2014 IEEE Bipolar/BiCMOS Circuits and Technology Meeting

(BCTM 2014)

Coronado, California, USA
28 September – 1 October 2014

IEEE Catalog Number: CFP14BIP-POD
ISBN: 978-1-4799-7231-9

Copyright © 2014 by the Institute of Electrical and Electronic Engineers, Inc
All Rights Reserved

Copyright and Reprint Permissions: Abstracting is permitted with credit to the source. Libraries are permitted to photocopy beyond the limit of U.S. copyright law for private use of patrons those articles in this volume that carry a code at the bottom of the first page, provided the per-copy fee indicated in the code is paid through Copyright Clearance Center, 222 Rosewood Drive, Danvers, MA 01923.

For other copying, reprint or republication permission, write to IEEE Copyrights Manager, IEEE Service Center, 445 Hoes Lane, Piscataway, NJ 08854. All rights reserved.

***This publication is a representation of what appears in the IEEE Digital Libraries. Some format issues inherent in the e-media version may also appear in this print version.**

IEEE Catalog Number: CFP14BIP-POD
ISBN 13:   978-1-4799-7231-9

Additional Copies of This Publication Are Available From:

Curran Associates, Inc
57 Morehouse Lane
Red Hook, NY 12571 USA
Phone: (845) 758-0400
Fax: (845) 758-2633
E-mail: curran@proceedings.com
Web: www.proceedings.com

TABLE of CONTENTS

A Low-Power SiGe Feedback Amplifier with Over 110GHz Bandwidth — 1

Leonardo Vera, John R. Long, B. Jeffrey Gross

Investigation of HBT layout impact on fT doubler performance for 90nm SiGe HBTs — 5

Vibhor Jain, Blaine Gross, John Pekarik, James Adkisson, Renata Camillo-Castillo, Qizhi Liu, Peter Gray, Aaron Vallett, Adam Divergilio, Bjorn Zetterlund, David Harame

An Active Frequency Doubler with DC-100GHz Range — 9

Leonardo Vera, John R. Long, B. Jeffrey Gross

A True-RMS Integrated Power Sensor for On-Chip Calibration — 13

Jonas Wursthorn, Herbert Knapp, Klaus Aufinger, Rudolf Lachner, Jidan Al-Eryani, Linus Maurer

Microwave Noise Properties of Heterojunction Bipolar Transistors — 17

Joseph Bardin

High-Resistivity SiGe BiCMOS Technology Development — 25

Anthony Stamper, Renata Camillo-Castillo, Hanyi Ding, James Dunn, Mark Jaffe, Vibhor Jain, Alvin Joseph, Ian McCallum-Cook, Kim Newton, Robert Rassel, Shyam Parthasarathy, Nicholas Schmidt, Srikanth Srihari, Randy Wolf, Michael Zierak

High-speed, waveguide Ge PIN photodiodes for a photonic BiCMOS process — 29

Stefan Lischke, Dieter Knoll, Lars Zimmermann, Alexander Scheit, Christian Mai, Andreas Trusch, Karsten Voigt, Marcel Kroh, Rainer Kurps, Pylyp Ostrovskyy, Yuji Yamamoto, Falk Korndörfer, Anna Peczek, Georg Winzer, Bernd Tillack

An SiGe heterojunction bipolar transistor with very high open-base breakdown voltage — 33

Thanh Viet Dinh, Tony Vanhoucke, Anco Heringa, Mahmoud Al-Sa'di, Ponky Ivo, Dick Klaassen, Peter Magnee

Examination of Horizontal Current Bipolar Transistor (HCBT) Reliability Characteristics — 37

Josip Žilak, Marko Koričić, Hidenori Mochizuki, So-ichi Morita, Katsumi Shinomura, Hisaya Imai, Tomislav Suligoj

Degradation and Recovery of High-Speed SiGe HBTs under Very High Reverse EB Stress Conditions — 41

Grazia Sasso, Niccolò Rinaldi, Gerhard Fischer, Bernd Heinemann

Device-to-Circuit Interactions in SiGe Technology: Challenges and Opportunities (DP - Invited Paper) — 45

John D. Cressler

Millimeter-Wave SiGe RFICs for Large-Scale Phased-Arrays — 56

Gabriel Rebeiz

A 2×2, 316 GHz SiGe Scalable Transmitter Array with Novel Phase Locking Method and On-die Antennas — 60

Saeed Zeinolabedinzadeh Namarvar, John Cressler, Aida Vera Lopez, Nelson Lourenco, Ahmet Cagri Ulusoy, Mehmet Kaynak, Mahmoud Kamarei, Bernd Tillack, John Papapolymerou

A broad-band BiCMOS transmitter front-end for 27-36GHz phased array systems   64

Yu Pei, Ying Chen, Domine Leenaerts

A Power-Efficient 4-element Beamformer in 120-nm SiGe BiCMOS for 28-GHz cellular communications   68

Anirban Sarkar, Kevin Greene, Brian Floyd

A Fast Precision Operational Amplifier Featuring Two Separate Control Loops   72

Derek Bowers

A BiCMOS 50 MHz Input Bandwidth, 1-to-16 Channelizer Optimized for Low Power Analog Signal Classification   76

Hao Li, Chris Thomas, Gert Cauwenberghs, Lawrence Larson

Linear low-power 13 GHz SiGe-Bipolar Modulator Driver with 7 Vpp differential Output Voltage Swing and on-Chip Bias Tee   80

Horst Hettrich, Michael Möller

An Integrated Transmitter for LED-Based Visible Light Communication and Positioning System in A 180nm BCD Technology   84

Zongyu Dong, Zongyu Dong, Fei Lu, Rui Ma, Li Wang, Chen Zhang, Gang Chen, Albert Wang, Bin Zhao

An Enhanced 180nm Millimeter-Wave SiGe BiCMOS Technology with fT/fMAX of 260/350GHz for Reduced Power Consumption Automotive Radar IC's   88

Jay John, Vishal Trivedi, Jim Kirchgessner, Dave Morgan, Ivan To, Pam Welch

A 90nm SiGe BiCMOS Technology for mm-wave and high-performance analog applications   92

John J. Pekarik, James Adkisson, Peter Gray, Qizhi Liu, Renata Camillo-Castillo, Marwan Khater, Vibhor Jain, Bjorn Zetterlund, Adam DiVergilio, Xiaowei Tian, Aaron Vallett, John Ellis-Monaghan, B. Jeffrey Gross, Peng Cheng, Vikas Kaushal, Zhong-Xiang He, Joseph Lukaitis, Kimberly Newton, Michael Kerbaugh, Edward Cahoon, David Harame

Device and circuit performance of SiGe HBTs in 130nm BiCMOS process with fT/fMAX of 250/330GHz   96

Vibhor Jain, Thomas Kessler, Blaine Gross, John Pekarik, Panglijen Candra, Peter Gray, Bodhisatwa Sadhu, Alberto Valdes-Garcia, Peng Cheng, Renata Camillo-Castillo, Kim Newton, Arun Natarajan, Scott Reynolds, David Harame

Low-Voltage Organic Field-Effect Transistors for Flexible Electronics   100

Hagen Klauk

Radio-Frequency Flexible Electronics: Transistors and Passives   107

Zhenqiang Ma, Jung-Hun Seo, Weidong Zhou

Transistor Laser for Optical Interconnect and Photonic Integrated Circuits   115

Milton Feng

Integration Challenges for High-Performance Carbon Nanotube Logic   123

James Hannon

Technologies for Very High Bandwidth Real-time Oscilloscopes   128

Peter Pupalaikis, Brian Yamrone, Roger Delbue, Amarpal Khanna, Kaviyesh Doshi, Balamurali

Bhat, Anirudh Sureka

Ultra-Wide-Bandwidth Oscilloscope Architectures and Circuits
136

Dan Knierim

A 27GHz, 31dBm Power Amplifier in a 0.25µm SiGe:C BiCMOS technology
143

Jaap Essing, Domine Leenaerts, Reza Mahmoudi

A Differential SiGe Power Amplifier Using Through-Silicon-Via and Envelope-Tracking for Broadband Wireless Applications
147

Jerry Tsay, Yu-Chun Donald Lie

W-band SiGe Power Amplifiers
151

Peter Song, Ahmet Çağrı Ulusoy, Robert Schmid, Saeed Zeinolabedinzadeh, John Cressler

A Low-Power and Ultra-Compact W-band Transmitter Front-End in 90 nm SiGe BiCMOS Technology
155

Taiyun Chi, Jong Seok Park, Robert Schmid, Cagri Ulusoy, John Cressler, Hua Wang

Comparison Between MOS and Bipolar mm-Wave Power Amplifiers in Advanced SiGe Technologies
159

Ayssar Serhan, Ayssar Serhan, Estelle Lauga-Larroze, Sylvain Bourdel, Jean-Michel Fournier, Nicolas Corrao

On-Wafer Small- and Large-Signal Measurement Systems at sub-THz Frequencies
163

Viktor Krozer

Temperature Impact on the In-Situ S-Parameter Calibration in Advanced SiGe Technologies
171

Andrej Rumiantsev, Ralf Doerner, Falk Korndoerfer

A Simple and Accurate Method for Extracting the Emitter and Thermal Resistance of BJTs and HBTs
175

Andreas Pawlak, Steffen Lehmann, Michael Schroter

A study on transient intra-device thermal coupling in multifinger SiGe HBTs
179

Rosario D'Esposito

A Low Phase Noise Signal Generation System for Ka-Band P2P Applications based on an Injection-Locked Frequency Tripler
183

Dwight Cabrera, Dwight Cabrera, Jean-Baptiste Begueret, Yann Deval, Olivier Tesson, Patrice Gamand, Olivier Mazouffre, Thierry Taris

A Digitally-Controlled Seven-State X-Band SiGe Variable Gain Low Noise Amplifier
187

Robert Schmid, John Cressler

Ultra-Low Noise and Low Power 18.7 GHz Radiometer LNAs in a 0.5 THz SiGe Technology Utilizing Back-Side Etched Inductors
191

Christopher T. Coen, Robert L. Schmid, John D. Cressler, Mehmet Kaynak, Bernd Tillack

K-Band Digitally Controlled Oscillator with Integrated Divide-by-16 Divider Chain for ADPLL Applications
195

Christopher Maxey, Sanjay Raman

A wide tuning triple-band frequency generator MMIC in 0.18um SiGe BiCMOS technology   199

Hechen Wang, Feng Zhao, Fa Dai, Guofu Niu, Bogdan Wilamowski

Small-Signal Modeling of the Lateral NQS Effect in SiGe HBTs   203

Shon Yadav, Anjan Chakravorty, Michael Schroter

An Electrothermal PIN Diode Model with Substrate Injection   207

Adam DiVergilio, John Pekarik, Vibhor Jain

An Investigation of fT/fmax Degredation due to Device Interconnects in 0.5 THz SiGe HBT   211
Technology

Cagri Ulusoy, Robert Schmid, Saeed Zainolabedinzadeh, Wasif Khan, Mehmet Kaynak, Bernd Tillack, John Cressler

Analysis of the Local Extraction Method of Base and Thermal Resistance of Bipolar   215
Transistors

Robert Setekera, Luuk Tiemeijer, Willy Kloosterman, Ramses van der Toorn

Welcome from the BCTM 2014 Chairmen

Welcome to the world-famous Coronado Island, San Diego, CA! On behalf of the IEEE BCTM'14 Executive Committee, we are honored and delighted to invite you to the 2014 IEEE Bipolar/BiCMOS Circuits and Technology Meeting (BCTM). We invite you to participate at the 2014 BCTM where the highlights include:

- Two keynote addresses from world-renowned speakers. The 1st one by the **Dean of College of Engineering and Professor Larry Larson of Brown University** on "The Next Era of Wireless Communications - Enabling Revolutions in Health Care, Transportation, Energy, and the Environment". The 2nd keynote will be by **Dr. Dave Harame, Chief Technical Executive Development and Enablement Value Foundry and Derivative Technology Development of IBM** on "Evolution of SiGe BiCMOS Technology adapting to the challenging wireless landscape"
- a day-long short course on "**SiGe BiCMOS Circuit Design for Wireless, Wireline and Radar Applications**", where Dr. Sean Nicolson from Broadcom will discuss "System specification and SiGe BiCMOS Circuit Design for W-Band Cellular Backhaul"; Prof. John Long on "The Design of RF Transceiver Front-Ends in SiGe BiCMOS"; Prof. Gabriel Rebeiz on "SiGe BiCMOS Circuit Design for Phase Array Radar Applications", and Prof. Sorin Voinigescu on "The Design of SiGe BiCMOS Digital Circuits for 60-100 Gbaud Wireline and Fiberoptics Communication Systems"
- a forward-looking **Emerging Technologies Session** with 4 invited speakers: Dr. Hagen Klauk (Max Planck Institute for Solid State Research) on "Low-Voltage Organic Field-Effect Transistors for Flexible Electronics"; Prof. Zhenqiang (Jack) Ma (University of Wisconsin-Madison) on "Radio Frequency Flexible Electronics: Transistors and Passives"; Dr. Milton Feng (University of Illinois) on "Transistor Laser for Optical Interconnect and Photonics Integrated Circuits"; and Dr. James Hannon, Manager, Carbon Electronics, IBM Thomas J. Watson Research Center on "Integration Challenges for High-Performance Carbon Nanotube Logic"
- invited papers exploring advances in process technology, device physics, wireless design, analog/mixed-signal, and modeling
- technical papers covering the latest advances in physics, design, performance, fabrication, characterization, modeling, and application of Si/SiGe/SiC bipolar, BiCMOS, and GaN ICs
- a luncheon & plenary speaker from Prof. Todd Martz: Scripps Institution of Oceanography, UCSD on the latest research in oceanography
- a fabulous evening dinner banquet

The IEEE BCTM is the world's premier forum focused on the needs and interests of the bipolar and BiCMOS community. If you are interested in leading edge bipolar/BiCMOS devices and technology, circuits, and applications, as well as networking with experts in these areas, please kindly join us this year at the beautiful Coronado Island at San Diego, CA, USA!

Prof. Donald Y.C. Lie
General Chair
Texas Tech University

Prof. Jean-Baptiste Begueret
Technical Program Chair
University of Bordeaux

2013 BEST PAPER AWARDS

The Best Paper Award for BCTM'13 is:

"A 12 bit 1.6GS/s BiCMOS 2x2 Hierarchical Time-Interleaved Pipeline ADC"

Manar El-Chammas, Xiaopeng Li, Shigenobu Kimura, Kenneth Maclean, Jake Hu, Mark Weaver, Matthew Gindlesperger, Scott Kaylor, Robert Payne, Charles Sestok, and William Bright, representing Texas Instruments, Dallas, USA.

The Best Student Paper Award for BCTM'13 is:

"TCAD Modeling of Accumulated Damage During Time-Dependent Mixed-Mode Stress"
Uppili S. Raghunathan, Partha S. Chakraborty, Brian Wier, John D. Cressler, Hiroshi Yasuda, and Philipp Menz, representing Georgia Tech, GA, & Texas Instruments, TX, USA.

Both of them will be honored during BCTM'14.

BCTM 2014 COMMITTEE MEMBERS

BCTM EXECUTIVE COMMITTEE
Donald Y.C. Lie (Texas Tech. Univ., USA, General Chair)
Jean-Baptiste Begueret (Univ. of Bordeaux, France, Technical Program Chair)
Niccolò Rinaldi (University of Naples, Italy, Vice-Technical Program Chair)
Niccolò Rinaldi (University of Naples, Italy, Publications Chair)
Doug Weiser (Texas Instruments, USA, Emerging Technology Chair)
Michael Schroter (Univ. Technology Dresden, Germany, Local Arrangements Chair)
Rob Rassel (IBM, USA, Finance Chair)
Foster Dai (Auburn University, USA, Short Course Chair)
Pete Zampardi (RFMD, USA, Vice-Short Course Chair)
Sorin Voinigescu (JSSC Guest Editor, University of Toronto, Canada)
Albert Wang (Univ. of California, USA, Publicity Chair)
Jan Jopke (CCS Associates, Conference Managing Consultant)
Catherine Shaw (C. Shaw, LLC, Atlanta Ga, USA, Conference Manager)
Sorin Voinigescu (University of Toronto, Canada, Analog/Mixed Signal Chair)
Andrej Rumiantsev (Brandenburg University of Technology Cottbus, Germany, Modeling/Simulation Chair)
Peter Magnee (NXP Semiconductors, The Netherlands, Device Physics Chair)
Edward Preisler (Jazz Semiconductor, USA, Process Technology Chair)
Wibo Van Noort (Texas Instruments, USA, Wireless Chair)

ANALOG/MIXED SIGNAL SUBCOMMITTEE
Sorin Voinigescu (University of Toronto, Canada, Chair)
Koichi Murata (NTT Photonics Lab, Japan)
Patrice Gamand (NXP, France)
Johann-Christoph Scheytt (University of Paderborn, Germany)
Michael Möller (Saarland University, Germany)
Laleh Najafizadeh (Rutgers University, USA)
Stephan Athan (DEC, USA)
Yang Lu (Maxim Integrated Products, USA)
Junxiong Deng (Marvell Semiconductor, USA)

MODELING/SIMULATION SUBCOMMITTEE
Andrej Rumiantsev (Brandenburg Univ. Tech., Germany, Chair)
Adam DiVergilio (IBM, USA)
Marco Bellini (ABB, Switzerland)
Michael Schroter (Univ. Tech. Dresden, Germany)
Breandán Ó hAnnaidh (Analog Devices)
Thomas Kessler (IBM, USA)
Vadim Issakov (Intel, Germany)

DEVICE PHYSICS SUBCOMMITTEE
Peter Magnee (NXP Semiconductors, the Netherlands, Chair)
Thomas Suligoj (University of Zagreb, Croatia)
Jake Steigerwald (Analog Devices, USA)
Peter Zampardi (RFMD, USA)
Hsien-Chang Wu (Texas Instruments, USA)
Tom Cheng (RFMD, USA)
Jiahui Yuan (SanDisk Corp, USA)
Tim Henderson (TriQuint Semiconductor, USA)
Vibhor Jain (IBM, USA)

PROCESS TECHNOLOGY SUBCOMMITTEE
Edward Preisler (TowerJazz, USA, Chair)
Rob Rassel (IBM, USA)
Julio Costa (RF Micro Devices, USA)
Tsugio Obata (Toshiba, Japan)
Jay John (Freescale Semiconductor, USA)
Grégory Avenier (STMicroelectronics, France)
Jack Pekarik (IBM, USA)
Alexander Fox (IHP, Germany)
Hiroshi Yasuda (Texas Instruments, USA)
Joost Melai (NXP, the Netherlands)
Josef Boeck (Infineon, Germany)

WIRELESS SUBCOMMITTEE
Wibo Van Noort (Texas Instruments, USA, Chair)
Gary Hau (Anadigics, USA)
Fa Foster Dai (Auburn University, USA)
Yang Xu (Illinois Inst. of Tech, USA)
John Rogers (Carleton University, Canada)
Ullrich Pfeiffer (University of Wuppertal, Germany)
Bruce Kim (University of Alabama, USA)
Hasan Gul (NXP, The Netherlands)
Albert Wang (University of California, Riverside, USA)
Mikhail Shirokov (Triquint, USA)

Anuj Madan (Skyworks, USA)
Tushar Thrivikraman (JPL, USA)
Hua Wang (Georgia Tech, USA)
Nils Pohl (Research Institute "Fraunhofer FHR", Germany)
LIN Fujiang (University of Science and Technology of China, China)
Thierry Taris (IMS Laboratory, France)

CALL FOR PAPERS
2015 BIPOLAR/BiCMOS CIRCUITS AND TECHNOLOGY MEETING
Boston, MA, USA
www.ieee-bctm.org

Tentative Dates: Short Course: Monday, September 28, 2015, Conference: Monday to Thursday, September 29 - October 1, 2015

The Bipolar/BiCMOS Circuits and Technology Meeting (BCTM) is a forum for technical communication focused on the needs and interests of the bipolar and BiCMOS community. Papers covering the design, performance, fabrication, testing and application of bipolar and BiCMOS integrated circuits, bipolar phenomena, and discrete bipolar devices are solicited. All papers must be suitable for a twenty-minute presentation. Text and figures must not have been presented at other conferences or published in any scientific or technical publications prior to BCTM.

Publication in the BCTM 2015 Proceedings does not preclude publication in an IEEE journal, and authors are encouraged to do so. A Special Issue of the *IEEE Journal of Solid-State Circuits* will include selected papers from BCTM 2015.

Papers are solicited in the following areas:

ANALOG / MIXED SIGNAL: Analog ICs: Mixed analog/digital ICs – Digital ICs - DACs and ADCs – Operational amplifiers - Voltage references and regulators - Integrated filters - Sensors - Networking ICs, MUX/DEMUX, Clock and data recovery, Decision circuits, Equalizers – Optical data links, Laser and modulator drivers - Gate arrays - Cell libraries -High-voltage ICs - Medical electronics - Motor controls – Analog subsystems within a VLSI chip - Packaging of high-performance ICs.

WIRELESS CIRCUIT DESIGN: Low Noise Amplifiers - Automatic gain control - Mixers – Voltage controlled oscillators - Frequency synthesizers - Power amplifiers – RF switches – Suppression of noise and distortion - Radio subsystems – RF Packaging - Integrated passives - Millimeter-wave circuits and systems.

DEVICE PHYSICS: New device physics phenomena in Si, SiGe, and III-V devices - Device design issues and scaling limits - Hot electron effects and reliability physics - Transport and high field phenomena - Noise - Linearity/Distortion – Novel measurement techniques - Operation in extreme environments (low and high temperatures, radiation effects).

MODELING/SIMULATION: Improved BJT and HBT models - Behavioural modelling techniques - Parameter extraction methods and test structures - De-embedding techniques - RF and thermal simulation techniques - Modelling of passives, interconnect and packages - Statistical modelling - Device, process and circuit simulation - CAD/modelling of power devices - packaging of power devices and ESD phenomena.

PROCESS TECHNOLOGY: Advances in processes and device structures demonstrating high speed, low power, low noise, high current, high voltage, etc. BiCMOS processes - Advanced process techniques – Si and Si-C homojunction bipolar/BiCMOS devices, III-V and SiGe heterojunction bipolar/BiCMOS devices. Manufacturing solutions related to Bipolar and BiCMOS yield improvements. Fabrication of high-performance passive components, including, MEMs. Process technology related to discrete and integrated bipolar/BiCMOS power devices, IGBT, RF power devices including DMOS. Wide bandgap bipolar devices (i.e. SiC, GaN, GaAs etc.) and related process technology.

STUDENT paper submissions are highly encouraged. Papers must be clearly marked as 'STUDENT SUBMISSION' in the abstract cover sheet to be eligible for the Best Student Paper Award.

If you know of people who may have a paper to contribute, please bring this Call for Papers to their attention.

IMPORTANT DEADLINES FOR AUTHORS
Monday, April 20, 2015 Deadline for receipt of abstract and summary
Monday, June 15, 2015 Notification of acceptance to be sent by email
Friday, July 17, 2015 Final proceedings manuscript due

SUBMISSION AND CONTACT INFORMATION
Visit the conference website: www.ieee-bctm.org, or contact: Catherine Shaw, Conference Manager, Phone 1-732 501-3334, e-mail: cshaw.cmpevents@gmail.com

2014 BCTM SPONSORS

GOLD SPONSORS

SILVER SPONSORS

BRONZE SPONSORS

BCTM is sponsored by the IEEE Electron Devices Society (EDS) in co-operation with the IEEE Solid-State Circuits Society (SSCS) and the IEEE Microwave Theory & Techniques Society (MTT).

A Low-Power SiGe Feedback Amplifier with Over 110GHz Bandwidth

Leonardo Vera[1], John R. Long[1] and B. Jeffrey Gross[2]

1. Electronics Research Laboratory/DIMES, Delft University of Technology, The Netherlands
2. IBM Microelectronics, Essex Junction VT

Abstract — The bandwidth of a single-stage, SiGe-HBT Darlington amplifier with feedback is improved 53% by cascoding and series peaking at the input, and by 25% for peaking alone. The cascode amplifier realizes 12-dB gain and better than 110GHz bandwidth (123GHz from simulation). Measured $|S_{11}|$ and $|S_{22}|$ are >10dB, group delay is ~6ps, and GBW/P_{dc} is 9.1GHz/mW. The 0.003mm^2 amplifier core is implemented in 90nm SiGe-BiCMOS and consumes 48mW from a 2.1V supply.

Index Terms — Broadband shunt-feedback amplifier, millimeter-wave amplifier, SiGe-HBT, Darlington amplifier, series peaking

I. INTRODUCTION

Broadband amplifiers are widely used in wireless and wireline communication systems. A general purpose gain block with input and output matched to 50Ω is cost-effective when conditioning signals between sub-systems, especially when needed in the later stages of system development. These amplifiers typically have an overall lowpass response (i.e., max. gain at DC), low and constant group delay, and a gain-bandwidth product exceeding the transistor f_T/f_{max}. Cascadable gain blocks designed for 50Ω systems should also be unconditionally stable and consume minimal power from a low-voltage supply.

A wideband, cascadable shunt feedback amplifier based on the Darlington pair is investigated in this paper. Characterized by its high input impedance and large current gain, Darlington-based amplifiers are widely used in broadband gain stages [1]. The objectives of this work are to benchmark the performance of a Darlington amplifier with shunt feedback designed in an advanced 90nm SiGe-BiCMOS technology, and to demonstrate circuit modifications that extend gain-bandwidth performance beyond the capabilities of the basic circuit topology. All of the amplifier prototypes developed in this work are input/output matched to 50Ω and designed for a low-frequency (S_{21}) gain of 12dB from a 2.1V supply. IBM 9HP SiGe-HBTs with f_T/f_{max} above 300GHz/350GHz [2] are used in this study.

The wideband shunt-feedback amplifier is analyzed in Section II of this paper, and modifications to extend the -3dB bandwidth of the circuit are proposed. Measurements of test circuits fabricated in a 90nm SiGe-BiCMOS technology are then presented and compared in Section III. Finally, a brief summary of the study is presented in Section IV.

II. Amplifier Design

The schematic of a broadband, gain block with shunt feedback in a 50Ω system is shown in Fig. 1. Power dissipation is minimized when supply current is sourced from a bias-T. Removing the bias-T and biasing the amplifier via the 50Ω load requires approximately 1V of headroom, and increases power consumption by approximately 50%.

Transistor Q_{b2} regulates the bias current of Q_1, while local

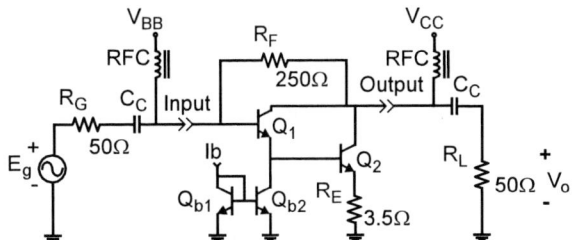

Fig. 1: Darlington amplifier with shunt feedback

feedback via R_E controls the DC bias of Q_2 and factors in its overall transconductance, g'_{m2}. The high input impedance of the Darlington pair minimizes loading on feedback network R_F and R_G compared to a single transistor gain stage. The gain-bandwidth (GBW) product is also close to a factor of 2 greater than a single device (i.e., f_T doubling), because the series connected base-emitters give the Darlington a lower input capacitance overall.

The input (Z_{in}) and output (Z_{out}) impedances of the amplifier in Fig. 1 at low frequency are given by

$$Z_{out} = Z_{in} = R_F/(1 - S_{21}) = 50\Omega, \qquad (1)$$

assuming that Q_1, Q_2 behave like an ideal transconductor, a 50Ω generator (R_G), and a 50Ω load (R_L). The value of R_F required to realize the desired S_{21} can be estimated from eq. 1. For $R_F = 250\Omega$, S_{21} is -4, or 12dB.

The transconductance required from the Darlington embedded in a 50Ω environment (as in Fig. 1) is determined from the small-signal gain equation

$$S_{21} = -g_{m-total}(R_F \| R_L) \approx -g'_{m2}(R_F \| 50\Omega). \qquad (2)$$

The degenerated transconductance g'_{m2} of transistor Q_2 is determined by transconductance g_{m2} and R_E, according to

$$g'_{m2} = \frac{i_o}{v_i} \approx \frac{g_{m2}}{1 + g_{m2}(R_E + r_{e2})} \quad , \qquad (3)$$

where i_o and v_i are the small-signal collector current and base-emitter voltage of Q_2, respectively. The extrinsic emitter resistance of Q_2, r_{e2}, is comparable to the typical value of R_E. For the amplifier of Fig. 1, eq. 2 predicts a g'_{m2} of 96mS. The total transconductance $g_{m-total}$ of the Darlington pair is approximately equal to g'_{m2}.

The relatively large transconductance and GBW required for the wideband 50Ω gain block give SiGe-HBTs a performance advantage over MOS transistors. A MOSFET with g_m~0.1S occupies more chip area and consumes more bias current than an HBT equivalent. Also, greater feedback from output to input via the MOS drain-gate parasitic capacitance limits its frequency response as an amplifier.

In order to simplify analysis of the small-signal equivalent for Fig. 1, parasitics of the Darlington are lumped into r_b (base resistance), C_μ (composite Miller capacitance, of which 32% is

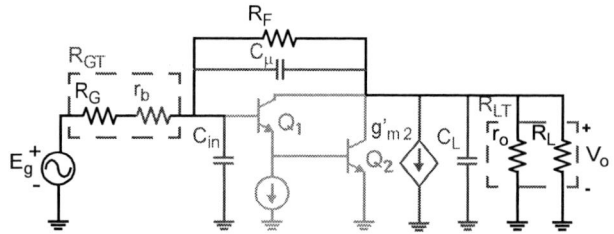

Fig. 2: Simplified small-signal circuit for the amplifier of Fig. 1 with key components highlighted in blue

contributed by Q_1 and 68% from Q_2), C_{in} (total input capacitance), and C_L (total collector-substrate capacitance) as shown in Fig. 2. Load (R_{LT}) is comprised of the output conductance paralleled with R_L, and R_{GT} is the series combination of r_b and the generator resistance.

Small-signal analysis of Fig. 2 predicts a frequency response that depends upon time constants τ_1 to τ_4, and a -3dB bandwidth that is inversely proportional to

$$\tau_i = C_{in}R_{GT} + (1 + g'_{m2}R_{LT})C_\mu R_{GT} + C_\mu R_{LT} + C_L R_{LT} . \quad (4)$$

The time constants listed in Table 1 are extracted from simulation of Fig. 1 where Q_1 and Q_2 are $4.5\times0.09\mu m^2$ and $7\times0.09\mu m^2$, respectively.

Table 1: Time constants as a percentage of the total from simulation of Fig. 2.

$\tau_1 = C_{in}R_{GT}$	$\tau_2 = (1+g'_{m2}R_{LT})C_\mu R_{GT}$	$\tau_3 = C_\mu R_{LT}$	$\tau_4 = C_L R_{LT}$	$\sum \tau_i$
46%	42.4%	2.9%	8.7%	100%

In order to reduce τ_1 and extend the amplifier bandwidth, a 72pH inductor (L_1) is added in series with the input, giving the series-peaked topology of Fig. 3. Inductor L_1 is series resonant with the input capacitance of the Darlington pair at 88 GHz. The multilayer shielded inductor used to implement L_1 occupies an area of just $12\times12\mu m^2$ and has a self-resonant frequency of 196 GHz. Series peaking increases the -3dB bandwidth of the original amplifier design by 25%.

A second variant, called the cascode amplifier (see Fig. 4), replaces Q_2 in Fig. 3 with cascode Q_2,Q_3 in order to reduce the Miller capacitance (C_μ in Fig. 2) and τ_2. Bias voltage V_{CB} may also be trimmed to optimize the f_T and f_{max} of Q_2 using the configuration of Fig. 4. It should be noted that neither the supply voltage nor the bias current increase when Q_3 is added. Thus, the trade-offs for the resulting 53% improvement in bandwidth compared to the reference design of Fig. 1, are a small increase in the overall circuit area, and a slight degradation in $|S_{12}|$ of the cascode amplifier.

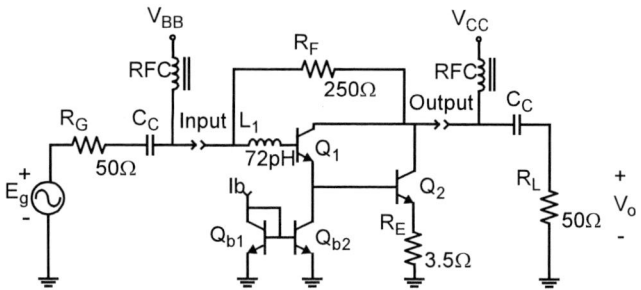

Fig. 3: Broadband series-peaked amplifier

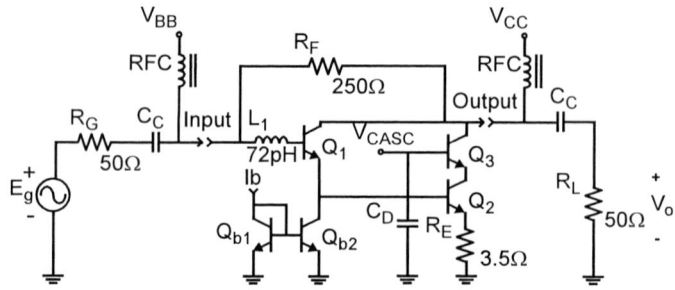

Fig. 4: Broadband cascode amplifier

Resistance in series with the base of Q_3 tends to destabilize the amplifier at high frequency. Therefore, a low extrinsic base resistance is desired for Q_3, and excellent decoupling of the base terminal to ground is needed to ensure wideband stability. The relatively low base resistance for HBTs in the 9HP technology provides better stability and reduces the amplifier noise figure. Transistor Q_3 is chosen identical in size to Q_2 (both $7\times0.09\mu m^2$). The bias voltage at the base of Q_3 is supplied off-chip, and is decoupled using a 5pF on-chip capacitor (C_D, see Fig. 4) and $1\mu F$ of off-chip decoupling capacitance.

Noise figure and linearity requirements for wideband gain blocks are modest, because they are typically used as post-amplifiers and not as preamplifiers or power stages. Simulated noise figure for the cascode amplifier is 5dB at low frequencies (dominated by thermal noise of R_F), rising to 8dB at 100GHz. The input-referred third-order intercept (IIP_3) from a two-tone simulation of the cascode amplifier at 5GHz is +7.1dBm, and 21dBm at 90GHz for a 10MHz tone difference. The amplifier is biased to draw 23mA (total) from a 2.1V supply in both noise and distortion simulations.

III. MEASUREMENTS

The reference (Fig. 1), series-peaked (Fig. 3), and cascode amplifiers (Fig. 4) were fabricated in IBM-9HP SiGe-BiCMOS technology. The micrograph of Fig. 5 shows the cascode amplifier configured for on-wafer probing using GSG probes (top and bottom), and DC bias connections on either side. Total test circuit area is $0.20mm^2$ (incl. pads), $0.003mm^2$ of which is occupied by active circuitry (i.e., transistors, peaking inductor and biasing). The cascode transistor is biased at $V_{CASC}=1.85V$ (see Fig. 4). All 3 amplifiers were biased to draw 23mA from a

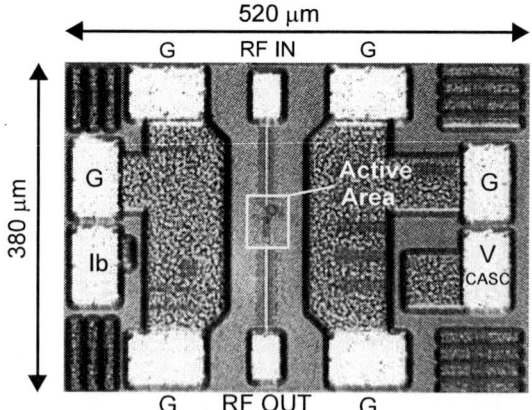

Fig. 5: Chip photomicrograph of the cascode amplifier

978-1-4799-7231-9/14 $31.00 © 2014 IEEE

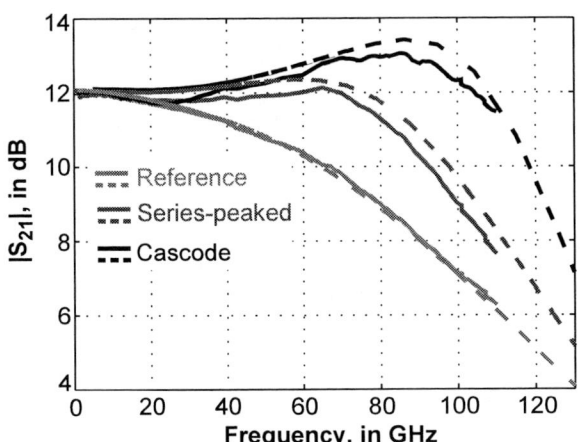

Fig. 6: Measured (solid line) vs. simulated (dashed) $|S_{21}|$ for the amplifiers of Figs. 1, 3 and 4

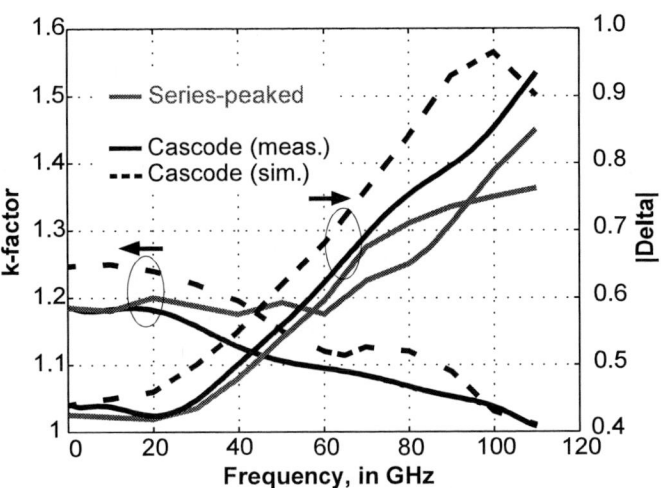

Fig. 8: Stability factors k and Δ extracted from measurements of the amplifiers shown in Figs. 3 and 4

single 2.1V supply during testing. Measurement data presented in this section includes the effects of pad parasitics and on-chip transmission lines connecting the input and output (i.e., no de-embedding).

Fig. 6 shows the measured and simulated forward transmission coefficient ($|S_{21}|$) across 1-110GHz for each amplifier. Post-layout simulations predict a bandwidth of 79, 106 and 123GHz for the reference (Fig. 1), series-peaked (Fig. 3) and cascode (Fig. 4) amplifiers, respectively, while the measured values are: 80 GHz (reference) and 100 GHz (series-peaked). The cascode amplifier has a -3dB bandwidth that exceeds the 110GHz measurement limit of the VNA. Feedback resistor R_F is 246Ω for the fabricated samples (measured separately) is just 1.6% below the design value of 250Ω. Thus, the low-frequency S_{21} of 12dB predicted from eq. 2 agrees very well with the measured data.

The reverse transmission coefficients ($|S_{12}|$ in Fig. 7) measured for the reference and series-peaked amplifiers are relatively flat, rising from -16dB at low frequency to -14dB at 110GHz. However, S_{12} for the cascode amplifier degrades more rapidly with increasing frequency due to extra phase shift in the forward gain path added by the cascode transistor. This

phase shift could be reduced and $|S_{12}|$ increased if Q_3 were sized larger to reduce its extrinsic base resistance, r_b. However, bias current for the cascode is fixed by Q_2, and therefore, bandwidth would suffer if the emitter area of Q_3 was made larger because f_T/f_{max} falls with decreasing current density.

The stability factor k and determinant (Δ) of the S-parameter matrix calculated from measured data are plotted for two amplifiers in Fig. 8. All 3 amplifiers are unconditionally stable (i.e., k>1 and $|\Delta|$<1), and can therefore be cascaded in a linear amplifier chain to increase gain. Stability improves as $|S_{21}S_{12}|$ increases, which could be realized by raising the forward gain at the expense of bandwidth.

The measured and simulated input (S_{11}) and output (S_{22}) reflection coefficients for the cascode amplifier are plotted on a Smith chart in Fig. 9. Measured input and output return loss across 1 to 110GHz are better than 10 dB for all amplifiers.

Single-stage amplifiers have low latency and cause minimal dispersion to the amplified signal. This is demonstrated by the group delay calculated from the measured S-parameter data for the 3 amplifiers, which is ~6ps across the 110GHz range (see Fig. 10). All 3 Darlington amplifiers developed in this work are therefore well-suited to high data rate system applications (e.g., post-amplification in a fiber-optic receiver).

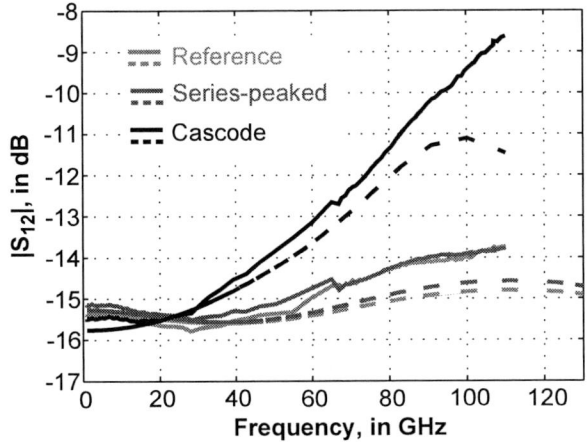

Fig. 7: Measured (solid line) vs. simulated (dashed) $|S_{12}|$ for the amplifiers of Figs. 1, 3 and 4

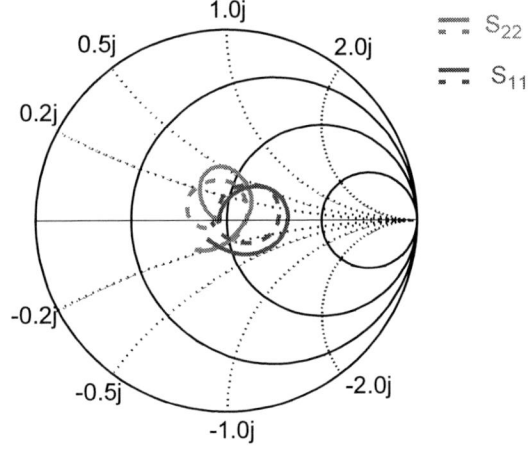

Fig. 9: Measured (solid line) and simulated (dashed) input and output reflection coefficients for the cascode amplifier

Table 2: Amplifier Comparison

Ref.	Technology/f_T	Gain (dB)	BW (GHz)	GBW (GHz)	Power Dissipation	GBW/P_{DC}	Area (mm²)	Topology
Fig. 1	90nm SiGe BiC-MOS/ 300GHz	12	79	316	48mW at 2.1V	6.58	0.152	Single-stage Darlington, single ended
Fig. 3			100	400		8.3	0.152	
Fig. 4			110 123(sim)	440 492 (sim)		9.1 10.25(sim)	0.197 0.003 (core)	
2007 [4]	0.18μm SiGe bipolar 200GHz	20	84.6	846	990mW at -5.5V	0.85	0.63	4 cascaded EF + 1 cascode, differential
2013 [5]	0.13μm SiGe BiC-MOS 200GHz	20	67 82(sim)	670 820(sim)	92mW at 2.7V	7.28 8.91(sim)	0.28 0.04 (core)	2-stage Darlington, single ended
2011 [6]	45nm CMOS SOI/ 350GHz	9	92	259	73.5mW at 1.2V	3.53	0.45	Distributed amplifier, single ended

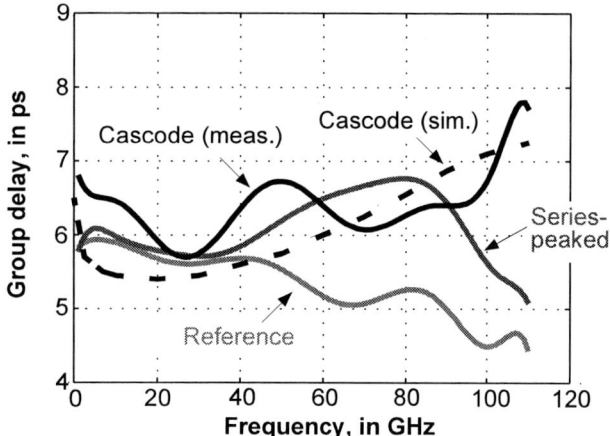

Fig. 10: Group delay extracted from S-parameters for the amplifiers of Figs. 1, 3 and 4

Table 2 compares the measured performance for the 3 amplifier prototypes with reported wideband amplifiers. The ratio of GBW to DC power consumption (P_{DC}) is a figure of merit (FoM) proposed for broadband amplifiers [3]. Amplifier [4] is a multi-stage design consisting of 4 emitter follower and 1 cascode stage that has a -3dB bandwidth of 84.6GHz. Its group delay (GD) is 5 times larger than GD for the amplifiers developed in this work (i.e., 40ps vs. 6ps) due to greater latency in a multi-stage design. Also, the GBW/P_{DC} of 0.85 is less than one-tenth the GBW/P_{DC} realized by the cascode amplifier. Example [6] is a distributed amplifier (DA) implemented in 45nm CMOS-SOI technology. The DA is capable of wide bandwidth, but it occupies more chip area (0.45mm²), consumes more power, and has a GBW/P_{DC} less than one-half that measured for the Darlington-based amplifiers developed in this work. For comparison with [5], a cascade of 2 Darlington amplifier stages was synthesized using the measured s-parameters for the cascode example in a simulation testbench. The resulting cascade realizes 23.9dB gain and a -3dB bandwidth of 116GHz. The predicted GBW is 1.82THz at 96mW dissipation, and GBW/P_{DC} is 19GHz/mW. This would exceed the GBW/P_{DC} of 8.9GHz/mW realized by the 820GHz GBW amplifier reported in [5].

Finally, the performance of Darlington gain blocks has progressed from the benchmark of 8dB gain and 17GHz bandwidth set by the 0.35μm SiGe bipolar technology reported in [7] (i.e., GBW=42.7GHz), to gain-bandwidth products that are over an order of magnitude greater in today's 90nm SiGe BiCMOS technology.

IV. Conclusions

Three low-power, cascadable broadband amplifiers matched to 50Ω and 12 dB of gain were demonstrated in a 90nm SiGe-BiCMOS technology. It has been shown that series peaking and cascoding of the g_m-driver transistor increases bandwidth of a single-stage Darlington amplifier by 53%. The resulting amplifier topology reaches the highest GBW/P_{dc} reported to date for a single-ended amplifier fabricated on silicon.

Acknowledgments

IBM Microelectronics for IC fabrication via the MOSIS service and A. Akhnoukh at TU Delft for technical support.

References

[1] J.F. Kukielka and C.P. Snapp, "Wideband monolithic cascadable feedback amplifiers using silicon bipolar technology," in *Monolithic Microwave Integrated Circuits*, IEEE Press, 1985, pp. 330-331.

[2] N.E. Lourenco, R.L. Schmid, K.A. Moen, et al., "Total Dose and Transient Response of SiGe HBTs from a New 4th-Generation, 90 nm SiGe BiCMOS Technology," in *2012 IEEE Radiation Effects Data Workshop*, Miami FL pp. 1-5, July 2012.

[3] Jun-De Jin, S.S.H. Hsu, "A 40-Gb/s Transimpedance Amplifier in 0.18μm CMOS Technology," *IEEE-JSSC*, vol. 43, no. 6, pp. 1449-1457, June 2008.

[4] S. Trotta, H. Knapp, K. Aufinger, et al., "An 84 GHz Bandwidth and 20 dB Gain Broadband Amplifier in SiGe Bipolar Technology," *IEEE-JSSC*, vol. 42, no. 10, pp. 2099-2106, Oct. 2007.

[5] Z. Xuan, R. Ding, T. Baehr-Jones, M. Hochberg, "A 92 mW, 20 dB gain, broadband lumped SiGe amplifier with bandwidth exceeding 67 GHz," Proc. of the IEEE-BCTM, Bordeaux FR, pp. 107-110, Oct. 2013.

[6] K. Joohwa and J.F. Buckwalter, "A 92 GHz Bandwidth Distributed Amplifier in a 45 nm SOI CMOS Technology," *IEEE Microwave and Wireless Components Letters*, vol. 21, no. 6, pp. 329-331, June 2011.

[7] J.R. Long, M.A. Copeland, S.J. Kovacic, D.S. Malhi, D.L. Harame, "RF analog and digital circuits in SiGe technology," *IEEE-ISSCC Tech. Dig.*, pp.82-83, Feb. 1996.

Investigation of HBT layout impact on f_T doubler performance for 90nm SiGe HBTs

Vibhor Jain*, B. J. Gross, J. J. Pekarik, J. W. Adkisson, R. A. Camillo-Castillo, Qizhi Liu, P. B. Gray, A. Vallett,
A. W. Divergilio, B. K. Zetterlund, D. L. Harame

IBM, 1000 River Road, Essex Junction, VT 05452 USA
*vibhorj@us.ibm.com

Abstract— **Peak f_T of 660GHz is reported for HBT f_T doubler designs in IBM 90nm SiGe BiCMOS technology 9HP. This high performance f_T doubler utilizes a longer HBT for output stage compared to the input stage HBT (length ratio 2:1) resulting in improved transconductance and lower thermal resistance. The impact of HBT layout on the circuit performance and trade-off between thermal resistance and f_T is also investigated. f_T doubler circuit can be used as a single transistor in several circuit applications like A/D converters and broadband circuits where higher performance is desired.**

Keywords— *BiCMOS, SiGe, HBT, f_T doubler*

I. INTRODUCTION

A high performance SiGe BiCMOS technology has potential high speed applications including RF transceivers, high bandwidth analog to digital converters and optical networks, terahertz imaging and sensing, automotive radars and instrumentation [1-4]. High frequency performance of SiGe heterojunction bipolar transistors (HBTs) has been consistently improving with several reports of simultaneous f_T and f_{MAX} greater than 300GHz [5-8]. Certain applications like radar and optical receivers still require devices with better AC performance and higher power at room temperature with no degradation in breakdown voltage. A f_T doubler circuit block can be used as a single device for analog designs having a higher power and a higher f_T than the single HBT. f_T doublers have been implemented in older SiGe HBT technologies for broadband applications including A/D converters, cellular, imaging systems and UWB [9-12]. This is the first instance of their implementation in 90nm SiGe HBTs.

In this paper, f_T doublers based on IBM's 90nm SiGe BiCMOS technology 9HP are reported [13]. High performance HBTs in this technology have simultaneous f_T/f_{MAX} of 300/360GHz. Peak f_T of 660GHz has been measured for a f_T doubler design having an input stage HBT with emitter area (A_E) of $0.1 \times 2 \mu m^2$ and output stage HBT with $A_E = 0.1 \times 4 \mu m^2$. f_T improvement is due to reduced input capacitance (C_π), improved transconductance (g_m) and lower thermal resistance (R_{th}) [9-10]. The impact of device layout on the circuit performance and trade-off between R_{th} and f_T is also investigated.

Introduction to the f_T doubler and basic circuit results are presented in Section II. Section III discusses the trade-off between performance and R_{th}. f_T doubler results with a longer output stage HBT compared to input stage are reported in Section IV followed by conclusions in Section V.

II. FT DOUBLER

A. Analysis

SiGe HBT f_T doubler design is based on a Darlington Pair (Fig. 1(a)) having two HBTs in series (T_{inp} and T_{out}) with a diode (T_{Dio}) connected in parallel to output transistor (T_{out}). The diode sets the bias currents. The entire circuit can be considered as a single transistor with three terminals – emitter, base and collector. A hybrid-pi equivalent circuit of a SiGe HBT is shown in Fig. 1(b) which can also be used for f_T doubler. For a single HBT, assuming negligible emitter and base resistances and high beta, f_T can be estimated from

$$f_\tau = \frac{g_m}{2\pi(C_\pi + C_{CB})} \qquad (1)$$

where g_m is the small signal transconductance, C_π is the input capacitance and C_{CB} is the collector base capacitance. For the f_T doubler, assuming identical input and bias transistors, T_{inp} and T_{Dio}, and longer T_{out}, f_T for the doubler can be estimated as

$$f_\tau = \frac{g_m(1+r)}{2\pi(C_\pi + 2C_{CB})} \qquad (2)$$

where r is the ratio of emitter length of T_{out} & T_{inp} and g_m, C_π and C_{CB} have their usual meaning for T_{inp} HBT. In reality, T_{inp} and T_{Dio} are not identical as $V_{CB} = 0V$ for T_{Dio} only. f_T improves for the doubler compared to single device due to series combination of T_{inp} and T_{out} which results in lower C_π and C_{CB} for the same collector current density. Series combination of T_{inp} and T_{out} reduces C_π by 50% but the presence of T_{Dio} in

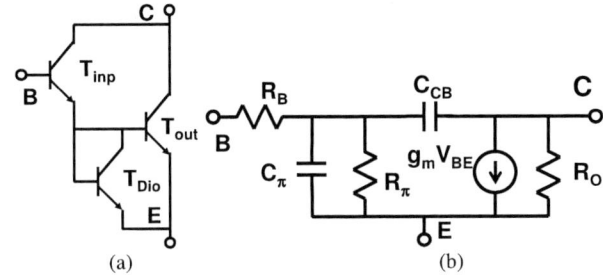

Fig. 1. (a) Schematic of f_T doubler (b) small signal equivalent circuit of SiGe HBT

parallel to T_{out} increases C_π. Thus f_T for the doubler is less than twice the value of a single HBT.

B. Results

Fig. 2 shows the DC gummel curves for a single $0.1 \times 2\mu m^2$ HBT in Collector-Base-Emitter (CBE) configuration in 90nm SiGe BiCMOS process (at $V_{CB} = 0.3V$) alongwith the gummel curves for a f_T doubler design ($V_{CB} = 0V$). The f_T doubler was designed using three HBTs in CBE configuration each having an emitter area (A_E) of $0.1 \times 2\mu m^2$. These devices have a BV_{CEO} of 1.7V and BV_{CBO} of 5.2V. Current gain (β) for the f_T doubler is higher than the single device as expected. Peak β for the single device on this wafer is measured as 780 which increases to 1700 for the doubler. In a f_T doubler, the output stage HBT is biased at V_{CE} higher than the BV_{CEO} value.

AC parameters for both single device and f_T doubler were extracted using two port S-parameter measurements with Load-Reflect-Reflect-Match (LRRM) calibration and open and short de-embedding to remove parasitic impedances associated with measurement cables, probes, wiring and pads. Fig. 3 shows the measured current gain H_{21} and Maximum Stable Gain *MSG* for the single HBT and f_T doubler biased at I_C corresponding to peak f_T performance. Measured R_B for single HBT is 60Ω which degrades to 105Ω for f_T doubler. Consequently, MSG for the doubler is similar to HBT despite an improvement in f_T.

Fig. 4 shows extracted f_T vs I_C curves for both the devices. f_T was extrapolated from H_{21} using a -20dB/dec slope at 20GHz. The HBT demonstrates a peak f_T of 294 GHz ($A_E = 0.1 \times 2\mu m^2$) at $I_C = 3.9mA$ which improves to 500 GHz for the f_T doubler at $I_C = 9.1mA$, $V_{CE} = 1.82V$. As expected, f_T for the

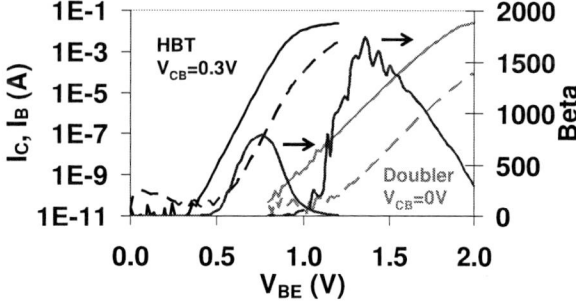

Fig 2. DC Gummel curves for SiGe HBT having $A_E = 0.1 \times 2\mu m^2$ at $V_{CB} = 0.3V$ and f_T doubler at $V_{CB} = 0V$

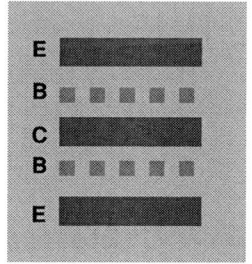

Fig. 5. Emitter-Base-Collector-Base-Emitter (EBCBE) compact layout where T_{out} and T_{inp} are combined into a single cell within the same isolation trench for low parasitics and high performance f_T doubler

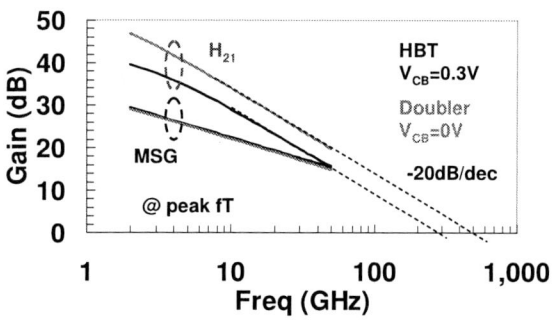

Fig 3. Measured H_{21} and *MSG* for HBT at $V_{CB} = 0.3V$ and f_T doubler at $V_{CB} = 0V$ biased at I_C corresponding to peak f_T. f_T for the HBT is 294GHz and for the doubler is 500GHz. *MSG* for the two devices is similar

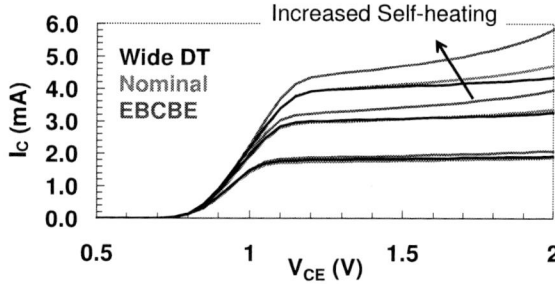

Fig. 6. Forced I_B Output curves (I_C-V_{CE}) for the three f_T doubler layouts – EBCBE, nominal CBE layout and Wide Emitter-DT space layout at $I_B = 2$, 4 and $6\mu A$; Due to high R_{th} for compact structure, thermal runaway occurs at lower I_C

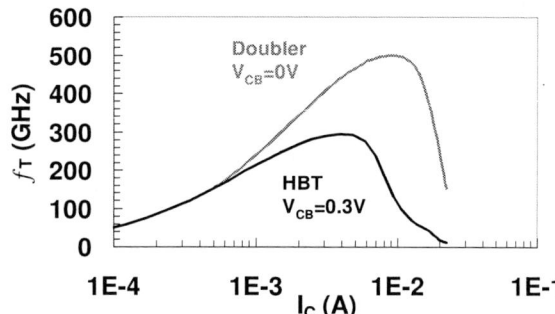

Fig 4. Extracted f_T for the HBT at $V_{CB} = 0.3V$ and doubler at $V_{CB} = 0V$; peak f_T improved from 294GHz to 500GHz for the doubler

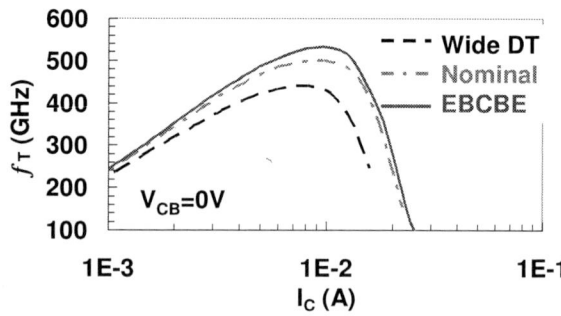

Fig. 7. Extracted f_T for the three f_T doubler layouts at $V_{CB} = 0V$; peak f_T improved from 440GHz for Wide emitter-DT space layout to 530GHz for EBCBE layout

978-1-4799-7231-9/14 $31.00 © 2014 IEEE

doubler is less than twice the f_T of a single HBT. Extracted C_π at peak f_T for the HBT is 26.4fF and C_π for the f_T doubler is 14.4fF. The presence of wiring parasitic also degrades f_T for the doubler.

III. PERFORMANCE AND THERMAL RESISTANCE TRADE-OFF

AC performance of the f_T doubler is dependent on the AC characteristics of a single HBT and on the internal wiring parasitics associated with the doubler layout as they are not removed by the de-embedding structure. Wiring parasitic can be reduced by utilizing a compact EBCBE layout shown in Fig. 5. In the EBCBE layout, the two HBTs – T_{out} and T_{inp} are combined into a single cell within the same isolation trench (EBCBE structure) while T_{Dio} is a separate CBE HBT. However, this compact layout degrades the thermal resistance of the doubler due to small footprint.

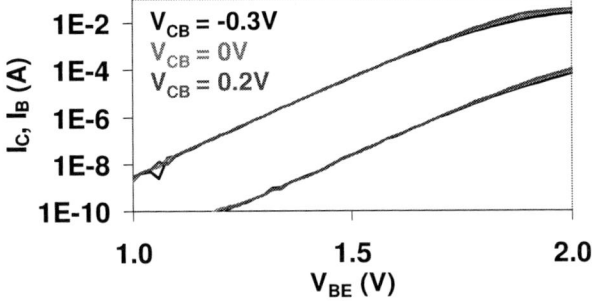

Fig. 8. DC Gummel curves for a f_T doubler with $T_{inp}/T_{Dio} = 0.1x2\mu m^2$ and $T_{out} = 0.1x4\mu m^2$ at $V_{CB} = -0.3V$, 0V and 0.2V

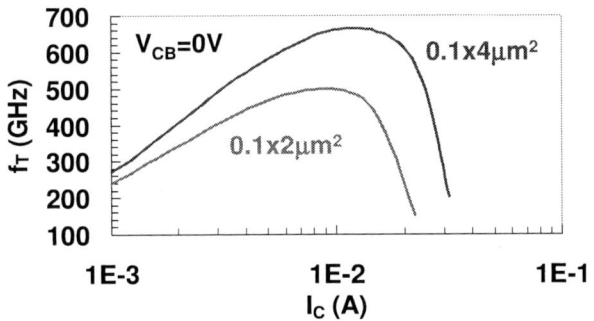

Fig. 9. Extracted f_T for the f_T doubler layouts with $T_{out} = 0.1x4\mu m^2$ and $0.1x2\mu m^2$ at $V_{CB} = 0V$; peak f_T improved from 500GHz to 660GHz

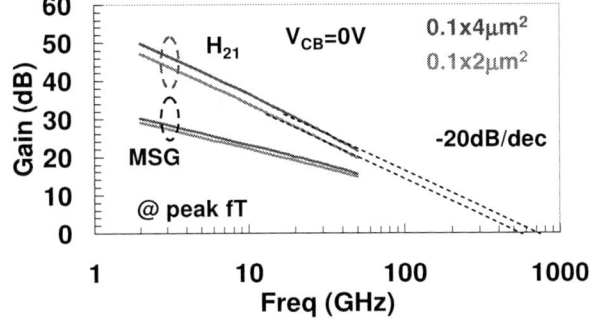

Fig 10. Measured H_{21} and MSG for f_T doublers at $V_{CB} = 0V$ biased at I_C corresponding to peak f_T.

In the f_T doubler design, since T_{out} is parallel to series combination of T_{inp} and T_{Dio}, it has twice the V_{CE} of a single HBT (V_{CE} is higher than BV_{CEO}). As a result, maximum self-heating occurs in the output stage. Self-heating in the devices can be reduced by increasing the Emitter-Isolation Trench (DT) space (Wide DT layout) resulting in lower R_{th} [14]. Forced I_B output curves (I_C-V_{CE}) for the f_T doublers designed using Wide DT HBT, compact EBCBE layout and nominal CBE HBTs are shown in Fig 6. Due to reduced R_{th} of the circuit with Wide DT layout, the circuit is thermally stable to a higher power as shown in the figure. However, as expected, f_T degrades due to higher wiring parasitic (Fig 7). For the nominal CBE f_T doubler, measured peak f_T is 500GHz (I_C = 9.1mA) which improves to 530GHz (I_C = 9.6mA) for the EBCBE layout and degrades to 440GHz (I_C = 7.2mA) for the f_T doublers with wide DT HBTs (all at $V_{CB} = 0V$). The peak f_T for the wide DT design is lower due to internal wiring parasitic and not due to degraded AC performance of a single wide DT HBT. There is no degradation in f_T or f_{MAX} for the single HBT in wide DT layout. I_C at peak f_T improves with peak f_T for the EBCBE doubler design due to lower internal wiring parasitics for the compact layout.

IV. DOUBLER WITH LONGER OUTPUT HBT

The f_T doublers discussed so far have the same size input and output stages and have a fundamental RF performance – thermal resistance trade-off. As mentioned earlier, HBT T_{out} has twice the V_{CE} of a single HBT for similar I_C bias, resulting in much higher junction temperature. Since R_{th} is a function of emitter length (lower R_{th} for longer HBT), performance-R_{th} trade-off can be improved by using a longer output HBT (T_{out}). Longer output HBT also has a higher g_m at a given V_{BE} which improves f_T. Increasing the emitter length by a factor of 2 reduces R_{th} by ~40% for HBTs in 9HP [14]. Since T_{out} has twice the V_{CE} of the 2µm long T_{inp}, its length was chosen to be 4µm for these designs.

Fig. 8 shows the DC gummel for f_T doublers with 4µm long HBT as T_{out} and 2µm long HBTs for T_{in} and T_{Dio} at different V_{CB}. Increase in I_B and I_C at high V_{BE} is prominent for higher V_{CB} due to self-heating. Extracted f_T for the two circuits (T_{out} = $0.1x2\mu m^2$ and $0.1x4\mu m^2$) is shown in Fig 9. Peak f_T for the 4µm T_{out} doubler improves to 660GHz (I_C = 12.8mA, V_{CE} = 1.83V) compared to 500GHz for the 2µm T_{out} doubler (I_C = 9.1mA, V_{CE} = 1.82V). The f_T doubler can now support higher current due to longer output device which also results in higher g_m at a given V_{BE}.

R_B for the long T_{out} device is lower than the nominal f_T doubler. As a result, *MSG* for this circuit is about 1dB higher at peak f_T bias. Figs. 11 and 12 show extracted C_π and g_m [15] for the three structures – single HBT, nominal f_T doubler with $0.1x2\mu m^2$ T_{out} and improved f_T doubler with $0.1x4\mu m^2$ T_{out}. For all three designs, T_{inp} and T_{Dio} are $0.1x2\mu m^2$ in size. As expected, C_π is lower for f_T doubler resulting in a higher f_T for the circuit block. C_π at peak f_T for the HBT is 26.4fF which reduces to 14.4fF for T_{out} = $0.2\mu m^2$ and 13.6fF for T_{out} = $0.4\mu m^2$. Similarly, g_m improves from 46.2mS to 58.8mS for the

978-1-4799-7231-9/14 $31.00 © 2014 IEEE

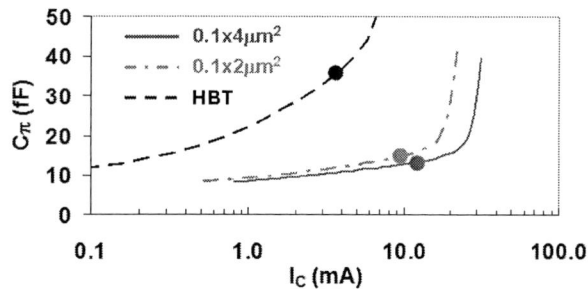

Fig. 11. Extracted C_π for the 3 devices – single HBT and doublers with $0.1x2\mu m^2$ T_{out} HBT and $0.1x4\mu m^2$ T_{out} HBT. C_π reduces for the f_T doubler improving f_T. Dots in the figure represent peak f_T point at $V_{CB} = 0V$

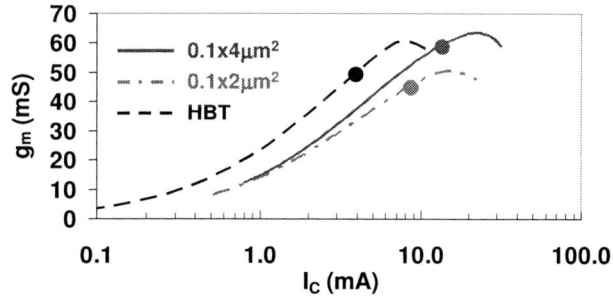

Fig. 12. Extracted g_m for the 3 devices – single HBT and f_T doublers with $0.1x2\mu m^2$ T_{out} HBT and $0.1x4\mu m^2$ T_{out} HBT. Dots in the figure represent peak f_T point at $V_{CB} = 0V$

$T_{out} = 0.4\mu m^2$ doubler resulting in higher f_T. R_B for the $T_{out} = 0.4\mu m^2$ doubler is lower at 93Ω due to longer HBT length.

Fig. 13 compares the measured and simulated H_{21} and *MSG* for the $T_{out}=0.4\mu m^2$ doubler design at peak f_T bias. Measured H_{21} and *MSG* are slightly lower due to absence of wiring parasitic in the circuit simulations.

V. CONCLUSIONS

In this paper, a f_T doubler in IBM's 90nm SiGe BiCMOS technology 9HP is demonstrated with peak f_T of 660GHz at room temperature and $V_{CB} = 0V$. In this high performance doubler design, output stage HBT has $A_E = 0.1x4\mu m^2$ and input stage HBT has $A_E = 0.1x2\mu m^2$. Thermal resistance of the longer HBT is lower which also improves the circuit's thermal

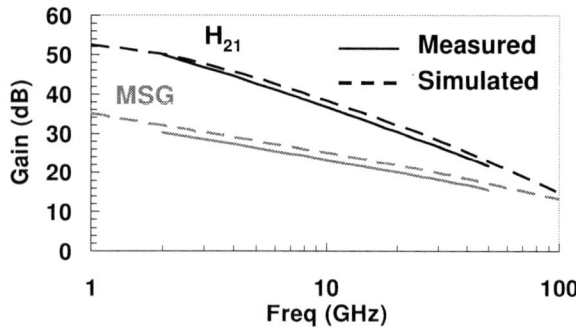

Fig. 13 Comparison of the measured H_{21} and *MSG* to simulations at peak f_T bias for the $T_{out} = 0.1x4\mu m^2$ doubler design

stability especially as the voltage drop across output stage is higher than BV_{CEO} and twice the V_{CE} of input stage for similar I_C bias. Due to the longer output HBT, g_m at a given V_{BE} is improved compared to nominal f_T doubler with similar input capacitance, thus pushing the f_T to greater than twice the f_T of a single HBT. R_{th} of the f_T doubler can also be improved by utilizing a wider emitter-DT space, however, with lower AC performance. All these f_T doubler circuit blocks can be used a single transistor cell in circuit designs for improved performance.

ACKNOWLEDGMENT

The authors would like to thank Adnan Beganovic for AC and DC measurements.

REFERENCES

[1] P. Candra et al., "A 130nm SiGe BiCMOS technology for mm-Wave applications featuring HBT with fT/fMAX of 260/320 GHz", IEEE Radio Freq Integrated Circuits Symp (RFIC), pp. 381-384, 2-4 Jun 2013

[2] J. W. May, and G. M. Rebeiz, "Design and Characterization of W - Band SiGe RFICs for Passive Millimeter-Wave Imaging", IEEE Trans on Microwave Theory and Tech., vol.58, no.5, pp.1420-1430, May 2010

[3] M. Chu et al., "A 40 Gs/s Time Interleaved ADC Using SiGe BiCMOS Technology", IEEE J of Solid-State Circuits, vol.45, no.2, pp. 380 - 390, Feb. 2010

[4] D. L. Harame et al., "Current Status and Future Trends of SiGe BiCMOS Technology", IEEE Trans on Electron Devices, Vol. 48, No. 11, pp. 2575, Nov 2001

[5] H. Rucker, B. Heinemann, and A. Fox, "Half-Terahertz SiGe BiCMOS Technology" et al., Silicon Monolithic Integrated Circuits in RF Systems, pp. 133-136, 2012

[6] R. A. Camillo-Castillo et al., "SiGe HBTs in 90nm BiCMOS technology demonstrating 300GHz/420GHz fT/fMAX through reduced Rb and Ccb parasitics", IEEE Bipolar/BiCMOS Circuits and Technology Meeting (BCTM), pp.227-230, Sept. 30 2013 - Oct. 3 2013

[7] A. Fox et al., "SiGe HBT module with 2.5 ps gate delay", Proc. IEEE IEDM, 2008, pp. 731-734

[8] P. Chevalier et al., "Towards THz SiGe HBTs", Proc. BCTM, pp. 57-65, 2011

[9] J. Yuan, J. D. Cressler, "Enhancing the speed of SiGe HBTs using fT-doubler techniques", Silicon Monolithic Integrated Circuits in RF Systems, SiRF, pp. 50-53, 2008

[10] M. Joodaki, "Small-signal characterization of SiGe-HBT fT-doubler up to 120 GHz", IEEE Trans on Electron Devices, vol.52, no.9, pp.2108 - 2111, Sept. 2005

[11] R. C. Walker, Kuo-Chiang Hsieh, T. A. Knotts, Chu-Sun Yen, "A 10 Gb/s Si-bipolar TX/RX chipset for computer data transmission", IEEE International Solid-State Circuits Conf, pp.302-303, 5-7 Feb. 1998

[12] Z. Xuan, R. Ding, T. Baehr-Jones and M. Hochberg, "A 92 mW, 20 dB gain, broadband lumped SiGe amplifier with bandwidth exceeding 67 GHz", IEEE BiCMOS Circuits and Technology Meeting (BCTM), pp.107 - 110, Sept. 30 - Oct. 3, 2013

[13] J. Pekarik et al., "A 90nm SiGe BiCMOS Technology for mm-wave and high-performance analog applications", submitted Proc BCTM, 2014

[14] V. Jain et al., "Study of mutual and self-thermal resistance in 90nm SiGe HBTs", IEEE Bipolar/BiCMOS Circuits and Technology Meeting (BCTM), pp.17 - 20, Sept. 30 - Oct. 3, 2013

[15] J. L. Olvera-Cervantes, et al., "A New Analytical Method for Robust Extraction of the Small-Signal Equivalent Circuit for SiGe HBTs Operating at Cyogenic Temperatures", IEEE Trans. on Microwave Theory and Techniques, , vol.56, no.3, pp.568-574, March 2008

An Active Frequency Doubler with DC-100GHz Range

Leonardo Vera[1], John R. Long[1] and B. Jeffrey Gross[2]

1. Electronics Research Laboratory/DIMES, Delft University of Technology, The Netherlands
2. IBM Microelectronics, Essex Junction VT

Abstract — A broadband, active doubler based on asymmetrically biased differential pairs delivers conversion gain (CG) from DC to 100GHz. Measured CG is >10dB up to 50GHz, >6dB up to 80GHz, and >0dB up to 100GHz. Second-harmonic (4x the input) suppression is -28dBc at 50GHz. Implemented in a 90nm SiGe technology, the 0.371mm^2 multiplier core consumes 25mA from a 4.5V supply.

Index Terms — frequency doubler, active multiplier, SiGe-HBT, SiGe-BiCMOS technology, millimeter-wave integrated circuits

I. Introduction

Integrated mm-wave frequency sources suffer from limited tunability and poor phase noise performance. A widely used alternative to fundamental synthesis is upconversion of a lower frequency source via a multiplier, or a chain of multiplier circuits.

Recent publications of silicon-based multipliers have reported impressive results, such as [1] which has a bandwidth of 220-275GHz. However, the conversion loss of passive, harmonic multipliers implies increased power consumption in high-frequency post-amplifier stages. They are also narrowband (e.g., 20% frac. BW in [1]), and are therefore tuned to a particular application.

A wideband frequency doubler capable of addressing multiple applications in the 1-100GHz range is described in this paper. The objective for this work is to demonstrate a low-power, active doubler with harmonic and spurious outputs better than -25dBc, and conversion gain across 100GHz. The use of on-chip resonators is avoided to improve bandwidth and portability between technology nodes with minimal chip area.

Section II describes the principle of operation used to realize doubling across a wide range from a low supply voltage, and design of the circuit demonstrator. Measured results of the fabricated chip's performance and comparisons to other frequency multiplier ICs are presented in Section III. Section IV summarizes the findings of this work.

II. Frequency Doubler Design

Harmonic frequency multipliers are typically comprised of a non-linear device followed by filtering. Active multipliers with conversion gain may be implemented using Gilbert's multiplier, but cascoding differential pairs consumes valuable supply headroom. An alternative based on unbalanced, or asymmetrical differential pairs proposed by Kimura [2] is selected for study in this work, because it is capable of operation from a supply below 2V in bipolar technology. A block diagram of the frequency doubler test circuit is shown in Fig. 1. An active balun converts the RF input from an external, single-ended source (terminated on-chip) to a differential signal, which drives the doubler core and generates a differential current (I+, I-) at twice the input frequency. Current from the core is fed to an active load, which suppresses DC

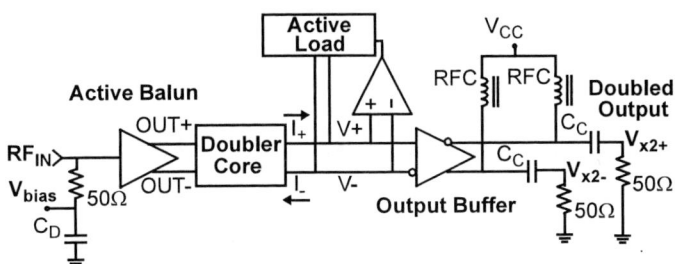

Fig. 1: Block diagram of the frequency doubler prototype.

offset between output voltages V+ and V- using a common-mode feedback (CMFB) loop. The output buffer provides the current gain required to drive a 50Ω load for testing and characterization.

A. The Kimura doubler

A schematic of Kimura's doubler is shown in Fig. 2. The differential output current (ΔI_{out}) as a function of the differential input voltage (V_{in}) is given by

$$\Delta Iout = \alpha_F I_0 [\tanh((V_{in}+V_K)/2V_T) - \tanh((V_{in}-V_K)/2V_T)], \quad (1)$$

where α_F is the common-base forward transfer ratio, V_T is the thermal voltage (kT/q), V_K is the applied offset voltage, and I_0 is the DC current biasing the differential pairs (ref. Fig. 2). Conversion gain (CG) is easily realized across a wide bandwidth using a resistive load at the output. Note that $V_K = ln(K)V_T$ for a bipolar transistor, where K is the ratio of emitter areas in an asymmetric differential pair. The factor K can be realized as an actual DC offset voltage, or using different emitter areas in each transistor pair. Simulations predict that the desired x2 output is optimized when V_K equals 60mV, or K=9.9.

Harmonic overtones, such as the quadrupled (x4) component in the output current, grow in proportion to the input amplitude. For example, simulations of the doubler core of Fig. 2 at 1GHz and V_K=60mV predict that the x4 component in ΔI_{out} is 70dB below the (desired) doubled output for 10mV input, while at 150mV input the difference between x2 and x4 components shrinks to 29dB. Simulations also show that the input amplitude must be constrained to <150mV in order to suppress x4 and

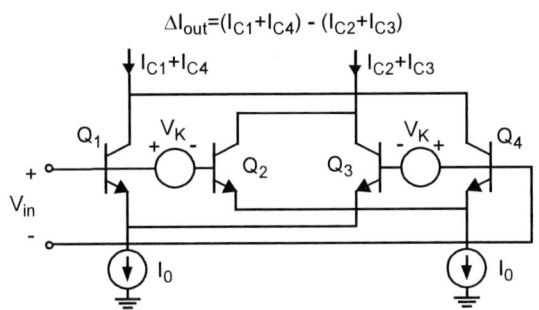

Fig. 2: The Kimura doubler stage [2].

978-1-4799-7231-9/14 $31.00 © 2014 IEEE

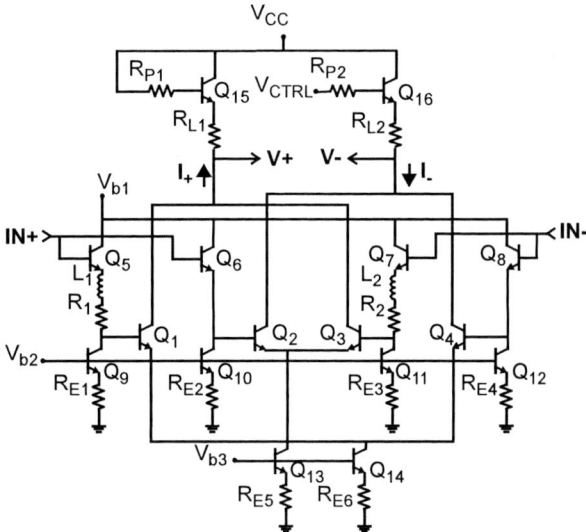

Fig. 3: Doubler and tunable active load.

other higher harmonics by better than -30dBc at the output.

A disadvantage of Kimura's multiplier arises from the offset between DC bias currents flowing through the collectors caused by V_K. In this work, a CMFB loop is used to suppress the DC offset at the output created by bias current asymmetry, and a wideband load is used to maximize output bandwidth.

B. Doubler core and tunable active load

The doubler circuit and active load are shown in Fig. 3. Followers Q_5-Q_8 buffer the input signal to matched differential pairs Q_1,Q_4 and Q_2,Q_3 with equal emitter areas. Identical area devices mitigate the effects of parasitic capacitances on the response seen at higher frequencies when mismatched pairs are used to form the multiplier core. The base-emitter voltage biasing Q_1 and Q_3 is 60mV less than V_{BE} biasing Q_2 and Q_4 due to voltage drops across resistors R_1 and R_2. Thus, quiescent currents of 0.25mA and 4.25 mA flow through Q_1,Q_3 and Q_2,Q_4, respectively. The load is comprised of a resistor (R_{L1}, R_{L2}) in series with an active inductor formed by a transistor (Q_{15}, Q_{16}) and a base resistor (R_{P1}, R_{P2}).

A single 4.5V supply was chosen to power all of the circuit blocks in the proof-of-concept prototype: 2V is applied to the

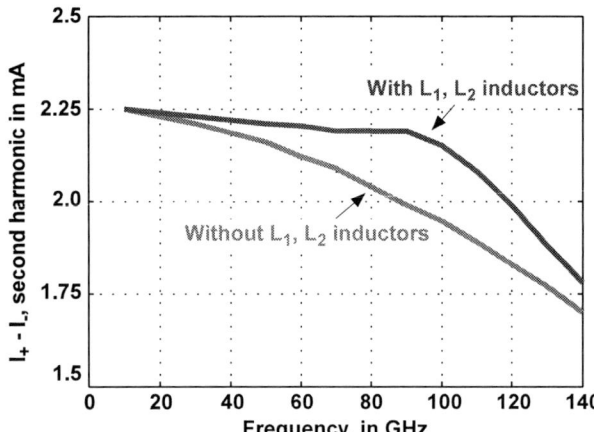

Fig. 4: Peaking the doubled output current using an active load.

active load, 1.3V is dropped across Q_1-Q_4, and 1.2V biases current sources Q_{13},Q_{14}. Operation from a 3V supply is possible, as the multiplier core requires just 1V bias across Q_1-Q_4. The supply headroom may be realized by substituting NMOS equivalents for sources Q_{13},Q_{14} and active loads Q_{15},Q_{16} (e.g., V_{DS}=0.5V across the active devices and 1V across R_{L1},R_{L2}) in the 90nm SiGe-BiCMOS technology.

The primary limitation on the frequency response of the multiplier core is the input time constant including R_1 and Q_1 (or R_2,Q_3). Input bandwidth is extended by adding inductors L_1 and L_2 (70pH each) designed to be series resonant with the input capacitance of Q_1 and Q_3, respectively. The double-frequency output current I_+-I_- drawn by the multiplier core of Fig. 3 was simulated using a short-circuited load to isolate the input and output loops. Adding inductors L_1 and L_2 peaks the response of the input loop and extends the frequency range of the doubler to ~100GHz, as shown in Fig. 4.

The impedance seen at the emitter of the active load formed by R_P and Q_L resonates with shunt parasitic capacitance at the output summing node to compensate for roll-off in currents I+ and I- with increasing frequency.

C. Op-amp

DC offset in the output current is sensed as a difference in voltage across the load of the multiplier core by an op amp (i.e., sensing V+ and V- in Fig. 3). This voltage is fed back to control the common-mode bias at the multiplier output by varying V_{CTRL} biasing the base of active load Q_{16}, as shown in Fig. 3. The 2-stage CMOS op amp (see Fig. 5) is designed using thick-oxide transistors and can handle common-mode inputs exceeding the supply voltage. It is Miller compensated by R_C and on-chip capacitor C_C.

D. Self-biased active input balun

The broadband input balun is shown in Fig. 6. Inductors L_1 and L_2 resonate with capacitance at the collectors of Q_1 and Q_2 to peak the frequency response and widen the output bandwidth. Capacitor C_1 compensates for error in the differential output current caused by b-e capacitive feedthrough. The areas of transistors Q_1 to Q_4 are also selected to reduce the phase error seen in the differential output (i.e., l_{eQ1}=2µm, l_{eQ2}=6µm, l_{eQ3}=6µm and l_{eQ4}=2µm). Simulations at 50GHz predict that the output amplitude difference reduces from 20% to 3% after these modifications, with an increase of just 0.8% (i.e., from 1.1 to 1.9%) in the phase difference between buffer outputs.

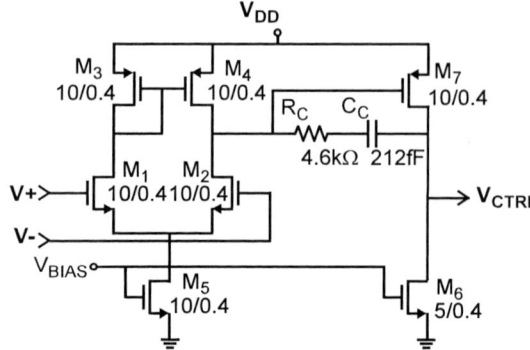

Fig. 5: Op-amp schematic.

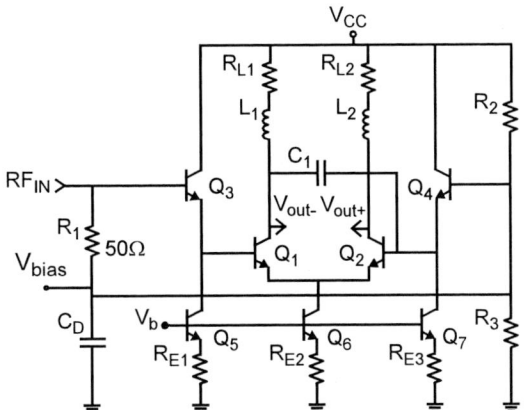

Fig. 6: Self-biased balun with amplitude and phase error compensation.

E. Output buffer

The output stage (see Fig. 7) buffers V+ and V- from the multiplier core to 50Ω loads for testing and characterization. Shunt feedback decreases the buffer output impedance, and peaking inductors L_{F1} and L_{F2} are added to increase the small-signal bandwidth from 70GHz to 130GHz. V_{CASC} is AC-grounded via a 500fF capacitor shunted by another 1.12pF capacitor in series with 3Ω on-chip. Further decoupling of 1μF‖22μF off-chip is added for testing and characterization.

III. Measurement Results

The multiplier prototype is shown in Fig. 8. Testchip area is 0.371mm² (incl. bondpads), where 0.072mm² is occupied by the doubler core and active load. The prototype consumes a total of 55.6mA from a 4.5V supply (i.e., 250mW): 25mA in the doubler core and active load, 9.5mA by the input balun, 12mA by the output buffer, and 8.5mA is dissipated in on-chip biasing networks.

On-die measurements of the prototype were performed across 10-50GHz using a spectrum analyzer (SA), and in the V- (i.e., 50-75 GHz) and W-bands (i.e., 75-100GHz) using mm-wave downconverting mixers feeding the SA. The RF power

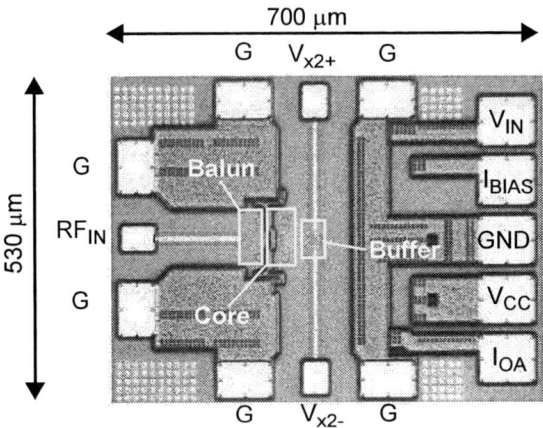

Fig. 8: Doubler testchip photomicrograph.

applied to the input and the power measured at the output are corrected for losses of the probes and cables comprising the test set-up. Fig. 9 shows the (overlapping) V- and W-band output spectra measured for 30, 35, 40, 45, and 50GHz inputs. Conversion gain (CG) varies from +12 to 0dB as the output frequency ranges from DC to 100GHz.

The relationship between input power (P_{in} at 40GHz) and doubler output power (P_{out} at 80GHz) is plotted in Fig. 10. The output power increases up to a P_{in} of -15dBm, beyond which output power is constant at approximately -7.5dBm. Measured (power) gain for an 80GHz output is 6.6dB when -15dBm applied to the input. Note that gain expansion is observed between -20 and -15dBm input power. The greatest fundamental suppression (28dBc) is realized when P_{in} is -14dBm. The 4th harmonic (i.e., 160GHz output) could not be measured due bandwidth limitations of the test set-up, but simulation data is included in Fig. 10 for comparison. The x4 component at the output increases rapidly above -15dBm input power (i.e., peak $V_{in} > \sim 6V_T$).

Measured and simulated output power vs. frequency are plotted in Fig. 11 from 10-110GHz. Suppression of the fundamental is also shown here, as the measured and simulated outputs at the fundamental are plotted on the same figure.

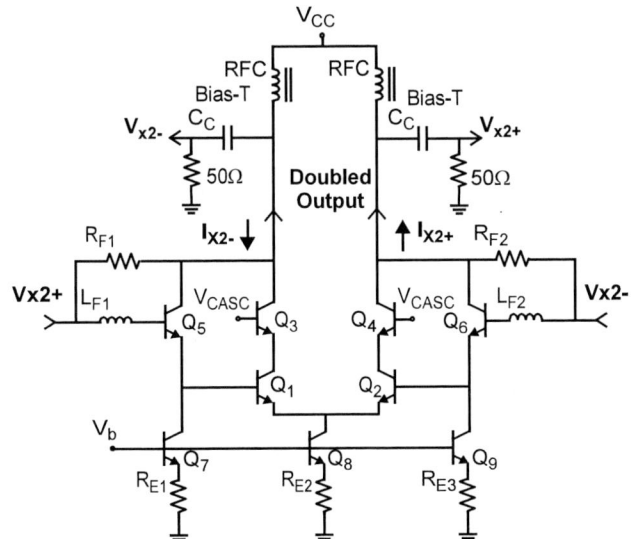

Fig. 7: Output buffer schematic.

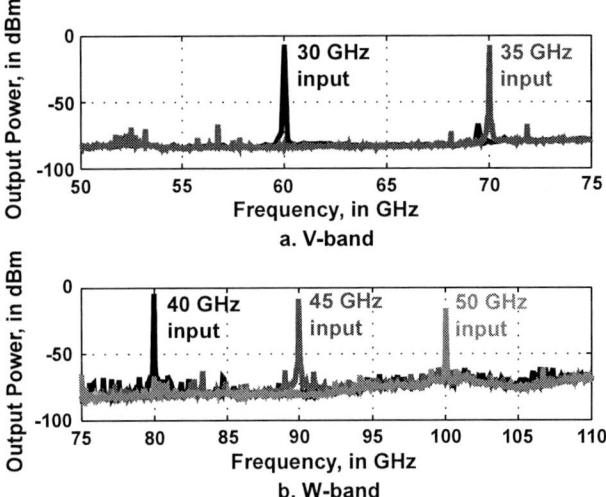

Fig. 9: Output spectrum in: a) V-band: 50-75 GHz, b) W-band: 75-100 GHz.

978-1-4799-7231-9/14 $31.00 © 2014 IEEE   11

Table 1: Frequency Multiplier Performance Comparison

Reference	Multiplication Factor	Max. CG (dB)	Output Bandwidth (GHz)	Input Power (dBm)	Max./Min. Suppression of f_{in} (dBc)	DC Power (mW)/ Supply Voltage	Technology
This Work	x2	12	DC-100	-16	28 / 22	250 / 4.5	90nm SiGe BiCMOS
[4], 2005	x2	1	DC-100	-12	30	150 / 4	0.2μm InP HBT
[5], 2008	x2	5.7	3-50	0	30/15	600 / -	GaAs pHEMT
[6], 2009	x2	10.2	36-80	-8	36 / 20	137 / 3.3	0.18μm SiGe BiCMOS
[7], 2009	x2	-10.0	110-125	8.5	12	- / -1.5	0.13μm CMOS

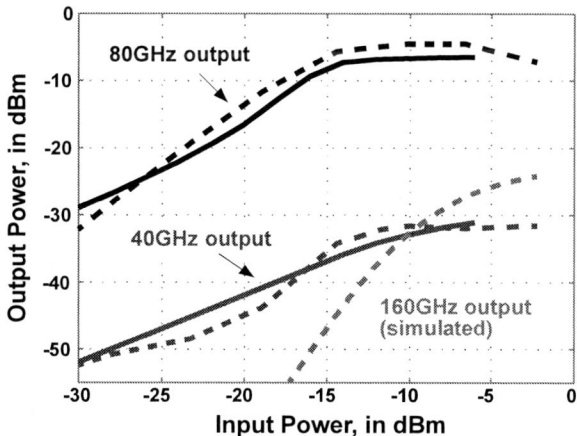

Fig. 10: Measured (solid line) and simulated (dashed) output power vs. input power at 40GHz input signal.

Across 10-110GHz, the fundamental and x4 components are suppressed by more than 28dBc and 30dBc, respectively (x2 output as the reference). Simulation and measurement trend closely, with simulation predicting 1-3dB higher output power than was measured across the entire range.

Table 1 compares the measured performance of the 90nm SiGe-HBT doubler to other examples from the literature.

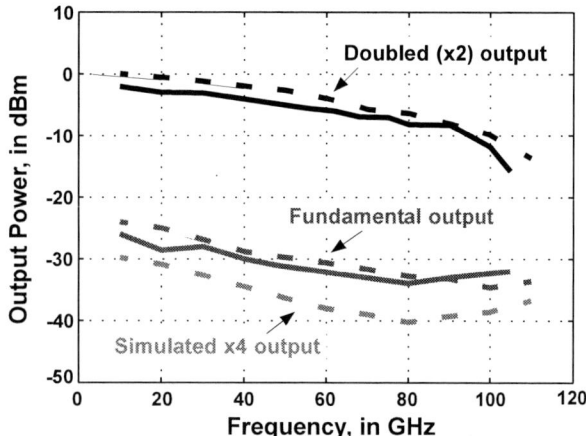

Fig. 11: Measured (solid line) and simulated (dashed) output power vs. output frequency for the wideband doubler.

Example [4] has comparable bandwidth, but the maximum CG of the InP-HBT doubler is 11dB lower than the gain of the SiGe-HBT circuit implemented in this work. Example [5] has a maximum conversion gain of only 5.7dB, one-half the bandwidth, and consumes 2.4x more power than the 90nm SiGe-HBT doubler prototype. Example [6] also uses SiGe-HBTs to achieve >10dB CG, but its operating range is restricted to 40-80GHz. The narrowband (110-125GHz) CMOS passive doubler from [7] has a conversion loss of 10dB and requires 8.5dBm input power in operation.

IV. Conclusions

A doubler suitable for applications in the 1-110GHz frequency range is proposed. It is realized in 90nm SiGe-BiCMOS technology using asymmetrically biased differential pairs, and a common-mode feedback loop to suppress DC offset at the doubled output. Measured conversion gain ranges from 12-0dB across DC-100GHz. Fundamental suppression at the output is 28 to 22dBc across the same frequency range. The 0.072mm^2 multiplier core dissipates 112.5mW from 4.5V.

Acknowledgments

IBM Microelectronics for IC fabrication facilitated by MOSIS, and W. Straver at TU Delft for technical support.

References

[1] O. Momeni, E. Afshari, "A Broadband mm-Wave and Terahertz Traveling-Wave Frequency Multiplier on CMOS," *IEEE-JSSC*, vol. 46, no. 12, pp. 2966-2976, Dec. 2011.
[2] K. Kimura, "A bipolar low-voltage quarter-square multiplier with a resistive-input based on the bias offset technique," *IEEE-JSSC*, vol. 32, no. 2, pp. 258-266, Feb. 1997.
[3] N.E. Lourenco, R.L. Schmid, et al., "Total Dose and Transient Response of SiGe HBTs from a New 4th-Generation, 90nm SiGe BiCMOS Technology," *2012 IEEE Radiation Effects Data Workshop*, Miami FL, pp. 1-5, July 2012.
[4] V. Puyal, A. Konczykowska, P. Nouet, et al., "DC-100-GHz frequency doublers in InP DHBT technology," *IEEE Transactions on Microwave Theory and Techniques*, vol. 53, no. 4, pp. 1338-1344, April 2005.
[5] Y. Liu, T. Yang, Z. Yang, J. Chen, "A 3–50 GHz Ultra-Wideband PHEMT MMIC Balanced Frequency Doubler," *IEEE Microwave and Wireless Components Letters*, vol. 18, no. 9, pp. 629-631, Sept. 2008.
[6] A.Y-K. Chen, et al., "A 36–80 GHz High Gain Millimeter-Wave Double-Balanced Active Frequency Doubler in SiGe BiCMOS," *Microwave and Wireless Components Letters*, vol. 19, no. 9, pp. 572-574, Sept. 2009.
[7] C. Mao, C.S. Nallani, S. Sankaran, E. Seok, K.K. O, "125-GHz Diode Frequency Doubler in 0.13-CMOS," *IEEE-JSSC*, vol. 44, no. 5, pp. 1531-1538, May 2009.

A True-RMS Integrated Power Sensor for On-Chip Calibration

Jonas Wursthorn[1,2], Herbert Knapp[1], Klaus Aufinger[1], Rudolf Lachner[1], Jidan Al-Eryani[2] and Linus Maurer[2]

[1]Infineon Technologies AG, Am Campeon 1-12, 85579 Neubiberg, Germany

[2]Universität der Bundeswehr München, 85577 Neubiberg, Germany

Abstract—An on-chip RF power sensor based on the bolometer principle is presented in this work. The sensor can be used for transmitter calibration purposes without using any RF equipment offering time and cost savings. Investigations on various layers of the f_T=170 GHz SiGe bipolar technology, which is used in this work, are performed and evaluated. Especially silicided polysilicon layers show a promising behavior as bolometer.

Measurements on a 76 GHz transmitter chip are performed to point out the feasibility of the on-chip absolute RF power sensor. The results show that this novel sensor concept is well suited for on-chip self-calibration.

I. Introduction

Because of higher data rates and better spatial accuracy, communication and radar systems tend to operate at increasingly higher frequencies. This trend is enabled by high performance silicon germanium bipolar processes like Infineons B7HF200 technology with maximum oscillation frequencies for transistors above 200 GHz [1]. The high speed transistors allow the circuit designer to realize the core parts of transceivers (VCOs, power amplifiers, mixers, ...) very well and with sufficient performance. Apart from these core parts there are design challenges for additional periphery like power detectors which do not benefit as much from high speed transistors as the core parts but still have to handle these high operating frequencies.

Diode-based or CMOS transistor-based power detectors do have an impressive sensitivity for the lower GHz frequency range [2], [3]. There are also CMOS transistor-based solutions up to frequencies around 60 GHz [4]. However, these approaches are limited to relative power measurements. To get a statement on the absolute power, one has to calibrate these detectors by measuring the actual RF power at e.g. a transmitter channel while monitoring the voltage at the diode-based detector. As the absolute power is of interest for example for radar applications where a certain output power may not be exceeded because of legal conditions, the solution with a diode-based detector only does not fulfill these requirements.

The power sensor presented in this work is based on a resistive layer in which RF power is dissipated. If the resistor is temperature dependent, it is a measure of dissipated power. The temperature coefficients of various layers are investigated in section II. Section III describes the implementation of the power sensor. The thermal time constant of the resistor forms a true-rms in terms of power from the RF signal applied to it. Another identical resistor is used as a reference and heated up with DC power in order to compensate the difference in resistance. This principle is further explained in section IV.

As the RF power needed to heat up the resistor significantly has to be relatively high, it is difficult to divert this amount of power from the actual RF transmitter path (as it can be done with diode-based detectors), while keeping the output power at the desired level. Nevertheless, the described sensor can be used to perform a one-time calibration of the sensitive diode-based detectors with a simple DC instead of an RF measurement. This method will be explained in section V. An existing transmitter chip has been extended with the described power sensor in a way that one can perform the mentioned calibration. The measurement results are discussed in section VI.

II. Investigation of Layers Suitable as Bolometer

The technology used offers four copper layers, differently doped polysilicon layers, which are used as base/emitter contacts or resistors, and an additional tantalum nitride layer which forms low-ohmic resistors. Regarding the fact that the layer needed for the power sensor has to have a significant temperature coefficient as well as a suitable sheet resistance, one can perform a preselection from the above mentioned layers. The metal stack does not suit the requirements because its sheet resistance is too small. A 50 Ω copper resistor would consume a lot of chip area and is therefore not suited for the application. The tantalum nitride layer is designed to have a temperature coefficient as small as possible which narrows the further investigation down to the polysilicon layers.

The polysilicon layers are available with two different doping levels resulting in different sheet resistances. To further increase the conductivity of the polysilicon layer there is a silicide option. This option is originally intended to be used for low-ohmic contacts of bipolar devices or short low-ohmic connections underneath the lowest copper layer.

Figure 1 shows the resistance of the various layers over temperature. The silicided poly resistor shows the largest temperature coefficient and is therefore further investigated.

In the application, the resistive silicided layer will be heated up by an RF signal applied to it. The power of these RF signals is limited by the on-chip power amplifiers and is expected to be in the range of 10-15 dBm at a 50 Ω load. In other words, the silicided poly resistor has to endure this power level without being destroyed from thermal heat-up. To investigate if the silicided resistor is capable to deal with the high temperature

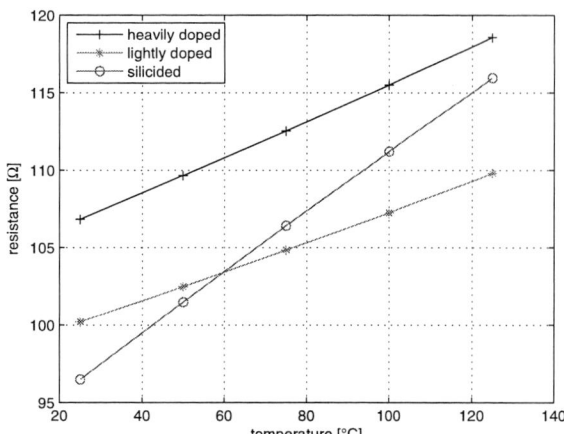

Fig. 1. Behavior of various polysilicon layers over temperature.

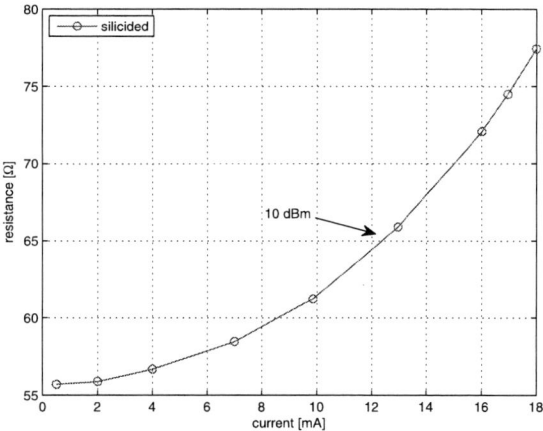

Fig. 2. Self-heating effect due to DC power disspated in the silicided resistor.

further measurements have been performed at a $50\,\Omega$ nominal value resistor. Due to process variation the resistance in Figure 2 is $56\,\Omega$ instead of $50\,\Omega$ for low DC currents. Figure 2 shows the resistor behavior when DC current is applied. As the self-heating of the resistor does not depend on the kind of power (DC or RF) this is a valid method to estimate how much power the resistor can handle. However, the resistance changes with increasing power and accordingly the impedance at the power amplifier's load changes as well. Please note that the measurements in Figure 1 and Figure 2 are performed on different test structures. Otherwise the resistance at room temperature in Figure 1 and the resistance at low currents in Figure 2 would be identical.

III. IMPLEMENTATION OF THE POWER SENSOR

Figure 3 shows a simple way to implement the power sensor on the chip. The RF signal provided by the power amplifier

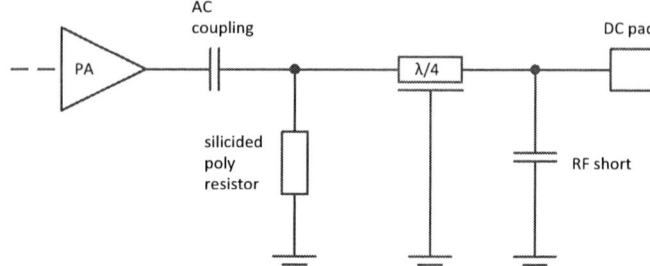

Fig. 3. Implementation of the power sensor at the power amplifier's (PA) output.

(PA) stage is AC coupled to the $50\,\Omega$ silicided resistor. The $\lambda/4$ transmission line transforms the RF short at the DC pad to an RF open. This prevents RF power from being radiated off the chip. For DC measurements the transmission line and the resistor are the only elements that have to be taken into account. A 4-wire resistor measurement can be performed using the DC pad and a ground pad.

IV. RF POWER COMPENSATION

The goal is to design a true-rms power sensor which allows absolute power level measurement of the RF signal. The simplest way is to compensate the difference in resistance caused by the RF power by applying a DC power to another resistor serving as reference. Assuming that the resistors are well matched, they show the same resistance if the same amount of power is dissipated. The DC current and voltage required to compensate the dissipated RF power can be measured and evaluated.

Evaluation might be performed automatically within an external microcontroller or even on-chip with a multiplier circuit. For this work the compensation is done manually. First, the resistance is measured when RF power is applied. In a second step the resistor is heated up with DC power only. If the measured resistance reaches the same value as for the RF heat-up before, the dissipated DC power equals the RF power.

V. CALIBRATION METHOD

State of the art RF transceivers with diode-based power detectors like the one described in [5] and [6] need an initial calibration for absolute power level monitoring. The output power has to be measured with an external RF power sensor while monitoring the diode-based power detector voltage. This leads to a voltage-power characteristic which allows to recalculate the RF power from the measured voltage during operation. An application for the presented power sensor is to replace the complex RF calibration measurement by a simple DC measurement.

Figure 4 shows an RF transmitter with two identical channels (TX1 and TX2) containing a common voltage controlled oscillator (VCO), a 1:n power splitter, a power amplifier (PA) and a diode-based power detector. The third channel is similar to the channels TX1 and TX2 and is needed for the calibration process. Instead of measuring the RF power at the transmitter

978-1-4799-7231-9/14 $31.00 © 2014 IEEE

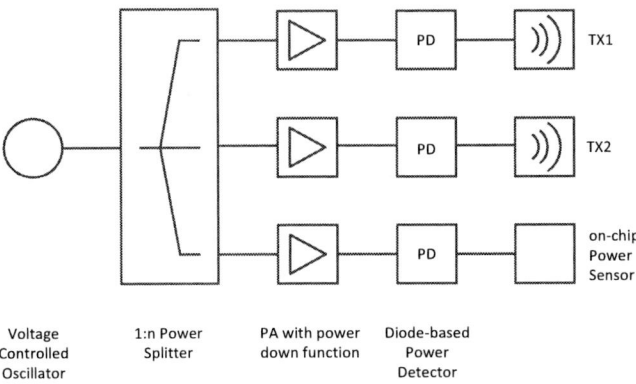

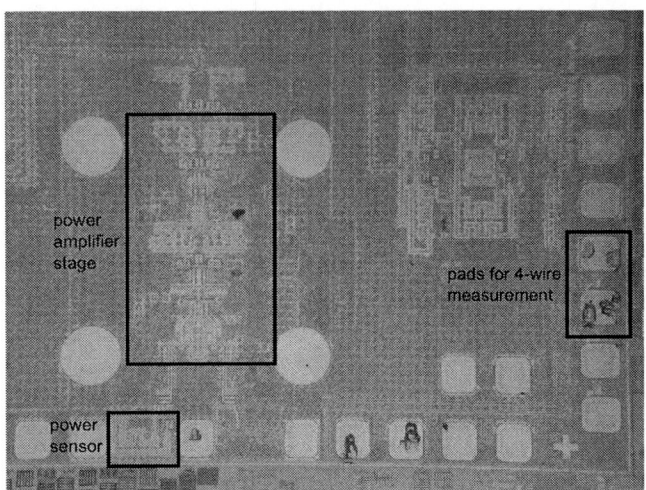

Fig. 4. Block diagramm of a three-channel transmitter chip where one channel is terminated with the on-chip power sensor serving as a calibration path. Even for multi-channel transmitters one calibration path is sufficient to perform the calibration process.

output (TX1 or TX2) the DC signal of the on-chip power sensor is evaluated. If the diode-based detector voltage is monitored at the same time, a voltage-power characteristic can be obtained (similar to the one described earlier). After performing the calibration process the calibration channel is switched to power down mode. If necessary, re-calibration is possible - even if the chip is mounted e.g. in a radar module. Assuming that the variation between the transmitter channels is low, one calibration path is sufficient even for multi-channel transmitter chips. The additional amplifier needed for the calibration path is a minor disadvantage, knowing that the minimum chip area is mainly limited by the required distance between the RF channels on the chip edges to keep the crosstalk at a tolerable limit.

VI. MEASUREMENT RESULTS

Measurements are performed on a modified two-channel transmitter chip which does originally not contain the additional calibration path shown in Figure 4. In one channel the original RF transmitter output is terminated by the presented power sensor while the other channel remains unchanged. The modified channel can now be used as the calibration path. Figure 5 shows the part of the chip with the modified channel. The chip area containing the power sensor has originally been occupied by an RF transmitter pad. The unchanged channel with the RF output is used to verify the measurement results gained from the presented power sensor. The power level measured by an external power sensor and the power detected by the presented power sensor are compared.

The power level provided by the power amplifiers can be set via a digital control interface. Unfortunately, the circuit does not provide equidistant power level steps. All measurements in this work are on-wafer measurements at room temperature if not denoted otherwise. The oscillation frequency of the VCO is set to 76 GHz.

The power level of both channels is swept simultaneously with the digital control interface. On the one hand, the actual RF power at channel two is monitored with an external power

Fig. 5. Part of the transmitter chip, where one RF pad is replaced by the presented power sensor. The 4-wire resistance measurement is performed on the marked pads.

sensor. On the other hand, the change in resistance at channel one is detected via a 4-wire measurement. In Figure 6 the change in resistance (R_{sense}, left y-axis) is plotted over the measured RF power (x-axis). The DC power required to introduce the same change in resistance as the RF power can be taken from Figure 2. Comparing the power levels of e.g. the 72 Ω point results in 8.9 dBm vs. 12.6 dBm for the RF and DC power level, respectively. The RF power level has to be corrected upwards by 1.5-2 dB because of insertion losses of the waveguide transitions. This results in a difference of 1.7 dB between the external RF power sensor measurement and the one on chip. This error is ascribed to the different loads the power amplifiers face at their outputs. One is loaded with the RF pads and the waveguide which is not a perfect 50 Ω match, the other is loaded with the on-chip resistive element changing its value with increasing power levels. The latter effect may be minimized by pre-heating e.g. a 35 Ω resistor with DC power which can be reduced if RF power is applied. This way the resistance is kept at 50 Ω. The decrease in DC power can be evaluated just the same way as an increase.

The right y-axis in Figure 6 shows the output voltage of the diode-based power detector (D_{sense}). As mentioned above it shows a non-linear behavior. Furthermore, no absolute power level can be obtained from this diode-based power detector.

VII. CONCLUSION

It has been shown that a silicided polysilicon layer which is available in the used technology can be used as a thermistor element. The change in resistance can be compensated with DC power, resulting in an on-chip RF power sensor. The error between an external RF power sensor measurement and the on-chip measurement is less than 2 dB and can be further minimized with little design effort. This novel sensor concept is promising for on-chip self-calibration of RF power sensors. The sensor principle is suitable for a wide bandwidth

978-1-4799-7231-9/14 $31.00 © 2014 IEEE

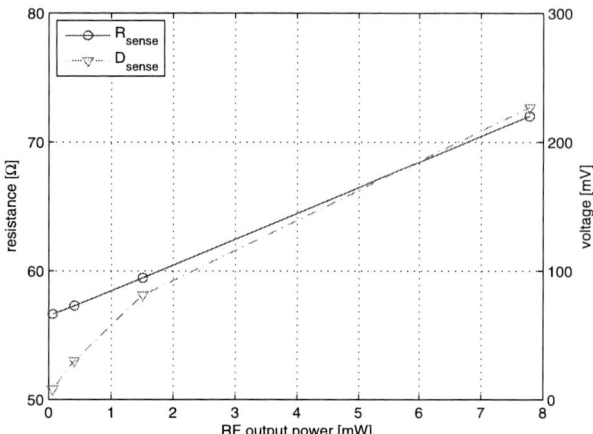

Fig. 6. The change in resistance at the output of channel one (left y-axis) scales linearly with the measured RF output power of channel two (x-axis) while the diode-based sensor delivers a non-linear output signal (right y-axis).

and can therefore be adapted to many other high-frequency applications.

ACKNOWLEDGMENT

The authors wish to acknowledge the DOTSEVEN project (316755) supported by the European Commission through the Seventh Framework Programme (FP7) for Research and Technology Development.

REFERENCES

[1] J. Böck, H. Schäfer, K. Aufinger, R. Stengl, S. Boguth, R. Schreiter, M. Rest, H. Knapp, M. Wurzer, W. Perndl, T. Böttner, and T. Meister, "SiGe Bipolar Technology for Automotive Radar Applications," in *Bipolar/BiCMOS Circuits and Technology, 2004. Proceedings of the 2004 Meeting*, Sept 2004, pp. 84–87.

[2] B. Francois and P. Reynaert, "A Transformer-Coupled True-RMS Power Detector in 40nm CMOS," in *Solid-State Circuits Conference Digest of Technical Papers (ISSCC), 2014 IEEE International*, Feb 2014, pp. 62–63.

[3] A. Valdes-Garcia, R. Venkatasubramanian, J. Silva-Martinez, and E. Sanchez-Sinencio, "A Broadband CMOS Amplitude Detector for On-Chip RF Measurements," *Instrumentation and Measurement, IEEE Transactions on*, vol. 57, no. 7, pp. 1470–1477, July 2008.

[4] J. Gorisse, A. Cathelin, A. Kaiser, and E. Kerherve, "A 60GHz CMOS RMS Power Detector for Antenna Impedance Mismatch Detection," in *Circuits and Systems and TAISA Conference, 2008. NEWCAS-TAISA 2008. 2008 Joint 6th International IEEE Northeast Workshop on*, June 2008, pp. 93–96.

[5] H. P. Forstner, H. Knapp, H. Jäger, E. Kolmhofer, J. Platz, F. Starzer, M. Treml, A. Schinko, G. Birschkus, J. Böck, K. Aufinger, R. Lachner, T. Meister, H. Schäfer, D. Lukashevich, S. Boguth, A. Fischer, F. Reininger, L. Maurer, J. Minichshofer, and D. Steinbuch, "A 77GHz 4-Channel Automotive Radar Transceiver in SiGe," in *Radio Frequency Integrated Circuits Symposium, 2008. RFIC 2008. IEEE*, June 2008, pp. 233–236.

[6] H. Knapp, M. Treml, A. Schinko, E. Kolmhofer, S. Matzinger, G. Strasser, R. Lachner, L. Maurer, and J. Minichshofer, "Three-Channel 77 GHz Automotive Radar Transmitter in Plastic Package," in *Radio Frequency Integrated Circuits Symposium (RFIC), 2012 IEEE*, June 2012, pp. 119–122.

Microwave Noise Properties of Heterojunction Bipolar Transistors

Joseph C. Bardin*, James Chingwei Li†, Ahmet Hakan Coskun*, Metin Ayata*, and Zachariah G. Boynton*
*Department of Electrical and Computer Engineering
University of Massachusetts Amherst, Amherst MA 01003–9292
Email: jbardin@ecs.umass.edu
†HRL Laboratories LLC, Malibu CA
Email: jli@hrl.com

Abstract—The broadband noise performance of silicon germanium (SiGe) and compound semiconductor (CS) heterojunction bipolar transistors (HBTs) is presented. The key noise mechanisms in HBTs are summarized to provide a framework through which the noise limitations of the devices can be understood. Process related details and transport physics of each class of device are then compared and contrasted with a focus on details relevant to the broadband noise performance. Fundamental noise performance limitations are presented for exemplary InP/InGaAs/InP BiCMOS DHBTs and SiGe HBTs. The state-of-the-art for SiGe and CS HBTs operating in the 1–300 GHz frequency range is reviewed and interpreted in the context of fundamental performance limitations. Finally, the paper concludes with a discussion of how the performance of future HBT technology generations should improve.

Index Terms—Noise, SiGe, InP/InGaAs/InP, low noise amplifier, noise figure, noise measure, HBT, cryogenic

I. Introduction

Low noise amplifiers (LNAs) are critical components in a variety of systems with applications ranging from terrestrial communications to quantum computing. These devices enable the sensitive systems that permit scientists to listen to microwave emissions from distant galaxies and are an essential component in nearly every radio receiver. The performance of a well designed low-noise amplifier is ultimately limited by the underlying intrinsic noise properties of the transistor technology. Therefore, decades of intense research has been focused on understanding and optimizing the noise performance of semiconductor devices.

In this paper, we review the broadband[1] noise properties of silicon germanium (SiGe) and compound semiconductor (CS) heterojunction bipolar transistors (HBTs), with an emphasis on fundamental frequency and temperature dependent performance limitations as well as the requirements for improving the state of the art. The outline of the paper is as follows:

1) A review of the key sources of noise in an HBT, and how they relate to different aspects of the device.
2) A summary of the key physical factors impacting the noise performance of CS and SiGe devices.
3) A discussion of the fundamental noise limits of devices from example SiGe and CS HBT processes.

[1]i.e. at frequencies above the $1/f$ noise corner.

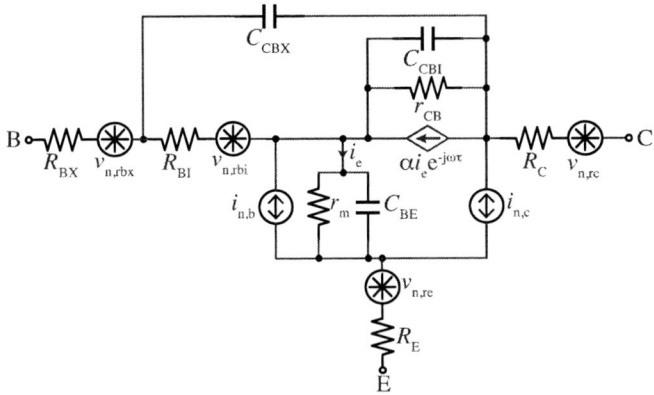

Fig. 1. Equivalent small-signal noise model for an HBT.

4) A summary of the state-of-the-art in HBT LNAs and a discussion of how this relates to fundamental limits.
5) A summary of how technology developments are likely to impact the broadband noise performance of future HBTs and a discussion of steps that need to be taken to overcome the limitations of currently available devices.

II. Broadband Noise Mechanisms in HBTs

The small-signal equivalent noise model for an HBT is shown in Fig. 1. The circuit model employs a tee representation of the intrinsic transistor and includes extrinsic resistances R_{BX}, R_C, and R_E to account for access resistances and extrinsic capacitance C_{CBX} to describe capacitive coupling between the extrinsic base and extrinsic collector. Thermal noise is considered for each physical resistance and the shot noise due to diffusion of carriers across the base–emitter junction is incorporated through a pair of shunt current noise sources connecting from the intrinsic base and collector nodes to the intrinsic emitter. The shot noise sources are partially correlated and their power and cross spectral densities are given as [1], [2]

$$\overline{|i_{n,b}|^2} = 2q\left(I_B + |1 - \exp\{-j\omega\tau_n\}|^2 I_C\right), \quad (1)$$

$$\overline{|i_{n,c}|^2} = 2qI_C, \quad (2)$$

978-1-4799-7231-9/14 $31.00 © 2014 IEEE

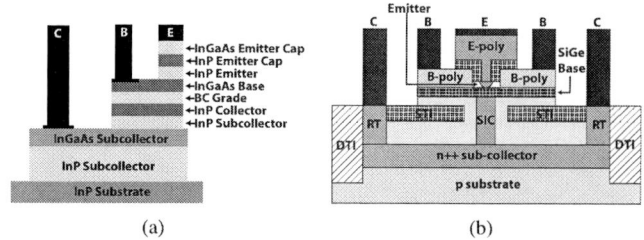

Fig. 2. Typical cross section of (a) InP and (b) SiGe HBTs, after [7] and [6], respectively. Drawings not to scale.

and

$$\overline{i_{n,b}i_{n,c}^*} = 2qI_C \left(\exp\left\{j\omega\tau_n\right\} - 1\right), \tag{3}$$

where I_B and I_C are the DC base and collector currents, q is the charge of an electron, ω is the angular frequency, and τ_n is a delay term known as the noise delay and is typically 50–65% of the forward transit time [2]–[4]. At low frequencies, the product of $\omega\tau_n$ is small and the shot noise sources simplify to the uncorrelated case. However, at higher frequencies, the correlation actually improves the achievable noise performance, much in the same way that correlation improves noise performance in field effect devices [3], [5].

Momentarily neglecting shot-noise correlation, the minimum noise temperature that can be achieved by an HBT can be estimated as [6]

$$T_{MIN} \approx T_a n_c \sqrt{\frac{1}{\beta_{DC}}\left(1 + 2\zeta\right) + 2\zeta\left(\frac{f}{f_t}\right)^2}, \tag{4}$$

where T_a is the ambient temperature, n_c is the collector current ideality factor, β_{DC} is the DC current gain, $\zeta = g_m\left(R_{BX} + R_{BI} + R_E\right)/n_c$, g_m is the transconductance, f is frequency, and f_t is the unity current gain cutoff frequency. Equation (4) sheds light on the properties that must be engineered to realize an HBT with optimized noise performance:

- In the low-frequency limit and for biases in the range that $2\zeta << 1$, the noise performance is primarily determined by the DC current gain, with larger values of β_{DC} resulting in improved noise performance.
- At higher current densities—and hence higher values of g_m/n_c—the term ζ can no longer be neglected. The low-frequency noise will be severely impacted as ζ grows greater than $1/2$ and the high frequency noise will grow with any increase in ζ. Therefore, to optimize the noise at all frequencies, it is critical that the base and emitter resistances be minimized.
- The corner frequency at which T_{MIN} begins to rise with frequency is given as $f_c = f_t T_{min,LF}/T_a n_c\sqrt{2\zeta}$, where $T_{min,LF}$ is the minimum noise temperature for $f << f_t$. Therefore, higher values of f_t lead to a larger value of f_c, and hence improved high-frequency noise performance.

In summary, achieving optimum low-frequency noise performance requires realizing large values of β_{DC} and minimizing the base and emitter resistances. Optimizing high-frequency

noise performance requires minimizing the base and emitter resistances while maximizing[2] f_t.

III. COMPARISON OF SiGe AND CS HBTs

In the design of a homojunction bipolar transistor, there is a severe trade-off between β_{DC}—proportional to the ratio of the emitter doping to the base doping—and the base resistance—proportional to the doping in the base. Since optimizing the noise performance of a bipolar transistor requires simultaneously maximizing β_{DC} and minimizing the base resistance, the lack of sufficient degrees of freedom to tune these parameters independently is a significant shortcoming.

In contrast, HBTs employ a heterojunction at the base–emitter interface as an additional degree of freedom to enable a heavily doped base and sufficiently large β_{DC}. Often, the base is a different material than both the collector and emitter, resulting in a second heterojunction at the base–collector interface; such a device is called a double heterojunction bipolar transistor (DHBT) as opposed to a single heterojuction bipolar transistor (SHBT), in which the base and collector are the same material. While both SiGe and CS HBTs employ band structure engineering, there are significant differences between these families of devices, both in terms of device structure and the underlying transport physics.

Schematic cross-sections for typical InP/InGaAs/InP and SiGe HBTs appear in Fig. 2. In general, an important distinction between CS and SiGe BiCMOS processes is the use of subtractive processing for CS devices and additive processing for SiGe transistors. The example CS device (Fig. 2(a)) is a triple mesa transistor that is defined from epitaxially grown layers on top of an InP substrate. A three layer InP/InGaAs/InP sub-collector is employed to allow for the definition of a bottom mesa layer through selective etching. The material composition employed in the sub-collector involves a careful trade-off between electrical and thermal properties [8], [9]. Above the sub-collector is the intrinsic device, which consists of four distinct layers: an InP intrinsic collector, a graded transition from InP to InGaAs, an InGaAs base, and an InP emitter. The base material is also InGaAs, allowing for selective etching to be used for definition of the base mesa. Finally, a two layer emitter cap is fabricated above the intrinsic emitter and an emitter mesa is defined using selective etching. While the epitaxial structure of Fig. 2(a) may appear complex, it is not uncommon for devices to employ additional layers to further enhance performance (e.g., [10]).

The SiGe HBT (Fig. 2(b)) is a dual mesa DHBT that is fabricated epitaxially on a lightly-doped silicon substrate. A buried subcollector is formed, either epitaxially [11]–[13] prior to processing the wafer or, less commonly, through ion implantation [14]. While the latter approach is lower cost, the performance of the epitaxially defined sub-collector is superior. The buried subcollector is contacted using reach-through implants. To reduce coupling to neighboring devices

---

[2]An alternative viewpoint is that one should maximize f_{max} to optimize high frequency noise.

due to capacitive coupling from the sub-collector to the doped substrate, deep-trench isolation is provided in most processes [11], [12]. Above the subcollector is the intrinsic device, which consists of a selectively-implanted collector, the SiGe base, and the emitter region. The SiGe film is deposited as a blanket layer, and the regions above mono-crystalline silicon and STI form the mono-crystalline SiGe intrinsic base and poly-crystalline SiGe base links, respectively. The poly-crystalline SiGe regions are patterned and then doped with additional boron and salicided to reduce the extrinsic base resistance [15]. Finally, the emitter mesa is capped using a polysilicon layer and a contact is formed at the top.

Typical doping profiles and the band structures for exemplary InP/InGaAs/InP and SiGe HBTs appear in Fig. 3. Impurity concentrations within the heavily-doped sub-collector and emitter cap regions are omitted from Figs. 3(a) and 3(b), so as to focus on the intrinsic device. Comparison of the doping profiles and band-diagrams reveals significant differences between the SiGe and CS devices. Referring to Fig. 3(c), the bandgap in InGaAs base region is significantly lower than that of the InP emitter. The net result is that electrons diffusing from the emitter to the base require exponentially less thermal energy to diffuse across the base-emitter conduction band barrier in comparison to holes being back injected from the base into the emitter. To achieve an acceptable base resistance despite the low hole mobility of doped InGaAs [16], heavy base dopant concentrations on the order of 1×10^{19} are required. The InGaAs base is also typically compositionally graded to create a base drift field that speeds the transit of minority carries towards the collector.

Since the electron mobility in the InP emitter is orders of magnitude higher than the hole mobility of the base [17], similar levels of doping are not required within the intrinsic emitter. Due to the large valence band offset and the heavily doped base, the base current is dominated by recombination in the neutral base and the space-charge regions rather than diffusion [18]. Previous studies have shown that the relationship of dopant concentration to the rate of recombination leads to a ratio of β_{DC}/R_{BIS}^2 that is a material related constant [19] (R_{BIS} is the intrinsic base sheet resistance). A base-collector grade region is employed to prevent the formation of an extra barrier that otherwise blocks carriers as they drift from the base to the collector [20]. The intrinsic collector is typically lightly doped and fully depleted near zero bias due the high electron mobility, low conduction band density of states, and need for low C_{BC}.

The doping profile of a SiGe HBT more closely resembles that of a classical bipolar device, with a very heavily doped emitter, a heavily doped base, and a lightly doped intrinsic collector. A graded SiGe alloy is employed in the base, with increasing Ge content in the direction towards the collector. The effect of the Ge is to reduce the bandgap of the alloy from that of silicon (1.12 eV) towards that of Ge (0.66 eV). As in the case of the InP/InGaAs/InP DHBT, the SiGe HBT benefits from a reduction in the barrier seen by electrons diffusing from the emitter to the base. Moreover, the graded Ge content causes

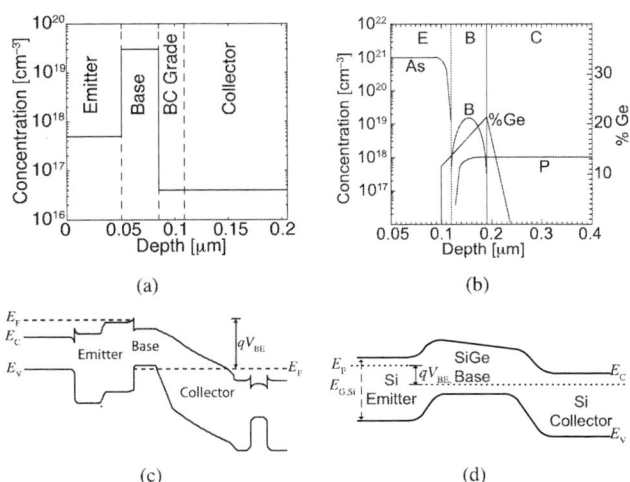

Fig. 3. Simplified doping profiles for (a) InP/InGaAs/InP DHBT and (b) SiGe HBT. Band diagrams for (c) InP/InGaAs/InP DHBT and (d) SiGe HBT. InP/InGaAs/InP HBT band diagram, after [21].

the bandgap to grow narrower in the direction of electron transport, which leads to a quasi-electric field that reduces transit times.

IV. EXAMPLE DEVICE PERFORMANCE

From a systems perspective, the noise performance of an amplifying device is impacted by both its noise figure and its available gain (G_{av}). In comparing devices, it is therefore important to use a metric that accounts for both noise and gain. Such a metric is Haus's noise measure [22], $M = (F - 1)/(1 - 1/G_{av})$. Similar to the noise figure, the noise measure is dependent upon generator impedance, bias condition, and, at a fixed bias, has a unique minima that can be found explicitly [23], [24]. It can be shown that the noise measure is a fundamental invariant of a network in the same way as Mason's gain; that is, the unique minimum value of M is invariant to any lossless imbedding, including feedback. Moreover, the product[3] of $T_0 M$ yields the noise temperature of an infinite chain of identical devices and is therefore an estimate of the achievable noise temperature for a device in a system application. Thus, one can define the cascaded noise temperature in terms of the noise temperature (T_e) [25]

$$T_{CAS} \equiv T_0 M = \frac{T_e}{1 - 1/G_{av}}, \qquad (5)$$

the global minima of which can be interpreted as the ultimate limit in the noise performance of a device. It should be noted that at frequencies well below f_{max}, where G_{av} is large, T_{CAS} is approximately equal to T_e.

A. Room Temperature Performance of CS HBTs

In a recent empirical study [26], the performance of 0.25 μm InP/InGaAs/InP DHBTs integrated with a 90 nm CMOS process was evaluated in terms of the cascaded noise

---

[3]$T_0 = 290$ K is the standard reference temperature.

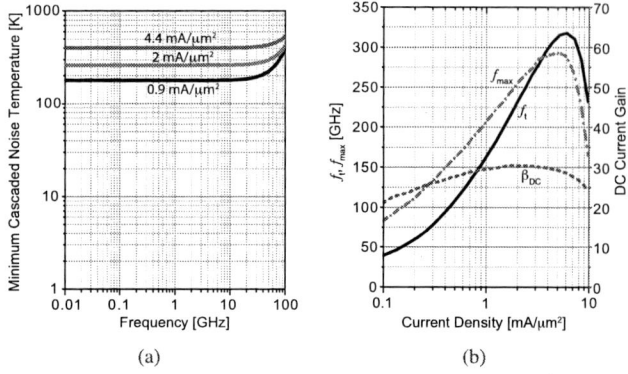

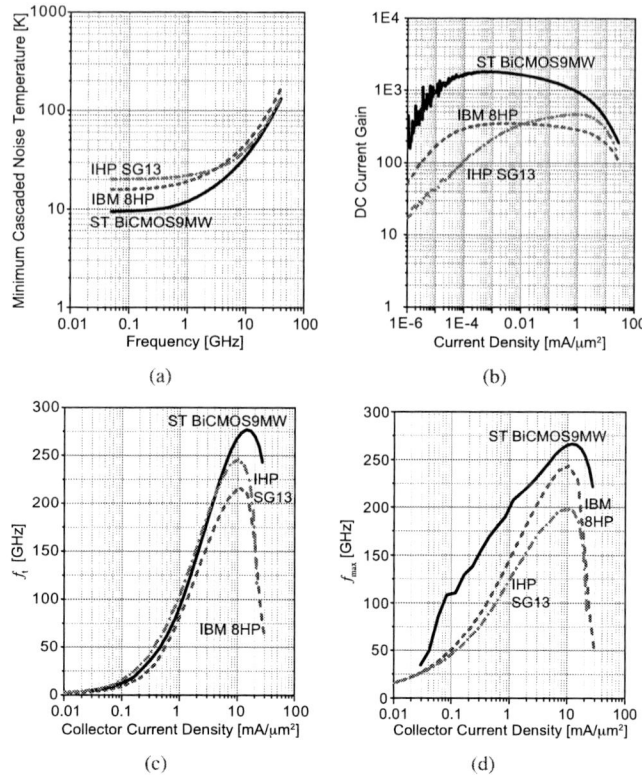

Fig. 4. (a) Cascaded noise temperature, mimimized as a function of generator impedance, and (b) key figures of merit as a function of current density for an InP/InGaAs/InP BiCMOS DHBT.

temperature. Process details can be found in [27]. Noise models were extracted using standard small-signal noise modeling techniques [1], [26], accounting for shot-noise correlation.

The extracted values of $T_{CAS,min}$ are plotted as a function of frequency in Fig. 4(a) and the bias dependence of β_{DC}, f_t, and f_{max} is provided in Fig. 4(b). The minimum cascaded noise temperature is independent of frequency into the tens of GHz. At higher frequencies, the available gain drops and the minimum cascaded noise temperature increases. This frequency dependence is easily explained by eq. (4) and the characteristics shown in Fig. 4(b). Since β_{DC} is in the range of 20–30, the low frequency noise floor is quite high. Moreover, due to the relatively weak dependence of β_{DC} on the bias point over the range of biases for which $T_{CAS,min}$ is shown, one would expect that the flat region of the $T_{CAS,min}$ curves would increase with bias (i.e., ζ), as is indeed the case.

Although the noise performance is limited by β_{DC} at lower frequencies, the millimeter wave performance is quite promising due to the high values of f_t and f_{max}. Moreover, with the recent demonstration of InP/InGaAs/InP devices having values of f_t and f_{max} exceeding 0.5 THz and 1.1 THz, respectively [28], it seems that there is considerable room to improve upon the high frequency results shown in Fig. 4(a). However, there is a significant caveat; increasing f_{max} requires reducing R_B, which in turn necessitates an increase in base doping. As the base recombination current is proportional to doping, the DC current gain will be reduced. The state-of-the-art DHBT having f_t/f_{max} values of 0.5/1.1 THz only achieved a β_{DC} of 17 [28]. In turn, this pushes the $T_{CAS,min}$ corner frequency (f_c) to higher frequencies due to the increased low-frequency noise floor.

While the noise performance of state-of-the-art InP/InGaAs/InP DHBTs is limited in the microwave frequency range due by low β_{DC}, this is not the case for all CS HBTs. For instance, a GaAs/AlGaAs HBT with room temperature DC current gains in excess of 600 has been demonstrated and used to implement an LNA in the low-GHz frequency range, which at the time achieved world record performance for any technology [29]. However, this

technology platform only achieved an f_t of 40 GHz, which is insufficient for the high-speed mixed-signal circuits that most CS HBT processes are tailored to. Other researchers have demonstrated high speed (f_t/f_{max} = 400/322 GHz) InP/GaAsSb/InP DHBTs with β_{DC} = 47 and a minimum noise figure of 1.2 dB (T_{MIN} = 92 K) at 20 GHz [30].

B. Room Temperature Performance of SiGe HBTs

A similar empirical study of noise performance has been carried out for SiGe devices [6]. In this case, the performance of several commercial technology platforms was experimentally modeled and the minimum cascaded noise temperature was extracted at each frequency with respect to both bias point and generator impedance. Shot noise correlation was ignored, so data are only provided to 40 GHz[4].

Example results appear in Fig. 5(a). Once again, these data are easily interpreted through eq. (4) and the data shown in Fig. 5. At low-frequencies, the noise depends strongly upon β_{DC} and noise temperatures in the 10 K range appear feasible for the ST BiCMOS9MW device in the sub-500 MHz frequency range. Due to the large values of β_{DC} achieved by the SiGe transistors, the noise corner frequency is in the 1–5 GHz frequency range. Therefore, larger values of β_{DC} will not

Fig. 5. Noise and terminal properties of representative SiGe HBTs at room temperature. (a) Minimum cascaded noise temperature. The values plotted are the minimum values with respect to both generator impedance and current density. (b) β_{DC}, (c) f_t, and (d) f_{max}.

[4]For studies of the millimeter wave noise performance of SiGe HBTs, see [31], [32]

978-1-4799-7231-9/14 $31.00 © 2014 IEEE

be terribly beneficial at microwave frequencies, as the primary effect will be to move the corner frequency proportionally lower in frequency.

On the other hand, improvements to f_t (and f_{max}) are expected to significantly enhance the millimeter-wave noise performance of these devices. The European DOTFIVE [33] program has already produced HBTs with f_t/f_{max} values as high as 300/500 GHz [34], [35] and the DOTSEVEN [36] program aims demonstrate SiGe HBTs with f_{max} above 700 GHz. Thus, one can expect the high-frequency noise performance of SiGe technology to continue to improve.

C. Cryogenic Performance of HBTs

Extremely low-noise cryogenically-cooled amplifiers are required for scientific systems used in a diverse set of fields ranging from Radio Astronomy to Quantum Computing. Historically, these needs have been met by CS HEMTs [37]. In fact, these amplifiers typically operate in the temperature range of 1–30 K, whereas homojunction bipolar junction transistors do not even work at temperatures below 30 K due to dopant-induced band narrowing in the heavily doped emitter region.

A number of research groups have studied the low-temperature DC current gain of CS HBTs. For AlGaAs/GaAs SHBTs, β_{DC} was found to initially increase at reduced temperatures before rapidly decreasing at temperatures below 100 K [38], [39]. AlInAs/GaInAs SHBTs were evaluated and little change was observed in β_{DC} from 300–77 K, although a downward trend was observed at the lower temperature range [40]. For InP/InGaAs SHBTs, a 25% drop in DC current gain was observed with cooling from 300–33 K [41] and for InP/InGaAlAs SHBTs, an increase in β_{DC} of 20% was observed with cooling from 300-73 K [42]. Finally, measurements of InP/InGaAs/InP DHBTs have shown reductions in β_{DC} of approximately 30% with cooling from 300 K to 9 K [43] whereas measurements of InP/InGaAs/InGaAsP DHBTs over the range of 300–90 K have found reductions in β_{DC} of over 80% [44].

The drop in β_{DC} observed in devices based upon Al-GaAs/GaAs heterostructures can be attributed to a positive temperature coefficient associated with the minority carrier lifetime in the base [45], with the best performing cryogenic devices being those whose base current is dominated by hole diffusion current [42]. It is believed that a similar mechanism limits the low temperature DC current gain of InP based HBTs [41]. Researchers have also investigated the temperature dependence of f_t and have found a 30% increase as AlGaAs/GaAs SHBTs are cooled to 110 K [46] and up to a 55% increase as InP/InGaAs SHBTs are cooled to 15 K [41], [47]. While one attempt to implement a cryogenic LNA using a high β_{DC} AlGaAs/GaAs process resulted in the demonstration of an DC–3 GHz amplifier that provided a noise temperature of approximately 14 K at 12 K operating temperature (limited by shot noise associated with base current) [48], applications of compound semiconductor HBTs in the ultra-low-noise application space have largely been limited by the degradation in β_{DC} discussed above.

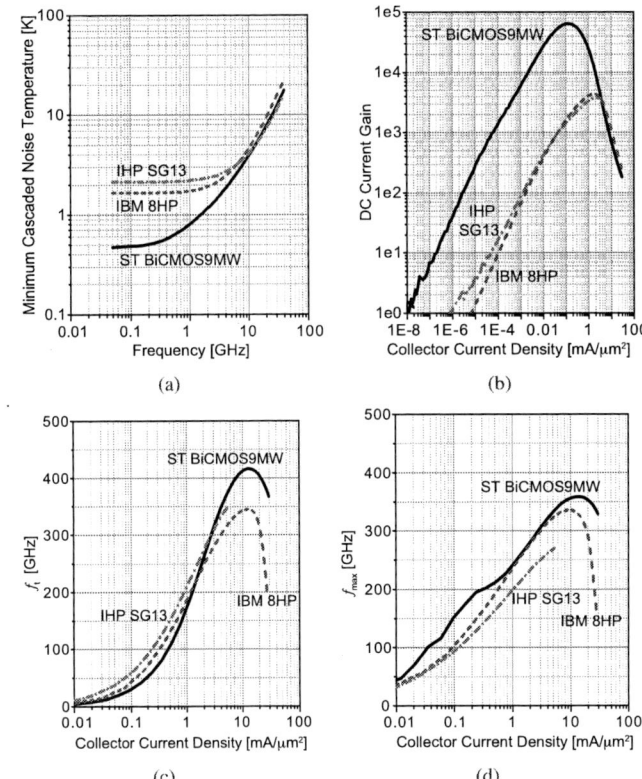

Fig. 6. Properties of three representative SiGe HBTs at 18 K physical temperature. (a) Cascaded noise temperature, simultaneously minimized as a function of bias and generator impedance, (b) DC current gain, (c) f_t, and (d) f_{max}.

SiGe HBTs, on the other hand, are an excellent choice for use in cryogenically-cooled amplifiers. As the base current in a SiGe device is dominated by diffusion current, due to the band offsets at the edge of the base–emitter space-charge-region, one expects an exponential enhancement of the DC current gain as the devices are cooled [49]. Moreover, as the critical regions are doped beyond the Mott transition, SiGe HBTs do not suffer from freeze-out effects; experimental evidence has shown that even first generation devices operate with excellent performance down to below 1 K [50].

Representative values of the minimum cascaded noise temperature for IBM, IHP, and ST technologies operating at 18 K appear in Fig. 6(a) [6]. Below 1 GHz, the ST device is capable of sub-K noise temperatures. The IBM and IHP performance is slightly worse. At higher frequencies, the noise increases, but remains below 20 K at 40 GHz, indicating that performance within a factor of 20 of the standard quantum limit, $T_Q = hf/2k$ [51], is achievable. This number can be reduced by further cooling of the devices, resulting in a decrease in thermal noise. It should be noted however that non-equilibrium transport dominates at these temperatures and both I_B or I_C are insensitive to temperature below approximately 30 K [6], [52].

The phenomenal noise performance of cryogenically cooled

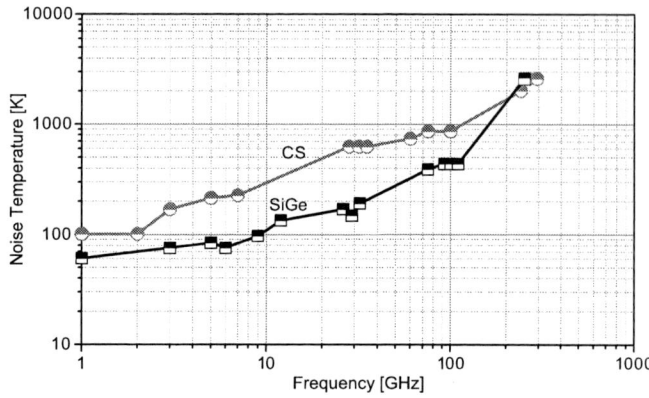

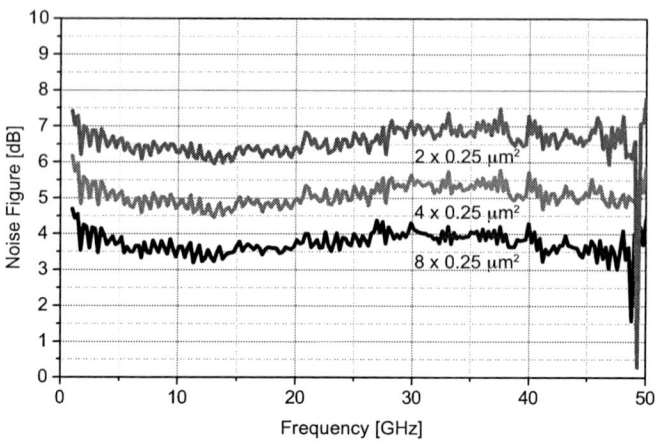

Fig. 7. State-of-the-art room-temperature noise performance for integrated-circuit low-noise amplifiers using CS and SiGe HBT technologies. Data taken from [29], [54]–[57], [69]–[77]. At frequencies below 200 GHz, the best reported SiGe amplifiers significantly outperform their CS counterparts. The CS amplifiers up to the 60 GHz frequency range were all implemented using GaAs-based SHBT technology (InGaP/GaAs or AlGaAs/GaAs), whereas higher frequency results were obtained for InP/InGaAs/InP DHBT technologies.

Fig. 8. 50 Ω noise figure for InP/InGaAs/InP HBTs heterogeneously integrated with CMOS substrate.

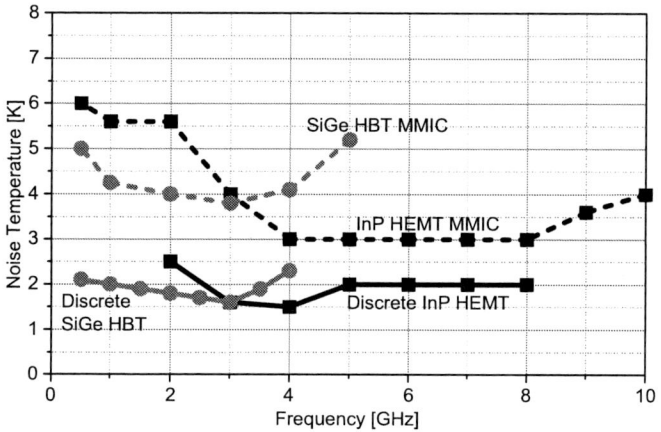

Fig. 9. State-of-the-art cryogenic noise performance for integrated circuit and discrete transistor amplifiers using InP HEMTs and SiGe HBTs. Data from [6], [54], [78]–[80]

SiGe HBTs is easily explained through the terminal characteristics (see Fig. 6). With cryogenic cooling, remarkable improvements in the DC current gain are observed for each of the three devices, with the ST device showing a peak current gain greater than 60,000 (see Fig. 6(b)). This translates to a drastic reduction in the low-frequency noise floor. Moreover, the significant improvement to the high-frequency noise is explained by an increase in f_t of approximately 50% that has consistently been observed across a wide range of high performance SiGe devices [6]. The example devices shown in Fig. 6(c) exhibit peak f_t values in the range of 350-415 GHz, as opposed to the room temperature values of 220-275 GHz. It is believed that this improvement can be attributed to an improvement in the transit time through the graded SiGe base [49], [53]. Increases in f_{max} averaging 35% were also observed with cooling (see Fig. 6(d)), although a high degree of variability was observed between different technologies, which is likely associated with process subtleties such as the extrinsic base structure [6].

V. STATE-OF-THE-ART AMPLIFIER RESULTS

Both SiGe and CS HBT technologies have been employed in low-noise amplifiers covering the range of DC–250 GHz and state-of-the-art noise results are collected in Fig. 7. With the exception of [54], the SiGe results are all for cutting edge technology platforms and were just reported within the past two years. On the other hand, the CS results span over 20 years of research and several different technologies. Referring to Fig. 7, it is apparent that the reported SiGe devices significantly outperform their CS competitors to frequencies exceeding 100 GHz.

Room temperature SiGe amplifiers have been demonstrated with noise temperatures better than 100 K at 9 GHz [55], 200 K at 32 GHz [56], and 450 K at 110 GHz [57]. In contrast,

results reported with CS HBTs are at least a factor of two worse over this frequency range. Thus, it may seem logical to conclude that CS HBT LNAs are not competitive with their SiGe counterparts in the microwave and millimeter-wave frequency regimes. However, limited work has been carried out in this area and comparison of the achievable noise limits of InP/InGaAs/InP DHBT technology (Fig. 4(a)) with the results shown in Fig. 7 indicates that the noise performance achieved by CS HBT LNAs can be improved by a factor of three at 32 GHz and a factor of two at 100 GHz; achieving these limits would bring the millimeter-wave noise performance of CS LNAs close to that of SiGe devices. To support this statement, the noise figure of three differently sized InP/InGaAs/InP BiCMOS DHBTs at a fixed current density of 3.6 mA/μm^2 is shown in Fig. 8. The 50 Ω noise figure measured for the unmatched 8×0.25 μm^2 device at 45 GHz corresponds to a noise temperature of 360 K, which is well below the state-of-the-art amplifier results shown in Fig. 7.

Several research groups have recently exploited the excel-

978-1-4799-7231-9/14 $31.00 © 2014 IEEE

lent cryogenic noise properties of SiGe HBTs to realize ultra-high-performance cryogenically cooled amplifiers (e.g., [58]–[64]). State-of-the-art results are compared to InP HEMT based amplifiers—the long standing gold standard for cryogenically cooled amplifiers [37]—in Fig. 9. The discrete and integrated circuit SiGe amplifiers achieve noise temperatures of approximately 1.8 K and 4 K in the low-GHz frequency range, where they outperform their InP HEMT based counterparts. Moreover, at cryogenic temperatures, the optimum generator impedance for noise match of SiGe HBT is of considerably lower quality factor than that of an InP HEMT, and very broadband matched amplifiers have already been demonstrated in SiGe technology [54], [65]. The excellent performance of these amplifiers has resulted in SiGe based devices replacing InP HEMT based amplifiers in a variety of low-GHz scientific instruments [66]–[68]. Referring to Fig. 6(a), it should be possible to realize SiGe amplifiers with noise temperatures within 1 K of the InP results in the 4–10 GHz range, so future research efforts should be focused on demonstrating competitive SiGe low noise amplifiers in the 4–10 GHz frequency range or at higher frequencies.

VI. CONCLUSIONS

The noise performance of HBTs is strongly dependent upon both the DC current gain and the high frequency figures of merit, f_t and f_{max}. At frequencies below 200 GHz, the best room temperature HBT low-noise amplifier results have been achieved using SiGe devices. However, this is not necessarily an indication that the noise performance of SiGe devices is intrinsically better than that of CS HBTs, but more likely a reflection of the lack of research efforts focused on the implementation of CS HBT low noise amplifiers.

Further improvements to the broadband room temperature noise performance of SiGe devices will primarily be driven by improvements to f_{max}, which will be correlated to reductions in the base resistance and base–collector capacitance. These improvements can also be expected to translate to improved cryogenic performance and should expand the frequency range over which cryogenically cooled SiGe LNAs are competitive with InP HEMT amplifiers.

On the other hand, improvements to the achievable broadband microwave performance of CS HBT LNAs will require research into device compositions that are able to drastically improve β_{DC} while maintaining high f_{max}. This will require doped base materials in which low rates of recombination can be maintained without compromising the low sheet-resistance that is required to achieve sufficiently high f_{max}.

VII. ACKNOWLEDGEMENTS

This work was partially supported by the DARPA/MTO DAHI/COSMOS program under AFRL contract FA8650-07-C-7714 and the National Science Foundation under CAREER grant CCCS-1351744. The views and conclusions contained in this document are those of the authors and should not be interpreted as representing the official policies, either expressly or implied, of the Defense Advanced Research Project Agency, the National Science Foundation, or the U.S. Government.

REFERENCES

[1] M. Rudolph et al., "An HBT noise model valid up to transit frequency," *IEEE Electron Device Lett.*, vol. 20, no. 1, pp. 24–26, Jan 1999.

[2] G. Niu et al., "A unified approach to RF and microwave noise parameter modeling in bipolar transistors," *IEEE Trans. Electron Devices*, vol. 48, no. 11, pp. 2568–2574, Nov 2001.

[3] M. Rudolph and P. Heymann, "On compact HBT RF noise modeling," in *Proc. IEEE IMS*, 2007, pp. 1783–1786.

[4] E. Ramirez-Garcia et al., "Intrinsic transit times and noise transport time study of Si/SiGe:C heterojunction bipolar transistors," in *Proc. IEEE EuMIC*, Oct 2012, pp. 175–178.

[5] T. C. Lim et al., "MOSFETs RF Noise Optimization via channel engineering," *IEEE Electron Device Lett.*, vol. 29, no. 1, pp. 118–121, Jan 2008.

[6] J. C. Bardin, "Silicon-germanium heterojunction bipolar transistors for extremely low-noise applications," Ph.D. dissertation, California Institute of Technology, 2009.

[7] J. C. Li et al., "Physical modeling of degenerately doped compound semiconductors for high-performance hbt design," *SSE*, vol. 50, no. 7, pp. 1440–1449, 2006.

[8] M. Dahlstrom et al., "High current density and high power density operation of ultra high speed InP DHBTs," in *Proc. IEEE IPRM*, 2004, pp. 761–764.

[9] J. C. Li et al., "Characterization and modeling of thermal effects in sub-micron InP DHBTs," in *Proc. IEEE CSIC*, Oct 2005, pp. 4 pp.–.

[10] V. Jain et al., "InGaAs/InP DHBTs in a dry-etched refractory metal emitter process demonstrating simultaneous," *IEEE Electron Device Lett.*, vol. 32, no. 1, pp. 24–26, 2011.

[11] B. Orner et al., "A 0.13 μm BiCMOS technology featuring a 200/280 ghz (ft/fmax) SiGe HBT," in *Proc. IEEE BCTM*, 2003.

[12] M. Racanelli et al., "Ultra high speed SiGe NPN for advanced BiCMOS technology," in *Proc. IEEE IEDM*, 2001, pp. 15–3.

[13] W. D. van Noort et al., "BiCMOS technology improvements for microwave application," in *Proc. IEEE BCTM*, 2008, pp. 93–96.

[14] H. Rucker et al., "SiGe:C BiCMOS technology with 3.6 ps gate delay," in *Proc. IEEE IEDM*, 2003, pp. 5–3.

[15] D. Harame et al., "Si/SiGe epitaxial-base transistors. ii. process integration and analog applications," *IEEE Trans. Electron Devices*, vol. 42, no. 3, pp. 469–482, Mar 1995.

[16] M. Sotoodeh, A. Khalid, and A. Rezazadeh, "Empirical low-field mobility model for III–V compounds applicable in device simulation codes," *JAP*, vol. 87, no. 6, pp. 2890–2900, 2000.

[17] "New semiconductor materials. characteristics and properties database."

[18] S. Spiegel, "Carrier recombination in single and double heterojunction bipolar transistors," in *Proc. Micro. Optoelec. Conf.*, vol. 2, Jul 1995, pp. 773–776 vol.2.

[19] C. Seabury et al., "Base recombination of high performance InGaAs/InP HBT's," *IEEE Trans. Electron Devices*, vol. 40, no. 11, pp. 2123–2124, Nov 1993.

[20] K. Kurishima et al., "Fabrication and characterization of high-performance InP/InGaAs double-heterojunction bipolar transistors," *IEEE Trans. Electron Devices*, vol. 41, no. 8, pp. 1319–1326, Aug 1994.

[21] J. C. Li, "Design considerations for 400 GHz InP/InGaAs heterojunction bipolar transistors," Ph.D. dissertation, University of California, San Diego, 2006.

[22] H. Haus and R. Adler, "Optimum noise performance of linear amplifiers," *Proc. IRE*, vol. 46, no. 8, pp. 1517–1533, Aug 1958.

[23] C. R. Poole and D. Paul, "Optimum noise measure terminations for microwave transistor amplifiers (short paper)," *IEEE Trans. Microw. Theory Tech.*, vol. 33, no. 11, pp. 1254–1257, Nov 1985.

[24] J.-C. Liu et al., "Comments, with reply, on "optimum noise measure terminations for microwave transistor amplifiers" by C.R. Poole and D.K. Paul," *IEEE Trans. Microw. Theory Tech.*, vol. 41, no. 2, pp. 363–364, Feb 1993.

[25] J. Bardin and S. Weinreb, "Experimental cryogenic modeling and noise of SiGe HBTs," in *Proc. IEEE IMS*, June 2008, pp. 459–462.

[26] J. Bardin et al., "Broadband noise performance of heterogeneously integrated InP BiCMOS DHBTs," *IEEE Electron Device Lett.*, in press.

[27] J. Li *et al.*, "Heterogeneous wafer-scale integration of 250nm, 300GHz InP DHBTs with a 130nm RF-CMOS technology," in *Proc. IEEE IEDM*, Dec 2008, pp. 1–3.

[28] M. Urteaga *et al.*, "130nm InP DHBTs with $f_t > 0.52$ thz and $f_{max} >$ 1.1 THz," in *Proc. DRC*, 2011, pp. 281–282.

[29] K. Kobayashi *et al.*, "Sub-1.3 dB noise figure direct-coupled MMIC LNAs using a high current-gain 1–μm GaAs HBT technology," in *Proc. IEEE GaAs IC*, Oct 1997, pp. 240–243.

[30] Y. Zeng *et al.*, "400-GHz InP/GaAsSb DHBTs with low-noise microwave performance," *IEEE Electron Device Lett.*, vol. 31, no. 10, pp. 1122–1124, Oct 2010.

[31] M. Deng *et al.*, "Millimeter-wave in situ tuner: An efficient solution to extract the noise parameters of SiGe HBTs in the whole 130–170 GHz range," *IEEE Microw. Wireless Comp. Lett.*, vol. PP, no. 99, pp. 1–3, 2014.

[32] K. H. K. Yau *et al.*, "Characterization of the noise parameters of SiGe HBTs in the 70–170-GHz range," *IEEE Trans. Microw. Theory Tech.*, vol. 59, no. 8, pp. 1983–2000, Aug 2011.

[33] [Online]. Available: http://www.dotfive.eu/

[34] R. Lachner, "Industrialization of mmWave SiGe technologies: Status, future requirements and challenges," in *Proc. SiRF*, Jan 2013, pp. 105–107.

[35] B. Heinemann *et al.*, "SiGe HBT technology with fT/fmax of 300GHz/500GHz and 2.0 ps CML gate delay," in *Proc. IEEE IEDM*, Dec 2010, pp. 30.5.1–30.5.4.

[36] [Online]. Available: http://www.dotseven.eu/

[37] M. Pospieszalski, "Extremely low-noise amplification with cryogenic FETs and HFETs: 1970-2004," *IEEE Microw. Mag.*, vol. 6, no. 3, pp. 62–75, Sept 2005.

[38] P. Enquist, L. Ramberg, and L. Eastman, "Comparison of compositionally graded to abrupt emitter-base junctions used in the heterojunction bipolar transistor," *JAP*, vol. 61, no. 7, pp. 2663–2669, 1987.

[39] N. Chand *et al.*, "Temperature dependence of current gain in AlGaAs/GaAs heterojunction bipolar transistors," *APL*, vol. 45, no. 10, pp. 1086–1088, 1984.

[40] M. Hafizi *et al.*, "Temperature dependence of DC and RF characteristics of AlInAs/GaInAs HBT's," *IEEE Trans. Electron Devices*, vol. 40, no. 9, pp. 1583–1588, Sep 1993.

[41] J. Kruse *et al.*, "Temperature dependent study of carbon-doped InP/InGaAs HBT's," *IEEE Electron Device Lett.*, vol. 17, no. 1, pp. 10–12, Jan 1996.

[42] W.-C. Liu *et al.*, "Temperature-dependent study of a lattice-matched InP/InGaAlAs heterojunction bipolar transistor," *IEEE Electron Device Lett.*, vol. 21, no. 11, pp. 524–527, Nov 2000.

[43] H. Wang and G.-I. Ng, "Electrical properties and transport mechanisms of InP/InGaAs HBTs operated at low temperature," *IEEE Trans. Electron Devices*, vol. 48, no. 8, pp. 1492–1497, Aug 2001.

[44] Z. Abid *et al.*, "Temperature dependent dc characteristics of an InP/InGaAs/InGaAsP HBT," *IEEE Electron Device Lett.*, vol. 15, no. 5, pp. 178–180, May 1994.

[45] C. van Opdorp *et al.*, "Temperature dependence of interface recombination and radiative recombination in (Al, Ga) As heterostructures," *APL*, vol. 42, no. 9, pp. 813–815, 1983.

[46] J. Laskar *et al.*, "Effect of reduced temperature on the f_T of AlGaAs/GaAs heterojunction bipolar transistors," *IEEE Electron Device Lett.*, vol. 12, no. 6, pp. 329–331, 1991.

[47] M. Agethen *et al.*, "Cryogenic temperature dependence and modelling of RF-noise parameters of carbon doped InP/InGaAs HBT," in *Proc. IEEE IPRM*, 2001, pp. 212–215.

[48] K. Kobayashi *et al.*, "A DC-3 GHz cryogenic AlGaAs/GaAs HBT low noise MMIC amplifier with 0.15 dB noise figure," in *Proc. IEEE IEDM*, Dec 1999, pp. 775–778.

[49] J. D. Cressler and G. Niu, *Silicon-germanium heterojunction bipolar transistors*. Artech house, 2002.

[50] L. Najafizadeh *et al.*, "Sub-1-K operation of SiGe transistors and circuits," *IEEE Electron Device Lett.*, vol. 30, no. 5, pp. 508–510, May 2009.

[51] A. R. Kerr, M. J. Feldman, and S.-K. Pan, "Receiver noise temperature, the quantum noise limit, and the role of the zero-point fluctuations," in *Proc. Eighth Intern. Space THz Technol. Symp*, 1997, pp. 101–111.

[52] D. M. Richey *et al.*, "Evidence for non-equilibrium base transport in Si and SiGe bipolar transistors at cryogenic temperatures," *SSE*, vol. 39, no. 6, pp. 785–789, 1996.

[53] J. Yuan *et al.*, "On the performance limits of cryogenically operated SiGe HBTs and its relation to scaling for terahertz speeds," *IEEE Trans. Electron Devices*, vol. 56, no. 5, pp. 1007–1019, May 2009.

[54] J. Bardin and S. Weinreb, "A 0.1–5 GHz cryogenic SiGe MMIC LNA," *IEEE Microw. Wireless Comp. Lett.*, vol. 19, no. 6, pp. 407–409, June 2009.

[55] T. Kanar and G. Rebeiz, "X- and K-band SiGe HBT LNAs with 1.2- and 2.2-db mean noise figures," *IEEE Trans. Microw. Theory Tech.*, vol. PP, no. 99, pp. 1–9, 2014.

[56] Q. Ma, D. Leenaerts, and R. Mahmoudi, "A 30GHz 2dB NF low noise amplifier for Ka-band applications," in *Proc. IEEE RFIC*, June 2012, pp. 25–28.

[57] A. Ulusoy *et al.*, "A 110 GHz LNA with 20dB gain and 4dB noise figure in an 0.13 μm SiGe BiCMOS technology," in *Proc. IEEE IMS*, June 2013, pp. 1–3.

[58] S. Weinreb, J. C. Bardin, and H. Mani, "Design of cryogenic sige lownoise amplifiers," *IEEE Trans. Microw. Theory Tech.*, vol. 55, no. 11, pp. 2306–2312, 2007.

[59] F. Voisin *et al.*, "Very low noise multiplexing with SQUIDs and SiGe heterojunction bipolar transistors for readout of large superconducting bolometer arrays," *Journal of Low Temp. Phys.*, vol. 151, no. 3-4, pp. 1028–1033, 2008.

[60] T. K. Thrivikraman *et al.*, "SiGe HBT X-band LNAs for ultra-low-noise cryogenic receivers," *IEEE Microw. Wireless Comp. Lett.*, vol. 18, no. 7, pp. 476–478, 2008.

[61] D. Prele *et al.*, "Development of superconducting NbSi TES array and associated readout with SQUIDs and integrated circuit operating at 2 K," *IEEE Trans. Appl. Supercond.*, vol. 19, no. 3, pp. 501–504, 2009.

[62] E. Bryerton, M. Morgan, and M. Pospieszalski, "Ultra low noise cryogenic amplifiers for radio astronomy," in *Proc. IEEE RWS*, 2013, pp. 358–360.

[63] N. Beev and M. Kiviranta, "Fully differential cryogenic transistor amplifier," *Cryogenics*, vol. 57, pp. 129–133, 2013.

[64] N. Wex, "Testing relativistic gravity with radio pulsars," *arXiv preprint arXiv:1402.5594*, 2014.

[65] S. Weinreb *et al.*, "Matched wideband low-noise amplifiers for radio astronomy," *RSI*, vol. 80, no. 4, p. 044702, 2009.

[66] P. Putz *et al.*, "NbTiN hot electron bolometer waveguide mixers on membranes at THz frequencies," *IEEE Trans. Appl. Supercond.*, vol. 21, no. 3, pp. 636–639, 2011.

[67] ——, "Terahertz hot electron bolometer waveguide mixers for great," *Astronomy and astrophysics*, vol. 542, 2012.

[68] M. Jung *et al.*, "Radio frequency charge sensing in InAs nanowire double quantum dots," *APL*, vol. 100, no. 25, p. 253508, 2012.

[69] P. Song *et al.*, "A high gain, W-band SiGe LNA with sub-4.0 dB noise figure," in *Proc. IEEE IMS*, June 2014, pp. 1–3.

[70] K. Schmalz *et al.*, "A 245 GHz LNA in SiGe technology," *IEEE Microw. Wireless Comp. Lett.*, vol. 22, no. 10, pp. 533–535, Oct 2012.

[71] K. Yamamoto *et al.*, "InGaP/GaAs HBT MMICs for 5-GHz-band wireless applications - a high P1dB, 23/4-dB step-gain low-noise amplifier and a power amplifier," in *Proc. IEEE IMS*, vol. 2, June 2004, pp. 551–554 Vol.2.

[72] K.-P. Ahn, R. Ishikawa, and K. Honjo, "Low noise group delay equalization technique for uwb ingap/gaas hbt lna," *IEEE Microw. Wireless Comp. Lett.*, vol. 20, no. 7, pp. 405–407, July 2010.

[73] S.-L. Chu *et al.*, "A Ka-band HBT two-stage LNA," in *Proc. IEEE GaAs IC*, Oct 1994, pp. 307–310.

[74] T. Morf *et al.*, "50 to 70 GHz InP/InGaAs HBT amplifier with 20 dB gain," in *Proc. IEEE IPRM*, 1999, pp. 431–434.

[75] S. Handa *et al.*, "60GHz-band low noise amplifier and power amplifier using InGaP/GaAs HBT technology," in *Proc. IEEE GaAs IC*, Nov 2003, pp. 227–230.

[76] P. Watson *et al.*, "A wide-bandwidth W-band LNA in InP/Si BiCMOS technology," in *Proc. IEEE IMS*, June 2014, pp. 1–4.

[77] K. Eriksson *et al.*, "Design and characterization of H-band (220–325 GHz) amplifiers in a 250–nm InP DHBT technology," *IEEE Trans. THz Sci. and Tech.*, vol. 4, no. 1, pp. 56–64, Jan 2014.

[78] A. Mellberg *et al.*, "Cryogenic 2-4 GHz ultra low noise amplifier," in *Proc. IEEE IMS*, vol. 1, June 2004, pp. 161–163 Vol.1.

[79] N. Wadefalk *et al.*, "Cryogenic wide-band ultra-low-noise IF amplifiers operating at ultra-low dc power," *IEEE Trans. Microw. Theory Tech.*, vol. 51, no. 6, pp. 1705–1711, June 2003.

[80] J. Pandian *et al.*, "Low-noise 6-8 GHz receiver," *IEEE Micro.*, vol. 7, no. 6, pp. 74–84, Dec 2006.

978-1-4799-7231-9/14 $31.00 © 2014 IEEE

High-Resistivity SiGe BiCMOS Technology Development

[1]Anthony Stamper, [1]Renata Camillo-Castillo, [1]Hanyi Ding, [1]James Dunn, [1]Mark Jaffe, [1]Vibhor Jain, [1]Alvin Joseph, [1]Ian McCallum-Cook, [1]Kim Newton, [2]Shyam Parthasarathy, [1]Robert Rassel, [1]Nicholas Schmidt, [2]Srikanth Srihari, [1]Randy Wolf, and [1]Michael Zierak

IBM Microelectronics Division
[1]Essex Junction, VT 05452, U.S.
[2]Bangalore, India

Email : astamper@us.ibm.com

Abstract — **IBM first qualified a 0.35μm generation 1000 Ω-cm high resistivity substrate (HiRES) SiGe BiCMOS technology in 2011. This technology was optimized for WiFi and cellular NPN power amplifier (PA), NPN low noise amplifier (LNA), and isolated CMOS NFET switch rf front-end-IC (FEIC) integration. It includes an optional through silicon via used as a low inductance ground path for NPN emitters. Data for 50 Ω-cm, 1st generation HiRES, and 2nd generation HiRES NPN PA, LNA, and CMOS NFET switch devices are reviewed.**

Index Terms - **High-resistivity substrate, SiGe BiCMOS, front-end-IC, switch, PA, LNA.**

I. INTRODUCTION

In most rf systems, the PA, LNA, and switch front-end components remain de-integrated due to the unavailability of a cost-effective single technology solution with competitive rf device specifications. In today's 2.4 and 5 GHZ band WiFi front-end, it is common to find multiple ICs typically comprising GaAs or SiGe for the PA and GaAs or SOI for the LNA and switch [1-2]. It is highly desirable to have a single-chip silicon IC solution, for cost reduction and for smaller device size.

The use of high resistivity substrates (HiRES) for improved devices has previously reported [3 – 5] and it is well known that HiRES silicon substrates improve the performance of inductors and transmission lines. In addition, the use of HiRES substrates reduces NPN collector-to-substrate capacitance [6]. HiRES handle substrates in an SOI technology have allowed for improved rf harmonics [7]. IBM L1 qualified 0.35μm generation SiGe BiCMOS 50 and 1000 Ω-cm substrate technologies targeting WiFi PA, cellular PA, and integrated front end IC (FEIC) applications in 2007 and 2011, respectively. In this paper, we discuss IBM's 1000 Ω-cm HiRES 0.35μm generation SiGe BiCMOS technology focusing on the WiFi PA, LNA, and switch enhancements that were L1 qualified in 2013-4.

II. TECHNOLOGY INTEGRATION

The process flow starts with a bulk 1000 Ω-cm HiRES silicon wafer. A low-resistivity region is defined after the sub-collector implant to maintain CMOS latch-up immunity and allow the use of standard FET logic libraries and models. Next, the standard process integration is followed: epi growth, STI, wells, poly gate, bipolar formation, final RTA, and Ti silicidation.

Prior to metallization, a trench isolation along with low mobility regions are defined around the active devices. The trench is filled and planarized with the oxide film that is also used for the pre-metal dielectric. Finally, the traditional metallization is completed to finish the wafer processing. The use of HiRES substrates adds 3 extra masks to the 50 Ω-cm base technology. Figure 1 shows the NPN device with the various elements integrated in this HiRES 0.35μm FEIC technology. Further information on the device offerings and process integration used in this technology can be found elsewhere [6, 8].

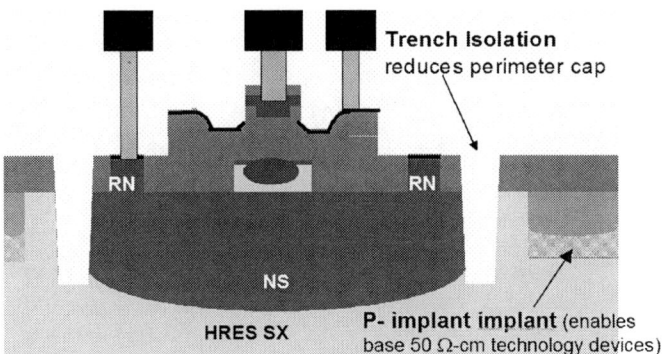

Fig. 1. Cross-sectional drawing showing the unique integration elements added around the NPN in the HiRES technology [6].

DEFECT DENSITY

A potential HiRES technology concern is degraded NPN defect density due to the use of a HiRES substrate and to the added masks and processes required for high resistivity substrate technologies. Fig. 2 shows defect density data for our standard LM test dicing channel SiGe BiCMOS multi-finger NPN test structures. These NPN yield chain test structures have an area larger than the typical NPN area on functional chip layouts and are tested for BE, BC, and EC leakage current and breakdown voltage. For the past year, the NPN yield chains had average yields of 99% for both the HiRES and 50Ω-cm versions of IBM's 0.35μm generation SiGe BiCMOS technologies. This demonstrates that the use of HiRES wafers does not degrade the baseline NPN defect density.

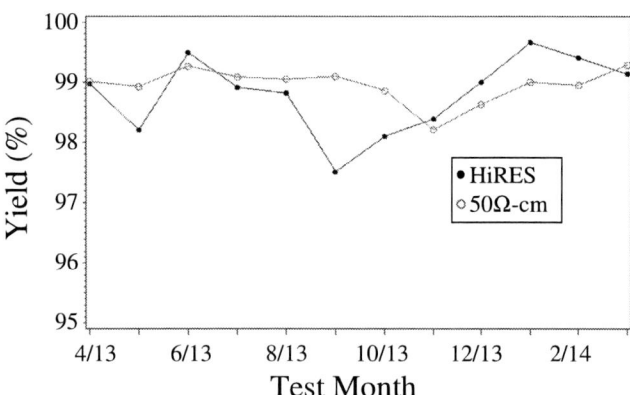

Fig. 2. 0.35μm generation BiCMOS technology NPN chain average monthly yields (data taken from several 10,000's of production wafers).

III. CMOS NFET SWITCH OPTIMIZATION

We previously reported HiRES substrate improvements for rf applications, characterized using a 4mm long metal-1 coplanar transmission line, for insertion loss and 2nd/3rd harmonics [6]. In this paper we report, for the first time, the key switch metrics Ron*Coff, insertion loss, and 2nd/3rd harmonics using 4 stack 1mm SPST NFET switch over HiRES substrates. The optimal method to extract Ron, and Coff from switch s-parameter measurements is to set the de-embed plan and use MAG data from the 'ON' state for Ron and Imag(Y12) from the 'OFF' state for Coff of the series wired switch. Fig.s 3-4 show insertion loss and Ron*Coff delay data for NFET switches vs. substrate resistivity. The dramatic insertion loss and Ron*Coff delay improvements for HiRES compared to 50 Ω-cm substrate NFET switches are due to reduced rf substrate losses. Not surprisingly, Coff is approximately constant vs. frequency and the Ron*Coff delay product increase at higher frequency is due to increasing Ron.

NFET switch rf properties can be improved by reducing the Lpoly width, which both decreases Coff and reduces Ron; and by reducing the contact width and contact to poly gate (PC) spacing, which reduces Coff. We L1 production qualified a NFET switch PC pitch reduction from 1.40 to 1.08μm in 2014. This PC pitch reduction was achieved by reducing Lpoly from 360 to 320nm; by reducing the contact width from 400 to 200nm; and by reducing the contact to PC spacing (Table 2). We measured incremental insertion loss reduction for the new 2014 HiRES layout, due to the shorter Lpoly; and substantially reduced Ron*Coff delay, due both to shorter Lpoly (Ron) and reduced switch area (Figs.3-4). Fig. 5 shows measured 2nd/3rd harmonic data for integrated HiRES NFET switches. This 4 stack switch using 3.3V DC rated NFETs easily achieves 32dBm of RF power without degradation in linearity or fundamental power. We conclude that these HiRES NFET switches meet the integrated WiFi FEIC switch insertion loss, 2nd/3rd harmonics loss, and power handling requirements.

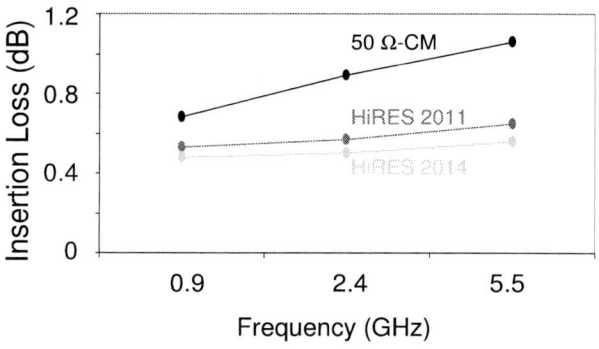

Fig. 3. Isolated 4 stack 1mm NFET switch insertion loss for devices on 50 Ω-cm and HiRES substrates. Data for the original (2011) and enhanced (2014) HiRES NFET switches are given.

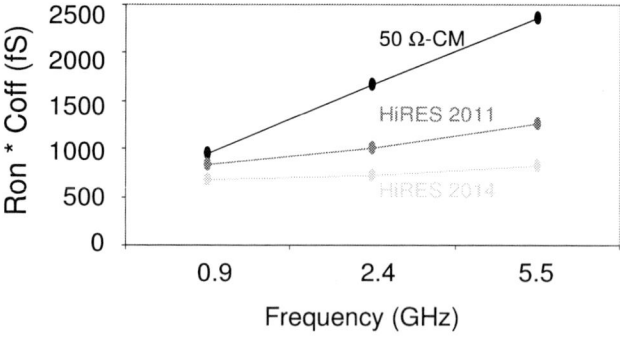

Fig. 4. Isolated 4 stack 1mm NFET switch Ron*Coff delay product for devices 50 Ω-cm and HiRES substrates. Data for the original (2011) and enhanced (2014) HiRES NFET switches are given.

TABLE 2: NFET switch layouts.

L1 date	PC pitch (μm)	Lpoly (nm)	Contact-PC space (nm)	Contact width (nm)	4 stack 1mm switch area (μm²)
2011	1.40	360	320	400	12,600
2014	1.08	320	280	200	10,800

Fig. 5. Isolated 4 stack 1mm HiRES SPST NFET switch 890MHz 2nd and 3rd rf harmonics electrical test data for the 1.4μm PC pitch device.

IV. PA OPTIMIZATION

IBM's 0.35μm generation SiGe BiCMOS technologies offer a high breakdown (HB) and high performance (HP) HBT NPN having BVcer of 8.3V and 6.3V, respectively [6, 8]. BVcer is the collector-emitter breakdown voltage at Ic=10uA for a forced Ib of 1nA. These 0.8μm or 1.2μm wide emitter devices are focused on WiFi and cellular PA applications. In late 2013, IBM L1 qualified an improved high efficiency (HE) NPN PA. To avoid changing the existing NPN, CMOS, or passive device process, we optimized the NPN selectively implanted collector process, which did not impact any other existing devices. To optimize the NPN, we focused on two metrics: improved Jc at peak fT (Jc,pk) normalized for same BVCER (Fig. 6) and the small signal performance, i.e. F_T plotted vs. Ic (Fig. 7). The HE NPN has a 0.2V reduced BVcer, normalized for beta, for the same HP 0.8x20x3

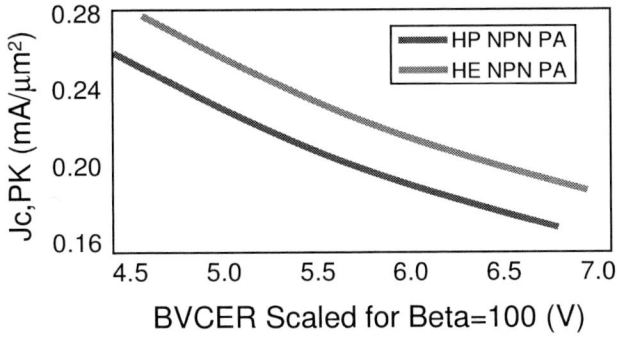

Fig. 6. $J_{C,PK}$ vs. collector-emitter breakdown voltage data at VCB=0V normalized to Beta = 100 for HiRES HP and optimized HE 0.8μm by 20μm 3 finger 48μm² SiGe HBT NPN PA.

device. It demonstrates an 7-19% improvement in Jc,pk, depending on collector-base voltage bias (V_{CB}), and a 1% increase in peak fT (fT,pk) compared to the HP NPN (Table 3).

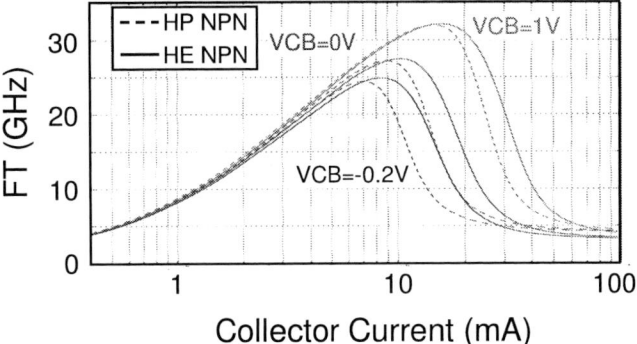

Fig. 7. Small-signal performance of the HiRES base HP and optimized HE 0.8μm by 20μm 3 finger 48μm² SiGe HBT NPN PA.

TABLE 3: 0.8X20X3 SiGe HBP NPN PA data summary.

Parameter	VCB=1V	VCB=0V	VCB=-0.2V
HE Ic,pk	15.9 mA	10.3	8.3 mA
HP Ic,pk	14.8 mA	8.9	7.0 mA
HE FT	32.2 GHz	27.5 GHz	24.8 GHz
HP FT	32.0 GHz	27.2 GHz	24.5 GHz

V. LNA OPTIMIZATION

One of the most important rf LNA metric is the noise figure (NFmin). It is well known that NFmin is a strong function of the base resistance [10]. To optimize LNA performance, we have systematically decreased the base resistance by optimizing the LNA layout. Fig. 8 shows a SEM cross-section of a NPN device highlighting the extrinsic base resistance path. Note that the collector implant conditions had little or no measurable impact on NFmin (data not shown).

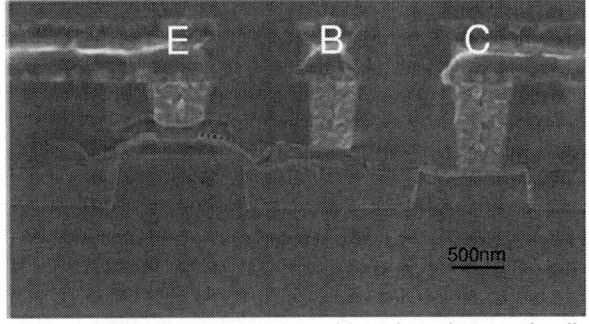

Fig. 8. NPN SEM cross-section with emitter, base, and collector labeled. The extrinsic base resistance path is shown in red (silicide) and blue (non-silicided) [7].

978-1-4799-7231-9/14 $31.00 © 2014 IEEE

For the LNA NFmin optimization discussed in this paper, we focused on decreasing the extrinsic base resistance, i.e. the resistance from the metal wiring through the base to the emitter opening. This extrinsic base resistance is dominated by the unsilicided base portion of the device, whose sheet resistance is >10X higher than the silicided portion of the base. Fig. 8 shows a NPN cross-section with the unsilicided base current path shown in blue. We reduced the resistance of this path by reducing the emitter poly edge to emitter-base opening width by ~30%. NFmin and gain data are shown in Fig. 9 and summarized in Table 4. This base resistance reduction translated into a 0.07dB (2.4GHz) to 0.11dB (5GHz) reduction in NFmin, and a ~0.5dB gain increase at 5GHz (2.4GHz gain data was not measured). Device dc characteristics like gummel plot and output curves were not degraded by this LNA layout optimization.

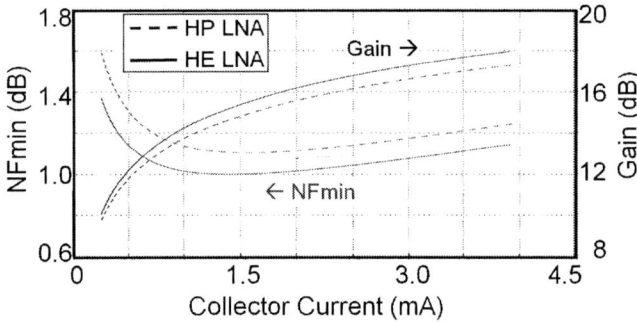

Fig. 9. 5GHz VCB=2V Noise Figure (NF) and gain vs. collector current data for the standard (HP) and reduced base resistance (HE) 0.44μm by 10μm 4 finger 17.6μm² HiRES SiGe HBT NPN LNA.

TABLE 4: 0.44X10X4 SiGe HBP NPN LNA data summary.
VBE=0.8V, VCB=2V.

Parameter	2.4GHz	5GHz
HE LNA NFmin	0.74dB	1.02dB
HP LNA NFmin	0.80dB	1.11dB
HE LNA Gain	-	15.19dB
HP LNA Gain	-	14.63dB

VI. CONCLUSION

A novel high-resistivity SiGe BiCMOS technology has been developed for industry-leading FEIC solutions. This technology allows for much improved substrate losses for active and passive devices. The HiRES NPN PA, NPN LNA and NFET switch devices were enhanced in 2013-4 to further support the integrated front end IC.

ACKNOWLEDGEMENT

The authors wish to gratefully acknowledge the assistance and support of the IBM Burlington manufacturing, manufacturing engineering, reliability, and modeling teams; and the IBM Bangalore modeling team for the success of this project. We especially acknowledge Santosh Sharma and Eric Johnson for their work on the HE NPN and NFET switch, respectively.

REFERENCES

[1] http://www.chipworks.com/blog/recentteardowns/2012/10/02/apple-iphone-5-the-rf/.

[2] http://www.triquint.com/products/p/TQP6M9017/.

[3] Weiming Yang, Chen Shi, MingJie Gao, Zhenguo Li, Jianxin Chen, **"The Microwave Performances of SiGe/Si HBT Based on the High Resistivity Substrate,"** *IEEE International Symposium on Microwave, Antenna, Propagation and EMC Technologies for Wireless Communications Proceedings*, pp.654-657, 2005.

[4] M. M. Kaleja, A. Grubl, F. X. Sinnesbichlerl, G. R Olbrich, K. M. Strohm, J.-F. Luy, E.M. Biebl,**"Si/SiGe HBT Active Integrated Antenna on High Resistivity Silicon Substrate,"** *IEEE MTT-S Digest*, pp. 1899-1902, 2000.

[5] Dietmar Beck, Michael Herrmann, and Erich Kasper, **"CMOS on FZ-High Resistivity Substrate for Monolithic Integration of SiGe-RF-Circuitry and Readout Electronics"** *IEEE Trans. On Elec. Dev.*, Vol. 44, No. 7, pp. 1091-1101, July 1997.

[6] A. Joseph, J. Gambino, R. M. Rassel, E. Johnson, H. Ding, S. Parthasarthy, V. Vanakuru, S. Sharma, M. Jaffe, D. Liu, M. Zierak, R. Camillo-Castillo, A., Stamper, A., and J. Dunn, **"A High-Resistivity SiGe BiCMOS Technology for WiFi RF Front-End-IC Solutions"**, *IEEE BCTM Proceedings*, in press, 2013.

[7] A. Botula, A. Joseph, J. Slinkman, R. Wolf, Z. -X He, D. Ioannou, L.Wagner, M. Gordon, M. Abou-Khalil, R. Phelps, M. Gautsch, W. Abadeer, D. Harmon, M. Levy, J. Benoit, J. Dunn, **"A Thin-Film SOI 180nm CMOS RF Switch Technology,"** *IEEE Topical Mg. on Si Monolithic Int. Cir. in RF Systems Dig.*, pp. 1-4, Jan 2009.

[8] Alvin Joseph, Qizhi Liu, Wade Hodge, Peter Gray, Kenneth Stein, Rose Previti-Kelly, Peter Lindgren, Ephrem Gebreselasie, Ben Voegeli, Panglijen Candra, Doug Hershberger, Ramana Malladi, Ping-Chuan Wang, Kim Watson, Zhong-Xiang He, and Jim Dunn, **"A 0.35 µm SiGe BiCMOS technology for power amplifier applications,"** *IEEE BCTM Proc.*, pp. 198-201, 2007.

[9] U. Gösele and T. Y. Tan, **"Oxygen Diffusion and Thermal Donor Formation in Silicon"**, *Appl. Phys.* A28, 1982 pp79-92.

[10] See, for example, J. D. Cressler, **"Fabrication of SiGe HBT BiCMOS Technology"**, CRC Press, 2007.

High-speed, waveguide Ge PIN photodiodes for a photonic BiCMOS process

S. Lischke[1], D. Knoll[1], L. Zimmermann[1], A. Scheit[1], C. Mai[1], A. Trusch[1], K. Voigt[2], M. Kroh[1], R. Kurps[1], P. Ostrovskyy[1], Y. Yamamoto[1], F. Korndörfer[1], A. Peczek[1], G. Winzer[1], and B. Tillack[1,2]

[1] IHP, Im Technologiepark 25, 15236 Frankfurt (Oder), Germany
lischke@ihp-microelectronics.com
[2] Institut für Hochfrequenz- und Halbleiter-Systemtechnologien
Technical University, Einsteinufer 25, 10587 Berlin, Germany

Abstract—**Waveguide-coupled, Ge lateral pin photodiodes featuring bandwidths of more than 50GHz and 40Gbps functionality are presented. Non-doping implantations are applied that allow one to reach this performance even under the effect of thermal steps acting when the diodes are integrated in a high-performance BiCMOS process. The effect of these implants is to lower the minority-carrier lifetime(s) and in this way, to reduce bandwidth degradation by minority-carrier diffusion in non-depleted, weakly doped regions.**

Keywords—*BiCMOS; Ge photodiode; Silicon photonics*

I. INTRODUCTION

Integration of photonic components such as waveguides, couplers, modulators, and photo detectors in microelectronics technologies is a key step towards complex receiver and transmitter devices for future communication applications with line rates exceeding 100Gbps. Compared to hybrid integration approaches, monolithic integration of detectors and modulators in the frontend of a Si-based integrated circuit technology allows for shortest possible interconnects between photonics and electronics from which high-speed performance of electronic-photonic integrated circuits can greatly benefit. Frontend integration of photonics has been pursued for some time, chiefly based on CMOS technologies [1-3]. Recently, we brought BiCMOS into play as integration platform, which promises significant advantages over advanced CMOS regarding cost and RF performance [4, 5].

Our photonic BiCMOS process offers a lateral pin photodiode, fabricated from a Ge layer, selectively grown on top of a Si waveguide [4]. Germanium epitaxy is carried out after the BiCMOS baseline source/drain anneal ($T_{Peak} > 1000°C$), but before the BiCMOS cobalt silicide module. In this way, melting of the Ge layer and mixing it with underlying Si material is prevented on the one hand, and on the other, any metal contamination of tools used for Ge epitaxy and pre-epitaxy wet cleaning is excluded. The B and P implantation steps applied to form the p and n regions of the pin diode are also carried out before silicide formation. To get low-ohmic $CoSi_2$ layers, silicide process modules typically include a final (second) anneal step with a temperature of ~700°C or even higher [6]. Under such conditions, implanted P profiles can significantly be smeared by diffusion [7]. Thus, fabrication of a diode with steep doping profiles necessary to get a bandwidth of 40GHz

or more, as already shown for Ge diodes fabricated with other integration schemes or without any frontend integration with electronics [8-10], becomes very challenging, in particular under the strict condition not too much to change those baseline modules which are critical for BiCMOS device parameter and yield.

Here, we will show for the first time that by additional implantation of non-doping elements, without changing the implantation conditions for the electrically active dopants, bandwidths of Ge lateral pin photodiodes can be significantly improved, even if the diodes are fabricated under effect of BiCMOS-given thermal steps. The effect of these implants is to lower the minority-carrier lifetime(s) and in this way, to reduce or even prevent bandwidth degradation by minority-carrier diffusion in non-depleted, weakly doped regions. The main results are as follows: (1) Diodes with non-doping implants show more than 50GHz bandwidth and 40Gbps functionality under use of a silicide module with 750°C anneal. This is a gain in bandwidth of about a factor of two, compared to diodes fabricated on the same wafer without these extra implantations. (2) Diodes show more than 0.65A/W responsivity at 1.55μm, with or without extra implants. (3) There is a dark current penalty of about a factor of ten for the diodes with highest bandwidth, compared to non-implanted devices. Conditions for the non-doping implantations can however be tuned in a way which enables bandwidths of around 45GHz without marked dark current increase.

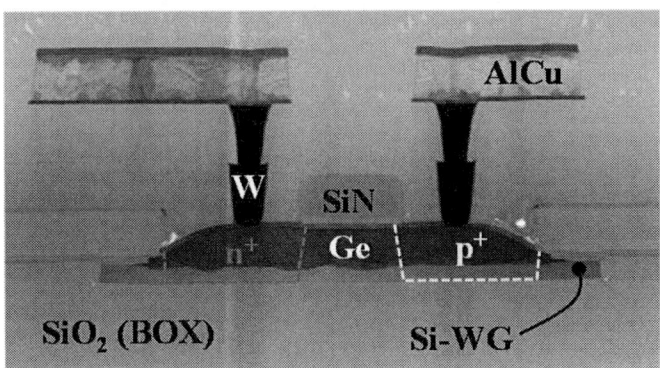

Fig. 1. TEM cross section of a Ge lateral pin photodiode, perpendicular to the direction of light entry. The SiN pedestal enables self-aligned implantations to form the diode n- and p-regions, and i-region as well. Doped regions are schematically marked in colors.

II. EXPERIMENTS

Fig. 1 shows essential structure features of the Ge lateral pin photodiode. The diode is evanescently-coupled to a 220nm thick silicon waveguide (Si-WG) rib on top of a 2μm thick SiO$_2$ layer. For light coupling to the waveguide, an enhanced grating coupler, as previously described [11], is used.

Fig. 2 illustrates the experiments carried out. On same wafers, identically designed diodes were fabricated without or with additional implantations of non-doping elements. The particular implant steps were carried out immediately after forming the Ge body and a Si-cap, before depositing the thick SiN pedestal layer. We tested two elements, anonymously called E1 and E2 here, varied implantation conditions for the second element (E2$_1$ or E2$_2$), and investigated also the combined effect of elements (E1 + E2$_1$ or E1 + E2$_2$).

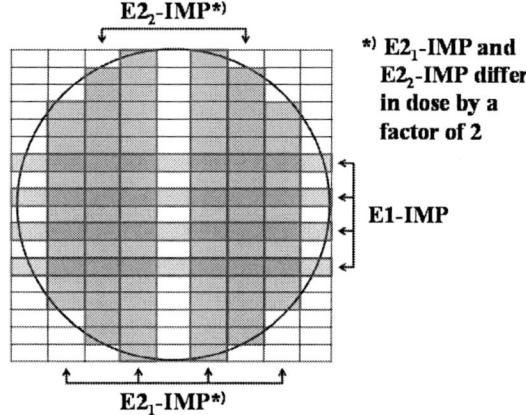

Fig. 2. Experiment overview: On 200mm wafers, Ge lateral pin photodiodes were fabricated with or without extra implantation of the non-doping elements E1 and E2. The particular implant steps were carried out for selected chip rows (E1-IMP) and columns (E2-IMP), respectively.

III. RESULTS AND DISCUSSION

Here, we focus on three cases: (a) reference devices, fabricated with no extra implantations, (b) diodes, fabricated with additional element E1 implantation alone, and (c) devices where both elements were additionally implanted (E1 + E2$_1$).

Fig. 3 shows the effect of post-implantation anneals on the diode bandwidth behavior for the case that no extra implants

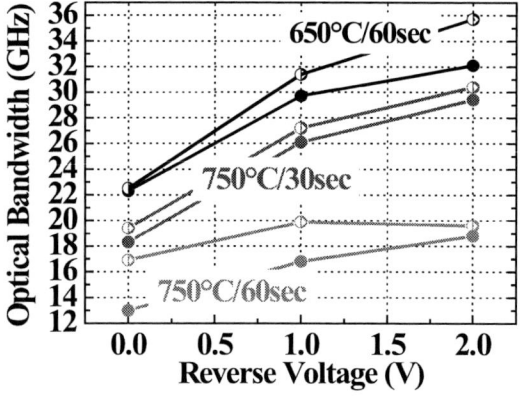

Fig. 3. (3-dB) bandwidth vs. reverse voltage of Ge lateral pin photodiodes fabricated without extra implantation of non-doping elements. Conditions for the 2nd CoSi$_2$ anneal were varied. Two diodes per variant were measured.

were applied. The critical, second anneal step of the cobalt silicide module was varied in temperature and time, between 650°C (low-ohmic CoSi$_2$ layer is just formed) and 750°C (close to BiCMOS baseline conditions). Strong performance degradation with increasing temperature and time is obvious and even with the lowest temperature applied in fabrication the diodes provide a bandwidth of about 35GHz only.

In following, we will compare diodes fabricated with 750°C, 30sec CoSi$_2$ anneal, with and without implantation of non-doping elements. Figures 4-6 show normalized response vs. frequency curves of the variants investigated here. While

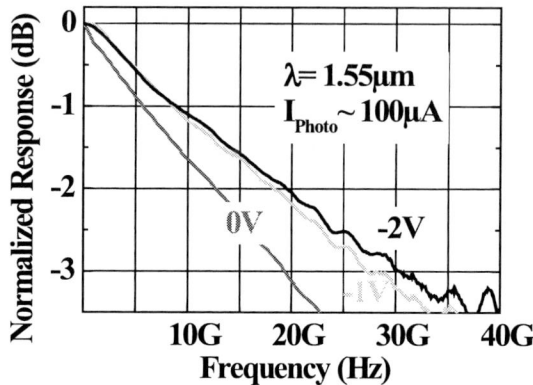

Fig. 4. Normalized s$_{21}$ response vs. frequency of a Ge lateral pin photodiode fabricated with 750°C, 30sec anneal, without extra implants.

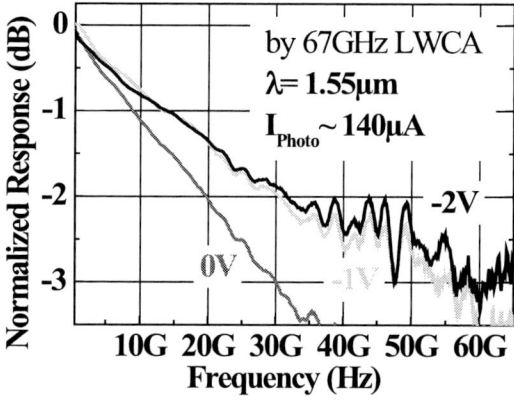

Fig. 5. Normalized s$_{21}$ response vs. frequency of a Ge lateral pin photodiode fabricated with both E1 and E2$_1$ implants and 750°C, 30sec anneal.

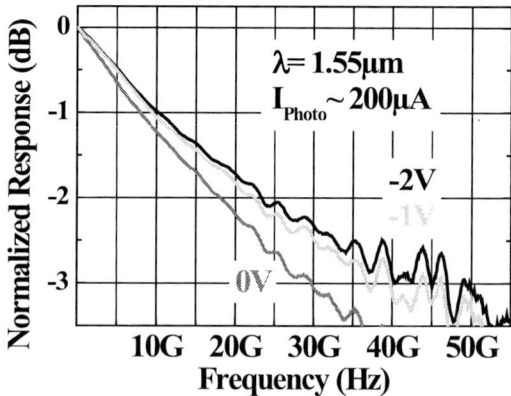

Fig. 6. Normalized s$_{21}$ response vs. frequency of a Ge lateral pin photodiode fabricated with E1 implantation only and 750°C, 30sec anneal.

978-1-4799-7231-9/14 $31.00 © 2014 IEEE

the reference devices provide about 30GHz bandwidth (Fig. 4), much better values of more than 50GHz are obtained for diodes implanted with both elements (Fig. 5). Diodes with element E1 implantation alone are in terms of bandwidth between, approximately, offering about 40-45GHz (Fig. 6). It is also obvious from figures 5 and 6 that the improvement by up to nearly a factor of two, compared to devices with no extra implants, fabricated on the same wafer, was reached without making use of special effects like inductive gain peaking [12].

Both the 40+GHz and, in particular, the 50+GHz diodes provide open 40Gbps eye diagrams, as shown in figure 7 for the latter, confirming the excellent high-speed performance of these devices.

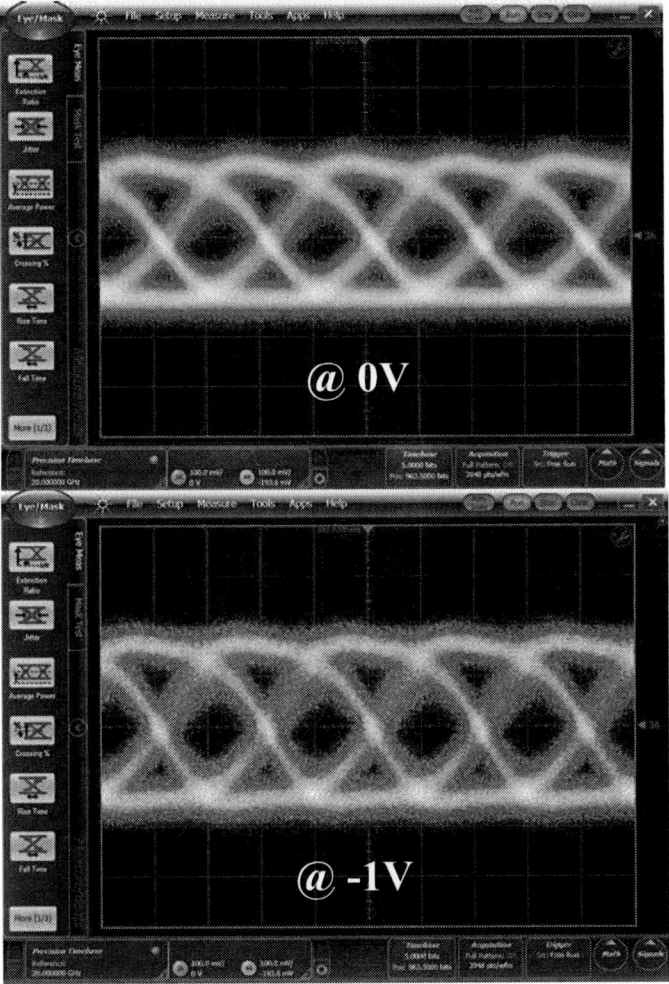

Fig. 7. 40Gbps eye diagrams, measured with 2^{31}-1 PRBS for a Ge lateral pin photodiode, fabricated with 750°C anneal and E1 and E2$_1$ implantations.

Interesting question was where the strong improvement in high-speed performance of diodes with additional implantation of non-doping elements, compared to reference devices without these implantations, comes from.

C-V curves, shown in figure 8, do not significantly differ between reference devices and diodes with extra implants. It proves that a change in the doping profiles due to the

particular non-doping implants applied here can very probably be ruled out as a cause. Big differences, however, show the DC characteristics, as can be seen from figure 9. Immediately, the higher forward currents of extra-implanted diodes are obvious. It let us conclude that forward current increase and bandwidth improvement has the same cause, which is reduction of minority-carrier lifetime(s) by the non-doping implants. Fig. 10 compares P vertical profiles in Ge after implantation and anneal. A diffusion tail, extending over some 100nm with more than 10^{17}cm^{-3}, can be observed. Assuming that lateral P diffusion resembles that behavior, our diodes

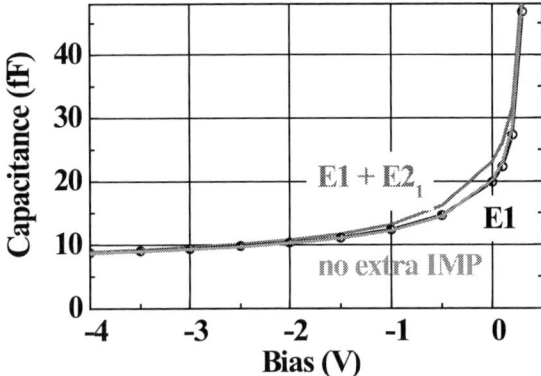

Fig. 8. Capacitance vs. voltage of Ge lateral pin photodiodes fabricated with or without extra implants, all with 750°C, 30sec anneal.

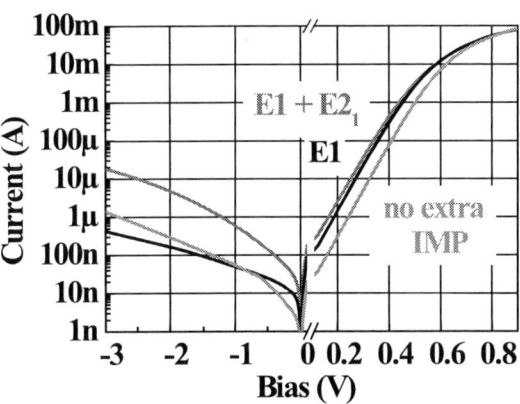

Fig. 9: DC characteristics of Ge lateral pin photodiodes fabricated with 750°C, 30sec anneal, with or without extra implants.

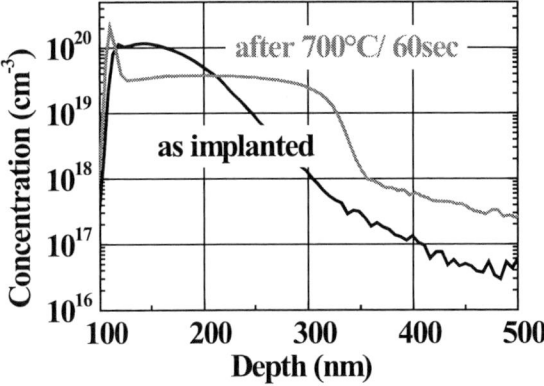

Fig. 10. P profiles in a Ge layer measured by SIMS after P implantation and 700°C, 60sec anneal, without extra implants.

very probably have a lateral P profile tail which can't fully be depleted at normal bias. Thus, a Ge layer with low n-doping remains where generated carriers can only move by slow diffusion. In result, bandwidth is no longer limited by charging or high-field drift time but partly by minority-carrier lifetime. Obviously, the non-doping implants applied here greatly suppress that effect by reduced lifetime(s), leading to the bandwidth gain observed.

Dark currents are also affected by the non-doping implants as can also be seen from figure 9. There is a penalty of about a factor of ten (@ -1V) for the diodes with highest bandwidth, compared to non-implanted reference devices. However, the dark current level is still better compared to other Ge pin photodiodes with more than 50GHz bandwidth [8]. For the 40+GHz devices, i.e. with extra implantation of element E1 alone, dark current behavior is much better, i.e. very similar to that of the reference devices.

Another question was why diodes with no extra implants degrade in bandwidth if conditions of the CoSi$_2$ module are changed (Fig. 3). C-V curves from diodes representing the particular annealing splits are shown in figure 10. They do practically no differ, which excludes a marked doping profile effect. However, the diodes differ in the forward current behavior, as can be seen from figure 11. Obviously, the stronger the annealing the lower the forward current is, indicating an increase in minority-lifetime(s) which is very probably responsible for the drop in bandwidth when anneal conditions become stronger.

Finally, no significant differences were found in the diode responsivity, which is in all cases higher than 0.65A/W with only a weak bias effect.

IV. SUMMARY AND CONCLUSIONS

Waveguide-coupled Ge photodiodes were presented showing more than 50GHz bandwidth and 40Gbps function. This excellent performance was reached with the help of additional implantations of non-doping elements which in particular allow us to reach such performance even under the effect of thermal steps acting when the diodes are integrated in a high-performance BiCMOS process. These results significantly improve the usability of our photonic BiCMOS process for high data rate communication applications.

ACKNOWLEDGMENT

The authors gratefully acknowledge support by German Ministry of Research and Education (BMBF), projects MOSAIC and RF2THzSiSoC.

REFERENCES

[1] B. Analui et al., "A Fully Integrated 20Gb/s Opto-Elec-tronic Transceiver Implemented in a Standard 0.13-µm CMOS SOI Technology," IEEE J. Solid State Circuits, Vol. 41, no. ?, p. 2945 (2006).

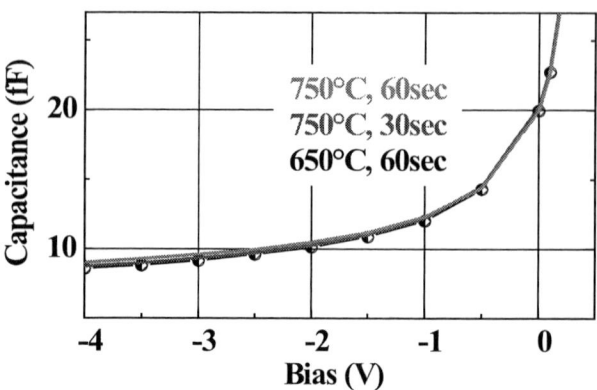

Fig. 10. Capacitance vs. voltage of Ge-PDs fabricated without additional implantation of non-doping elements, with different CoSi$_2$ anneals.

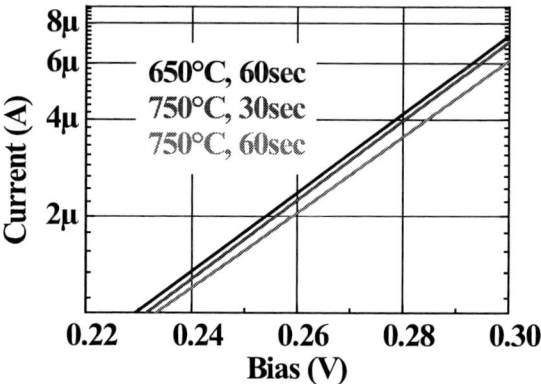

Fig. 11: Forward current characteristics of Ge-PDs fabricated w/o additional implantation of non-doping elements, with different CoSi$_2$ anneals.

[2] S. Assefa et al., "A 90nm CMOS Integrated Nano-Photonics Technology for 25Gbps WDM Optical Communications Applications," IEDM Tech. Dig., p. 809, San Francisco (2012).

[3] J. S. Orcutt et al., "Open Foundry Platform for High-Performance Electronic-Photonic Integration," Opt. Express, Vol. 20, no. 11, p. 12222 (2012).

[4] D. Knoll et al., "Monolithically Integrated 25Gbit/sec Receiver for 1.55 µm in Photonic BiCMOS Technology," Proc. OFC, Th4C.4, San Francisco (2014).

[5] L. Zimmermann et al., "Monolithically Integrated 10Gbps Silicon Modulator with Driver in 0.25µm SiGe:C BiCMOS," Proc. ECOC, We.3B.1, London (2013).

[6] A. C. Berti and S. P. Baranowski, "Self-Aligned Cobalt Silicide on MOS Integrated Circuits," US Patent, No. 5,736,461, April 7, 1998.

[7] T. Canneaux et al., "Modeling of Phosphorus Diffusion in Ge Accounting for a Cubic Dependence of the Diffusivity with the Electron Concentration," Thin Solid Films, Vol. 518 (2010) 2394-2397.

[8] L. Vivien et al., "Zero-Bias 40Gbit/s Germanium Waveguide Photodetector on Silicon," Opt. Express, Vol. 20, no. 2, p. 1096 (2012).

[9] C.T. DeRose et al., "Ultra Compact 45GHz CMOS Compatible Ge Waveguide Photodiode with Low Dark Current," Opt. Express, Vol. 19, no. 25, p. 24897 (2011).

[10] S. Klinger et al., "Ge-on-Si P-I-N Photodiodes with a 3-dB Bandwidth of 49GHz," IEEE Photonics Technology Letters, Vol. 21, no. 13, p. 920 (2009).

[11] S. Lischke et al., "High-Efficiency Grating Couplers for Integration into a High-Performance Photonic BiCMOS Process," Proc. ACPC, AW4A.1, Peking (2013).

[12] A. Novack et al., "Ge Photodetector with 60GHz Bandwidth Using Inductive Gain Peaking," Opt. Express, Vol. 21, no. 23, p. 28387 (2013)

An SiGe heterojunction bipolar transistor with very high open-base breakdown voltage

T. V. Dinh[1], T. Vanhoucke[1], A. Heringa[1], M. Al-Sa'di[1], P. Ivo[1], D. B. M. Klaassen[2], P. H. C. Magnee[3]

(1) NXP Semiconductors Central R&D, Interleuvenlaan 80, 3001 Leuven, Belgium
(2) NXP Semiconductors Central R&D, High-Tech Campus 46, 5656 AE Eindhoven, The Netherlands
(3) NXP Semiconductors Central R&D, Gerstweg 2, 6534 AE Nijmegen, The Netherlands

Abstract — An SiGe heterojunction bipolar transistor having a very high open-base breakdown voltage (BV_{CEO}), which is close to the hard breakdown voltage (BV_{CBO}), is introduced. This is achieved by draining the hot holes generated from impact ionization to the substrate. The carrier transport in those proposed devices is intensively investigated by device simulations which were confirmed by the electrical characteristics of our processed devices. The positive and constant base current over a large range of collector voltages (until BV_{CBO}) will improve the issues of electro-thermal reliability and distortion normally associated with a negative base current at voltages beyond BV_{CEO}, which is very attractive for RF power amplifiers.

Keywords—SiGe HBTs; BV_{CEO}; BV_{CBO}; hot carriers; impact ionization; power; cut-off frequency

I. Introduction

The RF power amplifier is an essential electronic component in RF systems which are now expanded remarkably to provide enough access and bandwidth for the incredibly increasing number of smart mobile phones. Thanks to their good RF performance and easy integration, SiGe HBTs are replacing expensive III-V devices currently used in those systems [1].

In order to improve further the capability of SiGe HBTs which are used for PA's, efforts have been done to create higher-breakdown SiGe HBTs following either traditional optimization in collector profiles (with breakdown voltages BV and cut-off frequency f_T trade-off, e.g. in [2-5]) or taking advantage of RESURF effects (by *junction* or *field plate RESURF*) to have performance breakthrough in term of BV×f_T, e.g. in [6-7]. Two types of breakdown voltages in HBTs are discussed briefly as follows.

In a bipolar, e.g. an NPN, the open-base breakdown voltage, BV_{CEO}, is defined as the voltage where the number of holes generated by impact ionization (due to the high electric field at the base-collector junction) equals the number of holes externally provided into the base contact making the external base current zero. This breakdown is not a hard breakdown since for collector-emitter voltages (V_{CE}) higher than BV_{CEO}, the device is still operating. The device breaks when V_{CE} is large enough to make the field at base-collector (B-C) junction larger than the critical field. This is the hard breakdown voltage, called open-emitter breakdown voltage BV_{CBO}. It is clear that BV_{CEO} is much smaller than BV_{CBO} in most cases, and in no means BV_{CEO} can be seen as a hard breakdown. However, for $V_{CE} > BV_{CEO}$ electro-thermal effects limit the maximum allowed voltage-current combination before experiencing instability problems described by the safe operating area [8-11]. This implicates issues with reliability (or degradation). Moreover, above BV_{CEO} the negative base current, which is strongly dependent on collector voltages, severely complicates the critical bias circuitry [12-13], and degrades the distortion behavior [12, 14]. Therefore, IC designs are often restricted to BV_{CEO} which limits the full potential of the transistor, especially for power amplifiers.

It is clear that BV_{CEO} depends on two factors: the electric field inside the device (determined by the B-C voltage V_{CB} and collector profile) and the current gain (determined by the amount of holes in the base needed to generate a specific collector current). Reducing the electric field will increase the breakdown voltage which can be done by for example reducing or reshaping the electric field in the B-C junction [6-7]. On the other hand, the negative current observed at the base terminal for given B-C and base-emitter (B-E) voltages can be reduced by reducing the current gain which is a topic studied in for example [15-16]. However, in these approaches, either BV_{CEO} has not been focused yet, or the BV_{CEO} improvement imposes disadvantages on other device characteristics (e.g. low current gain deteriorates high-frequency noise), or increases the process complexity dramatically.

In this paper, we demonstrate a method together with a device concept for increasing BV_{CEO} (close to BV_{CBO}) without affecting the current gain or adding complex process steps. High BV_{CEO} and BV_{CBO} are obtained by optimally redistributing the electric field, hence the impact ionization, within the collector to be able to have the highest breakdown voltages. In this concept, the region of high impact ionization, which is normally at the B-C junction, is shifted to the collector region close to the collector contact which is located further away from the B-C junction. Therefore, the base current is marginally affected by impact ionization since the generated holes do not flow to the base region (and therefore not observed at the base contact) but they are "drained" by the substrate. Details of the device concept and carrier transport will be discussed in Sec. II, supported by electrical measurements of the fabricated device shown in Sec. III.

II. Device Concept & Simulation

A. Traditional high voltage vertical SiGe HBTs

As mentioned in Sec. I, the most popular approach to obtain high-voltage SiGe HBTs is reduction of the collector dope by adjusting the collector profile. A typical device with this approach is shown in Fig. 1a, starting from a traditional vertical-collector device [5]. With this type of device, high electric fields appear at the B-C junction under high reverse voltages. Holes generated by field-induced impact ionization are forced by the electric field to the base region, which causes the base current to change its sign. The resulting hole-current distribution obtained from device simulation [17] is shown in Fig. 1b with the highest electric field and impact ionization concentrated in the narrow B-C region. Therefore with such device concept it is not trivial to shape the B-C field (i.e. limiting the impact ionization) while keeping f_T high and reduce the amount of generated holes at the B-C junction.

B. New high voltage lateral SiGe HBTs

In [18-19], the authors have introduced a double-emitter horizontal bipolar transistor, where 3D-collector charge sharing was used to fully deplete the collector and limit the electric field at the intrinsic base-collector junction to have a high BV_{CEO} (~BV_{CBO}); however a low f_T was shown, which can be attributed to a thick horizontal base. In another approach [20], the authors introduced a SiGe HBT device which is composed of a vertical base-emitter junction and a lateral collector drift region. The purpose of such a lateral collector is to spread the electric field over a larger (lowly-doped) region to obtain a higher breakdown voltage. The published device has higher f_T and a high BV_{CBO}, but it is still decoupled from BV_{CEO} resulting in $BV_{CBO} > 3 \times BV_{CEO}$.

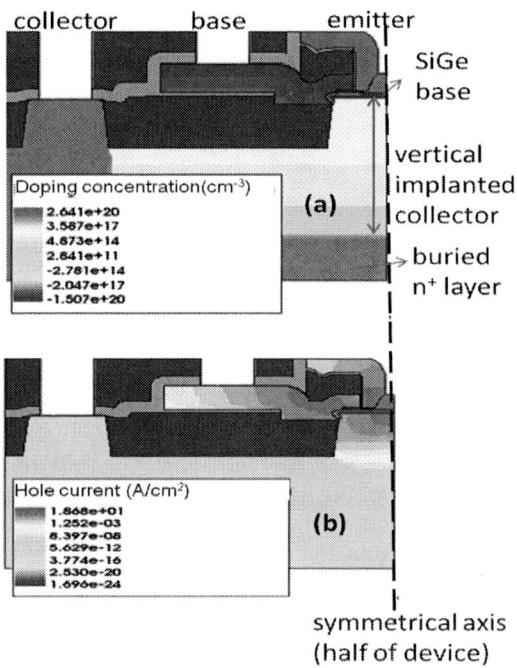

Figure 1: (a) A conventional high-voltage SiGe HBT device with a vertical collector profile following the traditional BV⇔f_T trade-off [5] (b) Hole current distribution within this device (obtained from device simulation) at the bias condition of V_{CE}=7V, V_{BE}=0.7V (BV_{CEO} of this device is 5V).

In this paper, we study in full details the carrier transport and impact ionization by device simulations and measurements, and come up with a similar lateral drift device concept having a more sophisticated engineering and optimization in the (lowly doped) n⁻ epitaxial collector drift region and the highly doped buried p⁺ layer underneath (device structure is shown in Fig. 2). The p⁺ layer, which can be

obtained in the default process flow from the P-well (and buried p⁺) for NMOS, creates a hard *junction* being used to introduce the RESURF effect [21]. An optimal RESURF, i.e. the field is optimally distributed to provide the highest BV_{CEO} (and BV_{CBO}), is obtained when the integral doping in n⁻ drift region is balanced with that of the buried p⁺ layer. Under that condition, compared to conventional devices (without junction RESURF, i.e. buried n⁺ instead of p⁺ layer under the n⁻ collector drift region), for the same V_{CE}, the field at the B-C junction is reduced by the RESURF effect, while the field at the substrate-collector junction increases (i.e. constant integral of the electric field). This field *redistribution* reduces the impact ionization at the B-C junction and drastically increases the voltage at which the base current changes its sign. The field distribution for the device with and without optimized p⁺ layer obtained from device simulations is shown in Fig. 3.

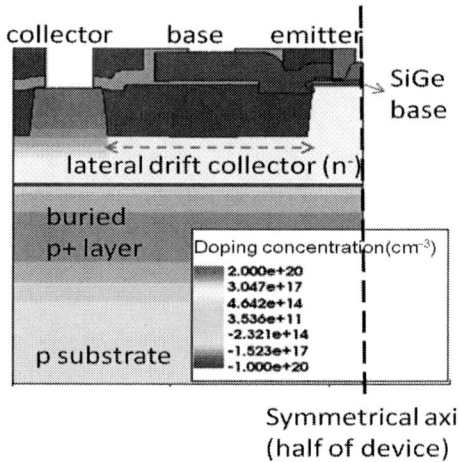

Figure 2: Structure of the optimized device obtained from TCAD with a lateral collector and a highly p doped buried layer.

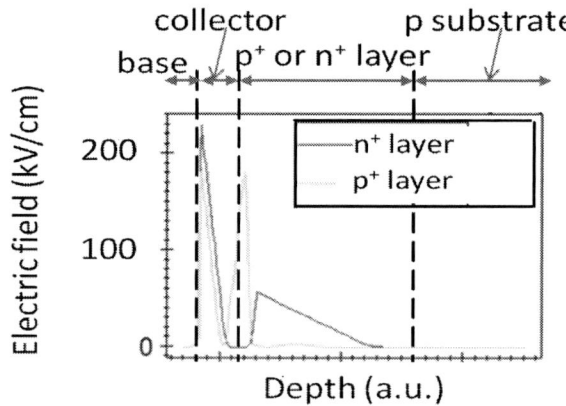

Figure 3: The electric field distribution along the vertical direction (emitter-base-collector-substrate, dashed line in Fig. 2) for two structures: *with buried p+ layer* and *without* (i.e. with n⁺ *layer*) at V_{BE}=0.6V and V_{CE}=5V (BV_{CBO} of the latter is ~20V).

The device with p⁺ layer (*i.e.* with *junction RESURF*) has a lower peak electric field at the B-C junction but an increased field at the collector-substrate junction. Note that due to the redistribution of the electric fields the area below both curves is identical and equals the same applied reverse voltage.

978-1-4799-7231-9/14 $31.00 © 2014 IEEE

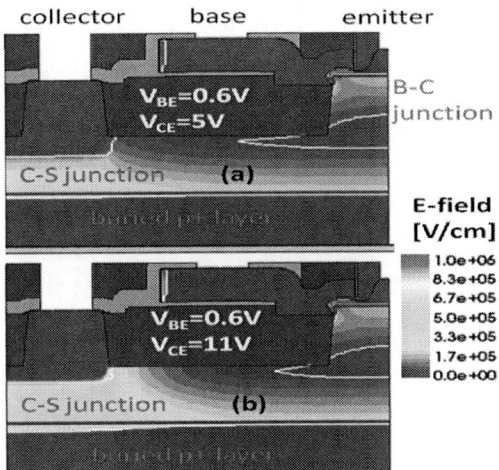

Figure 4: Simulations of the electric field in the lateral collector device (with *junction RESURF*) for (a) V_{CE} = 5 V and (b) VCE = 11 V (BV_{CEO} ~5V for the device *without junction RESURF*). The field at the B-C junction does not increase anymore for V_{CE} > 5V; instead only the field at the collector-substrate (C-S) junction increases.

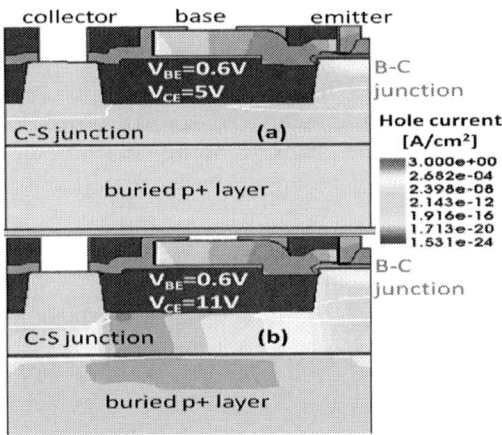

Figure 5: Hole current distribution in the optimized lateral drift device for (a) V_{CE}=5V and (b) V_{CE} = 11 V. At large V_{CE} (lower figure) the number of holes generated at the collector contact region and drained through the substrate terminal are much larger than the number generated at the B-C junction (which flow to the base contact).

More importantly, at a certain (high) reverse B-C voltage, the electric field at the B-C junction does not increase with voltage any more. Instead, the increasing field is pushed laterally to the (highly-doped) region at the collector contact as shown by the field distribution (in 2D) in Fig. 4. As a consequence, due to the redistribution of the electric fields most of the holes generated by impact ionization in the collector contact region go directly into the substrate and cannot flow towards the base anymore as is the case in the conventional devices. Therefore, the base current remains constant for increasing V_{CE} once the electric field at the B-C junction saturates until a junction breakdown happens (~BV_{CBO}). The intrinsic part of the device (i.e. the metallurgical base-collector and base-emitter junctions separated by the small neutral base region) does not "feel" the generated

avalanche current at the collector contact region and we have separated the intrinsic part from the avalanche generation region. With this concept, BV_{CEO} becomes independent of the current gain or in other words, the base current will not decrease (and become negative) anymore due to the generated holes for increasing V_{CE}.

The distribution of hole currents (holes generated by impact ionization) with increasing V_{CE} is shown in Fig. 5. It is seen that at low collector voltage, e.g. 5V, most of holes are generated at the B-C junction as usual. However, when V_{CE} is increased further, in accordance with the field distribution in Fig. 4, more holes are generated at the collector contact region and they are drained directly to the substrate terminal and therefore do not affect the base current.

III. DEVICE FABRICATION & MEASUREMENT

The devices based on the lateral drift concept described in Sec. II have been fabricated in the NXP QUBiC4 technology. As mentioned in Sec. II or with more details in [7, 21], optimal RESURF is obtained only at a certain window of collector epitaxial thickness and dope level (e.g. a dose of 2×10^{12} cm^{-2}), which is balanced by the same dose in the p$^+$ layer. Within the QUBiC4 technology, all those parameters are well tuned and accurately controlled. Fig. 6 shows a typical Gummel plot of these devices, from both measurement and simulation. The device has a current gain of ~1200, which is very high for high-voltage SiGe HBTs [1-5], but required for optimal high-frequency noise.

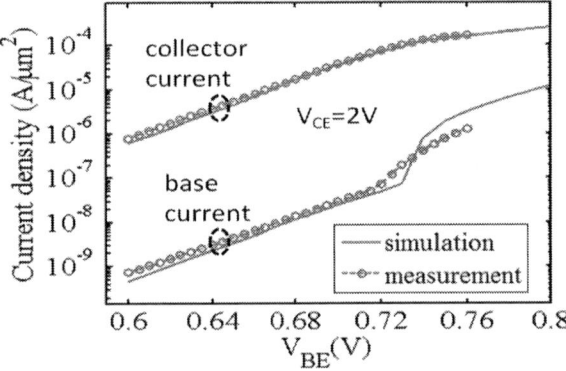

Figure 6: Measured and simulated base and collector currents for the device with p$^+$ layer at V_{CE}=2V.

The impact of the p$^+$ layer below the (lowly doped) n$^-$ epitaxial collector on breakdown voltages has been investigated. Fig. 7 shows the measured and simulated currents of two devices: one with p$^+$ layer and the other without this layer. It is seen that the point where the base current (I_B) changes sign (*i.e. BV_{CEO}*) is very different between devices with and without p$^+$ layer. Simulation results correspond well with the measured values. The large difference in BV_{CEO} between these two devices can be expected from the field and hole current distribution obtained from simulations in Fig. 4 and Fig. 5. Indeed, for the situation where no p$^+$ layer is present, the electric field redistribution is very ineffective and impact ionization takes place mostly at the B-C junction resulting in a low BV_{CEO} similar to traditional vertical-collector devices. In case of a p$^+$ layer however, the redistribution of the

978-1-4799-7231-9/14 $31.00 © 2014 IEEE

hole generation (i.e. electric field redistribution) results in very high BV_{CEO} values close to BV_{CBO} (~25 V) which is quite an advantage in application.

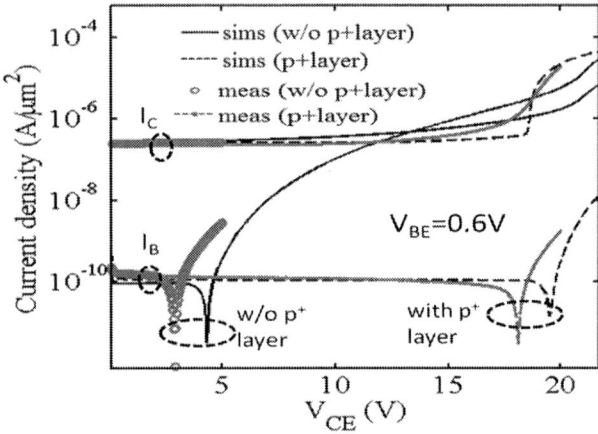

Figure 7: Comparison between BV_{CEO} by tracking the base current (I_B) and collector current (I_C) obtained from simulations (black) and measurements (red): the main difference in BV_{CEO} between "*with p+ layer*" and "*without p+ layer*" is due to the fact that in the p+ layer device, most of holes generated by impact ionization are drained by the substrate.

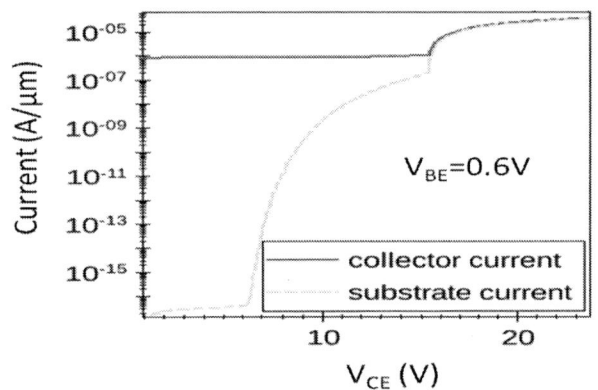

Figure 8: Collector and substrate current obtained from device simulations for the device with p+ layer (V_{BE}=0.6V).

The draining of the generated holes by the substrate is clearly shown through the behavior of substrate current (I_S) at low and high V_{CE} in Fig. 8. At low V_{CE}, I_S is low since the impact ionization is weak and mostly located at the B-C junction. When V_{CE} is high (e.g. V_{CE}>10V), the increase of I_S indicates that most of generated holes flow to the substrate while generated electrons add to the collector current. For very high V_{CE}, the collector current equals the substrate current meaning that both currents are fully determined by the avalanche at the collector contact region.

Fig. 9 shows the RF performance of the device with p+ layer. A product $f_T \cdot BV_{CEO} \approx 400\,\text{GHz}\cdot\text{V}$ was obtained, which is very high compared to reported high-voltage SiGe HBTs [2-6, 18-20]. The p+ layer has an impact on the collector resistance which might deteriorate RF performance [20]; this

resistance can be reduced significantly by layout and process optimization.

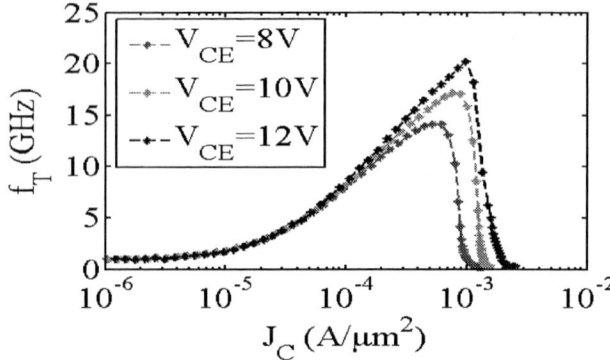

Figure 9: Measured f_T at various collector voltages for the device with p+ layer (BV_{CEO} of 18V, shown in Fig. 7).

IV.   CONCLUSION

A lateral drift device, which has a junction RESURF together with a fourth terminal to drain the holes generated from the impact ionization process, has been proposed and fabricated. This device has an optimized collector and substrate dope and thickness resulting in a very high open-base breakdown voltage BV_{CEO}, which could be as high as the open-emitter breakdown voltage BV_{CBO}. This high BV_{CEO}-device gives IC designers more flexibility in choosing a load-line which can bring the best performance to a RF power amplifier.

ACKNOWLEDGMENT

We would like to acknowledge Jan Slotboom, Fred Hurkx, Fred van Rijs and Domine Leenaerts for valuable discussions, and the BiCMOS team in Nijmegen for processing all these devices.

REFERENCES

[1] K. Nellis *et al.*, IEEE Journal of Solid State Circuits, pp. 1746-1754, vol. 39, 2004.
[2] D. R. Greenberg *et al.*, IEDM Proceedings, pp.799-802, 1997.
[3] E. J. Preisler *et al.*, BCTM Proceedings, pp. 202-205, 2007.
[4] B. Geynet *et al.*, SiRF Proceedings, pp. 210-213, 2008.
[5] H. Mertens *et al.*, BCTM Proceedings, pp. 158-161, 2011.
[6] J. Melai *et al.*, ISPSD Proceedings,pp. 33-36, 2004.
[7] R. Hueting *et al.*, IEEE Trans. Electron Devices, pp. 1108-1113, vol. 51, 2004.
[8] M. Rickelt *et al.*, IEEE Trans. Electron Devices, pp. 774-783, vol. 48, 2001.
[9] N. Rinaldi *et al.*, IEEE Trans. Electron Devices, pp. 1683-1697, vol. 53, 2006.
[10] T. Vanhoucke *et al.*, BCTM Proceedings, pp. 2.1-2.4, 2006.
[11] J. Kraft *et al.,* BCTM Proceedings, pp. 33-36, 2005.
[12] T. Leitner, Microwave Conference APMC, pp. 1160-1163, 2009.
[13] T. Leitner, US patent US7714659.
[14] G. Niu *et al.*, BCTM Proceedings, pp. 9.1-9.4, 2006.
[15] J.J.T.M Donkers *et al.*, IEDM Proceedings, pp. 243-246, 2004.
[16] P-M. Mans *et al.*, Active and Passive Electronics Components, pp. 1-6, 2010 (doi:10.1155/2010/542572).
[17] Synopsys Sentaurus Manual, 2013.
[18] M. Koricic *et al.*, BCTM Proceedings, pp. 5-8, 2011.
[19] M. Koricic *et al.*, IEEE Trans. Electron Devices, pp. 3647-3650, vol. 59, 2012.
[20] R. Sorge *et al.*, SiRF Proceedings, pp. 223-226, 2012.
[21] J.A. Appels and H.M.J. Vaes, IEDM Proceedings, pp. 238-241, 1979.

Examination of Horizontal Current Bipolar Transistor (HCBT) Reliability Characteristics

J. Žilak, M. Koričić, H. Mochizuki*, S. Morita*, K. Shinomura*, H. Imai* and T. Suligoj

Department of Electronics, University of Zagreb, HR-10000, Zagreb, Croatia
Tel: (385) 1 612 9898, Fax: (385) 1 6129653, e-mail: tom@zemris.fer.hr
*Asahi Kasei Microdevices Corp. 5-4960, Nobeoka, Miyazaki, 882-0031, Japan

Abstract – The reliability characteristics of HCBT are examined for the first time by employing reverse emitter-base (EB) and mixed-mode stresses. Three HCBT structures with different *n*-collector doping profiles and different oxide etching parameters before polysilicon deposition are measured, exhibiting different behavior at each stress test. Due to the specific HCBT structure, the traps generation causing I_B degradation, occurs at different regions, i.e. at EB *pn*-junction near both the top and the bottom EB oxide for reverse EB stress and only at the bottom EB oxide for mixed-mode stress, as discovered by TCAD simulations. Pre-deposition oxide etching conditions turned out to be critical for I_B degradation after reverse EB stress, whereas the *n*-collector vertical doping profile mostly impacts the trap generation after the mixed-mode stress. The 1/f noise characteristics also show the highest degradation for HCBT structures with the highest stress damage.

Keywords—Horizontal Current Bipolar Transistor, reliability, reverse emitter-base stress, mixed-mode stress, 1/f noise.

I. INTRODUCTION

The Horizontal Current Bipolar Transistor (HCBT) is developed as a compact bipolar structure with high-frequency characteristics suitable for wireless communication circuits [1]. The integration with CMOS technology; with only 2 or 3 additional masks, together with the availability of zero-cost high-voltage transistors [2], results in a flexible low-cost BiCMOS technology platform. Recently, the down-converting mixer is developed as the first HCBT RF circuit demonstrating the suitability of HCBT technology for the ISM band communication circuits [3]. In order to verify the suitability of the HCBT technology for the production, the device reliability has to be tested, which is the purpose of this paper.

Both reverse emitter-base (EB) stress and mixed-mode stress are applied to HCBT structures with different fabrication sequences. The reverse EB stress is traditional approach for reliability analysis that leads to a degradation of the transistor performance through the increase of the base recombination current. Such degradation is caused by the impact ionization in the emitter-base depletion region and consequent hot carrier injection into the emitter-base spacer oxide due to the high EB electric filed [4]. More recently, the mixed-mode stress is developed where the transistor is biased simultaneously with constant emitter current (I_E) and high collector-base voltage (V_{CB}) such that collector-emitter voltage (V_{CE}) is higher than breakdown voltage BV_{CEO} [5-7]. In this way, the hot carriers are generated in the collector-base (CB) depletion region and

can be injected both in the EB and CB oxides. Hence, the mixed-mode stress also causes the increase of the base recombination current, but with a different origin of hot carrier generation.

In the specific HCBT structure, the active transistor region is fabricated at the sidewall of the shallow trench isolation and the position of impact ionization region (as the source of hot carriers) both in reverse EB and mixed-mode stresses is different than in the conventional vertical-current bipolar devices. The HCBT base is always one-side contacted and oxide-silicon interface subject to hot carrier injection is both above and under the emitter n^+ polysilicon region, as shown in Fig. 1. The position of impact ionization region is at different distance from the oxide-silicon interface and those interfaces experience different treatments in fabrication than in vertical-current transistors. Hence, the purpose of this study is to examine the characteristics of HCBT structure under reverse EB and mixed-mode stresses.

The impact of various *n*-collector doping profiles and oxide etching step parameters before the polysilicon deposition on device reliability and 1/f noise is examined by measurements. The physical mechanisms behind the measured characteristics are then explained by TCAD simulations.

II. HCBT MEASUREMENTS

A. Examined HCBT Structures

Due to the lateral current flow, the HCBT characteristics can be optimized by designing the *n*-collector vertical doping profile [8]. Hence, the HCBTs with three different *n*-collectors, the same as those used in the study of mixer linearity in [3] are

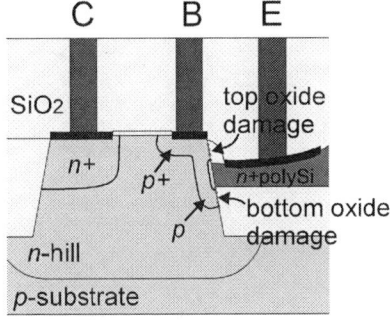

Fig. 1. Cross section of single-poly HCBT structure examined in reliability study with marked top and bottom oxide-silicon interface damaged by hot carrier injection.

978-1-4799-7231-9/14 $31.00 © 2014 IEEE

TABLE I.   MEASURED ELECTRICAL PARAMETERS OF HCBT 1
(STEEP N-HILL), HCBT 2 (UNIFORM N-HILL) AND HCBT 3 (CMOS N-WELL)

	HCBT 1 Steep *n*-hill	HCBT 2 Uniform *n*-hill	HCBT 3 CMOS *n*-well
Emitter area	0.1 x 1.8 µm²		
Peak β	153	158	155
f_T (GHz)	36	38	47
f_{max} (GHz)	67	66	65
BV_{CEO} (V)	3.9	3.3	2.8
f_TBV_{CEO} (GHzV)	140	125	132
Re (Ω)	120	108	155
pre-deposition HF dip	short	long	short

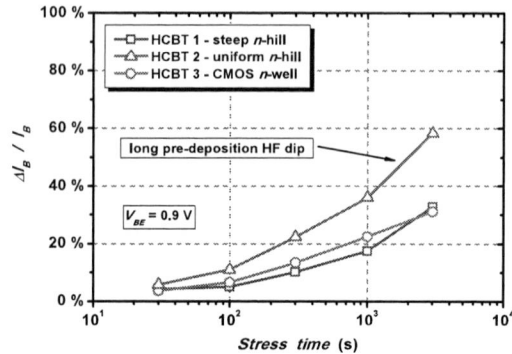

Fig. 3. Base current degradation for three HCBT structures showing the effects of reverse-bias emitter-base stress as a function of stress time. Stress conditions: $I_{EB,stress}$ = 0.5 µA, open collector.

compared. Measured electrical parameters of analyzed HCBT structures are summarized in Table I. The highest f_TBV_{CEO} product of 140 GHzV, is achieved for HCBT 1 (steep *n*-hill) since it suppresses the charge sharing between the extrinsic and intrinsic base regions most efficiently [8]. The HCBT 2 (uniform *n*-hill) achieves the best high-current linearity performance in RF mixer among analyzed structures as shown in [3]. Finally, the HCBT 3 (CMOS *n*-well) uses the implantation steps from CMOS *n*-well process, which saves one mask in fabrication, resulting in the lowest-cost process. Since all three HCBTs have certain advantages for commercial applications, their reliability characteristics are tested.

In addition to different *n*-collector regions, the silicon sidewall surface is treated differently before the emitter polysilicon deposition. A longer HF dip is used in HCBT 2 as compared to HCBT 1 and 3 resulting in its lower emitter resistance (R_e) due to a thinner emitter interface oxide. However, this also means that the interface between silicon sidewall and top oxide spacer presumably has a stronger H-termination, which should result in a higher concentration of interface states at the end of the fabrication and is more susceptible to hot carrier damage.

B. Reverse-bias emitter-base stress

The three HCBT structures from Table I are used in the reverse-bias EB and mixed-mode stress measurements. Reverse emitter-base stress is performed at a constant emitter-base reverse current of 0.5 µA with an open collector at room temperature. The resulting reverse EB voltage is around 3 V producing a high electric field sufficient for the strong impact ionization. The forward Gummel characteristics are measured at the collector-base voltage V_{CB} = 0 V after different stress times for all examined HCBT structures. The stress times are from 10s to 3000s. Fig. 2 shows the measured Gummel characteristics for the HCBT 2 and the identical measurements are done for the other two HCBTs.

Base current degradation for all three HCBT structures is shown in Fig. 3. HCBT 1 (steep *n*-hill), HBCT 2 (uniform *n*-hill) and HCBT 3 (CMOS *n*-well) exhibit base current degradation of 33 %, 58 % and 31 %, respectively, after 3000s of stress at V_{BE} = 0.9V, which is the bias point around peak f_T. HCBT 1 and HCBT 3 show similar I_B degradation due to the same pre-deposition HF dip, whereas HCBT 2, where the longer pre-deposition HF dip is used, has a higher I_B degradation. These results mean that the pre-deposition surface treatment is critical for the trap generation by the reverse EB stress. Expectedly, the *n*-collector doping profile does not affect the I_B degradation, as shown by almost equal post-stress characteristics of HCBT 1 and HCBT 3.

Different behavior of HCBT structures after reverse EB stress can also be observed by low-frequency noise measurements. Base current noise spectral densities (S_{IB}) of examined HCBT structures measured before and after reverse EB stress are shown in Fig. 4. Measurements are done at base

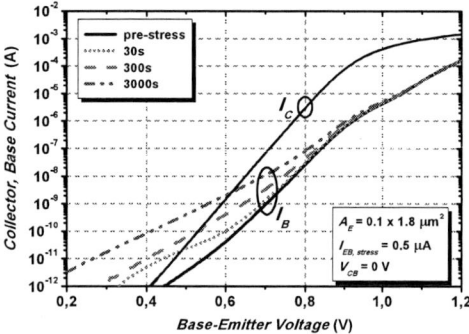

Fig. 2. Measured forward Gummel characteristics of the HCBT 2 (uniform *n*-hill) before and after the reverse EB stress.

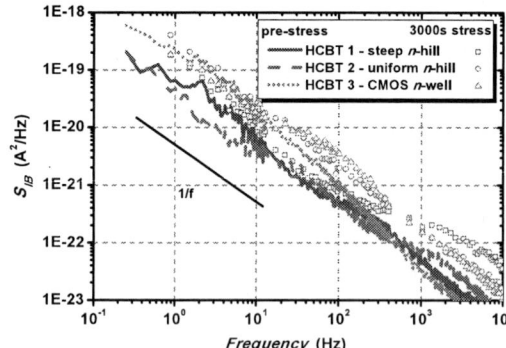

Fig. 4. S_{IB} spectra of pre- and post-stress (reverse EB stress) low-frequency noise for three HCBT structures, A_E =0.1 x 1.8 µm². Bias : I_B = 1 µA, V_{CE} = 1.5 V.

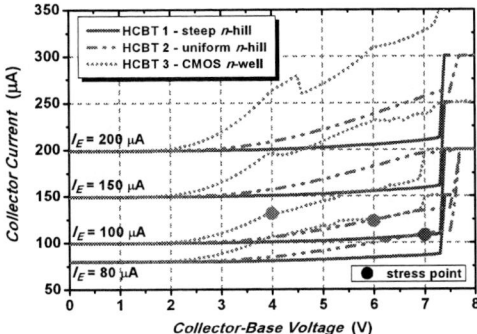

Fig. 5. Output characteristics with constant I_E for all three types of HCBT structures. Large dots indicate defined mixed-mode stress conditions for each HCBT structure.

current $I_B = 1\ \mu A$ and stress time is 3000s. The HCBT 2 (uniform n-hill) exhibits the highest post-stress S_{IB} increase at the lowest frequencies, due to the highest concentration of generated traps, as already shown by I_B degradation in Fig. 3, while the other two structures show similar degradation of S_{IB} which corresponds to the results shown in Fig. 3. HCBT 1 and HCBT 3 structures show smaller degradation of S_{IB} at lowest frequencies, e.g. below 10 Hz which correspond to the results shown in Fig. 3. Extracted values of K factor, defined as product of flicker-noise coefficient (K_f) and emitter area (A_E), in the range of 10^{-8} to 10^{-10} place the HCBT among the best low-noise performance of npn silicon BJTs and SiGe HBTs[9].

C. Mixed-mode stress

In order to determine mixed-mode stress conditions for three HCBT structures the output characteristics with constant emitter current (I_E) are measured and shown in Fig. 5. Increase of the collector current due to avalanche multiplication in the CB region occurs in the range between the breakdown voltages BV_{CEO} and BV_{CBO}. Due to variations in collector region design various stress conditions are used. The emitter current is fixed at 100 μA for all three structures. HCBT 3 (CMOS n-well) enters the instability region first and it is stressed by $V_{CB} = 4$ V. Collector current increase for HCBT 2 (uniform n-hill) has a milder slope and it is stressed with $V_{CB} = 6$ V. HCBT 1 (steep

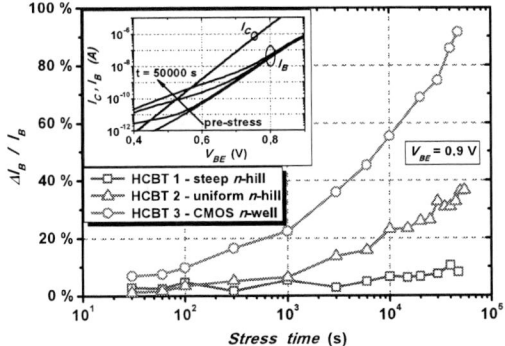

Fig. 6. Base current degradation for three HCBT structures showing the effects of mixed-mode stress as a function of stress time. Stress conditions: $I_{E,stress} = 100\ \mu A$, $V_{CB} = 7$ V (HCBT 1), $V_{CB} = 6$ V (HCBT 2) and $V_{CB} = 4$ V (HCBT 3). Measured forward Gummel characteristics of the HCBT 2 (uniform n-hill) is in inset.

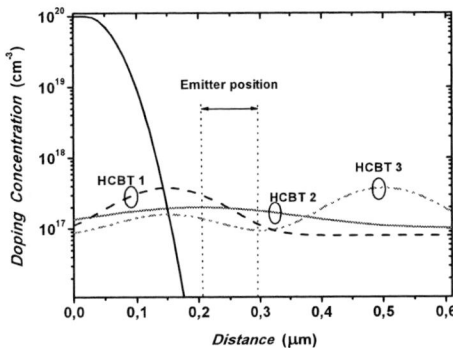

Fig. 7. Vertical doping profiles of simulated HCBT structures.

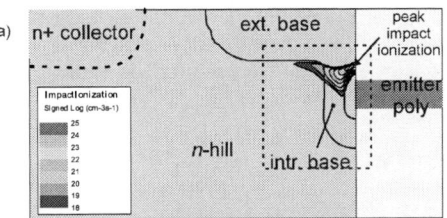

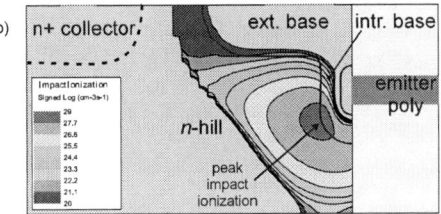

Fig. 8. Simulated impact ionization rates for HCBT 2 (uniform n-hill) for: a) reverse-mode stress ($V_{EB} = 3$V, $I_C = 0$), b) mixed-mode stress ($V_{CB} = 8.2$ V, $I_{E,stress} = 50\ \mu A/\mu m$).

n-hill) does not exhibit instabilities and it is stressed with largest collector-base voltage $V_{CB} = 7$ V. During stress measurements, the HCBTs showed a high immunity to the mixed-mode stress so the stress time is increased up to 60000s.

Base current degradation under mixed-mode stress for all three HCBT structures is shown in Fig. 6. Although it is stressed with largest collector-base voltage, the HCBT 1 (steep n-hill) is least affected by the stress, e.g. 8 % after 50000s of stress. HCBT 2 (uniform n-hill) exhibits 36 % base current increase for the same stress time and smaller V_{CB}, while the greatest base current degradation, i.e. 91 % occurs for HCBT 3 (CMOS n-well) despite the fact that it is stressed by the smallest collector-base voltage $V_{CB} = 4$ V.

III. SIMULATONS AND DISCUSSION

In order to explain and gain a deeper insight into measured reverse EB and mixed-mode stress results, HCBT structures are analyzed by 2D device simulations. Simulated structures correspond to the ones analyzed by the measurements. Therefore, the main focus of the analysis is the collector doping profile impact on the reliability properties. Vertical doping profiles in simulations are shown in Fig. 7.

All three HCBT structures are simulated at the same stress conditions, i.e. $V_{EB} = 3$ V with an open collector in the case of reverse EB and $I_E = 50$ µA/µm and $V_{CB} = 8.2$ V in the case of mixed-mode stress. The results of the impact ionization rates for HCBT 2 (uniform *n*-hill) under reverse and mixed-mode stresses are shown in Fig. 8. Under the reverse EB stress, peak impact ionization occurs at the top portion of EB junction next to the top EB oxide. The maximum electric field is placed at the curvature of EB junction. Simulations show that hot electron current injected into the insulator has two peaks which are placed next to the top and bottom EB oxides as shown in Fig. 9, with a higher peak current at the top EB oxide. Simulation results show that there is no significant difference in the injected current between three structures. This is in a good agreement with the measured results from Fig. 3 for HCBT structures 1 and 3. According to simulations, the similar characteristics would be expected for HCBT 2, which is not confirmed by measurements. Therefore, in case of HCBT 2, degradation of the base current after stressing is attributed to the lower quality of oxide-silicon interface at the top EB oxide caused by the longer pre-deposition HF dip.

Simulated impact ionization rate for the mixed-mode stress in case of HCBT 2 (uniform *n*-hill) is shown in Fig. 8.b. Peak impact ionization is located at the bottom part of the CB junction. Due to the charge sharing between extrinsic and intrinsic base acceptors, current crowding occurs at the bottom part of the base. This effect is partly compensated by the optimized steep *n*-hill profile in HCBT 1, as explained in [8]. Simulated hot electron currents injected into the insulator for all three structures are shown in Fig. 9. The maximum injection current is placed at the bottom part of the base and it can be assumed that interface traps are introduced mainly at the bottom EB oxide. Peak values of the hot electron injection currents for HCBT 1 (steep *n*-hill), HCBT 2 (uniform *n*-hill) and HCBT 3 (CMOS *n*-well) are 0.15, 0.9 and 7 pA/µm, respectively. For higher collector concentration in the bottom part (see Fig. 7) the electric field is higher and higher peak hot electron current is obtained. It can be also observed that the peak value is not placed directly at the BE junction, but in the neutral base. However, a part of the hot electrons are injected into the oxide within EB depletion region, especially for smaller V_{BE}, where EB depletion region is wider and contribution to recombination I_B is stronger. Certain amount of hot electron current is injected from the top portion of the EB junction in the case of HCBT 2 and 3. However, this current is several orders of magnitude smaller than the one associated with the bottom peak and is considered to be negligible. Since the bottom EB oxide has the same physical properties for all three structures, hot electron injection depends only on the collector profile. The higher the collector concentration in the bottom part, the higher the electric field and more recombination centers are introduced resulting in the higher I_B increase. This is in a good agreement with the measured results in Fig. 6.

IV. CONCLUSION

Different locations of oxide-silicon interface damage after reverse EB and mixed-mode stresses result in a different degree of I_B and S_{IB} degradation for various HCBT designs. The results clearly show the guidelines for HCBT fabrication to improve its reliability. To make it more immune to the reverse EB stress, the oxide-silicon interface above the n^+

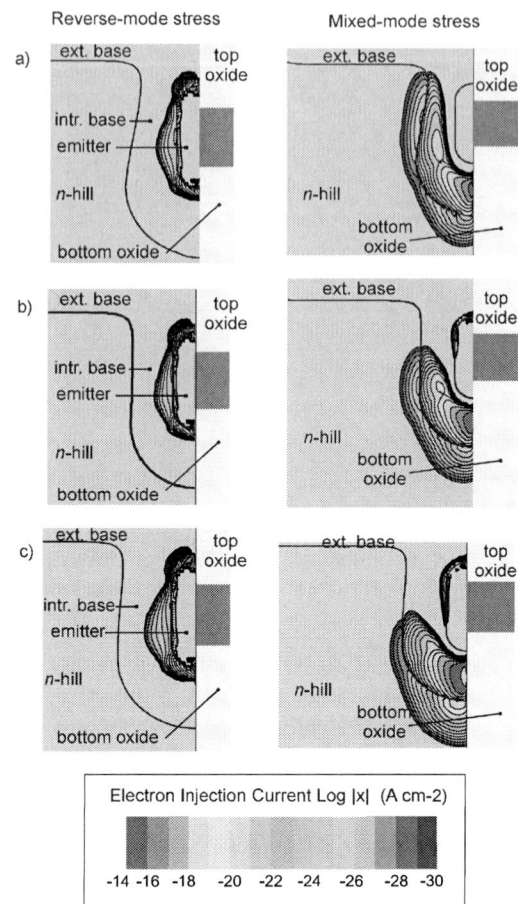

Fig. 9. Simulated hot electron injection current, which is several orders of magnitude larger then the hole injection current in the used model, at reverse emitter-base and mixed-mode stress for a) HCBT 1 (steep *n*-hill), b) HCBT 2 (uniform *n*-hill) and c) HCBT 3 (CMOS *n*-well). Portion of intrinsic transistor marked in Fig. 8.a is shown.

polysilicon region must have a good quality, requiring a shorter pre-deposition HF dip, such as for HCBT 1 and 3. On the other hand, the electric field at the bottom part of the CB junction has to be reduced in order to minimize the damage from mixed-mode stress, which favors a steep *n*-hill profile, such as in HCBT 2. Hence, the most reliable device would be HCBT with a steep *n*-collector profile with a short pre-deposition HF dip. The 1/f noise characteristics are examined for the first time showing respectable low-frequency noise performance of HCBTs and S_{IB} degradation follows the I_B degradation.

REFERENCES

[1] T. Suligoj, et. al., *IEEE EDL*, vol. 31, pp. 534-536, 2010.

[2] M. Koričić, et. al., *Proc. BCTM*, 2011, pp. 5-8.

[3] T. Suligoj, et. al., *Proc. BCTM*, 2013, pp. 13-16.

[4] J. Tang, et. al., *Proc. BCTM*, 2013, pp. 163-166.

[5] G.G. Fischer, et. al., *Proc. BCTM*, 2012, pp. 1-4.

[6] G. Zhang, et. al., *IEEE TED*, vol. 49, pp. 2151-2156, 2002.

[7] J.D. Cressler, et. al., *IEEE TDMR*, vol. 4, pp. 222-236, 2004.

[8] T. Suligoj, et. al., *Proc. BCTM*, 2010, pp. 212-215.

[9] P. Cheng, et. al., *Proc. BCTM*, 2010, pp. 165-168.

Degradation and Recovery of High-Speed SiGe HBTs under Very High Reverse EB Stress Conditions

G. Sasso, N. Rinaldi

DIETI - Dept. of Electrical Eng. and Information Tech.
University Federico II, via Claudio 21, 80125 Naples, Italy
grazia.sasso@unina.it

G. G. Fischer, B. Heinemann

IHP - Innovations for High Performance Microelectronics
Im Technologiepark 25, Frankfurt (Oder), Germany

Abstract—Hot-carrier induced degradation and post-stress recovery of 240/300 f_T/f_{MAX} SiGe HBTs are investigated. The investigation includes the impact of stress conditions and lateral scaling. Self-heating is employed as a means to investigate thermal-induced recovery activation and rate.

Keywords—SiGe HBT; reliability; reverse EB stress; recovery; self-heating.

I. INTRODUCTION

In the ongoing development of high performance RF circuits, silicon-germanium (SiGe) heterojunction bipolar transistor (HBT) technologies have maintained their competitiveness against advanced Si CMOS and compound semiconductor millimeter-wave circuits since more than 20 years thanks to its excellent RF performance combined with monolithic integration [1]. To keep this competiveness in the market of millimeter-wave systems, the European project DOTSEVEN [2] supports the development of SiGe HBTs with maximum oscillation frequencies f_{MAX} of 700 GHz. This development is enabled by aggressive scaling of lateral and vertical device dimensions together with new concepts for reduced transistor parasitics [3]. The aggressive operating conditions, required to achieve the maximum operating frequency, combined with the highly-scaled dimensions of SiGe HBTs, make device reliability an increasingly important topic in circuit and system design. Hot-carrier damage is typically categorized into reverse emitter-base (EB) stress, forward stress and mixed-mode stress (MM) and many efforts have been spent to investigate the responses of SiGe HBTs under stress conditions [4]. However, available studies on state-of-the-art HBTs mainly focus on MM stress effects and relatively little attention has been paid to the reversibility of the EB stress degradation by annealing. Hot-carrier induced damage may often be eliminated, at least in part, by subjecting the devices to post-stress thermal annealing. The physical background of the annealing processes provides insight into the stress mechanisms and allows a careful treatment in the evaluation of the degradation during and after the application of the stress, in order to avoid that the measurements affect the results of the degradation itself. Thus, the understanding of both effects, degradation and annealing, is of great importance for proper device design, specification and real-world simulation of circuit malfunction due to degradation.

In this work, we intend to contribute to the current understanding by reporting detailed investigations of the reverse EB stress induced degradation and recovery on the DC and RF performance of SiGe HBTs, including its implication to lateral scaling.

II. TECHNOLOGY AND REVERSE EB STRESS RESULTS

The investigated SiGe HBTs are fabricated in the IHP 0.13 µm SiGe BiCMOS technology SG13S and exhibit peak cut-off frequency f_T of 240 and f_{MAX} of 300 GHz. Details of the fabrication process and performance can be found in [5]. In this study, SiGe NPNs with single collector and single base layout (CBE configuration) have been analyzed; emitter effective sizes W_E and L_E, areas $A_E=n\cdot(W_E\cdot L_E)$ and perimeters $P_E=2\cdot n\cdot(W_E+L_E)$ are summarized in Table I, together with the applied stress conditions $V_{EB,stress}$. All the stressed DUTs are multi-transistors composed by 4 parallel HBTs (n=4). The stress and unstressed measurements have been conducted on wafer at a temperature of 300 K. In order to check the AC performance reliability as well, we investigated RF device configurations with an RF test set. A stress voltage $V_{EB-stress}$ has been applied at the BE junction with $V_{CE}=0$, i.e., $V_C=V_E=0$ V. This bias condition was verified to be equivalent to the open collector (OC) stress bias, since the collector current I_C is negligible during stress, and an uncontrolled forward biasing of the substrate-collector junction was avoided. The samples have been stressed at $V_{EB-stress}$ of 3.75, 4.00, 4.25, and 4.50 V, hence, always above the reverse open-collector breakdown voltage BV_{EBO}, whose process target is 1.7 V. The applied stress conditions, rather extreme and very far from the usual device bias conditions, cause the acceleration of the degradation and, thereby, allow the estimation of the time-to-failure (TTF) of a transistor and integrated-circuit fabrication technology within a reasonable measurement time. The stress has been periodically interrupted for monitoring any change in the device characteristics. In order to prevent any variation of the results of the degradation due to the monitoring measurements, the evaluation of the impact of reverse EB stress on the DC and AC performance has been conducted separately on different samples. In the following, reliability results concerning one chip will be reported, but DC and AC stress and recovery effects have been evaluated on three dies confirming all the observed trends and results.

The Gummel-plot of HBT #2 is depicted in Fig. 1 before and after 1000s of applied stress and shows that the collector current I_C remains unchanged after stress, while the base current I_B increases with increasing stress time and, consequently, the current gain $\beta_F=I_C/I_B$ decreases. The I_B degradation strongly shrinks in the high-bias region and the high-injection β_F is marginally affected by degradation. Base current degradation $\Delta I_B(t)=I_B(t)-I_B(0)$ as a function of the stress time t is depicted in Fig. 2 under different stress conditions. Constant voltage stressing results indicate that I_B degradation, for a given stress duration, increases with

978-1-4799-7231-9/14 $31.00 © 2014 IEEE

increasing reverse EB bias, since larger reverse voltage yields an increasingly larger amount of current flowing through the EB junction. This reverse current, which is primarily confined to the emitter perimeter region, under the influence of localized electrical fields can generate hot carriers. These hot carriers, in turn, can damage the Si-SiO$_2$ interface by

TABLE I.   HBTs LAYOUT AND STRESS CONDITIONS

HBT	W_E [μm]	L_E [μm]	A_E [μm²]	P_E [μm]	$V_{EB,stress}$ [V]
1	0.13	0.88	0.4576	8.08	3.75÷4.50
2	0.16	0.88	0.5632	8.32	3.75÷4.50
3	0.19	0.88	0.6688	8.56	3.75÷4.50

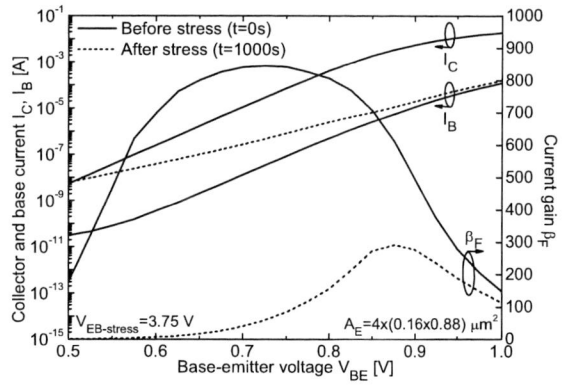

Fig. 1. Currents (left) and current gain (right) measured at V_{CB}=0 V for HBT #2 before and after applied stress of $V_{EB\text{-}stress}$=4.0 V for 1000s.

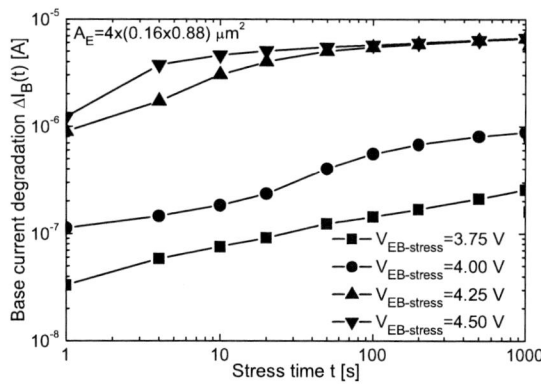

Fig. 2. Base current increment measured at V_{BE}=0.7 V and V_{CB}=0 V as a function of stress time for HBT #2 under different stress conditions.

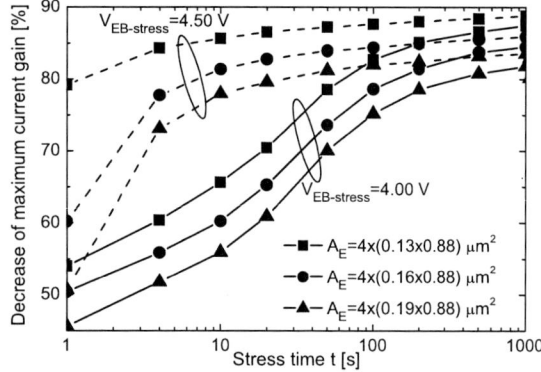

Fig. 3. Maximum gain degradation measured @ V_{CB}=0 V as a function of the stress time for all the investigated geometries under stress conditions of $V_{EB\text{-}stress}$ 4.0 V (full lines) and 4.5 V (dashed lines).

increasing interface trap density and, if energetic enough, they surmount the Si-SiO$_2$ barrier giving rise to an increase in fixed oxide charge density and oxide trap charge density. Maximum gain degradation $\Delta\beta_{F\text{-MAX}}$ is reported in Fig. 3 for the different emitter geometries at two different stress conditions. For a given emitter layout, maximum gain degradation follows the same trends of $\Delta I_B(t)$ shown in Fig. 2. For $V_{EB\text{-}stress}$< 4 V, the degradation is almost linear with stress time. When the stress voltage increases, a very fast degradation occurs in the first few seconds; then, the rate of the increase of I_B reduces with increasing stress time and a saturation trend can be observed. A simply explanation is that saturation is reached when the generation process has exhausted the available electrically contributing surface states. The density of available states and, thus, the size of the induced leakage is sensitive to the stress bias. However, although a higher $V_{EB\text{-}stress}$ increases the energy of the hot carriers, so that the number of potential surface states to be generated should increase, the saturation level reached at different high stress voltages was found to be almost independent of the $V_{EB\text{-}stress}$. Therefore, further investigations supported by device simulation are necessary. Besides, the performance of the device with the smallest emitter width W_E degrades more rapidly, since the scaling of the emitter size to smaller dimensions inevitably increases the perimeter-to-area ratio of the emitter. Fig. 4 reveals that the base current degradation normalized to perimeter P_E is almost comparable for different device sizes and these results confirm that the hot-carrier damage induced by the stress is mostly located around the emitter perimeter, adjacent to the space charge region between the emitter and the base.

Stress tests have been repeated to monitor the AC performance degradation. The chosen stress conditions and device layouts are the same as described above, but different samples have been considered in order to (i) avoid that RF measurements could affect the effects of degradation and (ii) assure that measurements between stress steps are as fast as possible. Junction capacitances, AC current gain H_{21}, and f_T have been extracted from the scattering parameters S. Despite of the very strong applied stress conditions, no significant variations has been detected for RF parameters (with the exception of low-frequency values of H_{21}, which follow the degradation of DC performance). It's worthwhile noting that when RF performance are evaluated, the complete stress cycle should take less than one hour and a dummy unstressed sample should be measured simultaneously, in order to monitor any possible shift in the S parameters not ascribable to the applied stress.

III. NATURAL AND SELF-HEATING RECOVERY EFFECTS

An in dept examination of recovery phenomena should account for three types of recovery. Namely, in addition to the thermal recovery, natural and forward bias relaxation effects should be considered. To the authors' knowledge, detailed studies of recovery mechanisms after reverse EB stress, including activation and rate, are widely available for Si BJTs [6], [7], while such investigations are still lacking for SiGe HBTs. We observed that the stress-induced base current increment ΔI_B decreases (and β_F increases) immediately after the removal of the stress voltage. As depicted in Fig. 5, this relaxation transient originates a not negligible decrease of ΔI_B which takes place over few minutes apparently approaching a

978-1-4799-7231-9/14 $31.00 © 2014 IEEE

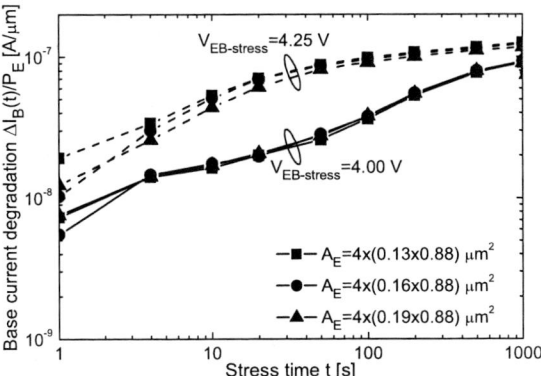

Fig. 4. Excess base current on perimeter ratio measured @ V_{BE}=0.7 V and V_{CB}=0 V as a function of stress time for different geometries under stress conditions of $V_{EB\text{-}stress}$ = 4.0 V (full lines) and 4.25 V (dashed lines).

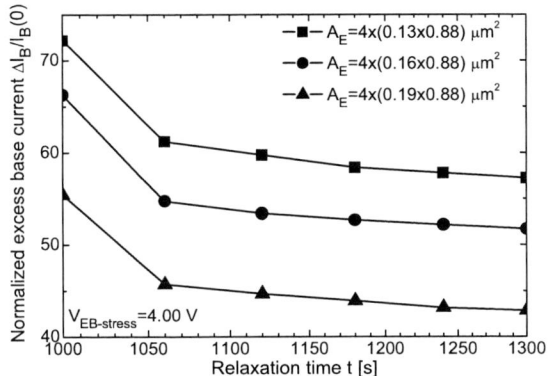

Fig. 5. Excess base current measured at V_{BE}=0.7 V and V_{CB}=0 V vs. relaxation time after applied stress of $V_{EB\text{-}stress}$=4.0 V for 1000s.

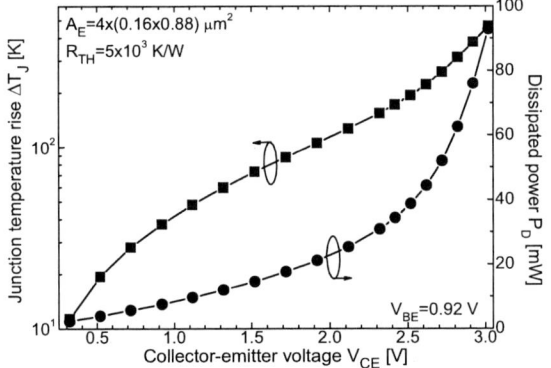

Fig. 6. Dissipated power (right) and junction temperature rise (left) as obtained for the unstressed HBT #2 by sweeping V_{CE} at V_{BE}=0.92 V (this V_{BE} value gives the peak f_T at V_{CE}=1.2 V).

steady-state condition. HBTs have been previously stressed at $V_{EB\text{-}stress}$=4.0 V for 1000s. Then Gummel-plots at V_{CB}=0 V are repeatedly measured with a sample rate of 60s, and ΔI_B and β_F changes are compared to the fresh (unstressed) values. The collector is shorted to the base during forward-bias measurements, in order to avoid uncontrolled thermal annealing induced by self-heating. Consistently to the measured stress results, the $\Delta I_B(t)$ variation rate is comparable for different geometry at a given stress voltage and the emitter layout does not affect the duration of the relaxation. After few

minutes the relaxation approaches a saturation, indicating that this recovery is spontaneous and no forward bias recovery occurs. The effects of the natural recovery remain irrespective of time and subsequent bias. Therefore it is more likely due to the detrapping of oxide trapped carriers than to a fast change in occupancy of interface states [6]. Although the reasons for natural recovery are unclear, the results suggest that DC or AC measurements should be performed immediately after the applied stress and as fast as possible, in order to avoid that natural recovery affects the stress results.

Recovery of the degradation was further investigated in terms of applied stress and lateral scaling. The kinetics and temperature dependences of the annealing of reverse bias stress-induced changes can provide important information on hot-carrier induced degradation. A standard procedure to investigate annealing and recovery activation consists in setting an annealing time and varying the annealing temperature with steps; after each annealing cycle at a given temperature, the device should be fully cooled down to quantify the effect of recovery. However, the heating and cooling needs a large amount of time and would accordingly give rise to ambiguous results. A possible alternative way to overcome this issue consists in setting the annealing temperature equal to the one applied during the measurement. However, this approach strongly limits the investigation and impedes the evaluation of the activation energy and annealing rates with temperature. In this work a different approach is proposed, in that we use self-heating as a means to accurately study the thermal recovery activation and rate. Since the DC thermal behavior of DUTs (i.e. the thermal resistance R_{TH}) is known, controlled self-heating conditions can be applied to reliably evaluate thermal and kinetics properties of recovery. The chuck temperature is always kept to T_A=300 K and annealing steps of constant duration and temperature are obtained by setting the dissipated power P_D; if the thermal resistance of the device is known, the annealing temperature corresponds to the junction temperature T_J, and can be calculated from the dissipated power as $T_J=T_A+R_{TH}\cdot P_D$ [8]. At preselected times, the annealing experiments are interrupted and the recovery rate is evaluated. In this way, the sequence of annealing and evaluation cycles is simply obtained by switching the applied voltage bias, and the controlled self-heating at higher dissipated power allows increasing recovery temperature of one's choice. Firstly, for all the geometries the output characteristics of a fresh sample are measured, in order to set the bias points for the annealing test and define the power dissipation function to be applied for the recovery. The bias function and the resulting dissipated power and junction temperature rise $\Delta T_J=T_J-T_A$ are shown in Fig. 6 for HBT #2, as obtained by sweeping V_{CE} at V_{BE}=0.92 V; thermal resistance of the intermediate device is reported as well. The application of each bias point has the duration of 10s and has been alternated to the measurement of the Gummel-plot at V_{CB}=0 V to evaluate the effect of the applied self-heating. This procedure allows monitoring recovery effects as a function of dissipated power, cumulated energy, cumulative heating time, and junction-to-ambient temperature increase. Recovery results, described by the base current unrecovered fraction $f_I(t)=I_B(t)/I_B(0)$, are depicted in Fig. 7a for HBT #2 as a function of cumulated heating time t. The annealing process is

978-1-4799-7231-9/14 $31.00 © 2014 IEEE

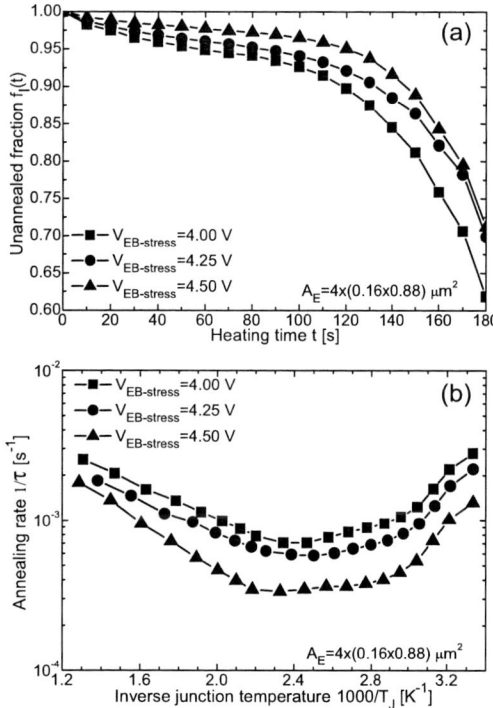

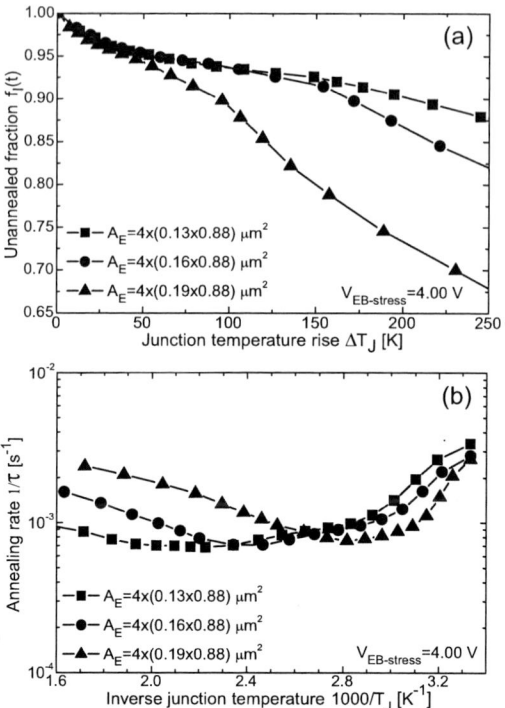

Fig. 7. (a) Base current unannealed fraction and (b) annealing rate vs. (a) heating time and (b) inverse junction temperature for HBT #2, measured at V_{BE}=0.7 V and V_{CB}=0 V after 1000s of different stress voltages.

Fig. 8. (a) Base current unannealed fraction and (b) annealing rate vs. (a) junction temperature rise and (b) inverse junction temperature measured at V_{BE}=0.7 V and V_{CB}=0 V for different layouts, after applied stress of $V_{EB\text{-}stress}$=4.0 V for 1000s.

accelerated by increasing the annealing temperature. After a given annealing time, i.e. at a given annealing temperature, base current recovery is lower when the applied stress voltage was higher, as already deducible from natural recovery results. These results are confirmed by the corresponding annealing rates $1/\tau$ depicted in Fig. 7b as a function of the inverse junction temperature and computed from f_I as $f_I(t)=exp(-t/\tau)$ [7]. Recovery results obtained for different layouts, preliminary stressed at $V_{EB\text{-}stress}$= 4.0 V, are compared in Fig. 8a as a function of ΔT_J, as computed for each geometry according to its specific thermal resistance value. Corresponding annealing rates are depicted in Fig. 8b. The annealing data show two distinct regions: initially, the recovery starts with a higher rate for smaller device and the recovery rate reduces as the temperature increases; in this period, the easily annealed states are removed and the detrapping of oxide trapped carriers occurs. Later, when the temperature increases, the thermally activated annihilation of interface states prevails, the recovery rate increases with temperature and, accordingly, the activation energy can be derived. The largest device, affected by the lowest degradation, is the fastest to recover, in agreement with results in Fig. 7b. The higher the degradation is, i.e. the trap density, the slower will be the recovery and the annealing rate is actually not proportional to the recombination current. Hence, the identification of defects nature and recovery is not foregone and further investigations are needed. However, results obtained for the recovery are promising and reveal that, even though a very high reverse EB stress condition is applied, the activation and significant effects of the recovery are reached by applying self-heating for a few hundred seconds.

IV. CONCLUSIONS

We studied in detail hot-carrier induced degradation in SiGe HBTs with different emitter geometries. Results confirm that the oxide/silicon interface traps generated by electrical stressing are located around the emitter perimeter, but a saturation effect of the degradation with increasing applied stress voltage was observed. Natural and thermal recovery effects were investigated. By means of self-heating, an accurate analysis of the activation and dynamics recovery was possible. Despite of the very strong applied stress conditions, no significant variation was detected for RF performance.

ACKNOWLEDGMENT

The authors wish to acknowledge the DOTSEVEN (316755) project supported by the European Commission through the Seventh Framework Program (FP7) for Research and Technology Development.

REFERENCES

[1] J. D. Cressler, *Proc. IEEE SiRF*, pp. 81–83, 2013

[2] http://www.dotseven.eu

[3] P. Chevalier et al., *Proc. IEEE BCTM*, pp. 57–65, 2011

[4] J. D. Cressler, *IEEE Trans. in Device Materials and Reliability*, pp. 222–236, 2004

[5] H. Rücker et al., *IEEE Journal of Solid-State Circuits*, pp. 1678–1686, 2010

[6] A. Neugroschel et al., *IEEE Transaction on Electron Devices*, pp. 792–800, 1997

[7] H. Wurzer et al., *IEEE Transaction on Electron Devices*, pp. 533–538, 1994

[8] V. d'Alessandro et al., *Proc. IEEE BCTM*, pp. 137–140, 2010

Device-to-Circuit Interactions in SiGe Technology: Challenges and Opportunities

John D. Cressler

School of Electrical and Computer Engineering, 777 Atlantic Drive, N.W.
Georgia Institute of Technology, Atlanta, GA 30332-0250 USA
(cressler@ece.gatech.edu)

Invited Paper

Abstract— **The tight coupling between the nuanced physics of silicon-germanium (SiGe) heterojunction bipolar transistors (HBTs) and the circuits in which they are utilized in many ways represents the "final frontier" for research in technology optimization, device physics, compact modeling, circuit design, and system implementations. As relevant examples of the inherent complexities associated with such "device-to-circuit interactions" within the SiGe world, I examine two distinct scenarios: 1) Our ability to accurately predict the end-of-life reliability of actual SiGe HBT circuits; and 2) Our ability to mitigate transient radiation effects in SiGe HBT circuits. In each example, I address the scope of the problem, the challenges faced in trying to solve them, and the opportunities presented if and when that success comes.**

Keywords— *silicon-germanium, SiGe, heterojunction bipolar transistor, HBT, circuits, device-to-circuit interactions, modeling, TCAD, reliability, radiation, aging*

I. MOTIVATION

As an integrated circuit paradigm, SiGe BiCMOS technology has enjoyed great success over the 2+ decades of its existence. The seamless combination of bandgap-engineered SiGe HBTs, with their superior analog/RF properties (transconductance, output conductance, noise, gain, linearity, etc.), with standard CMOS in a high yield, 100% silicon-manufacturing compatible platform, has proven to be a compelling solution for a diverse variety of performance-constrained analog, digital and RF applications, for both wired and wireless systems ranging from DC to sub-mm-wave operational frequencies [1,2].

At least four distinct generations of SiGe technology presently exist in foundries in the US, the EU, and Asia, and small-signal intrinsic device performance (i.e., f_T and f_{max}) has routinely exceeded multi-hundred GHz speeds while maintaining usable breakdown voltages (BV_{CEO} and BV_{CBO}). It is also apparent that we are not in any danger of approaching the ultimate limits of scaling for (manufacturable) SiGe HBTs, and THz SiGe HBTs seem increasingly like a real possibility [3]. This all bodes well for the future, of course, especially as the world's appetite for mm-wave and sub-mm-wave systems begins to coalesce into commercial realities that can drive production volume at the state-of-the-art.

Given this situation, we might be naively tempted to assume that with the relentless march of technology scaling in the SiGe world, that all is known, that the issues we face are well understood and in-hand, that the remaining problems, while certainly non-trivial, nevertheless invite straightforward analysis and systematic solutions. Far from it! Even in the realm of materials, our cherished SiGe strained layer epitaxy, many open questions remain, and clearly the robust fabrication of manufacturable SiGe platforms under the constraints of decreasing thermal cycle budgets is increasingly problematic. The same holds true for multiple aspects of the basic physics of the SiGe HBTs themselves, and how one best models that physics (see, for instance, [4]).

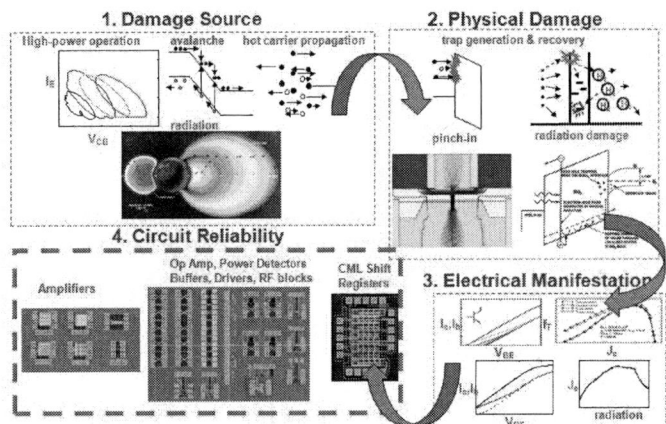

Fig. 1. The reliability "circle," leading from damage sources, to operative damage mechanisms, to resultant transistor damage, and ultimate circuit impact (after [5,7]).

More challenging still is the intersection between the SiGe transistors we design, fabricate, measure, and model, and the circuits and systems that are ultimately constructed from them. A reminder: it is the latter that ultimately pays the bills! I will refer to this domain of intersection as the realm of "device-to-circuit interactions." Of particular interest in this context is the inherent coupling between the physics of the transistors and the intended (and often unintended!) response of the circuit (or system) in question. The complexities associated with such device-to-circuit interactions may be obvious (e.g., the circuit utilizes a feedback loop around the transistor which measurements of single devices cannot, even in principle, reflect), or they may be more subtle (the nodal impedances on transistor are time varying within the circuit, impacting the

978-1-4799-7231-9/14 $31.00 © 2014 IEEE

physics of the device itself as it responds to some external stimulus). It is my own view that this device-to-circuit interaction sphere represents the final frontier in SiGe research, presenting many interesting challenges, certainly, but also many opportunities that should bear fruit for the sustained evolution of our field.

It is worth pointing out the often-unstated obvious: there exists a deeply-embedded, legacy disconnect between device technologists that design vertical profiles using TCAD and then figure out how to fabricate the devices at high yield, the device modelers that must accurately measure and then build a model to represent the electrical behavior of the resultant transistors post-fabrication, and the circuit designers, the end-users, who often would prefer to simply treat the transistor as a perfectly modeled "black box." This disconnect makes dealing with device-to-circuit interactions challenging, since most circuit designers are not especially versed in the many nuances of devices and technology or compact modeling, and vice versa. When final circuit measurements inevitably do not agree perfectly with pre-fabrication simulations, bi-directional finger pointing necessarily ensues. This scenario will generally grow worse with technology evolution, since transistor fabrication and high fidelity modeling grow more challenging with dimensional scaling, and circuit designers increasingly desire to "push" the transistors to their limits to squeeze every last ounce out of the available device performance without compromising reliability. Such is life.

To explore these emerging complexities of device-to-circuit interactions, I will choose two unique example problems with broad applicability that have occupied my research team for some time: 1) the prediction of the end-of-life reliability of SiGe HBT mixed-signal circuits (Fig. 1), and 2) the mitigation of transient radiation effects in SiGe HBT mixed-signal circuits. As will be seen, both topics offer both significant challenges and compelling opportunities. To add additional insight, I will highlight the fact that there are fundamental differences between the way circuit designers and the way device designers and modelers view these types of problems. For brevity, I will only give a cursory overview of the various results, and refer the interested reader to the appropriate references should they care to know more detail (see, for instance, [5-12] for reliability and [13-19] for radiation effects).

II. PREDICTING CIRCUIT-LEVEL RELIABILITY

Obtaining robust reliability is essential for any viable integrated circuit (IC) technology, and entire organizations within any given company are devoted to measuring and predicting the reliability of their IC technology offerings. When we think about IC reliability, most of us naturally think about transistor reliability, and as discussed below, there are good reasons for this. In truth, however, it is circuit reliability that ultimately matters most. (I will, for brevity, ignore other important aspects of the problem – e.g., packaging reliability, interconnect reliability, etc.).

There are several relevant questions that **circuit designers** are keen to know the answers to: 1) What happens to the circuit I have meticulously designed, simulated, fabricated, and measured as a natural consequence of the technology "aging" process – that is, under normal (or even extreme) use conditions, how will the characteristics of my circuit change over time due to reliability stress, and will it continue to satisfy my system specifications as long as it is used? 2) To what voltages and currents can I safely bias my device (both DC and AC) within my circuit and "get away with it," without breaking something or negatively affecting circuit yield? 3) Can I, as a circuit designer, "look" at the end of use characteristics of my circuit within a familiar circuit design environment (e.g., Cadence, ADS, etc.), and do that in a believable way. This last question represents the "holy grail" in this field, at least from the viewpoint of the circuit designer. Their desire? **A button in Cadence that allows them, as circuit designers, to quickly and efficiently look at end-of-life response of the circuit as it operates, and which captures all possible reliability concerns, and makes accurate downstream projections of circuit performance.**

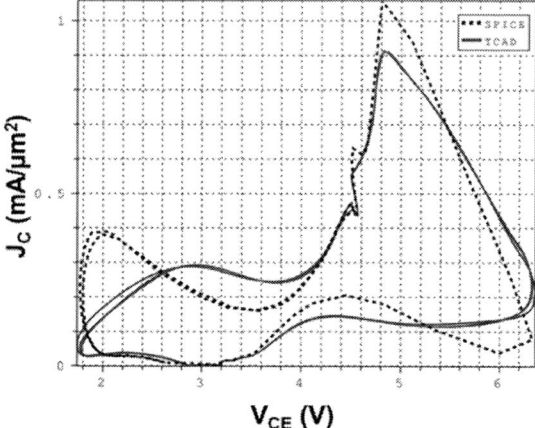

Fig. 2. Example of a load-line for a relevant SiGe HBT circuit (after [9]).

That such a tool does not presently exist should be obvious to most, but a number of research teams, including my own, have been working hard in that direction for some time now. Progress is being made, and I remain optimistic about achieving our end goal; but there is a long way to go still.

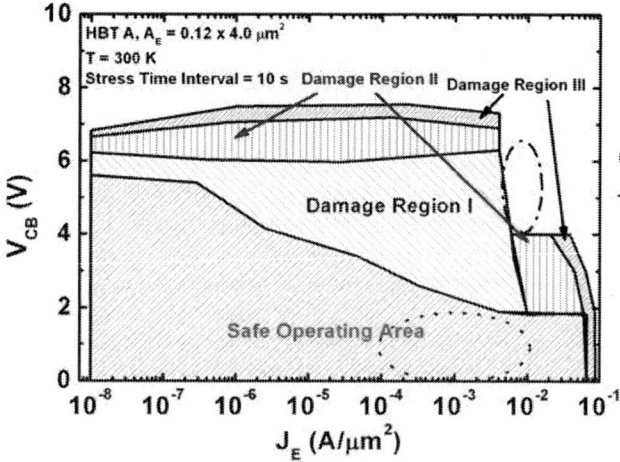

Fig. 3. Example of the output stress plane for a SiGe HBT, showing multiple damage regions and the classically defined safe operating area (after [5,7]).

978-1-4799-7231-9/14 $31.00 © 2014 IEEE

If such a circuit-level reliability prediction tool COULD be brought to bear, its payoff would be potentially enormous. Circuit designers could not only see what their circuits will do over time, but, more importantly, they could then leverage that knowledge to design better circuits. For instance, if I, as a circuit designer, can gain a performance advantage by biasing my transistors in ways that the device and modeling and reliability teams advise me NOT to, and yet remain confident that I have not compromised the net reliability of my circuit in doing so, then I can and I will and I should do that. Alas, as it stands presently, most circuit designers simply "play by the rules," blindly building their circuits to satisfy the performance/reliability corners and design ground-rules of the process design kits (PDKs) that have been provided to them.

Post-fabrication they measure their "baby", pump the air with their fist when it works as intended, and then shed a tear as they hand their circuit over to the reliability team that will then subject it to some sort of nasty accelerated stress to see if it is going to hold up under use (or over-use) conditions as a function of time. Perhaps it will, perhaps it won't, but the rule of the day is inevitably to play it overly conservative (an excellent recent example of this can be found in [12]). This legacy approach is VERY inefficient, and in general will leave additional performance "on the table," not in their circuits. While this "thrown-away" performance might not have mattered a decade ago, it most certainly matters now.

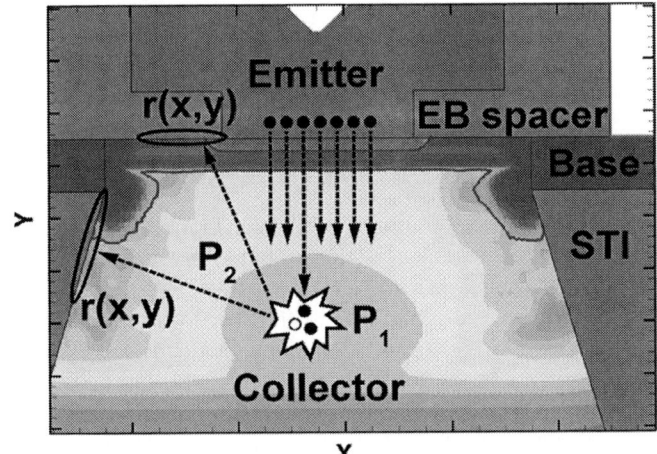

Fig. 4. Cross-section of 2-D device model, annotated with the basic process of the mixed-mode degradation mechanism (after [9]).

What makes this dream of circuit reliability prediction so challenging to realize? Several things, actually. For one, the near-infinite variety of circuits in play (analog, digital, RF through sub-mm-wave) makes circuit-level reliability prediction a tall order to fill, even in principle. One useful way to look at this is to imagine the dynamic load line of the circuit in question – not only for its output, but for each node around each transistor within the circuit (Fig. 2). For many/most/all circuits that are time-varying in nature, each device will sweep through a different set of V-I points in the output plane, with some unique time dependence and duty cycle.

This would be complicated even if there was only one type of damage physics occurring in the device at any given instant. That is not the case. Truth be told, the physics of reliability

damage of the transistors themselves is almost unbearably complicated, and inevitably involves multiple operative damage mechanisms (Fig. 3). This is especially true for SiGe HBTs (even more so than for CMOS, in my opinion). The voltages and currents in a given transistor within a given circuit will thus sweep through multiple regions of fundamentally different damage physics, each with different dependencies on time, voltage, current, temperature, terminal impedance, etc. This creates a very messy picture for any comprehensive understanding, even at the single device level.

In this context, circuit designers are inevitably tempted to say dumb things to device and modeling folks, such as: "What is the lifetime of the transistor?" Or, "How much current (I) can I put through it?" Or, "How much voltage (V) can I put across it?" Any decent device reliability person will roll their eyes (or curse under their breath!). There ARE no simple answers to such questions, other than, "It depends." Stalemate ensures between circuit designers and device technologists, often translating to walls built between opposing camps; which is a shame, and in the end, counter-productive.

Reliability groups deal with these inherent complexities by choosing the only manageable path: focus on the transistor itself, and use a variety of overstress conditions to see how and where the transistor falls apart; tease apart its death throes to better understand the underlying mechanisms. When done, CONSERVATIVELY spec the safe-operating-area (SOA) of the device, the upper-bound conditions under which you will allow circuit designers to handle your precious devices, OR ELSE! In the old bipolar days, this type of SOA definition was done for reverse emitter-base (EB) stress, high forward current stress, and perhaps thermal runaway. Most places today also now add "mixed-mode" stress (Fig. 4), a (more circuit relevant) cross-section of high V and high I applied to the device during stress, and perhaps pinch-in induced bias instabilities [6,7]. Still, these are (statistical) data based on measurements of single isolated transistors, and typically involve monitoring changes to the current gain or perhaps a few other DC parameters. Modelers take those net results and then code this information into the PDK. Business as usual.

But if the dream of a viable circuit-level reliability prediction tool is to be realized, there is a logical (to me, at least), if challenging, approach to follow.

1) Completely characterize the isolated SiGe HBT response to applied electrical stress (e.g., all combinations of DC V-I for a given stress time series – e.g., to 10,000 seconds) to understand the boundaries in the so-called "output stress plane" where the different damage mechanisms are operative. CW RF stress is a different matter altogether [6-8].

2) Use a calibrated (to pre-stress) hydrodynamic TCAD model of the SiGe HBT and then code-in the physics associated with damage (e.g., the classical trap "reaction-diffusion" damage kinetics model) as TCAD wrap–around stress module, which are then solved self-consistently under each stress condition, at each time step (Figs. 5 and 6). I view using TCAD as an essential step, since it accurately captures the physics of the problem (local fields, current vectors, impact ionization, carrier energy and transport, etc.).

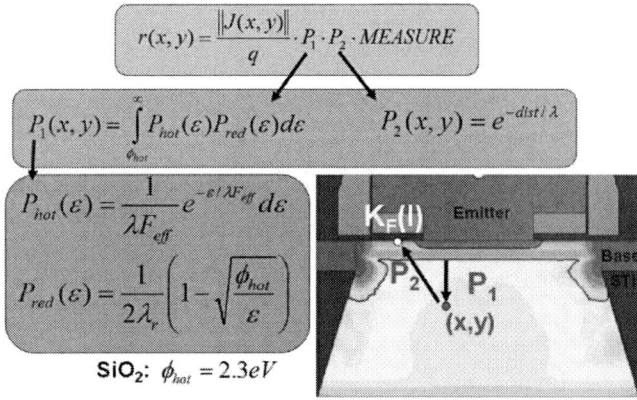

Fig. 5. The physical basis of the stress degradation model implemented in TCAD (after [11]).

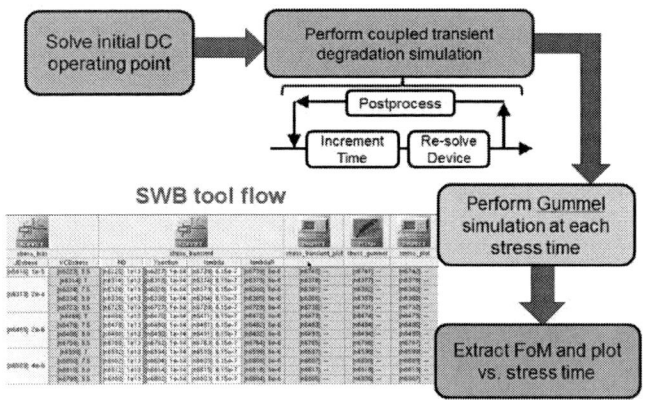

Fig. 6. Stress degradation model implemented in TCAD (after [11]).

3) Compare measured stress data to TCAD results in a particular region of the output plane where only one damage mechanism is known to be operative (e.g., mixed-mode damage will dominate at "high" V and "modest" I). One can now virtually predict the stress response of the device (e.g., excess base current (or gain degradation) as a function of time).

4) Calibration of the trap model wrap-arounds is next (this is non-trivial), and now one can easily begin to test the predictive capability of the TCAD reliability model at V-I points beyond step 3. Getting the impact ionization coefficients correct is obviously key since it will drive much of the physics. For this calibration step, carrier mean free path length is the key parameter. Additional relevant calibration steps include appropriate accounting for temperature effects (e.g., thermally-induced trap self-annealing and self-heating). Large volume 3-D meshes are often required in this thermal context. This calibration process also has the advantage of allowing you to do what TCAD does so beautifully: See how the stress results change when I make structural or profile changes to the transistor. Powerful information, since one can now in principle optimize a device for reliability, not just performance!

5) Now, continue to "move" in the V-I plane to expose where the TCAD model is accurate and where it begins to fail (Fig. 7). Adding temperature (e.g., self-heating, which can induce trap self-annealing) to the damage kinetics needs to happen, as well as other damage mechanisms that may be operative (e.g., Auger generation dominates at "low" V and "very high" I stress [10]) – different physics.

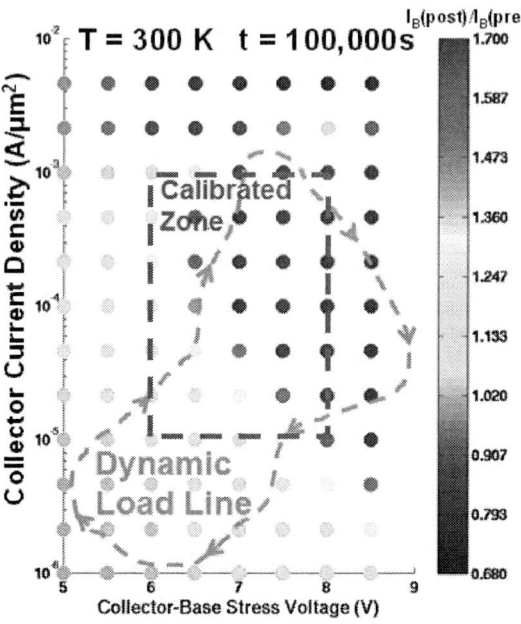

Fig. 7. Example of TCAD stress degradation simulation, showing excess base current across the output stress plane (after [11]).

7) We can in principle now use this TCAD model to simulate moving along a load line in the output stress plane (Fig. 7). This is a key step closer to what circuits actually do. The goal here is to traverse the load line in the stress plane in the way the circuit would, such that various mechanisms are engaged and disengaged as the location along the load line evolves in time. The path taken along that load line may or may not matter, depending on the circuit. You would decide on the specific shape of the load line by running a simulation within Cadence for the circuit you have in mind. Comparisons to data for V-I plane path trajectory would follow, to help establish capabilities and limitations of the model. Initial results have been encouraging (Fig. 8, top).

8) As the best test of the efficacy of the model, you might then choose to run "mixed-mode" TCAD [16], in which the TCAD simulation runs self-consistently within a Cadence environment. One step closer still to circuits. This is not for the faint of heart given the inherent computational complexity and difficulty in convergence, but it will contain the best physics. You could then compare this to **circuit stress data**, hopefully to find good agreement.

978-1-4799-7231-9/14 $31.00 © 2014 IEEE   48

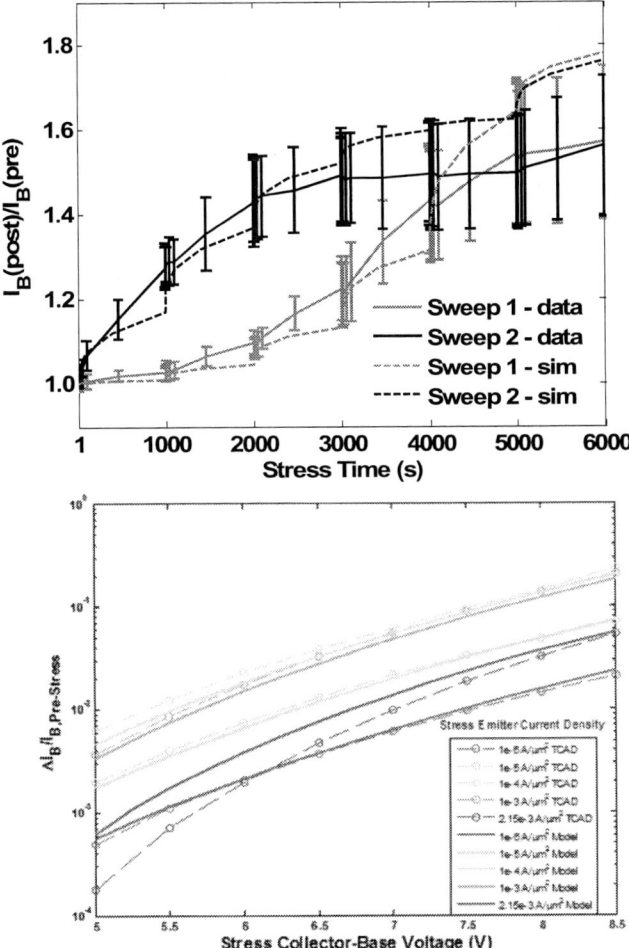

Fig. 8. (Top) Example of TCAD stress degradation simulation along a load line trajectory in the output stress plane, compared with data. The bifurcation in the damage results from different trajectories in a stress sequence. Both end at a similar point (after [11]). (Bottom) A comparison between the analytical formulation used in a compact model implementation of electrical stress and calibrated TCAD results.

9) The computational complexity associated with mixed-mode TCAD is prohibitive for getting a circuit designer to actually use this approach. To make things numerically efficient, you then need to run a slew of stress-calibrated TCAD simulations, and now begin to build a stress damage wrap-around for the **compact model** (HICUM, VBIC, MEXTRAM). This can be physics-based (i.e., equations), but necessarily contains only a shadow of the complex, operative physics the TCAD solution gets at. A preliminary result showing agreement between calibrated TCAD and the compact model is shown in Fig. 8 (bottom).

10) Once the compact model exists, you have basically reached the goal, the holy grail; a push button in Cadence that now swaps in a compact "reliability stress model" for the SiGe HBTs in the circuit in question. As the circuit operates, it "ages" the transistor, which in turn changes the circuit

response (in complex ways). Run it forward virtually 10 years with the push of a button and see what your circuit does as a result of that aging; i.e., where it ends up specification wise. Depending on the answer, the circuit designer can then iterate their design to squeeze additional performance out without compromising circuit reliability. Nothing left on the table!

Done! Pretty slick, huh? And easier said than done, of course. How far have we come to date? Well, TCAD models with calibrated stress models in them have been built [9], and for at least the regime of mixed-mode damage in the output stress plane, seem to give decent agreement between simulation and reality: meaning, it is predictive [9,11]. Extending this to cover other stress regimes [10] is in process, as well as temperature dependence. Finally, building the corresponding compact models to automate the process in a Cadence environment is also now underway.

III. MITIGATING TRANSIENT RADIATION EFFECTS

For example number two of complex device-to-circuit interactions in SiGe circuits, I will turn next to transient radiation effects. Because this topic is less familiar to many readers, I will begin with some background to define the context to the problem. "Extreme environment" electronics represents an important niche market in the $1T+ dollar global electronics industry, and spans the operation of electronic circuits and systems in surroundings lying outside the domain of conventional commercial or military environmental specifications [13]. Extreme environments might include, for instance: 1) Operation down to very low temperatures (e.g., to 77 K or even 4.2 K or below); 2) Operation up to very high temperatures (e.g., to 200C or even 300C or higher), 3) Operation across very wide and/or cyclic temperature swings (e.g., -230C to +120C night-to-day, as found on the lunar surface); 4) Operation in a radiation-rich environment (e.g., in space while orbiting the Earth, traveling to an outer planet, or even inside the detectors of a terrestrial particle physics collider (e.g., CERN)), or 5) at worst case, all four occurring simultaneously. I will focus here on scenario #4, but the other extreme environments also represent interesting examples of device-to-circuit interactions. Interested readers are referred to [13] for a comprehensive treatment.

It has been recognized for some time now that the unique bandgap-engineered features of SiGe HBTs offer considerable potential for simultaneously being able to cope with all four of these extreme environments, potentially with little or no process modifications, ultimately providing compelling advantages at the integrated circuit and system level with respect to size, weight and power (SWaP) constraints, across a wide class of envisioned commercial and defense applications [13]. Such an "environmentally-invariant" (i.e., the circuitry can operate unattended and without protection in any environment in which it finds themselves) electronics technology platform would allow mission designers and vehicle architects to re-imagine how space systems could and should be designed and operated, and thus SiGe represents a candidate "game-changer" in this important field.

Within this context, we now focus in on radiation effects in SiGe HBT circuits as another important and exceptionally challenging device-to-circuit interaction problem. Ionizing radiation is ubiquitous to operation in Earth orbit, for instance, where the planet's magnetosphere traps high energy particles that are toxic to conventional electronics. These radiation fields produce two major types of damage in devices: 1) total ionizing dose (TID), where the radiation source (high energy electrons, protons, gamma rays, X-rays, etc.) creates, among other things, trap states at dielectric interfaces (think emitter-base spacer, shallow trench edge); and 2) single event effects (SEE), in which high energy particles (e.g., GeV cosmic rays) pass right through the spacecraft and the IC, depositing electron-hole pairs as it traverses the silicon at nearly the speed of light, and which is then collected by the various device junctions, translating to very fast, time-dependent current "pulses" on the device terminals. By definition, SEE are a transient phenomenon, but they can have substantial impact at the circuit and system level, even to the point of lethality to a satellite system. Earth orbit is home base to an increasing fraction of the global communications infrastructure, as well as weather radars, climate and resource mapping systems, observation platforms, and science observatories, GPS, and many other vital systems - read: radiation matters a lot.

It is well-established that SiGe HBTs have a very favorable built-in total ionizing dose (TID) tolerance, to (extreme) multi-Mrad levels [14]. For reference, a dose of a few hundred rads of radiation is lethal to humans. While the actual TID tolerance of SiGe HBTs varies some with generational scaling node, and from company-to-company, the same claim remains essentially true: SiGe HBTs are multi-Mrad TID tolerant as-built. The general observation for TID damage in SiGe HBTs is a very modest degradation in the base current, producing an excess base leakage that degrades current gain in low injection. This base current degradation is identical to that well known in Si BJTs (albeit at far smaller magnitudes for SiGe) and is the consequence of radiation-induced trap states located at the emitter-base (EB) spacer interface. Such traps act as recombination centers when located within the EB space charge region, producing parasitic base current leakage and current gain degradation. To first order, collector current remains unchanged, except at high injection levels where small changes to the parasitic base and emitter resistance can sometimes be seen.

This built-in TID tolerance in SiGe HBTs is unique among semiconductor technologies, and has understandably garnered significant interest from the space community [14]. One might logically wonder about the source of this attractive built-in TID tolerance. The simple answer is that it is not Ge related at all, but is rather a free "perk" associated with the device structure required to robustly embed a strained SiGe alloy inside the base region of a bipolar transistor to produce a high-speed SiGe HBT. The classical weak-link for radiation tolerance of bipolar transistors is associated with the emitter-base (EB) spacer oxide. In the case of the SiGe HBT we have three structural features in our favor: 1) the EB spacer is very

thin (e.g., 100 nm) and importantly is contained within a heavily doped region of the epitaxial base, such that any interface damage is effectively "caged-in" by the high doping (the trap-induced modulation of the local space charge region cannot easily occur) and parasitic leakage is thus suppressed; 2) the base itself is very thin (e.g., 100 nm) and heavily doped, minimizing the effects of displacement damage, and 3) the shallow trench isolation in the collector-base (CB) junction is thin and located well away from the carrier transport path of the transistor. Each is a fortuitous but effective combination of self-mitigation schemes. We emphasize that ALL epitaxial base SiGe HBT structures, no matter who is building it, embody these features, explaining the universality of the observed built-in TID tolerance. Examples can be seen in Figs. 9-10.

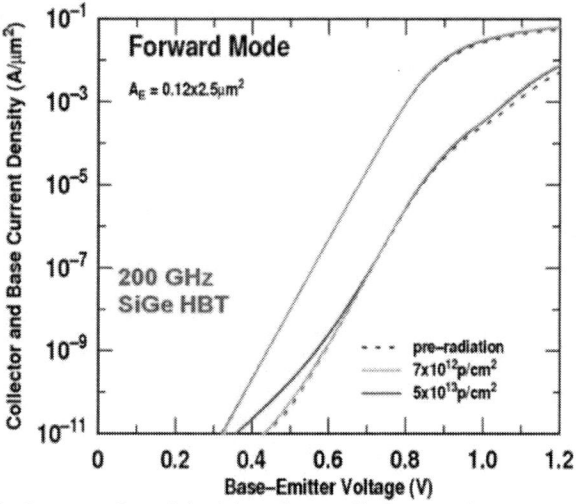

Fig. 9. Impact of total ionizing dose radiation on the DC Gummel characteristics of a third-generation SiGe HBT. For reference, 5×10^{13} protons/cm^2 particle flux corresponds to 6.7 Mrad(SiO$_2$) of total dose (after [14]).

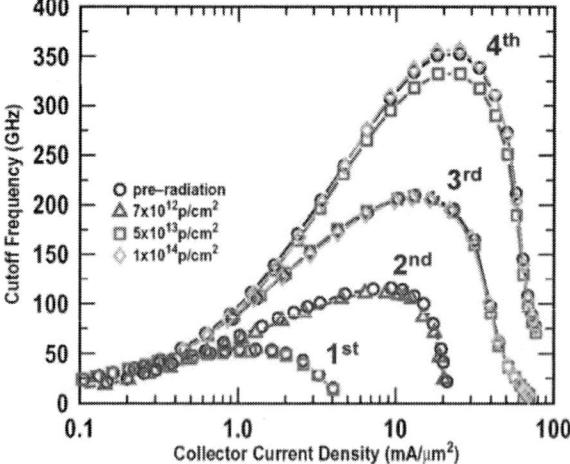

Fig. 10. Impact of total ionizing dose radiation on the AC small-signal performance of multi-generation SiGe HBTs. For reference, 5×10^{13} protons/cm^2 particle flux corresponds to 6.7 Mrad(SiO$_2$) of total dose (after [14]).

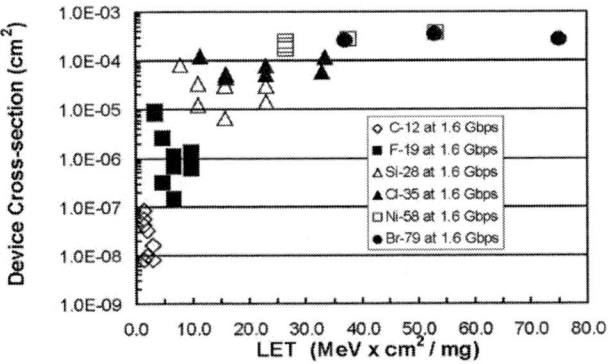

Fig. 11. Measurements by Marshall *et al.* on first-generation SiGe HBT SEU performance (after [19]).

Alas, nature did not smile on the SiGe HBT when it comes to single event effects (a.k.a., single event transients (SET), or in the case of a pure digital system, single event upset (SEU) - i.e., bit flips in digital logic). It has been appreciated for some time now [19] that SiGe HBT digital circuits are SEU sensitive, meaning low LET (linear energy transfer – deposited charge per unit path length, pC/μm) heavy ions (or even protons) can easily produce bit upsets in fast SiGe digital circuits (Fig. 11). This is not hard to understand. Two things are working against us (Fig. 12). First, for an ion strike located within the bounds of the active intrinsic device, the high density of charge deposition from the heavy ion effectively "shorts out" the emitter, base and collector of the transistor, producing large quasi-resistive transients on the E,B,C terminals. In addition, for any ion strike located within the deep trench (DT) isolation "ring," the reverse-biased n+ to p-sub-collector to substrate junction forms an ideal collection point to swallow virtually all of the deposited charge associated with a heavy ion strike in the substrate. In this case there is a fast field-driven transient, and then a longer diffusion-driven tail, often with complex structure to it due to the device 3-D geometry. Because this junction is electrically tied to the collector node, THE sensitive point for inducing a bit-flip in a conventional digital latch, the problem is compounded. Fig. 13 shows the results from calibrated TCAD simulations, illustrating the two dominant SEE mechanisms (the fast, high magnitude transient ion-shunt effect (with only modest total collected charge -- the integral of current over time), and the slower, lower magnitude substrate diffusion transient (with much large total collected charge). Even to the eye, the current transients produced by SEE phenomena is complex looking. It gets worse.

While it is tempting to prioritize the importance of both drift and diffusion mechanisms, we emphasize that caution is in order. BOTH mechanisms can matter, depending on the circuit type in question. For example, for a high speed latch, the fast high magnitude transient matters most for generating bit flips. For this reason, we point out that moving to a thick film SOI platform to eliminate the long substrate ion-induce diffusion tail, while generally helpful for SEE mitigation, is not a "magic bullet" for solving SEU in fast digital circuits (see [14] and references within).

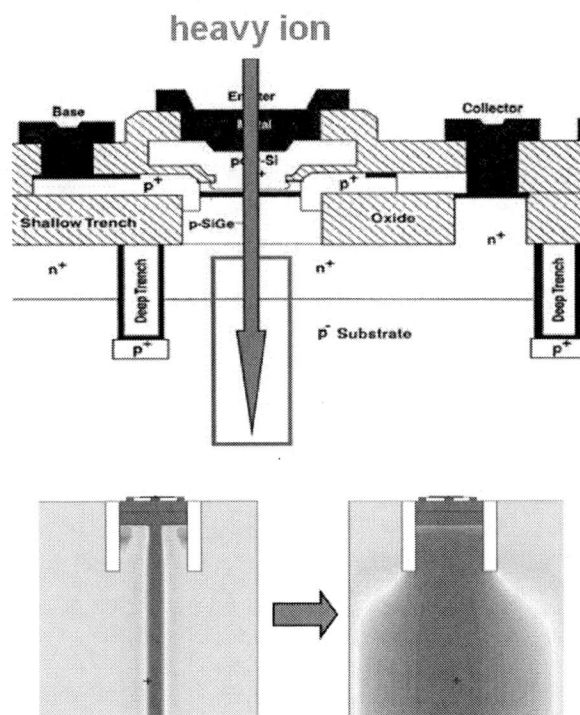

Fig. 12. Illustration of a heavy ion strike of a SiGe HBT, and the resultant charge deposition and drift-diffusion process (after [14]).

It is worth pointing out single event latchup (SEL), which can be lethal to circuit and systems, is a non-issue in SiGe HBTs (not so for CMOS!), due to the self-isolating nature of the heavily-doped patterned sub-collector used in mainstream SiGe platforms. SEL has never been observed in SiGe, clearly a nice feature for space applications. It is also important to realize that SEU in Gb/sec digital logic should be considered worst case for SEE. In RF systems, for instance, which are inherently frequency domain and not time domain, SEE is generally less of an issue, but often more nuanced.

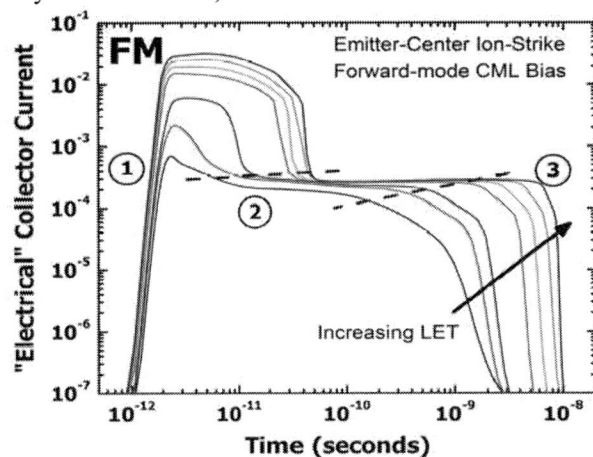

Fig. 13. TCAD simulated electrical collector (physical collector) current transients for a bulk 90nm SiGe HBT operating in the **forward-mode** regime. Transients for multiple ion LETs spanning from 0.89 (10 MeV boron) to 30.23 (10 MeV krypton) are overlaid (after [15]).

978-1-4799-7231-9/14 $31.00 © 2014 IEEE

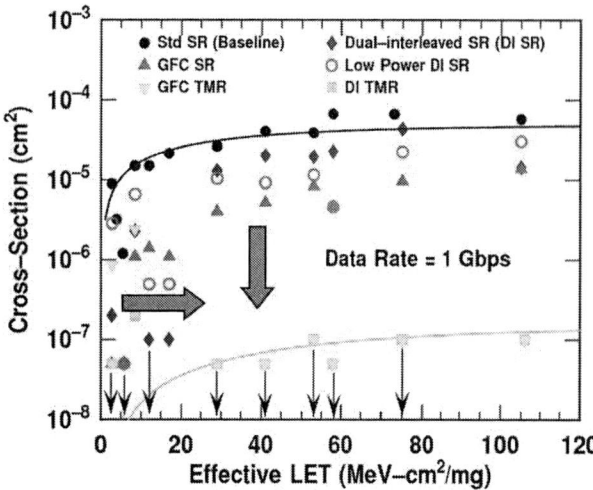

Fig. 14. Measured SEU error event cross-section in a shift register as a function of ion LET (charge deposition per unit path length) for a control and several RHBD versions, showing that SEU mitigation is in fact possible. For reference, DI stands for dual-interleaved, GFC stands for gated-feedback cell, and TMR stands for triple modular redundancy. Data points with downward arrows refer to "limiting cross-section" – zero errors observed (after [14]).

Nevertheless, SEU in SiGe digital circuits exists, is serious, and must therefore be dealt with, at both the circuit level and the system level (the latter, for example, via error correction). Substantial research over the past decade has been focused on first understanding via a combination of measurements and standalone TCAD simulations the quite complicated physics of SEE in SiGe HBTs, and then leveraging that knowledge to attempt to mitigate SEE by a variety of clever means [14]. Some progress has been made. As can be seen in Fig. 14, a variety of radiation hardening by design (RHBD) approaches, including simple transistor layout changes (but no process changes), improved latch designs such the dual-interleaved (DI) or gated-feedback cell (GFC) latches (Fig. 15 shows the DI latch topology), the use of triple-modular redundancy (TMR), and hardened clock trees (e.g., with GFC latches), were combined to demonstrate the first SEU-hard (no bit upsets to LET's above say 60-70) multi-Gb/s digital logic in SiGe. This was accomplished in third-generation SiGe technology. This is an important existence proof that SEU-hardening can be accomplished in SiGe, albeit at substantial area and power overhead, which is clearly undesirable.

Since the first demonstration of SEU sensitivity in fast SiGe digital logic, focus has been logically placed on finding practical mitigation schemes. Attempts to suppress substrate charge collection via the use of various "n-rings" and substrate ties which can be arranged to shunt charge away from the sensitive collector node, have yielded modest success in SEU mitigation. The most appealing of those n-ring approaches are the external n-ring, placed is a ring around the DT isolation since the collection volume of inside the DT is not affected and it can be easily implemented in design kits with no design rule changes (i.e., pure RHBD). Even so, the use of n-rings and substrate ties is not an SEU "magic bullet," since it does

not strongly affect the ion shunt effect associated with the intrinsic transistor (which remains unchanged in these mitigation schemes).

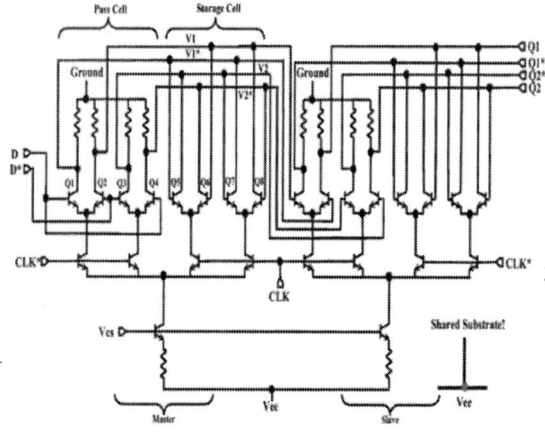

Fig. 15. Circuit schematic for the dual-interleaved RHBD latch with internal redundancy to help mitigate SEU (after [14]).

A more recent interesting approach is the "inverse mode cascode" SiGe HBT, which attempts to decouple the collector and emitter nodes by adding an additional cascode transistor operating in inverse mode (emitter and collector swapped) [14]. At third-generation SiGe nodes, this double transistor IMC SiGe HBT occupies no additional area over a standard SiGe HBT, and while some performance degradation is observed in inverse mode, optimization can easily allow multi-Gb/sec logic in IMC logic. The SEU benefits are real, but still in the category "modest." Most recently, focus has been placed on simply using the SiGe HBT in inverse mode for the sensitive transistors within the latch (e.g., the switching pair). Calibrated TCAD solutions (Fig. 16) suggest that substantial leverage in SEU can be gained in this manner, since the fast, high magnitude SEU transients, while still present in the inverse SiGe HBT, are a) much shorter, and b) lower in magnitude (compare Fig. 13 and Fig. 16 [17]). Careful optimization of the inverse mode SiGe HBT is required, but that procedure is known [14], and importantly is naturally favored by technology scaling. One advantage of this mitigation scheme is that the inverse mode SiGe HBTs could be used for SEE sensitive nodes, while the standard SiGe HBTs would be used everywhere else, representing an ideal division of labor. Figs. 17-18 show recent SET data on fourth-generation device and circuit hardware, comparing forward and inverse mode transients [15]. The results are again, encouraging, but not overwhelmingly positive since performance is degraded in inverse mode, even with careful optimization.

We note, finally, that technology node scaling in SiGe naturally favors improved SEU tolerance, primarily because the transistor size (volume within the DT isolation ring) quickly decreases, providing a natural reduction of SEU sensitive volume. In addition, the DT isolation, typically 5-10 μm deep provide intrinsic protection against angular strikes, since the DT effectively truncates the charge track passing

978-1-4799-7231-9/14 $31.00 © 2014 IEEE

laterally through this device. This is a good thing, obviously, although it can be challenging to model.

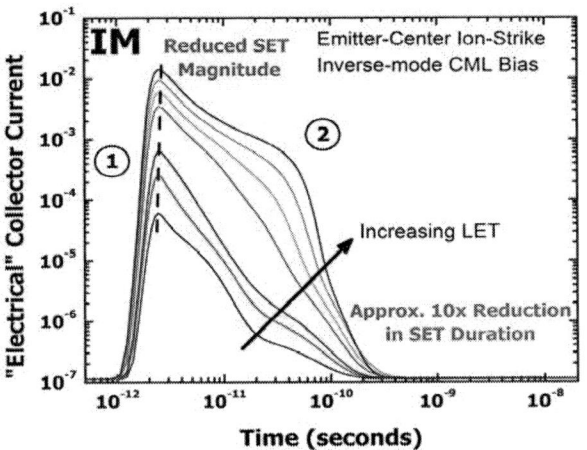

Fig. 16. TCAD simulated electrical collector (physical emitter) current transients for a bulk 90nm SiGe HBT operating in the inverse-mode regime. Transients for multiple ion LETs spanning from 0.89 (10 MeV boron) to 21.17 (10 MeV krypton) are overlaid (after [15]).

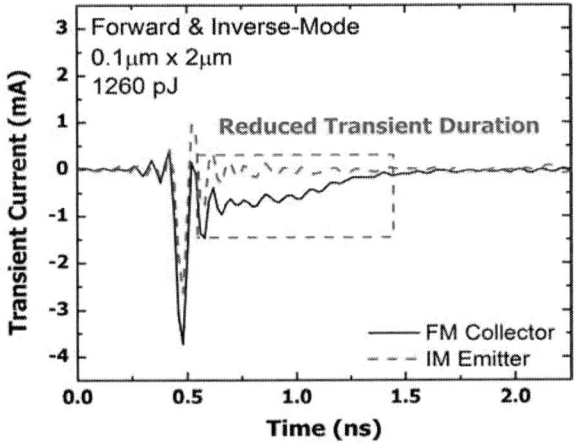

Fig 17. Measured forward-mode and inverse-mode electrical collector current transients for a 9HP (0.1 μm x 2 μm) SiGe HBT (after [15]).

This is pretty much where we stand today with respect of transient radiation effects in SiGe. What comes next? There are two major goals: 1) In the realm of (worst case) SEU in digital logic, can we create digital circuits that have modest LET threshold (say 10-20) without having to exploit TMR, with its substantial area and power overhead? That is, is there a magic bullet for mitigating SEU in SiGe digital logic? 2) How important is SEE in others types of circuits found in space systems (e.g., analog, RF) and how it is best mitigated for these types of circuits?

It is healthy to step back from this problem and ask: What makes this transient radiation problem so challenging? The answer has several instructive layers: 1) The physics of the charge deposition process itself is exceptionally complex,

given that we are dealing with the interaction of ionizing radiation with matter within a 3-D multi-layer composed of metals, dielectrics, and semiconductor(s), each of which respond differently to that impinging radiation; 2) This radiation physics couples to both of carrier transport mechanisms in the semiconductor (drift and diffusion – both matter); 3) What happens to the deposited charge after the ion strikes is a complicated time-dependent problem which is a strong function of the properties of the collecting junctions themselves, which in turn have 3-D field distributions, non-constant doping profiles, Ge within Si, etc.; and 4) The dynamics of the transient radiation strike **within the device**, and ultimately what currents are induced on what terminals (this is what matters most to the circuit), depend not just on local bias (V and I) and temperature, but also on the terminal impedances (R, C, and L as a function of frequency), which in many relevant circuits are time varying (e.g., the digital latch). Said another way, this is a device-to-circuit interaction problem of extreme difficulty.

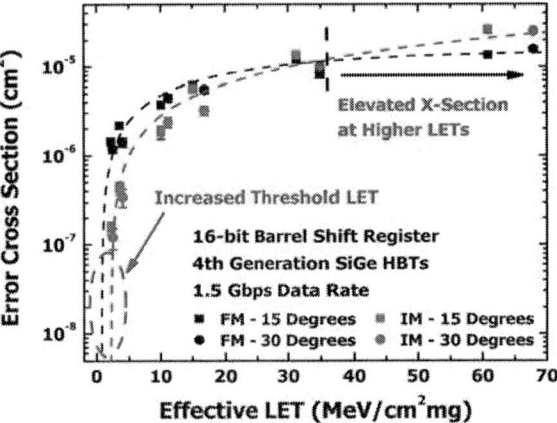

Fig 18. Measured SEU cross-section as a function of LET for forward-mode and inverse-mode SiGe HBTs in a 90 nm SiGe platform (after [15]).

So what is one to do to crack this nut? The classical approach is to simulate the ion strike on a transistor with standalone TCAD under a given bias DC condition (static voltage sources on their terminals, at high input impedance), and capture the current transients on all four terminals (E,B,C, SX). This is not a trivial thing to do in 3-D for a SiGe HBT, and appropriate modeling of the ion-induced charge column also needs to be taken to account. Large simulation meshes are required, since the ions are deeply penetrating and charge collection can stretch tens of microns into the substrate and outward laterally beyond the bounds of the deep trench. An example result is shown in Fig. 13. From these transient simulations, we can build a look-up table (or fitting function), and then add supplemental current sources to our circuit schematic within Cadence to "inject" the ion-induced current pulses on the device terminals at a given time step when the ion "strikes" our circuit within the Cadence environment. This is typically first done one transistor at a time to gauge which device dominates the response, and is referred to as a "current injection" approach to SEE simulation. While useful, it is generally NOT predictive; meaning, rarely does it agree with data.

Such a current injection approach of SEE is appealingly straightforward, but demonstrably in error. Why? For one thing, the shape and duration of the transients themselves depend on the local nodal voltages and the local nodal impedances. Change those and the device transients change, often radically. Even for a DC circuit (e.g., a voltage reference or a current mirror), internal circuit feedback can impact the local charge collection and transient waveforms [16], and certainly in a time dependent circuit (a latch) this will inevitably be the case. Options?

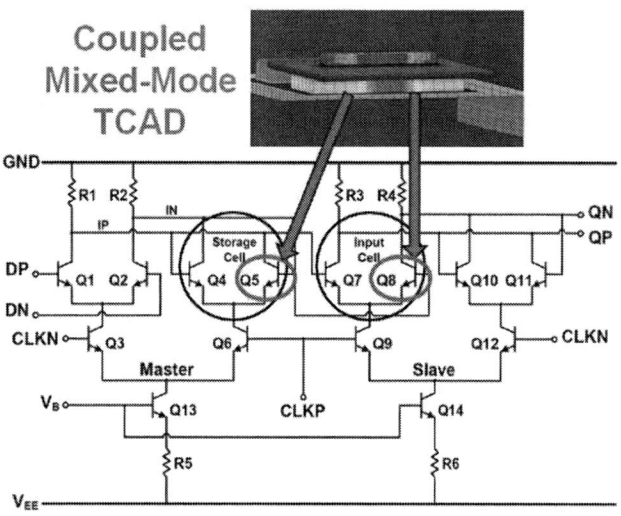

Fig. 19. Illustration of mixed-mode TCAD in a standard master-slave latch (after [16]).

Well, we must necessarily revert to so-called "mixed-mode TCAD," in which there is a dynamic linkage between the 3-D TCAD solver and the Cadence circuit simulation environment. A SiGe HBT (or even multiples of HBTs) is self-consistently solved with the circuit simulation under the Cadence manifold. Sound numerically challenging? It is! The merit is that the voltages, impedances, and internal feedback paths are free to adjust dynamically as they will during circuit operation, and the TCAD then adjusts its solution(s) to match. To date we have successfully done this for a bandgap voltage reference [16] and a master-slave latch [18]. In both cases, the most accurate solutions compared with data are the full-bore mixed-mode TCAD solutions. Not surprising, really. An example of the differences between the two approaches are shown in Figs. 19-20 for a latch.

We are presently extending this methodology to RF blocks (LNA, VCO, etc.) and other mixed-signal circuits (comparator, current mirror, etc.), and comparing the results to data (which, by the way, are DIFFICULT to obtain; an art form really). The payoff in doing this more cumbersome TCAD-within-Cadence mixed-mode approach is that: 1) It gets the physics right, thereby adding insight to the problem and its solution; 2) With this added insight, the hope is that

more advanced circuit mitigation schemes can be invoked; and 3) One can also then make changes to the TCAD deck (layout, vertical profile) to assess directly the SEE impact of a process change **at the circuit level**, which would be exceptionally useful. Time will tell if these potential payoffs are borne out, but I am hopeful.

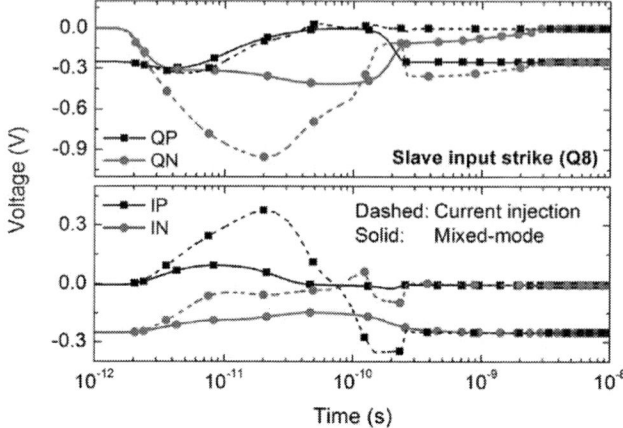

Fig. 20. Illustration of the differences mixed-mode and current injection models for the impact of an ion strike on the internal voltage waveforms of a standard master-slave latch (after [16]).

On a passing note, one might wonder if a similar scheme as that outlined for the circuit reliability device-to-circuit interaction problem could be attempted here. That is, can we use the TCAD results as a basis to then build a new "ion strike" compact model for making this complex problem "easy" for circuit designers to iteratively simulate (i.e., a push button in Cadence). Perhaps, though that seems to me more of a stretch in the case of transient radiation effects, since the TCAD results so intimately depend on the dynamics of the local circuit environment; meaning, the coupling of the device to the circuit is substantially stronger.

IV. SUMMARY

The tight coupling between the physics of SiGe HBTs and the circuits constructed from them, so-called device-to-circuit interactions, represents both extreme challenges to our understanding of SiGe technology, as well as the "final frontier" for new research in technology optimization, device physics, compact modeling, circuit design, and system implementations. Two device-to-circuit examples with broad applicability were explored to highlight the challenges and opportunities in these types of difficult problems: 1) our ability to accurately predict the end-of-life aging characteristics of actual SiGe HBT circuits; and 2) our ability to mitigate transient radiation effects in SiGe HBT circuits.

ACKNOWLEDGMENT

The author is grateful to the members of his Georgia Tech research team, both past and present, particularly: P. Chakraborty, N. Lourenco, U. Raghunathan, B. Wier, M. Oakley, S. Phillips, and K. Moen. The contributions of A. Joseph, D. Harame, J. Pekarik, M. Kaynak, B. Tillack, H. Yasuda, P. Menz, K. Green, S. Jordan, J. Babcock, P.

978-1-4799-7231-9/14 $31.00 © 2014 IEEE

Amouzou, L. Cohn, and K. LaBel are gratefully acknowledged. This work was supported by a NASA, SRC, DARPA, Texas Instruments, IBM, Tower Jazz, and IHP.

REFERENCES

[1] J.D. Cressler and G. Niu, *Silicon-Germanium Heterojunction Bipolar Transistors*, Artech House, Boston, MA, 2003.

[2] J.D. Cressler (Editor), *Silicon Heterostructure Handbook – Materials, Fabrication, Devices, Circuits, and Applications of SiGe and Si Strained-Layer Epitaxy*, CRC Press, Taylor & Francis Group, Boca Raton, FL, 2006.

[3] P.S. Chakraborty *et al.,* "130 nm, 0.8 THz f_{max}, 1.6 V BV_{CEO} SiGe HBTs Operating at 4.3 K," *IEEE Electron Device Letters*, vol. 35, pp. 151-153, 2014.

[4] J.D. Cressler, "A Retrospective on the SiGe HBT: What We Do Know, What We Don't Know, and What We Would Like to Know Better," *Proceedings of the 2013 IEEE Topical Workshop on Silicon Monolithic Integrated Circuits in RF Systems*, 2013, pp. 81-83.

[5] J.D. Cressler, "Emerging Reliability Issues for SiGe HBTs for Mixed-Signal Circuit Applications," *IEEE Transactions on Device and Materials Reliability,* vol. 4, pp. 222-236, 2004.

[6] C.M. Grens *et al.,* "Reliability of SiGe HBTs for Power Amplifiers -- Part I: Large-Signal RF Performance and Operating Limits," *IEEE Transactions on Device and Materials Reliability*, vol. 9, pp. 431-439, 2009.

[7] P. Cheng *et al.,* "Reliability of SiGe HBTs for Power Amplifiers -- Part II: Underlying Physics and Damage Modeling," *IEEE Transactions on Device and Materials Reliability*, vol. 9, pp. 440-448, 2009.

[8] P. Cheng *et al.,* "An Investigation of DC and RF Safe-Operating-Area of npn + pnp SiGe HBTs on SOI," *IEEE Transactions on Electron Devices*, vol. 58, pp. 2573-2581, 2011.

[9] K.A. Moen *et al.,* "Predictive Physics-Based TCAD Modeling of the Mixed-Mode Degradation Mechanism in SiGe HBTs," *IEEE Transactions on Electron Devices*, vol. 59, pp. 2895-2901, 2012.

[10] B.R. Wier *et al.,* "Base Current Degradation Mechanisms in NPN SiGe HBTs Subjected to High Current Stress," *Proceedings of the 2013 IEEE International Solid-State Device Research Symposium*, pp. 1-2, 2013 (on CDROM).

[11] U. Raghunathan *et al.,* "TCAD Modeling of Accumulated Stress Damage DuringTime-Dependent Mixed-mode Stress, " *Proceedings of the 2013 IEEE Bipolar/BiCMOS Circuits and Technology Meeting*, pp.179-183, 2013.

[12] M.A. Oakley *et al.,* "On the Reliability of SiGe HBT Cascode Driver Amplifiers, *Proceedings of the 2014 IEEE MTT-S RFIC Symposium,* to appear.

[13] J.D. Cressler and H.A. Mantooth (Editors), *Extreme Environment Electronics*, CRC Press, Boca Raton, FL, 2013.

[14] J.D. Cressler, "Radiation Effects in SiGe Technology," *IEEE Transactions on Nuclear Science*, vol. 60, pp. 1992-2014, 2013. (A complete set of references for SiGe radiation effects is contained within.)

[15] N.E. Lourenco *et al.,* "An Investigation of Single-Event Effects and Potential SEU Mitigation Strategies in 4th Generation 90nm SiGe BiCMOS," *IEEE Transactions on Nuclear Science,* vol. 60, pp.4175-4183, 2013.

[16] K.A. Moen *et al.,* "Accurate Modeling of Single Event Transients in a SiGe Voltage Reference Circuit," *IEEE Transactions on Nuclear Science*, vol. 58, pp. 877-884, 2011.

[17] S. Phillips *et al.,* "Single Event Response of the SiGe HBT Operating in Inverse Mode," *IEEE Transactions on Nuclear Science,* vol. 59, pp. 2682-2690, 2012.

[18] K.A. Moen *et al.,* "Establishing Best-Practice Modeling Approaches for Understanding Single Event Transients in Gb/sec SiGe Digital Logic," *IEEE Transactions on Nuclear Science,* vol. 59, pp. 958-964, 2012.

[19] P. Marshall, M. A. Carts, A. Campbell, D. McMorrow, S. Buchner, R. Stewart, B. Randall, B. Gilbert, and R. Reed, "Single event effects in circuit hardened SiGe HBT logic at gigabit per second data rates," *IEEE Transactions on Nuclear Science*, vol. 47, pp. 2669–2674, 2000.

Millimeter-Wave SiGe RFICs for Large-Scale Phased-Arrays

(Invited Talk)

Gabriel M. Rebeiz

University of California, San Diego, La Jolla, CA, 92093-0407, USA

Abstract — **This talk will present our latest work on silicon RFICs for phased-array applications with emphasis on very large chips with built-in-self-test capabilities, and for wafer-scale integration. SiGe is shown to be ideal for mm-wave applications due to its high temperature performance (automotive radars, base-stations, defense systems, etc.) and lower power consumption. These chips drastically reduce the cost of microwave and millimeter-wave phased arrays by combining many elements on the same chip, together with digital control. The phased-array chips also allow a much easier packaging scheme using either a multi-layer PCB or wafer-level packages. Examples from 30 GHz to 140 GHz will be presented, together with their system-level performance (imaging radars, packaging, antenna patterns, etc.)**

I. INTRODUCTION

Silicon phased-array chips are now ubiquitous for millimeter-wave operation with many companies having developed SiGe and CMOS solutions at 60 GHz, and some even up to 100 GHz. These chips are especially important at millimeter-wave frequencies where there is very little space for electronics in a two-dimensional phased array with an antenna spacing of 0.5λ (λ=3 mm at 100 GHz). One way to build such arrays is to integrate as much as possible of the SiGe (or CMOS) chip, and this not only includes the phase shifters and VGAs, but also, the entire transmit/receive phased-array modules, Wilkinson power combining network, digital and SPI control, bias circuitry, and in some cases, entire up/down-converters. Such chips become very complex to test and therefore, it is essential to design a complete built-in-self-test (BIST) system in order to test these chips in a low-cost manner.

Another important aspect is the packaging of such chips. This has been done using bond-wires, flip-chip techniques and C4 bumps, but requires expensive millimeter-wave substrate which contains a low-loss distribution network between the chip and the antennas, and low-loss antennas. Such substrates are typically expensive to manufacture since they must be composed of multiple layers of low-loss Teflon-based or LTCC substrates. This is what the industry has typically done at 60-80 GHz [1-3], but it is hard to extend it to 90 GHz and above.

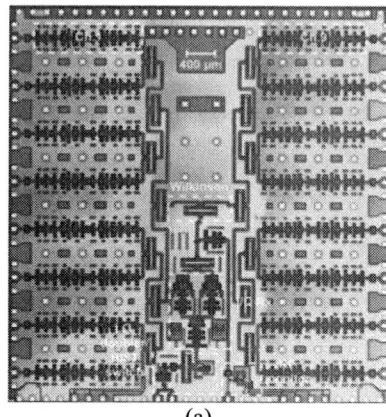

(a)

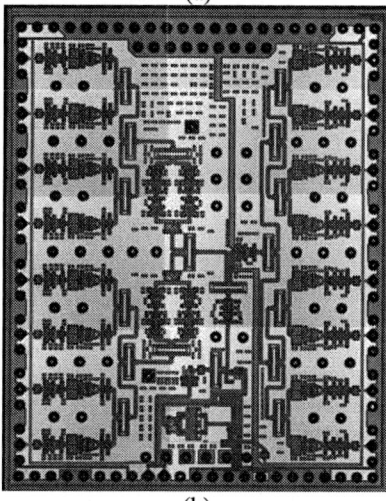

(b)

Fig. 1. (a) 16-element 76-85 GHz SiGe BiCMOS phased array receiver chip (5.5 x 5.8 mm^2) with down-converter and BIST, and (b) 8-Tx/8-Rx 76-85 GHz SiGe BiCMOS phased array chip with down-converter, up-converter, Rx BIST, Tx BIST.

One radical solution, proposed by UCSD, is to integrate everything required for a phased array on a single chip, not only the electronics, but also the high-efficiency antennas leading to a wafer-scale implementation [4]. Also, the entire phased array should be integrated on a single piece of silicon so that no mm-wave interconnect occurs between different substrates. The wafer-scale phased array can then be placed on a low-cost printed-circuit board where the input/output data signals (up to Gbps), control and power are placed. The entire mm-wave

978-1-4799-7231-9/14 $31.00 © 2014 IEEE

functionality is contained on the chip and the wafer-scale implementation is therefore easy to implement, and requires no RF design from the "end user". It is a self-contained solution.

This paper will present some of these ICs for mm-wave radar and communication systems [1-7].

II. SiGe Phased-Array Chips for Automotive Radars

Fig. 1 presents two different chips developed at UCSD for automotive radar applications [1-3]. The first chip is a 16-element phased array with receive capabilities at 76-85 GHz and BIST circuitry, while the second chip is a 76-84 GHz 8-element Tx/8-element Rx with up- and down-converters and BIST for the transmit and receive sections. The 16-element receive array is designed for bond-wire packaging, while the 8-Tx/8-Rx array is designed for flip-chip packaging with C4 bumps due to the required low ground inductance and the high isolation (> 50 dB) between the Tx and Rx. The 16-element Rx chip consumes 1.1 W with a system NF of 17 dB at 100 kHz

offset, and a P1dB of -22 dBm per channel. The system NF includes the mixer, IF amplifier and the 1/f noise from the SiGe transistors. The 8-Tx/8-Rx chip has much higher linearity (-10 dBm/channel), a transmit power of +3 dBm per channel, and a system NF of 17 dB at 100 kHz offset. In fact, both of these chips only require an external LO at 38-42 GHz for proper operation, which comes from an external chip and is fed in a differential fashion to the chips (at the bottom).

The 16-element receive chip was assembled with 16 antennas as shown in Fig. 2. The scanning is done on the azimuthal plane and the antennas are designed as vertical elements using series-fed microstrip arrays space at 0.6λ apart. The antenna results in a directivty of 31 dB and a gain of 28 dB. The chip is connected to the 16-element antenna array using G-CPW lines on the RO3003 susbtrate, and the different line lengths are compensated using phase settings in the 16-channels in the SiGe chip.

Fig. 2b presents the measured patters at 79 GHz for the 16-element receive array. The measured E-plane patterns (azimuth scan) show $\pm50^\circ$ scanning with an average sidelobe level of < -16 dB up to 40°. The measured patterns show a 3-dB beamwidth of 5.5°–7° at 0–40°. To our knowledge, this is the first demonstration of a phased array pattern at > 60 GHz using SiGe chips and high density integration.

II. 108-114 GHz Wafer Scale Phased Array Transmitter

Fig. 3 presents a SiGe wafer-scale phased array transmitter at 108-114 GHz transmitter with high-efficiency on-chip antennas [4]. The 4x4 array is based on an RF beamforming architecture with an equi-phase distribution network and phased shifters placed on every element. The differential on-chip antennas are implemented using a 100 µm thick quartz superstrate and with a simulated efficiency of 45% at 110 GHz (Fig. 4).This type of antenna, pioneered at UCSD, requires the use of a 50-100 µm thick quartz (or low-loss dielectric) superstrate above the silicon wafer so as to increase the volume for the electric field under the antenna and increase its efficiency [5]. The phased array is also designed with low mutual coupling between the antenna elements and results in very stable active antenna impedances versus scan angle. The phased array is built in the Jazz SBC18H3 SiGe BiCMOS process, and is 6.5x6.0 mm².

Measurements show two-dimensional pattern scanning capabilities up to 30 degrees, limited only by the two bit phase shifter used on each cell (0, 90, 180, 270 degrees) and the size of the array (Fig. 5), with a directivity of 17.0 dB, an array active gain of 26.5 dB at 110 GHz, and an EIRP of 23–25 dBm at 108–114 GHz. The power

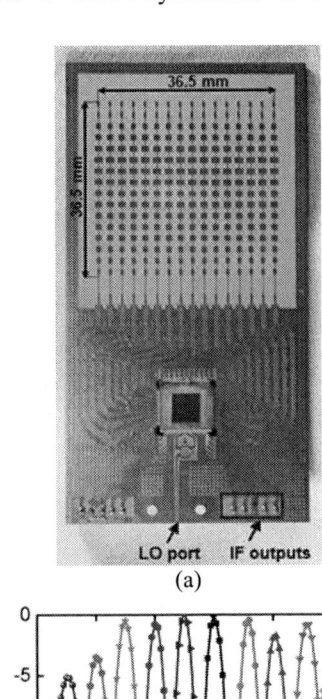

(a)

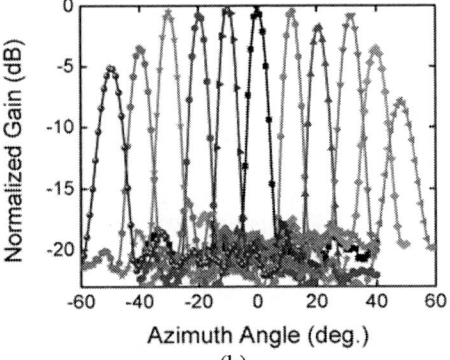

(b)

Fig. 2. (a) The SiGe chip packaged with a 16-element antenna for azimuth scanning. (b) Measured scanned patterns.

978-1-4799-7231-9/14 $31.00 © 2014 IEEE

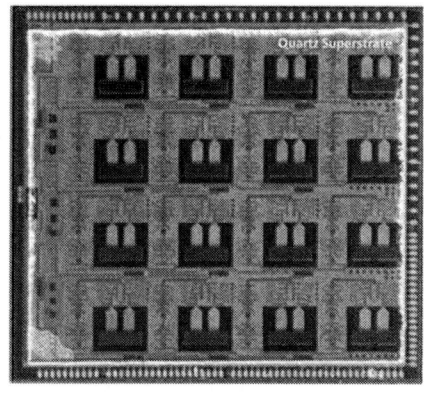

(a)

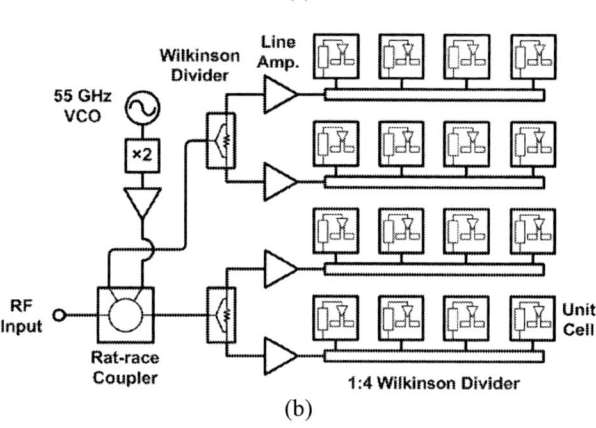

(b)

Fig. 3. (a) A 108-114 GHz 16-element transmit wafer-scale phased- array with integrated high efficiency differential dipole antennas. (b) Block diagram showing the integration on the chip.

consumption is 3.4 W from a 1.9 V supply. To our knowledge, this work represents the first wafer-scale phased array to-date. The application areas are in point-to-point communication systems in the 100–120 GHz range.

III. 90-100 GHz 4x4 POLARIMETRIC RADAR CHIP

Fig. 6 presents a 4x4 transmit/receive (T/R) SiGe BiCMOS phased-array (IBM8HP) chip at 90–100 GHz with vertical and horizontal polarization capabilities, 3-bit gain control (9 dB), and 4-bit phase control [6,7]. The 4x4 phased array fits into a $1.6 \times 1.5 \text{mm}^2$ grid, which is required at 94 GHz for wide scan-angle designs. The chip has simultaneous receive (Rx) beam capabilities (V and H) and this is accomplished using dual-nested 16:1 Wilkinson combiners divider with high isolation. The phase shifter is based on a vector modulator with optimized design between circuit level and electromagnetic simulation and results in 1 dB and 7° rms gain and phase error, respectively, at 85–110 GHz. The V

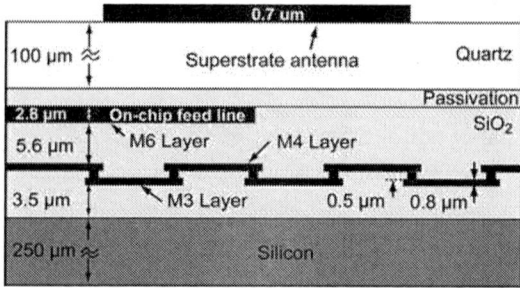

(a)

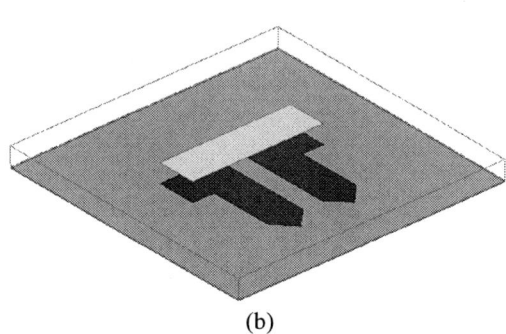

(b)

Fig. 4. (a) Superstrate antenna on top of a silicon RFIC. The antenna is placed on a 100 μm thick quartz substrate for improved efficiency. (b) Differential dipole antenna (yellow) on the top side of the quartz and the feed line (dark blue) on the top side of the silicon substrate.

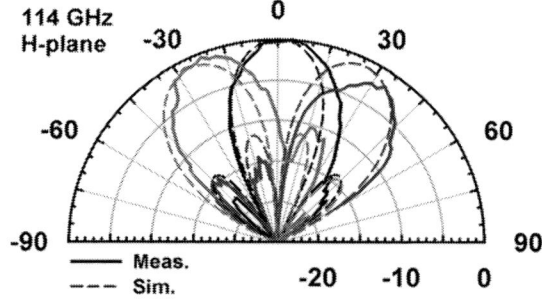

Fig. 5. Measured patterns at 114 GHz for the 16-element wafer-scale phased array transmitter.

and H Rx paths result in a gain of 22 and 25 dB, respectively, a noise figure of 9–9.5 (at max. gain), and 11 dB (at min. gain) measured without the T/R switch, and an input P1dB of -31 to -26 dBm over the gain control range. The measured output Psat is +1 dBm per channel, limited by the T/R switch loss. The chip consumes1100 mA from a 2-V supply in both the Tx and Rx modes.

Measurements show +/-6° and +/-0.75 dB variation between the 4x4 array elements in the Tx mode and Rx mode, respectively, and -40 dB coupling between the different channels on the chip.

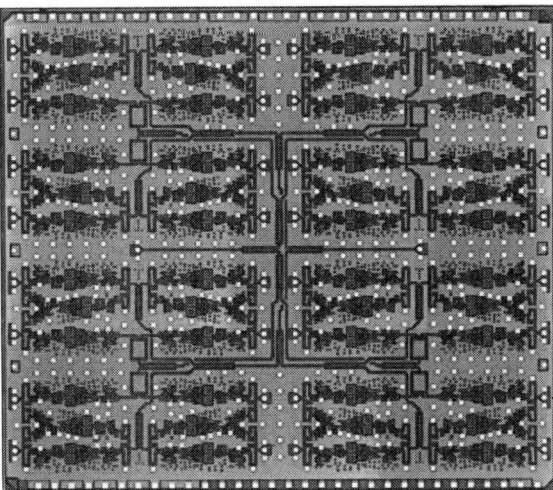

Fig. 6. A 90-100 GHz 16-element transmit/receive phased array with simultaneous receive beams and dual nested Wilkinson couplers. This chip contains 48 different phased array channels and is 6.6x5.8mm². The channels fit into a λ^2 antenna grid which is placed on top of the chip using polyimided RDL (redistribution layers).

In this design, the antennas are not integrated on chip, and 5-10 µm thick polyimide redistribution layers are used on top of the cihp in order to attach the internal antenna ports on the RFIC (32 of them, 16 for V and 16 for H) to the external dual polarized antenna. This is still a wafer-scale implementation since the completed product will contain the RFIC and the antennas on a single chip, and no mm-wave transitions are used (assuming that an up/down converter is used on chip). To our knowledge, this chip has 48 independent T/R channels and two 16:1 Wikinson combiners, and is the largest phased-array chip ever developed.

IV. CONCLUSION

This paper presented several silicon RFIC chips with very complex capabilities. In the opinion of the author, this trend is going to continue and even larger chips and with more capabilities will be designed and built in the near future, thereby lowering the cost of millimeter-wave phased array systems.

ACKNOWLEDGEMENTS

This work was supported by the DARPA GRATE, DARPA MFRF, Intel and Toyota. The author acknowledges the work of his students and collaborators.

REFERENCES

[1] Sang Young Kim, Ozgur Inac, Choul-Young Kim, Donghyup Shin and Gabriel M. Rebeiz, "A 76-84 GHz 16-Element Phased Array Receiver with a Chip-Level Built-In Self-Test System," *IEEE Transactions on Microwave Theory and Techniques,* vol. 61, no. 8, pp. 3083-3098, August 2013.

[2] B. Ku, P. Schmalenberg, S. Kim, C. Kim, O. Inac, J. Lee, K. Shiozaki, and G.M. Rebeiz, "A 16-Element 77–81-GHz Phased Array for Automotive Radars with ±50o Beam-Scanning Capabilities," *IEEE Int. Microwave Symposium,* June 2013, pp. 1-4.

[3] Bon-Hyun Ku, Ozgur Inac, Michael Chang and Gabriel M. Rebeiz, "A 75-85 GHz Flip-Chip Phased Array RFIC with Simultaneous 8-Transmit and 8-Receive Paths for Automotive Radar Applications," *IEEE Radio Frequency Integrated Circuits Conf. (RFIC),* June 2013, pp. 1-4.

[4] W. Shin, B. Ku, O. Inac. Y.C. Ou, and G.M. Rebeiz, "A 108-116 GHz 4x4 Wafer-Scale Phased Array Transmitter with High-Efficiency On-Chip Antennas," *IEEE Journal Solid-State Circuits,* vol. 48, no. 9, pp. 2041-2055, September 2013.

[5] J. M. Edwards and G. M. Rebeiz, "High-Efficiency Elliptical Slot Antennas With Quartz Superstrates for Silicon RFICs," *IEEE Trans. Antennas and Propagation,* vol. 60, no. 11, pp. 5010-5020, November 2012.

[6] Fatih Golcuk, Tumay Kanar, and Gabriel M. Rebeiz, "A 90-100 GHz 4x4 SiGe BiCMOS Polarimetric Transmit-Receive Phased Array with Simultaneous Receive-Beams Capabilities," *IEEE Transactions on Microwave Theory and Techniques,* vol. 61, no. 8, pp. 3099-3114, August 2013.

[7] Ozgur Inac, Fatih Golcuk, Tumay Kanar, and Gabriel M. Rebeiz, "A 90-100 GHz Phased-Array Transmit/Receive Silicon RFIC Module with Built-In Self-Test," *IEEE Transactions on Microwave Theory and Techniques,* vol. 61, no. 11, pp. 3374-3782, October 2013.

978-1-4799-7231-9/14 $31.00 © 2014 IEEE

A 2×2, 316 GHz SiGe Scalable Transmitter Array with Novel Phase Locking Method and On-die Antennas

Saeed Zeinolabedinzadeh, Aida L. Vera López, Nelson E. Lourenco, Ahmet Cagri Ulusoy, Mehmet Kaynak[2], Mahmoud Kamarei[3], Bernd Tillack[2], John Papapolymerou, and John D. Cressler

[1]*School of Electrical and Computer Engineering, Georgia Tech, Atlanta, GA 30332-0250 USA*

[2]*IHP Microelectronics, Frankfurt-Oder, Germany*   [3]*University of Tehran*

Abstract—**We demonstrate a fully integrated 2×2 SiGe transmitter array at 316 GHz. A novel method is introduced to phase-lock the high frequency signal sources without any additional circuitry. This method suppresses the other possible oscillatory modes automatically. The power of the locked signal sources are combined in the air by the 2×2 on-chip antenna array which are fed with the locked sources. This array is scalable to larger sizes in the same layout style. This array transmitter utilizes Colpitts oscillators as high frequency signal sources; however, this method can be applied to any other oscillator architecture to perform the frequency and phase locking. The frequency of this transmitter follows the simple equations for a single Colpitts oscillator, thereby facilitating the design of the circuit.**

Keywords—*Array, Colpitts, on-chip antenna, oscillator, sub-mmW, SiGe, Terahertz.*

I. INTRODUCTION

Advances in silicon-based technology have enabled the serious consideration of constructing millimeter-wave (mmW), sub-millimeter wave (sub-mmW), and terahertz (THz) circuits on silicon. One of the main challenges for implementing THz circuits in silicon is the difficulty in generating signals with sufficient power levels. Since the f_T and f_{max} of current Si-based processes are still not adequate to make fundamental signal sources at THz frequencies, higher harmonics of the fundamental sources are utilized to generate signal powers at frequencies higher than the f_T and f_{max} of the transistors.

There are two main approaches used for this kind of implementation. One method is the use of a multiplier chain and amplifiers with a lower frequency input signal; while the other method is to use harmonic oscillators. The multiplier-based approach has the flexibility of utilizing high purity external signals with more control on the generated frequency and higher output power. On the other hand, the harmonic oscillator has the advantage of compact design with less power consumption, which makes it suitable for integrated array implementations. This latter approach will result in a narrow band system with less control on the tuning frequency; however, with current EM simulators, the oscillation frequencies can be predicted within the available tuning range of the oscillators. Nevertheless, the combination of these two approaches can also be employed. Both design approaches have been utilized recently to create signal power beyond the operational frequency limits of the utilized transistors [1] - [9].

Due to the compact size and lower power consumption, as well as its ease for creating large size arrays, we have designed and implemented harmonic oscillators in this paper to generate strong second harmonic power at 316 GHz. Four oscillators were designed and phase-locked using a novel technique, with each oscillator radiating via on-chip patch antennas. The proposed technique locks the oscillators strongly together without any additional circuitry. The array was designed and optimized in such a way that its net combined power (into the air) approaches a 6 dB of theoretical power improvement compared to a single transmitter. Although this work uses a 2×2 array for demonstration purposes, the concept and layout are easily scalable to any custom 2-D array sizes.

II. CIRCUIT DESIGN

Several approaches have been reported recently to create more power at frequencies beyond the peak f_T of the transistors. Power combining is among those that can potentially be scaled for larger sizes. One solution is to use a central signal source with multiple amplifier-multiplier units with a combiner (e.g., a Wilkinson combiner). The generated powers are inherently phase locked in this approach. However, such a design requires complex routings at extremely high frequencies, as well as high DC power consumption, and potential instability concerns, making this approach more challenging. On the other hand, any power combiner has a certain insertion loss associated with it, which limits the number of combinatory networks for very large array sizes. The other solution could be the use of local signal sources which are placed close to radiating antennas and directly feed them.

The main issue of this latter approach is the effective synchronization between the signal sources. If the signal sources were not frequency and phase-locked, even with very small phase or frequency deviations the resultant combined power will vary in time with the rate equal to the difference in the phase or frequency of the sources. At one instant, and in the certain point in space, the power will reach a peak value and then drop to zero as time evolves and changes the phase.

In [2], the signal sources are injection-locked to the central VCO to make a 4×4 radiating array at 280 GHz (and requires

978-1-4799-7231-9/14 $31.00 © 2014 IEEE

complex high frequency routing). Each cell of the array itself contains 4 cross-coupled oscillators. The limited locking range of the injection locking approach, as well as its looser locking mechanism, may result in the system failure in the presence of any frequency shift, due to process variations or any environmentally-induced deviation from the design frequency. In the present work, we propose another locking method, which we call a "hard-lock" approach, to lock four oscillators strongly together and make the 2×2 transmitter array. Since Colpitts oscillators out-perform other oscillator configurations at millimeter-wave frequencies, it is highly desired to find a way to lock these types of oscillators together for use in the array transmitters. We show how this can be accomplished.

Fig. 1 shows the schematic of the single Colpitts oscillator [1]. This oscillator has the best signal purity among the other silicon-based oscillators at the relevant frequencies. The oscillator utilizes a push-push Colpitts configuration which extracts the second harmonic signal from the common-base connection. The circuit configuration is a high-order system with multiple possible solutions, the so-called "modes". The circuit has been designed in such a way to only excite the desired mode and dampen out the other possible modes. For instance, the two left and right half-branches of the circuit can oscillate either in a differential mode or in a common mode, each of which can be considered a possible solution to this complex high-order system. Where in common-mode operation, each branch will oscillate exactly at the same frequency and phase as the other branch. While in differential mode, the two branches will have anti-phase oscillations at the same frequencies. The losses introduced in the common nodes of the circuit will effectively dampen the common-mode solution and thus only the differential mode solution will survive.

Consider that we want to lock four oscillators together. The best approach is to create an electrical locking mechanism in which the resultant oscillation frequency follows the oscillation frequency of the single oscillator. In other words, the oscillation frequency of these four oscillators is desired to be the same as the oscillation frequency of a single oscillator, so that designing the system can be reduced to the design of a single oscillator cell.

The proposed technique for this locking mechanism is shown in Fig. 2 and Fig. 3. We first divided the single oscillator circuit into two half branches at node B. Each branch in this new oscillator should be connected to the neighbor branch of the neighbor oscillator, as seen in Fig. 3. This procedure should be followed until the second branch of the last oscillator is connected to the second branch of the first oscillator, in order to make a loop. The resistors at the common-node connections will effectively dampen any common-mode solution between each connected branches. By inspecting this circuit one can easily see the solution of the single Colpitts oscillator is also a solution to this complicated oscillatory network. It can be shown that because of the generated loop, this solution is the only solution to this complex high-order system. The same approach can be used to lock any other type of oscillators together. The first advantage

of this proposed solution is the possibility of predicting the oscillation frequency by simply finding the oscillation frequency of the single oscillator cell, thus reducing the complexity of the design of this network to the design of single cell oscillator.

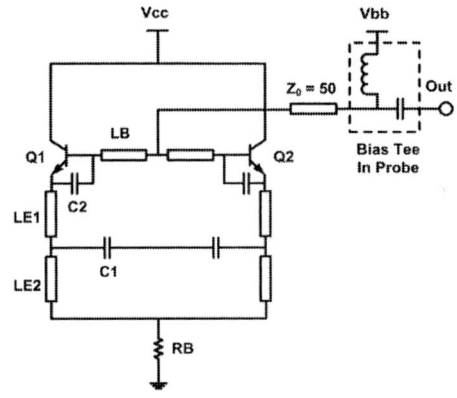

Fig. 1. Push-Push Colpitts oscillator.

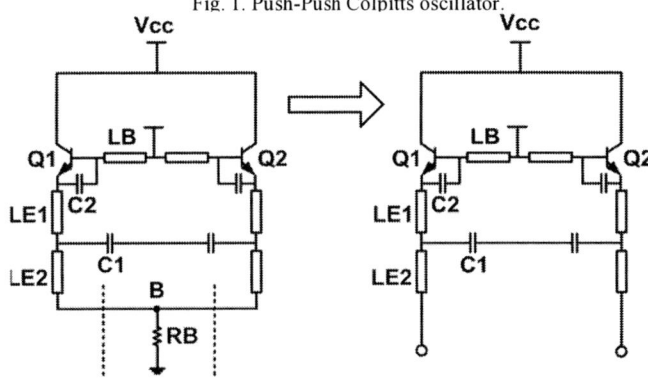

Fig. 2. Breaking the oscillator circuit from its common node.

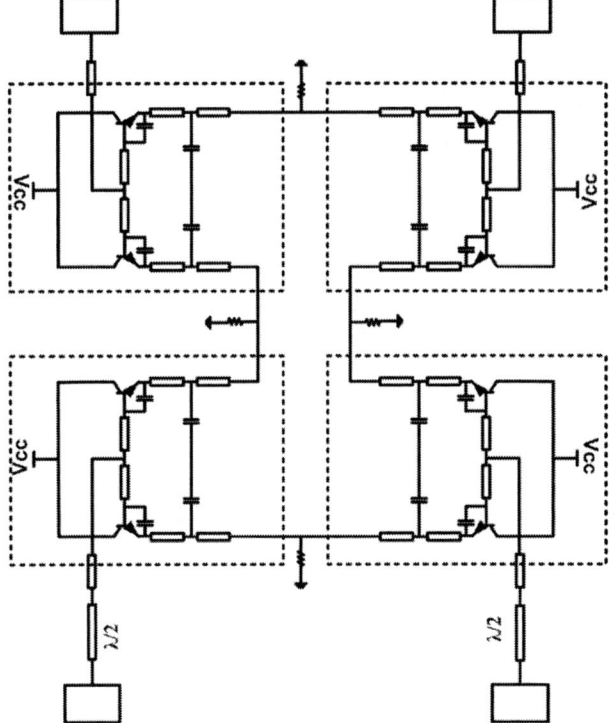

Fig. 3. Proposed 2×2 phase-locked oscillatory array.

The second advantage is its easy scalability to any larger 2-D array. The LE2 inductors in Fig. 2 were utilized as the interconnections between the neighbor oscillators while serving as an inductor for the single oscillator cell. Since the oscillatory network will be placed inside the array of antennas, it is important to pay attention to the direction in which the circuit feeds the antennas so that radiation fields can combine constructively in the air. In our layout, 180 degree phase shifter lines have been added to the bottom cells to make the radiated power from the array combine in phase.

On-chip patch antennas were designed for this 2×2 transmitter array, since this type of antenna automatically suppresses the surface wave excitation through the shielding GND plane. However, this type of antenna has a narrow bandwidth and the circuit needs to be carefully EM simulated to predict the oscillation frequency within the bandwidth of the antennas. The distance between the array elements is around λ/2. The dimensions and spacing of the antennas were optimized to get the maximum gain of the array network.

III.   MEASUREMENT RESULTS

The proposed circuit was implemented in a SiGe HBT BiCMOS technology with the peak f_T and f_{max} of 300 GHz and 450 GHz, respectively (the IHP SG13G2 SiGe platform). This process technology has seven metal layers and the top layer has the thickness of 3 μm, which is suitable for high frequency transmission lines and antennas. The silicon substrate has a thickness of around 300 μm and no substrate thinning was performed on the die. Fig. 4 shows the die photo of the circuit which occupies 1.3×1.4 mm². The measured phase noise of the single Colpitts oscillator is shown in Fig. 5.

In order to measure the radiated power and frequency of the transmitter, the die was mounted on a board and placed in front of the receiver. The receiver setup consists of a WR2.8 horn antenna with 26 dBi of gain and WR2.8 EHM mixer. The mixer is a 20th harmonic mixer which downconverts the received signal to the baseband. The distance between the receiver and the transmitter array was well in the far-field of the receiver antenna. To determine the far-field distance of the receiver antenna, we varied the distance between the transmitter and receiver between 120 mm to 155 mm. It was observed that the signal loss follows the free-space path loss, meaning the transmitter was correctly placed in the far-field of the receiver antenna. Fig. 6 shows the photo of the custom setup for measuring the frequency and power of the transmitted signal. The detected signal was amplified after down conversion with a 20th harmonic mixer and the detected IF signal is shown in Fig. 7. The baseband amplifier provides around 24 dB of gain to the IF signal. The distance between the transmitter antenna and the receiver antenna was set to 155 mm for this data. Based on the power of the detected signal, and by calculating the free-space path loss as well as the loss of the receiver chain, the effective radiated power from the transmitter was predicted to be -0.3 dBm at 316 GHz.

It is worth noting that the antennas were originally designed for 340 GHz and there is around 3 dB of gain loss because of the frequency shift of the transmitter circuit. More accurate

EM simulations predict the oscillation frequency with less than 2% error.

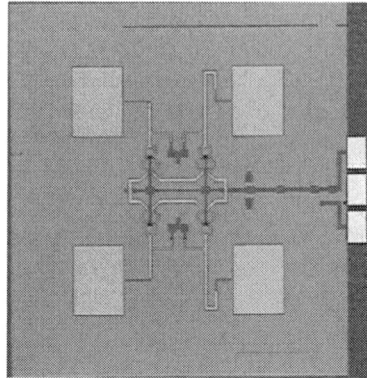

Fig. 4. Die micrograph of the 2×2 array circuit.

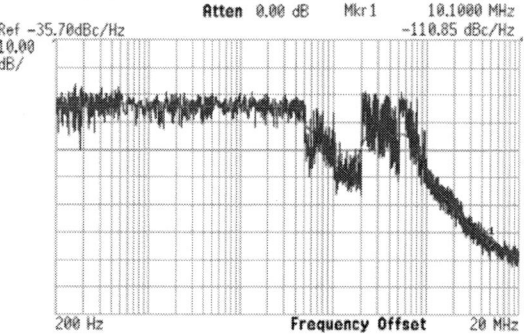

Fig. 5. Measured phase noise of the single cell Colpitts oscillator after down-conversion with a 20th harmonic mixer.

Fig. 6. Measurement setup for detecting the transmitted signal.

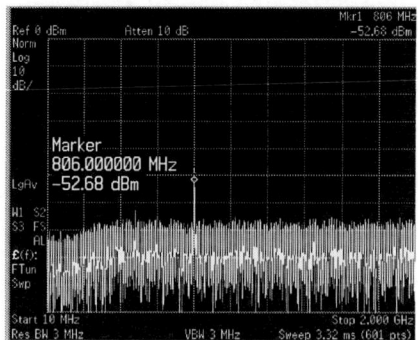

Fig. 7. The spectrum of the detected signal after down conversion with a 20th harmonic mixer and base-band amplification.

978-1-4799-7231-9/14 $31.00 © 2014 IEEE

TABLE I. PERFORMNCE SUMMARY

Reference	Frequency (GHz)	Radiated Power EIRP (dBm)	Array Size	DC Power consumption (mW)	Process
[2]	280	+9.4	4×4	820	45nm SOI CMOS
[7]	420	+3	2×4	700	45nm SOI CMOS
[8]	338	+17.1	4×4	1540	65nm CMOS
This work	**316**	**-0.3**	**2×2**	**225**	**130nm BiCMOS**

Fig. 8 shows the measurement setup and the measured radiation pattern of the antenna-array at 316 GHz. The gain was estimated by considering the mixer's conversion loss from the data sheet of the mixer. Absorbers were used for this measurement to absorb additional reflections. There is a small gain difference around zero degrees between the measured E-plane and H-plane which comes from the small misalignment between the horn antenna and transmitter array in the H-plane setup.

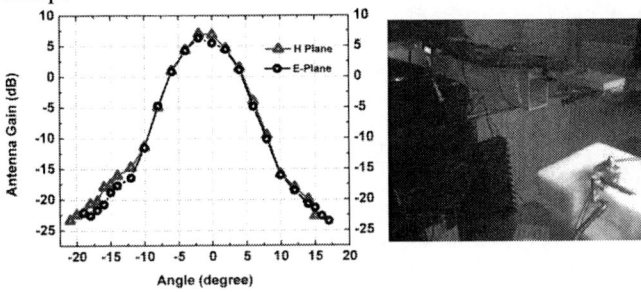

Fig. 8. Measured E-Plane and H-Plane pattern cut (left), and measurement setup (right).

Another simple test was also carried out to prove the phase-locking condition of the designed array. Simulations show that a low frequency tone is generated at the output in the presence of strong mismatch in array oscillators. The frequency of this tone is equal to the difference between the frequencies of the array elements. This tone is the result of modulation between the different oscillator tones with different frequencies in the array architecture in the presence of strong mismatch.

A test circuit consists of the 2×2 oscillatory array without antennas was also fabricated. Fig. 9 shows the die photo of this test circuit. The experiment showed that by probing only one of the array elements (50 Ω at the output) and leaving the other pads open, we can induce a mismatch between the oscillator cells. As a result of that mismatch, a low frequency tone appears at the output. However, by probing the output of the other array elements that tone disappears. Since at matched termination at the output of the oscillator cells, we don't observe any low frequency tone, thus verifying the frequency locking of the array oscillators.

The performance of this array transmitter is compared with the state-of-the-art in Table I. Considering the smaller array size and lower power consumption, this transmitter has the competitive performance.

IV. SUMMARY

A 2×2 scalable SiGe transmitter array with on-chip antennas was demonstrated. A novel technique to phase lock the on-chip signal sources was also described and demonstrated. The array consumes 225 mW of DC power while radiating -0.3 dBm at 316 GHz.

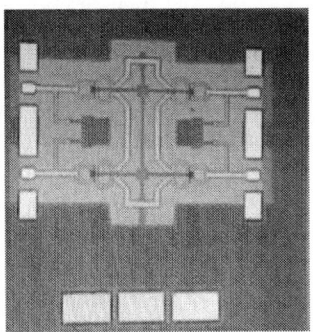

Fig. 9. Die photo of the test circuit of the locked oscillators.

ACKNOWLEDGEMENT

The authors would like to acknowledge IHP Microelectronics for chip fabrication, and the members of SiGe team at Georgia Tech for their contributions.

REFERENCES

[1] S. Zeinolabedinzadeh et al., "Low Phase Noise and High Output Power 367 GHz and 154 GHz Signal Sources in 130 nm SiGe HBT Technology," *Proc. Int. Micro. Symp.*, pp. 1-4, 2014.

[2] K. Sengupta et al., "A 0.28 THz power-generation and beam-scanning array in CMOS," in *IEEE ISSCC Dig.*, pp. 256–258, 2012.

[3] Y.M. Tousi et al., "A 283 to 296 GHz VCO with 0.76 mW Peak Output Power in 65nm CMOS," *ISSCC Tech. Dig.*, pp. 258-259, 2012.

[4] E. Ojefors et al., "Subharmonic 220- and 316-GHz SiGe HBT receiver front-ends," *IEEE Trans. Microw. Theory Tech.*, vol. 60, no. 5, pp. 1397–1404, May 2012.

[5] J-D Park et al., "A 0.38 THz fully integrated transceiver utilizing quadrature push-push circuitry," *IEEE Symp. VLSI Circuits*, pp. 22-23, June 2011.

[6] U. Pfeifer et al., "A 0.53THz reconfigurable source array with up to 1mW radiated power for terahertz imaging applications in 0.13μm SiGe BiCMOS," *ISSCC Tech. Dig.*, pp. 256-257, 2014.

[7] F. Golcuk et al., "A 0.39–0.44 THz 2x4 Amplifier-Quadrupler Array with Peak EIRP of 3 - 4 dBm," *IEEE Trans. Microw. Theory Tech.*, vol. 61, no. 12, pp. 4483–4491, Dec. 2013.

[8] Y. Tousi et al., "Scalable THz 2D phased array with +17dBm of EIRP at 338GHz in 65nm bulk CMOS," *ISSCC Tech. Dig.*, pp. 258-259, 2014.

[9] S. Zeinolabedinzadeh et al., "A 314 GHz, Fully-Integrated SiGe Transmitter and Receiver with Integrated Antenna," *Proc. IEEE RFIC.*, pp. 361-364, 2014.

978-1-4799-7231-9/14 $31.00 © 2014 IEEE

A Broad-band BiCMOS Transmitter Front-end for 27-36GHz Phased Array Systems

Yu Pei, Ying Chen, Domine M.W. Leenaerts

NXP Semiconductors
Eindhoven, Netherlands
yu.pei@nxp.com

Abstract— A 27-36 GHz wide-band phased array TX front-end is demonstrated in a 0.25um SiGe:C BiCMOS process. The TX front-end presents a saturation power more than 12.5dBm across 9GHz bandwidth. The front-end provides variable phase shift from 0°~360° with ~10° resolution, and the relative phase shift remains constant in the desired band. A 2-bit amplitude resolution is available for advanced beamforming algorithms. The wide-band PA can be applied in saturation mode and in linear mode due to its high linearity with an OIP3 over 21dBm.

Keywords— beam steering; broadband amplifiers; phased arrays; phase shifters; power amplifiers.

I. INTRODUCTION

Ka-band (26.5 GHz to 40 GHz) transmitters are broadly used in professional applications such as communication satellite uplink, base station back haul point-to-point (P2P) communication, and very small aperture terminal (VSAT) communication. For social security equipments some frequencies in the Ka-band are used for radars aboard of airplanes, vehicle speed detection, weather detection, etc [1]. By using a phased array approach, the transmit beam can be steered electronically rather than mechanically, which reduces mechanics hardware complexity, weight and the deployment cost. Furthermore, the radiated power of several elements can be combined to relieve the output power burden of each power amplifier (PA). Where in the past these RF PAs and RF vector modulators, needed for the phased array approach, were realized in III-V technologies, the latest improvements in silicon-based SiGe:C bipolar technologies enable the realizing of these circuits in a low-cost silicon technology [2-4, 6-8].

For VSAT communication, which uses high order QAM modulation, linearity is a crucial parameter. On the other hand, radar transmitters operate in saturation mode, where the saturation power becomes more important. Compared with phase shifting in the RF path, LO path phase shifting has less influence on the transmitter linearity and the RF output power, as long as the mixer is hard switched by the LO signal. However, a clipped LO signal cannot provide amplitude tuning for the phase shifter; additional VGAs in the IF or RF path can offer such function, but will influence the TX linearity. LO phase oversampling vector modulators (LO-POVM) [4] replace VGAs with simple switches in the LO path to keep the advantages of LO phase shifting, and in addition, to realize amplitude variation to the RF output signal.

In this paper, a wideband TX front-end is proposed to cover in linear mode the VSAT communication band (27.5-31GHz)

and in saturated mode the radar band at 35GHz. The wideband TX provides high linearity and output power for both applications. In section II the system architecture of the wideband phased array TX front-end is introduced; Section III discusses the RF PA; and then in section IV the line-up of the PA and the wideband LO-POVM phase shifter are briefly described. Detailed measurements are provided in section V together with a comparison with state of the art. We conclude in section VI.

II. SYSTEM ARCHITECTURE

The architecture of the Ka-band phased array front-end is shown in Fig. 1. The phase shifting up-conversion is realized with an LO-POVM block. The LO-POVM was designed to provide 2-bit amplitude variation and 10 degrees phase resolution at each TX channel. Previous results of a similar stand-alone LO-POVM showed that the relative phase shift of the RF output signal can remain flat in a 5GHz bandwidth [4]. The focus of the paper is on an improved wide band LO-POVM combined with the RF PA.

Rather than a CMOS technology, BiCMOS is chosen for the implementation, as the application is in the professional market, where high performance and reliability are required, while cost, die size and power consumption have lower priority. A bipolar transistor has a higher breakdown voltage than a MOS device, thus same output power can be achieved with a higher voltage swing and a lower current swing; the optimal load-line matching is easier to realize. Moreover, a

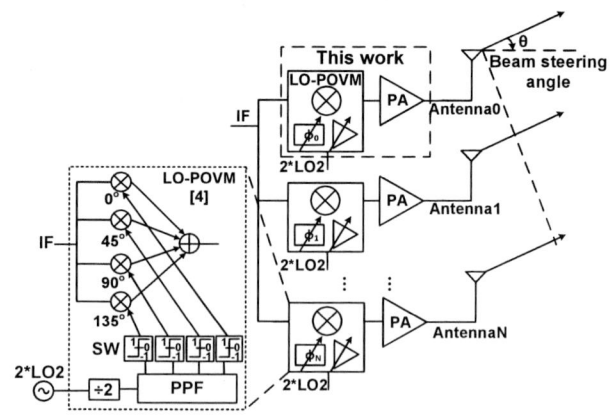

Fig. 1. Block diagram of the phased array transmitter architecture.

moderate output power is achievable with a single path power amplifier instead of a multi-path power combining technique as is common in CMOS PAs. Avoiding a power combining circuit will improve the total power efficiency.

III. WIDEBAND POWER AMPLIFIER DESIGN

As linearity is an important parameter for VSAT applications, a 2-stage Class-A PA has been designed (Fig. 2). The PA line-up consists of a medium power amplifier (MPA), a driver (DR) and three matching networks (MN). The MPA is the most critical stage for the linearity of the transmitter chain [5], and must also provide the desired output power level. The driver operates as an interface between the transmitter and the MPA, and improves the overall power gain and power matching. The matching networks MN_in, MN_inter and MN_out provide desired impedance transformation at the source, between the two stages and at the load, respectively. MN_in and MN_out also work as baluns for single-ended to-differential (S2D) and differential-to-single-ended (D2S) transformation.

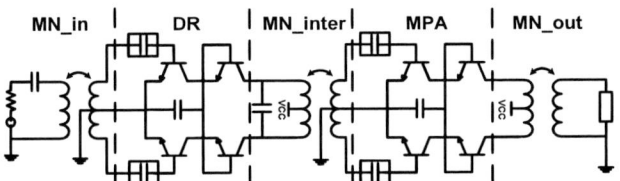

Fig. 2. Simplified schematic of the 2-stage wide-band power amplifier.

A. Active stages

Since the circuit is driven by a 3.3V power supply, a cascode topology for both stages has been chosen to allow both devices to operate within the breakdown voltage limitation. A cascode design also improves output-to-input isolation. The common-base transistors are high-voltage NPN devices with a 2.8V collector-emitter breakdown voltage with open base (BVceo) which allows higher voltage swing than standard transistors with 1.45V BVceo. The common-emitter transistors are standard NPN devices to maintain the high gain. The DC current is supplied from a 3.3V power supply via the center tap of the transformer's primary winding, and the winding width is selected to handle the total current.

B. Matching networks

The complete PA (Fig. 2) is designed for a 50Ω source and load impedance off-chip. Assuming the PA is loaded with a 50Ω single-ended antenna, the output matching network (MN_out) transfers the differential output signal of the RF PA to the single-ended load. MN_out also creates the optimal load-line matching at the MPA output at 33GHz to compensate the gain roll-off at high band. Similarly, MN_inter ensures that the driver stage provides highest gain at 35GHz, such that the driver linearity is not limiting the total linearity within the desired band. The 50Ω source is conjugate matched with the driver input at 32GHz center frequency. Note that the capacitive output impedance of the driver and the MPA are taken into account when designing the MN_inter and MN_out.

Transformers are used in the matching networks as they have a wider bandwidth than single stage LC networks.

IV. WIDEBAND TX FRONT-END

For the wide-band TX front-end in the dashed box in the top of Fig. 1, LO2 frequency is fixed at 20GHz, and the IF frequency is switched between 10GHz and 15GHz. In this way the VSAT band around 30GHz and the radar band at 35GHz can be reached. Fixing LO2 allows for an easier design of the LO-POVM, meanwhile designing broadband IF circuits between 10-15GHz is not too difficult. The 20GHz differential LO2 signals with 0°-135° phases are generated by an integrated 40GHz frequency divider and a poly-phase filter (PPF). Variable complex gain is tuned digitally with the switched amplifiers (SW), see Fig. 1 [4].

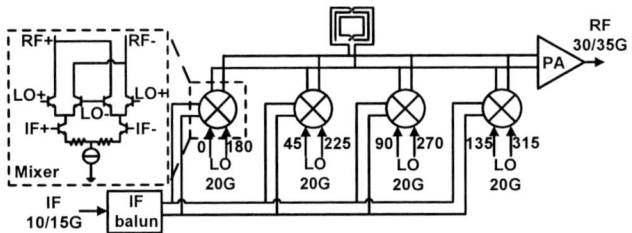

Fig. 4. Block diagram of the LO-POVM mixer and the PA line-up.

The line-up of the LO-POVM mixer and the PA (Fig. 3) shows the improvement of the LO-POVM compared with the previous design. The focus of [4] was on the phase shifting property rather than the output power, thus the LO-POVM output power is kept at a moderate level with a single buffer. With the implementation of the PA, the LO-POVM is now loaded with the input of the driver without MN_in in Fig. 2. Wideband power matching network between the LO-POVM and the driver consists of the current combining inductor and serial capacitors. The power transfer is optimized at 32GHz center frequency of both bands. By proper design of the mixers' biasing, the nonlinearity of the mixer is not degrading the line-up's total linearity, which is still decided by the PA.

V. MEASUREMENT RESULTS

The test chips were fabricated in a 0.25um SiGe:C BiCMOS process. The NPN device features a 210GHz cut-off

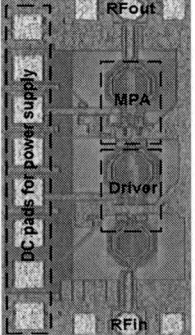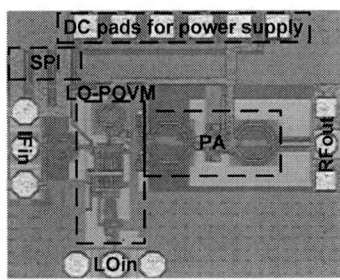

Fig. 3. Die photo of the fully integrated power amplifier left) and TX front-end (right).

978-1-4799-7231-9/14 $31.00 © 2014 IEEE

frequency and a DC current gain of 2000. The micrograph of the fabricated PA is shown in Fig. 4 (left). The area of the core circuit excluding pads is 200um × 640um. The power amplifier consumes 337mW DC power from a 3.3V power supply. The micrograph of the TX front-end is shown in Fig. 4 (right). The core circuit excluding pads occupies 1000um × 670um area and consumes 529mW DC power from a 3.3V power supply.

A. Small signal test of the stand-alone wide-band PA

First, the stand-alone wide-band PA is measured with a network analyzer and the small signal S-parameters are shown in Fig. 5. From 27 to 36 GHz, the transducer gain |S21| is higher than 15dB. |S22| reaches a minimum value of -4dB due to load-line matching at the output node. The input matching is centered around 32GHz and |S11| is better than -15dB from 27 to 36 GHz. |S12|, which is not plotted in the figure, is lower than -45dB.

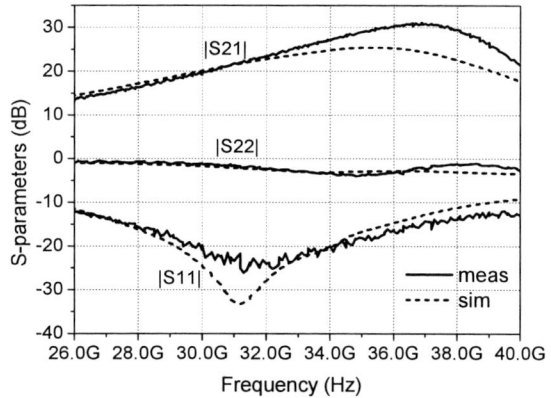

Fig. 5. Measured S-parameters of the stand-alone PA.

B. Large signal test of the PA and the TX front-end

AM-AM responses of the stand-alone PA and the TX front-end are measured from 27-36GHz (Fig.6). The saturation power of the PA in the VSAT band is better than 15dBm and in the radar band it is still above 12.5dBm. The 3dB Psat drop in the radar band will easily be covered by more than doubled number of antennas in the phased array approach for the radar. The PAE of the PA is higher than 13% in the VSAT and radar

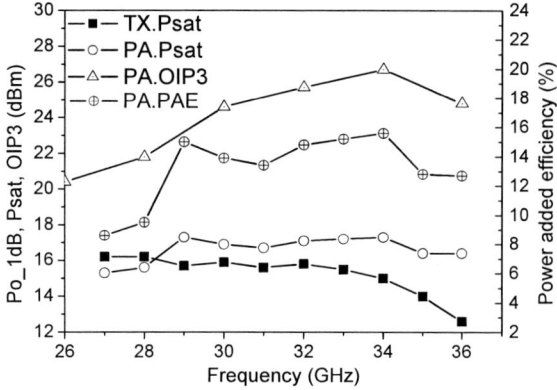

Fig. 6. Measured large signal frequency response of the stand-alone PA and the TX front-end.

frequency bands. The OIP3 of the PA is higher than 23dBm in the VSAT band. For VSAT communication with 16-QAM modulation schemes, the PA can operate at 1.5-6 dB back-off from saturation region to achieve higher linearity, signal to noise ratio (SNR) and lower symbol error rate (SER) [9], when the output power is still above 10dBm; while for radar applications, the PA will operate in saturated mode for high output power and power efficiency.

C. Wideband phase shift test

The measured normalized complex gain constellations (Fig. 7) at 30GHz and 35GHz match well with the calculated constellation. 81 gain points are available with 10° phase resolution and 2-bit amplitude resolution. The 10° phase shift resolution can be translated to a 3° beam steering resolution, assuming a phased array with a half wavelength antenna spacing, and such resolution is adequate for most of the applications. The 2-bit amplitude resolution can be applied for advanced beamforming algorithms, which improves the radiated beam pattern by suppressing the sidelobes, generating a nulling direction, bi-directional transmission, and etc. Several phase shifter settings are selected to demonstrate flat relative phase shifts from 27 to 36GHz (Fig. 8). The phase errors of these settings are less than ± 9° (Fig. 9).

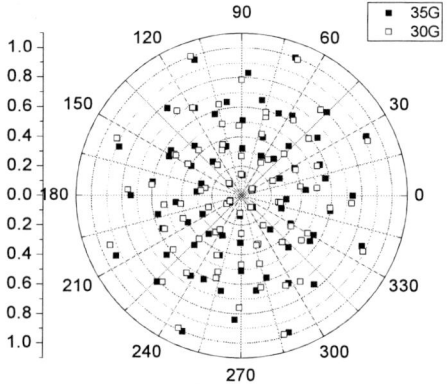

Fig. 8. Measured complex gain constellation of the front-end at 30GHz and 35GHz.

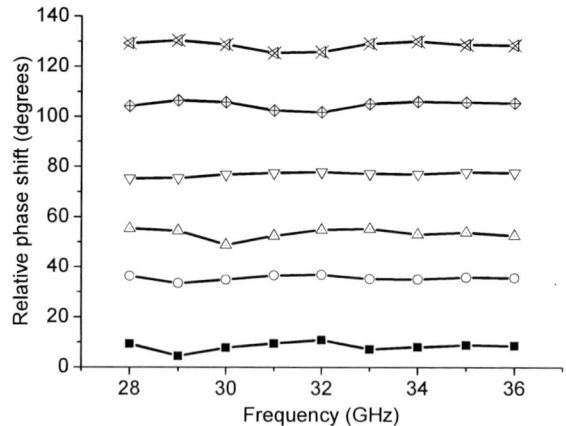

Fig. 7. Measured relative phase shifts of several settings within 27GHz to 36GHz band.

978-1-4799-7231-9/14 $31.00 © 2014 IEEE

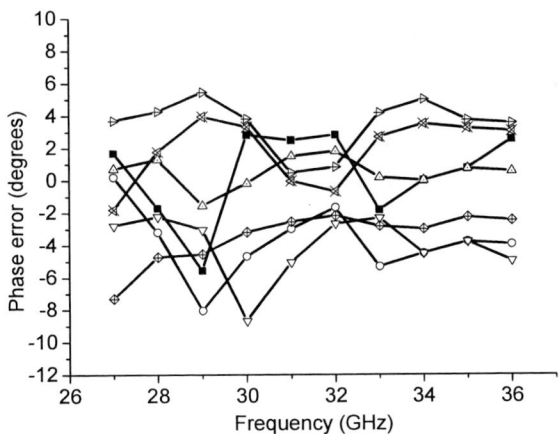

Fig. 9. Measured phase error of several settings within 27GHz to 36GHz band.

The performance of the phased array TX front-end is compared with published works in the 30-50GHz frequency range with BiCMOS technolgies (TABLE I). The proposed phased array TX front-end achieves broadest operation bandwidth with highest linearity in this frequency range. This work achieves a significant improvement in the output power, which is more than 6dB higher than the reported data. The high output linearity is advantageous for VSAT transmitters that utilize high order QAM modulations. The phase/amplitude resolution is in line with state- of-the-art.

VI. CONCLUSIONS

In this paper, a wideband phased array TX front-end configuration is proposed to cover Ka-band VSAT communication band in linear mode and radar band in saturated mode. The TX front-end stands out with high output power, power added efficiency and linearity within 28% of bandwidth. Furthermore, accurate complex gain constellation is achieved at both 30GHz and 35GHz bands. The phase shift levels remains relatively constant within desired band. The PA and the TX front-end both perform as qualified candidates for Ka-band advanced multi-mode/multi-band phased array systems.

ACKNOWLEDGMENT

This research is conducted as part of the Sensor Technology Applied in Reconfigurable systems for sustainable Security (STARS) project. The authors would like to thank J. Bergervoet from NXP for the valuable support during the measurements.

REFERENCES

[1] C.K. Chong, D. Layman, R.H. Le Borgne, M.L. Ramay, R.J. Stolz, and Xiaoling Zhai, "Development of High-Power Ka/Q Dual-Band and Communications/Radar Dual-Function Helix-TWT," *IEEE Trans. Electron Devices*, vol. 52, no. 5, pp. 653-659, May 2005.

TABLE I. PERFORMANCE COMPARISON

	[6]	[7]	[8]	[4]	This work
freq(GHz)	44	40-45	36-38	27.5-32.5	27-36
BiCMOS tech.	0.18um	0.18um	0.13um	0.25um	0.25um
core area (mm^2)	0.658	~0.52*	~1.28*	0.135	0.67
gain(dB)	+18	+12.5	0	+10	> +15
Psat(dBm)	-1.7	-2.5±1.5	~ +6	~ -5	> +12.5
OIP3	~ +6**	~ +5**	~ +14**	+3	>+20
P_{DC}(mW)/ V_{dd}(V)	786/4.4	~225/5*	43/1.8*	393/3.3	558/3.3
phase resolution	Cont.	22.5°	11°	~10°	~10°
amplitude resolution	-	-	5-bit	2-bit	2-bit
remarks	up-conv. LO VM	TX, RF IQ VM	TX, RF LC network	up-conv. LO-POVM	TX LO-POVM

** OIP3 estimated from 1dB compression point.
* per array element.

[2] P.J. Riemer, J.S. Humble, J.F. Prairie, J.D. Coker, B.A. Randall, B.K. Gilbert, and E.S. Daniel, "Ka-Band SiGe HBT Power Amplifier for Single-Chip T/R Module Applications," *IEEE MTT-S Intl. Microwave Symp. Dig.*, June 2007, pp. 1071-1074.

[3] M. Chang and G.M. Rebeiz, "A 26 to 40GHz wideband SiGe balanced power amplifier IC," *IEEE RFIC Symp. Dig.*, June 2007, pp. 729-732.

[4] Y. Pei, Y. Chen, D.M.W. Leenaerts, and R. Mahmoudi "A phase-shifting up-converter for 30GHz phased array applications," *IEEE RFIC Symp. Dig.*, June 2012, pp. 499-502.

[5] B. Razavi, *RF Microelectronics*. Prentice Hall, 1997.

[6] S. Kim, L. E. Larson, "A 44-GHz SiGe BiCMOS phase-shifting sub-harmonic up-converter for phased-array transmitters," *IEEE T-MTT*, vol. 58, no. 5, pp. 1089-1099, May, 2010.

[7] K.-J. Koh, J. W. May, G. M. Rebeiz, "A millimeter-wave (40-45 GHz) 16-element phased-array transmitter in 0.18-um SiGe BiCMOS technology," *IEEE JSSC*, vol. 44, no. 5, pp. 1498-1509, May. 2009.

[8] D.-W. Kang, J.-G. Kim, B.-W. Min, G. M. Rebeiz, "Single and four-element Ka-band transmit/receive phased-array silicon RFICs with 5-bit amplitude and phase control," *IEEE T-MTT*, vol. 57, no. 12, pp. 3534-3543, Dec. 2009.

[9] H. AbdulHussein Al-Asady and M. Ibnkahla, "Performance evaluation and total degradation of 16-QAM modulations over satellite channels," *CCECC*, vol.2, May 2004, pp. 1187-1190.

978-1-4799-7231-9/14 $31.00 © 2014 IEEE

A Power-Efficient 4-element Beamformer in 120-nm SiGe BiCMOS for 28-GHz cellular communications

Anirban Sarkar, Kevin Greene and Brian Floyd
Department of Electrical and Computer Engineering
North Carolina State University, Raleigh, NC 27606
Email: {asarkar3, kbgreen2, bafloyd}@ncsu.edu

Abstract—**A 4-element beamformer designed in 120-nm SiGe BiCMOS technology for 28-GHz mobile millimeter-wave broadband system is presented in this paper. Each element of the beamformer consists of a 4-bit active phase shifter and a two-stage Power Amplifier (PA). A two-stage PA design with a Class-C pre-driver and a 2nd-harmonic-tuned Class-AB driver stage is adopted for high gain and high efficiency at both peak and backed-off power levels. The active phase shifter employs in-phase/ quadrature phase current steering and digital control of transconductance (Gm). Measurement results show a 33-dB gain, 16.5-dBm saturated output power, 15.7-dBm oP_{1dB}, 27.5% peak PAE and 8.2% 7-dB back-off PAE at 27 GHz for a single element. The minimum (maximum) RMS gain and phase errors across the 27-29 GHz band were 0.5 dB (3 dB) and 1.5°(12°). The beamformer also includes a 1:4 power splitter and a serial interface for digital control and occupies a die area of 5.32mm^2.**

Index Terms—**phased array, 28-GHz, SiGe, millimeter-wave.**

I. INTRODUCTION

The ever-increasing demand for high speed mobile data has encouraged the use of millimeter-wave frequency bands for future cellular communication systems. In [1] and [2] the feasibility of a 1-Gbps data rate mobile millimeter-wave broadband(MMB) system using the LMDS bands around 28 GHz was studied. The uplink (handset to base-station) budget suggests the necessity of moderate beamforming, high output power, power efficient and linear transmitter circuitry. Although phased array transmitters have been demonstrated in Ka-band, there has been little effort on improving the efficiency of the overall system at backed-off power levels. This paper demonstrates a fully integrated 4-element array working in the 27-29 GHz band that achieves efficiency and linearity performance that is suitable for use in a mobile handset with a compact area.

We target an EIRP greater than 20 dBm for a 1-Gbps link at 500 m with a 4-element array assuming 0 dBi unit antenna gain. Considering 7-dB back-off for good linearity performance, this translates to a 16-dBm 1-dB compressed output power (oP_{1dB}) per element. To limit the total dc power consumption of the array to 320 mW, a front-end efficiency greater than 10% at 7-dB back off is required. To achieve the oP_{1dB} and PAE targets, we adopt a two stage PA design with a 2nd-harmonic-tuned Class-AB driver stage delivering a high saturated output power with good Power-Added Efficiency(PAE) and a class-C pre-driver stage improving the

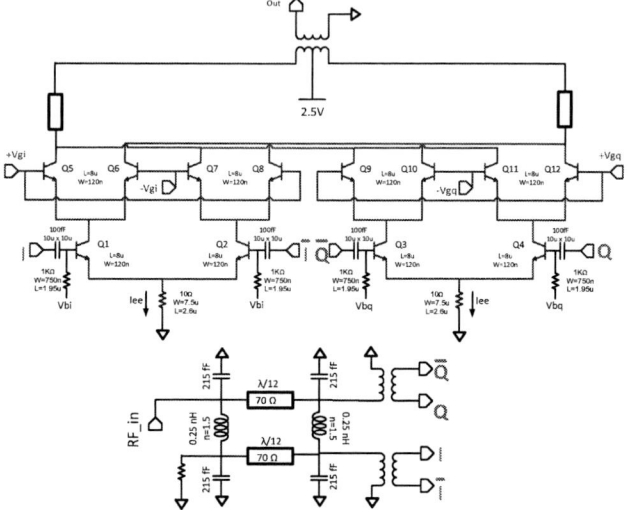

Fig. 1: Phase Shifter schematic.

linearity and gain of the PA while having a minimum effect on PAE. An active vector interpolator based topology is chosen for the phase shifter in order to achieve full 360° phase shift, low gain and phase errors, additional gain and high oP_{1dB} to linearly drive the PA. The next section presents the circuit design techniques adopted to achieve high performance. Section IV presents the measurement results followed by a summary.

II. CIRCUIT DESIGN

The beamformer has three stages: a power splitter, a phase shifter and a two-stage PA. The 1:4 power splitter consists of two levels of cascaded 1:2 Wilkinson power splitters. The quarter-wave resonators of the Wilkinson splitter are implemented as single section T-networks of two series MIM capacitors and one shunt spiral inductor for compact area and low loss.

A. Phase Shifter

The phase shifter is realized using a vector-interpolator topology, as shown in Fig. 1. In-phase and quadrature-phase RF signals are created using a lumped-element equivalent of a branchline coupler. The $\sqrt{2}Z_0$ quarter-wave lines have been realized with 70-Ω $\lambda/12$ series transmission lines and 115-fF shunt capacitors. The Z_0 quarter-wave lines have been realized

978-1-4799-7231-9/14 $31.00 © 2014 IEEE

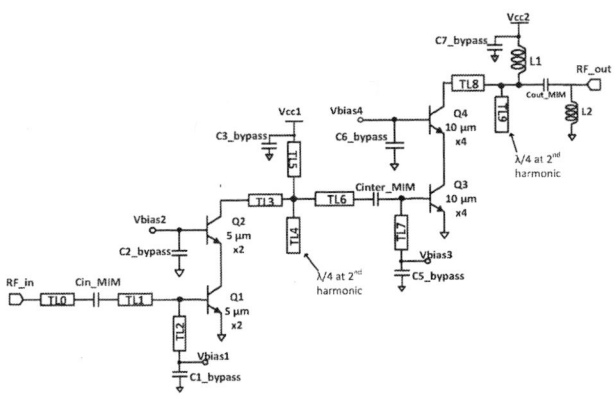

Fig. 2: Two stage PA schematic.

with 0.25-nH series inductors and 100-fF shunt capacitors. Differential I and Q signals are created using two single-turn transformers each with 150-μm diameter. The coupler was simulated using a combination of SONNET [13] and design-kit models, and showed less than 2.5° rms phase error and < 0.3-dB rms gain error across 27-29 GHz. The total area of the differential quadrature hybrid is 430x530-μm^2, and this area includes space for all of the active circuitry within the vector interpolator.

The active portion of the phase shifter uses a differential current-steered (or Gilbert-cell) design [3],[4]. A *tanh*-based current steering function is implemented using cross-coupled cascode devices. These cross-coupled devices are controlled through an inverse-*tanh* pre-distortion circuit driven by sin-weighted DACs targeting 4-bit accuracy. The input transconductance (Gm) is ideally fixed to provide constant impedance loading to the coupler; however, Gm can be compensated to calibrate the gain and phase response across phase settings and frequency if needed. Simulations for the phase shifter indicate 8.8-dB gain, 11-dB noise figure, and -6.6 dBm oP_{1dB}, while consuming 36 mW from 2.5-V supply.

B. 2-stage Power Amplifier

A simplified schematic of the two-stage PA is shown in Fig. 2. The driver stage is similar to the single-stage PA presented in [5] with improvements to the output match and base bypass network of the common-base transistor. A cascode amplifier is used with devices having an emitter length of 10 μm and 4 fingers biased in class-AB mode with an approximate conduction angle of 240°. The optimum harmonic terminations at fundamental $(70 + j25)\Omega$, 2nd harmonic $(8 - j60)\Omega$ and 3rd harmonic $(-j10)\Omega$ were determined using a multi-harmonic load pull as described in [5] and indicate a continuous mode class-B/J operation as the optimum one for maximum PAE. In this work the output matching network is designed to have the ability to tune the 2nd-harmonic impedance independently of the fundamental impedance. This is accomplished by using a 2nd-harmonic-quarter-wave open stub(TL9) [8]. The 2nd harmonic impedance can be tuned using the series microstrip line TL8; the fundamental can be tuned using the portion of the network to the right side of TL9. The fundamental and second harmonic impedances provided by the matching

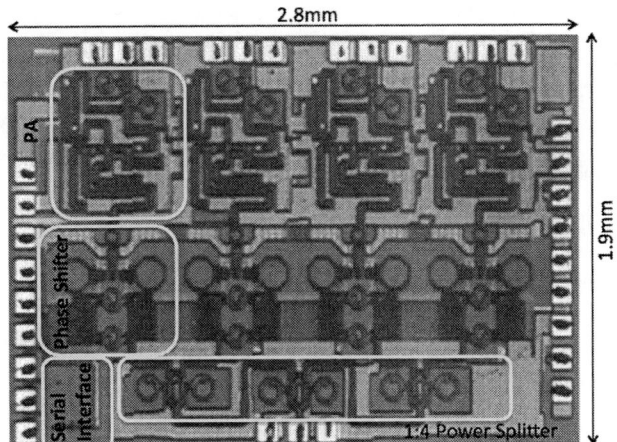

Fig. 3: Chip microphotograph.

network are $(76-j8)\Omega$ and $(40-j83)\Omega$ respectively. Note that the fundamental and second harmonic impedances provided by this network have a different relationship compared to the continuous modes discussed in [6]-[7] because both have a capacitive reactance part. This causes the voltage swing to move into the knee region which would theoretically result in loss of linearity, output power and efficiency[7]; however the efficiency and output power improve because the network has a higher quality factor compared to the one with inductive fundamental reactance.

A class-C pre-driver stage was added to improve the overall gain and linearity of the PA. The pre-driver stage is also a cascode amplifier with four times smaller devices, half the collector supply voltage compared to the driver stage and an approximate conduction angle of 170° to minimize dc power consumption and have less impact on overall efficiency. The class-C PA shows a tunable gain expansion characteristic versus swept input power which can be used to partially compensate for the gain compression of the driver-stage PA and improve the oP_{1dB}. The compensation is imperfect because the overall gain in the compensated region is not flat but shows some expansion/compression limited to +/-1dB. As such the PA would benefit from digital pre-distortion to improve its EVM in this region. The input of the pre-driver is conjugately matched to 50-Ω whereas the interstage matching network design between the pre-driver and the driver stage involves a trade-off between gain, bandwidth and PAE. We design the interstage matching network for maximum PAE at 28 GHz with the restriction that the gain variation in the 27-29 GHz band is limited to one decibel. The optimization of PAE while maintaining a certain gain flatness involved the simultaneous use of load pull contours and gain circles.

Simulation results of the PA show a 30-dB gain, 18-dBm saturated output power (P_{SAT}), 17-dBm oP_{1dB}, 37% peak PAE and 15% 7-dB-back-off PAE at 28 GHz. Simulation results of the array show a 33-dB gain, full 360° phase shift, RMS gain and phase errors less than 0.3 dB and 3° in the 27-29 GHz band for 4-bit phase shift, 17-dBm oP_{1dB}, 32% peak PAE and 12% 7-dB-back-off PAE.

978-1-4799-7231-9/14 $31.00 © 2014 IEEE

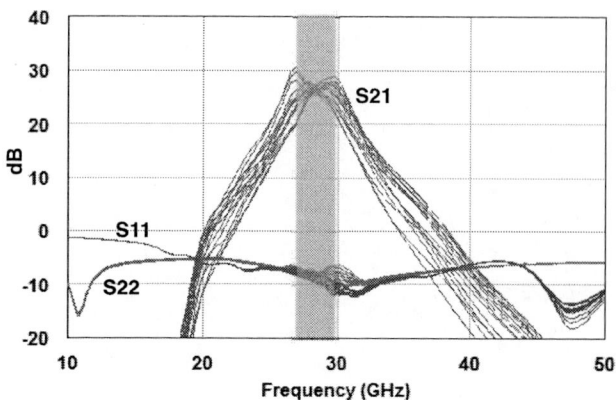

Fig. 4: Measured S-Parameters of a single beamformer element.

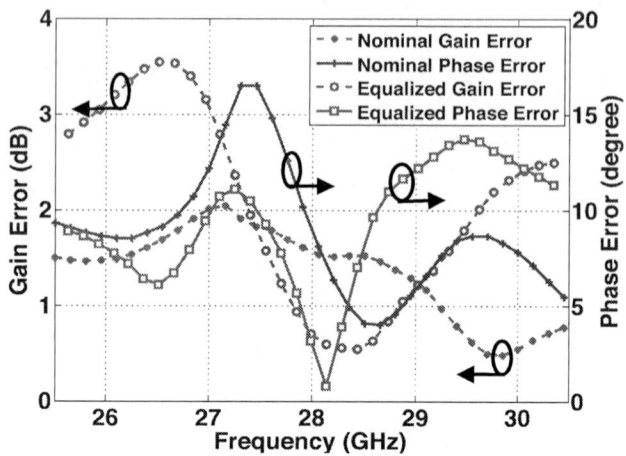

Fig. 6: Measured RMS gain and RMS phase errors for nominal and equalized cases.

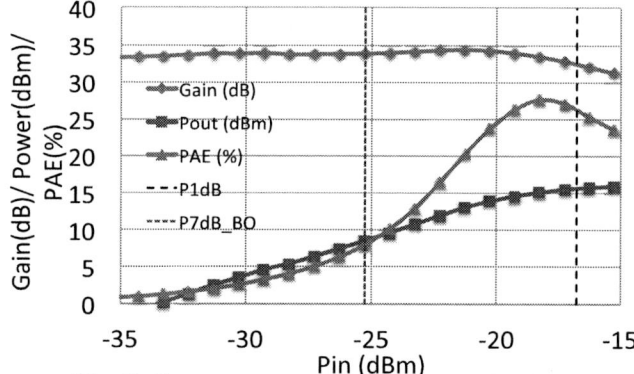

Fig. 7: Swept power measurements at 27 GHz.

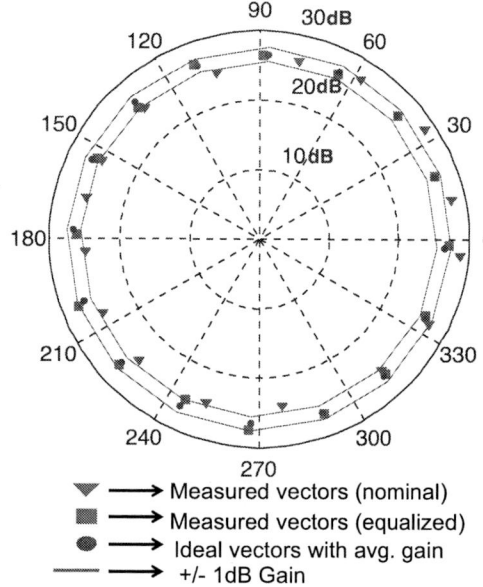

▼ ⟶ Measured vectors (nominal)
■ ⟶ Measured vectors (equalized)
● ⟶ Ideal vectors with avg. gain
— ⟶ +/- 1dB Gain

Fig. 5: Measured vectors of a single beamformer element at 28.14 GHz.

III. LAYOUT AND MEASUREMENT RESULTS

The design is implemented in IBM 120-nm SiGe BiC-MOS 8HP technology featuring NPN transistors with peak f_T/ f_{MAX} of 200/ 265 GHz, BV_{CEO} = 1.5 V, and BV_{CBO}= 5.5 V. The full chip meets electromigration current density requirements at 100°C. Passive elements used for matching networks were microstrip lines, MIM capacitors and spiral inductors. Additional loss was incorporated into the design kit models of microstrip lines and spiral inductors based on EM simulations. The base bypass capacitances at the common base transistors were realized using a combination of MOS capacitors and custom designed Metal-Oxide-Metal capacitors for low resistance and high capacitance density. The die photo is shown in Fig. 3. The chip measures 2.8mm by 1.9mm including RF and dc pads.

All measurements were performed with collector supply voltages of 2.5 V, 1.8 V and 3.6 V for the phase shifter,

pre-driver and driver stages respectively. S-parameter measurements of the beamformer showed a gain variation across frequency and phase setting, indicating a quadrature gain/phase error within the phase shifter. Revised EM simulations of the transformers used in the quadrature generation circuit in HFSS [14] indicated a nearly 3-dB gain difference between the differential components of both I and Q vectors. Using the additional control settings within the phase shifter Gm and current-steering DACs, it is possible to calibrate out the I/Q gain error at a particular frequency to obtain an equalized phase shifter gain and phase response. Fig.4 shows the s-parameters with these equalized phase-shifter settings for 28.14 GHz. The polar plot in Fig. 5. shows the measured vectors for nominal and equalized settings along with the ideal vectors at 28.14 GHz. RMS gain error and phase error across the band are presented in Fig. 6. The beamformer achieves a 27-dB end-to-end gain, full 360° phase shift, 3-dB bandwidth of 26-30 GHz, RMS gain error of 0.5-3 dB and RMS phase error of 1°-11.5° in the 27-29 GHz band. Measured gain variation across the four-elements of the array for a given phase and frequency was +/-3dB, which is due to a cross-chip variation of a supply voltage used within the bias circuit of the PAs.

Swept power measurement results of a single element at

27 GHz and 337.5° phase shift are shown in Fig. 7. The beamformer element achieves a 33.2-dB gain, 16.5-dBm P_{SAT}, 15.7-dBm oP_{1dB}, 27.5% peak PAE, 25.6% 1-dB-compression PAE and 8.2% 7-dB back-off PAE. Swept power performance across the band for a phase shift of 337.5° is summarized in Table I. P_{SAT} and oP_{1dB} show less than 1-dB variation, whereas PAE metrics show less than four percentage point variation. Measurements across the 16 phase settings for a fixed frequency show less than 1.5-dB variation in power metrics and less than four percentage point variation in PAE metrics. Measurements for 2 other chip samples show similar results.

IV. SUMMARY

The 4-element beamformer presented in this work has shown the capability to generate an EIRP greater than 21 dBm assuming 0 dBi unit antenna gain with a total dc power consumption less than 380 mW at 7-dB-back-off using a commercial SiGe BiCMOS technology. The circuit design techniques mentioned in Section II, namely, optimum harmonic terminations, analog pre-distortion using pre-driver stage enabled high performance. The efficiency and linearity achieved makes it an useful solution for mobile devices using MMB for Gbps communications. A performance comparison with some of the phased array results from different millimeter-wave frequency bands is shown in Table II. To the best of our knowledge, this beamformer achieves the highest peak and back-off PAE among the millimeter-wave beamformers/ phased arrays designed in Silicon-based technologies.

ACKNOWLEDGMENT

This work was partially funded by Samsung Electronics. The authors wish to thank F. Aryanfar, Z. Pi for helpful discussions and Y. Takeda, T. Fujibayashi for their assistance in layout and measurement.

REFERENCES

[1] Z. Pi and F. Khan, "An introduction to millimeter-wave mobile broadband systems," *IEEE Comm. Mag.*, June 2011, pp. 101-107.
[2] F. Khan, Z. Pi and S. Rajagopal, "Millimeter-wave Mobile Broadband with Large Scale Spatial Processing for 5G Mobile Communication," *Proc. 50th Annual Allerton Conference on Communication, Control, and Computing*, Oct. 2012, pp.1517-1523
[3] Ming-Da Tsai and A. Natarajan, "60GHz passive and active RF-path phase shifters in silicon," *IEEE RFIC Symp. Dig. Tech. Papers*, June 2009, pp.223-226
[4] Kwang-Jin Koh and G. M. Rebeiz, "0.13-μm CMOS Phase Shifters for X-, Ku-, and K-Band Phased Arrays"*IEEE Journal of Solid State Circuits*, Nov 2007, vol.42, no.11, pp.2535-2546
[5] A. Sarkar and B. Floyd, "A 28-GHz Class-J Power Amplifier with 18-dBm output power and 35% peak PAE in 120-nm SiGe BiCMOS," *IEEE SiRF Dig. Tech. Papers*, Jan. 2014
[6] Steve C. Cripps, "RF Power Amplifiers for Wireless Communications," 2nd ed. Boston, MA: Artech House, 2006.
[7] T. Canning, P. J. Tasker and S. C. Cripps, "Continuous Mode Power Amplifier Design Using Harmonic Clipping Contours: Theory and Practice,"*IEEE Transactions on Microwave Theory and Techniques*, Jan. 2014,vol.62, no.1, pp.100-110
[8] J. Moon, J. Lee, R. S. Pengelly, R. Baker, and B. Kim, "Highly Efficient Saturated Power Amplifier", *IEEE Microwave Magazine*, Jan.-Feb. 2012, vol.13, no.1, pp.125-131
[9] Kwang-Jin Koh, J. W. May, G. M. Rebeiz, "A Millimeter-Wave (40-45 GHz) 16-Element Phased-Array Transmitter in 0.18-μm SiGe BiCMOS Technology", *IEEE Journal of Solid State Circuits*, May 2009, vol.44, no.5, pp.1498-1509
[10] M. Tabesh et. al.,"A 65 nm CMOS 4-Element Sub-34 mW/Element 60 GHz Phased-Array Transceiver",*IEEE Journal of Solid State Circuits*, Dec 2011, vol.46, no.12, pp.3018-3032
[11] A. Valdes-Garcia et. al., "A 16-element, SiGe BiCMOS phased-array transmitter for 60-GHz communications",*IEEE Journal of Solid State Circuits*, Dec 2010, vol. 45, no. 12, pp. 2757-2773
[12] S. Y. Mortazavi et. al., "A Class F^{-1}/F 24-to-31GHz Power Amplifier with 40.7% Peak PAE, 15dBm OP1dB, and 50mW Psat in 0.13μm SiGe BiCMOS", *IEEE ISSCC Dig. Tech. Papers*, Feb. 2014
[13] Sonnet ver. 12.56, Sonnet Software Inc., Syracuse, NY, 1986-2011.
[14] Ansoft HFSS ver. 15, Ansoft Corporation, Pittsburgh, PA, 2013.

TABLE I: Performance of the beamformer across band

Frequency (GHz)	27	28	29
Gain (dB)	33.2	30.6	29.8
P_{sat} (dBm)	16.5	16.5	16.9
oP_{1dB} (dBm)	15.7	15.8	16.9
Peak PAE (%)	27.5	24.3	26.2
PAE 1-dB comp. (%)	25.6	22.3	26.1
PAE 7-dB-back-off (%)	8.2	8.5	10.3

TABLE II: Performance comparison with millimeter-wave beamformers

Reference	Frequency (GHz)	Gain (dB)	RMS Gain Error (dB)	RMS Phase Error (°)	oP_{1dB} per element (dBm)	Peak PAE (%)	PAE at 7-dB-back-off(%)	Technology
This work	27-29	33.2	0.5-3	1.5-11.5	15.7	27.5	8.2	SiGe 120nm
[9] Koh et. el.	40-45	12.5	<1.5	<8.8°	-3.5	-	-	SiGe 180nm
[10] Tabesh et. el.	57.9-65.6	-	-	-	-1.5	20	-	CMOS 65nm
[11] Valdes-Garcia et. el.	58-65	35	1dB	<5	15	8	-	SiGe 120nm
[12] Mortazavi et. el. (PA only)	26-30	9	-	-	15	40	5	SiGe 120nm

978-1-4799-7231-9/14 $31.00 © 2014 IEEE

A Fast Precision Operational Amplifier Featuring Two Separate Control Loops

Derek F. Bowers

Analog Devices Incorporated, San Jose, California 95134, USA

Abstract — **An operational amplifier is described which uses separate loops to control the output voltage and the error voltage between its inputs. To a large extent this architecture combines the high-speed characteristics of "current feedback" amplifiers with the low input referred errors of precision architectures.**
The technique has been applied to produce an amplifier with precision characteristics comparable to OP-07 type amplifiers but with over 100 times the bandwidth and slew-rate. Additionally, the power consumption has been reduced by a factor of three.

Index Terms — **Silicon bipolar, precision amplifier, operational amplifier, current feedback, analog circuits.**

I. INTRODUCTION

Self-contained monolithic precision bipolar op-amps have been around since the early 1970's and generally feature two or more stages of gain and a differential-pair input stage running at around 10μA or so per side to minimize input bias current. Such designs are not noted for having high speed characteristics – the popular OP-07 [1] has a gain-bandwidth product of 600kHz and a slew-rate of 0.2V/μS. Such amplifiers are often used in high closed-loop gain, where the low bandwidth becomes particularly troublesome.

The highest speed op-amps often use a "current feedback" architecture where the differential input pair is replaced by a high-speed buffer and the output is controlled by the supply currents of the buffer via (usually) a single gain stage. These amplifiers can have very high bandwidths (>1GHz) and slew-rates (>1000V/μS) but have generally inferior DC characteristics (and often high power consumption).

The amplifier presented here is not intended to compete with existing high speed amplifiers, but is a new architecture intended to bring an element of speed to precision amplifiers without significantly sacrificing the DC performance.

II. THE CURRENT FEEDBACK OP-AMP

Most "current feedback" op-amps are derived from the topology shown in fig.1.

Q1-Q4 form a buffer which keeps the voltage at the inverting input close to that at the non-inverting one. Any current flowing through the feedback resistor (Rfb) will flow in either Q3 or Q4 and end up charging the compensation capacitor (Cc) via one of the Wilson current mirrors at top and bottom. The

loop will stabilize when the current in Rfb is zero and the output voltage will satisfy the usual gain equations for op-amps [2].

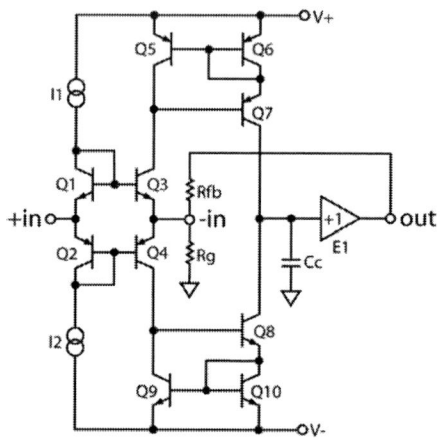

Fig.1. Basic Form of a "Current Feedback" Op-Amp

If the output impedance of the buffer were zero, under dynamic conditions all charging of Cc (and therefore all output voltage change) would result from the current through Rfb, which would form a time constant with the compensation capacitor. This time constant would determine the bandwidth (regardless of gain setting) and also imply no slew-rate limiting. In practice, when the current in Rfb is high, not all the current will reach Cc and slew limiting will occur, but at a much larger value than with conventional op-amps. Also, the finite output impedance of the buffer will form a potential divider with the gain setting resistor (Rg) and will cause bandwidth reduction at high gains; but once more at a much higher level than conventional amplifiers.

The major disadvantage of such amplifiers lies in the DC characteristics.

A multitude of effects including transistor mismatch, alpha errors and Early effect cause the difference in currents in Q7 & Q8 to be non-zero when the buffer current is zero. This is corrected by a current flowing in Rfb which appears at the output as an offset voltage (independent of gain). This can be substantial and also a significant function of temperature.

Additionally, offset errors in the buffer due to Early voltage and alpha errors appear as an input offset term which becomes multiplied by the closed loop gain before appearing at the output.

Another problem is that the gain of the input buffer is rarely more than 0.999 or so, and this limits CMRR to about 60dB.

Other deficiencies if the "current-feedback" amplifier include high input and supply currents, though to be fair this is mostly due to the quest for highest possible speed and not necessarily a drawback of the architecture,

III. A BALANCED CURRENT FEEDBACK TOPOLOGY

The idea behind the design presented here is to find a "current feedback" topology which is inherently balanced with respect to static currents to reduce output referred errors to a minimum. The input buffer then forms part of a precision voltage feedback amplifier to improve its offset and gain error.

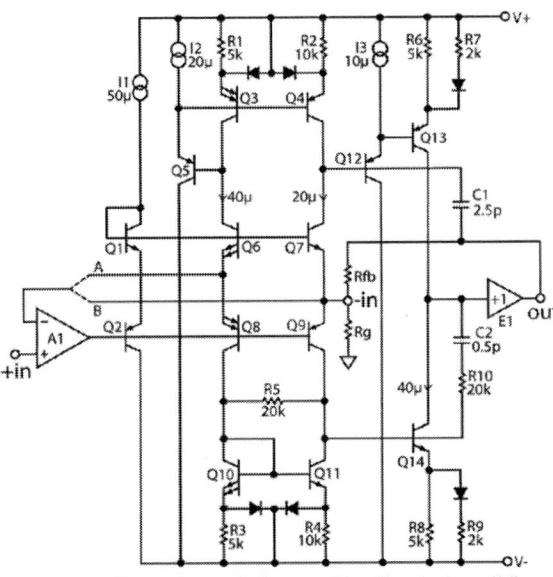

Fig.2. Fully-Balanced Current Feedback Amplifier Topology

Referring to fig.2, in this case the output stage of the input buffer consists of Q1, Q2, Q7 & Q9. In parallel with the buffer output transistors (Q7 & Q9), Q6 & Q8 have been added and have twice the emitter area, so the nominal currents in them under DC conditions will be exactly double[1] of those in the buffer output transistors. Q3-Q5 form a current mirror with a very accurate gain of 0.5, yielding nominally equal currents in Q4 & Q7 at balance. Any imbalance in these currents is sensed by a second gain stage consisting of Q12 & Q13 which drive the overall output away from zero to correct this. The combination of these two gain stages is intrinsically balanced so the only net output offset is due to random mismatches in components;

[1] The reason for choosing the current to be double on one side is that it improves the noise performance without adding additional DC errors.

mainly Q6-Q9 which are carefully laid out to minimize this effect. Additionally, laser trims on R1 & R2 are further used to minimize this error current, and thus also minimize the output offset voltage.

Superficially, the circuitry below Q6 & Q7 looks similar to the gain stage above it, but the presence of R5 limits the gain of this stage so that for small perturbations Q14 approximates a current source.

Loop feedback ensures that under DC conditions the current in Q13 is also equal to this. Under slewing conditions, sufficient voltage is developed across R5 to allow the current in Q14 to rise to whatever is necessary to support the resulting slew-rate.

The result is that like the "current-feedback" op-amp of fig.1, the slew currents in the compensation capacitors C1 & C2 are derived from the current in Rfb (nominally 10kΩ in this design), and can be much larger than any of the quiescent stage currents.

III. THE INPUT BUFFER

The amplifier inverting input (which is also the output of the input buffer – point B) is forced to equal that of the non-inverting one by A1, a precision voltage feedback op-amp. It is desirable to force this directly by closing the loop from point B directly back to A1. Such a connection would result in minimum noise and offset. This causes an interesting AC problem however, since A1's frequency compensation forces the output of the buffer to appear inductive at high frequencies. At low gains, this inductance is swamped by Rg, but interaction between this inductance and the frequency response of the main amplifier results in instability at high closed-loop gains, a very bad feature for a precision op-amp. An alternative feedback connection is from the emitter of Q6 (point A) since it settles out to the same voltage as point B but has no direct interaction with Rfb. Unfortunately, this now places the output impedance of the input buffer outside the feedback loop of A1, which causes severe bandwidth loss above gains of 20. It also increases the input referred noise since the noise contributions of Q6-Q9 are no longer inside the loop either.

Given a choice of feedback connections which alternately produce high-gain peaking or roll-off, this suggests that there should be some intermediate point where the response is optimum. This could be implemented as a tap point on a resistive divider stretched between points A & B, but a low overall divider resistance would severely reduce the gain of the current feedback stage and a high resistance would add greatly to the input referred noise. In practice, there is no useful compromise here. Fig.3 shows details of the correction amplifier with an alternative solution to splitting the feedback.

Q21-Q24 form the input stage with feedback via both Q23 & Q24. Because Q23 is twice the size of Q24,

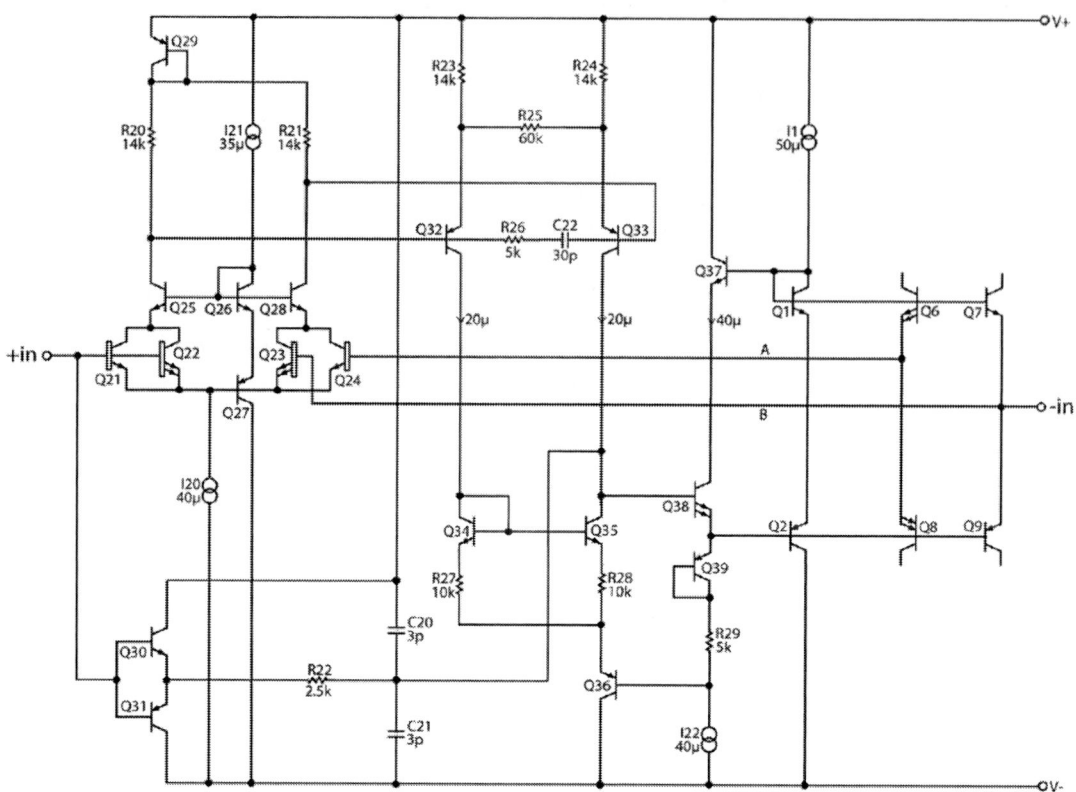

Fig.3. Details of the Closed-Loop Input Buffer Showing the Interface to the Overall Amplifier

one-third of the feedback is taken from point A and the other two-thirds from point B. There is nothing fundamental about this apportionment, but simulation indicated that this was a good compromise between maximum bandwidth and minimum peaking over the full gain range. Symmetry is restored to the differential pair by making the non-inverting side the same, but with both transistors in parallel. All four transistors are "superbeta" types', resulting in an input bias current of around 10nA. This is further reduced to about 2nA by a cancellation circuit (not shown). The input stage currents pass trough a differential current mirror, and are reflected into the active load formed by Q34 & Q35. This load is bootstrapped by Q36 to achieve high gain and CMRR from the single gain stage, which is then double-buffered by Q38 and Q9 before arriving at the inverting input.

The input offset is adjusted by laser trims on R20 & R21, and the use of proportional-to-temperature biasing throughout ensures that drift is minimized by the offset trim.

Dominant-pole compensation is afforded by C20 & C21 (split into two for PSRR reasons) and the resulting slew-rate of the buffer is about 6Vμ/S. This is effectively multiplied by the closed-loop gain by the time it reaches the output, and thus would not limit the overall slew-rate in (non-inverting) gains above eight. At balance, the compensation node and the input are at approximately the same potential, so a simple feedforward network consisting of Q30, Q31 & R22

can be used to precharge the capacitors to within 0.6V or so of final value at the lower gains. This maintains a high overall slew-rate (50V/μS) over the full gain range.

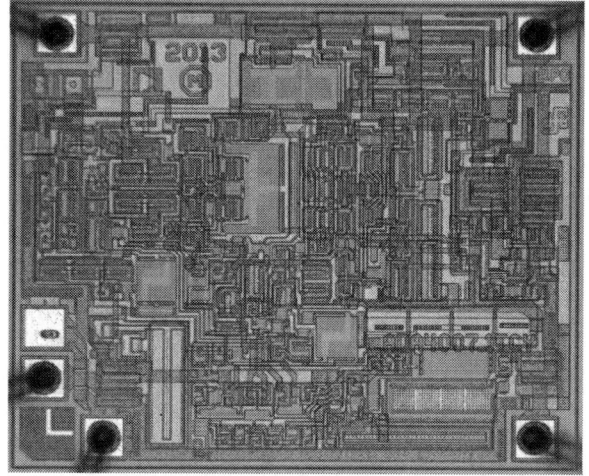

Fig.4. Photomicrograph of the Amplifier

IV. PROCESS

The process used is a 36 volt oxide-isolated complementary bipolar process including superbeta NPN transistors and 2kΩ/□ CrSiC laser-trimmable

thin-film resistors. Two levels of Al metallization were used.
A photomicrograph of the 0.99mm² die is shown in fig.4.

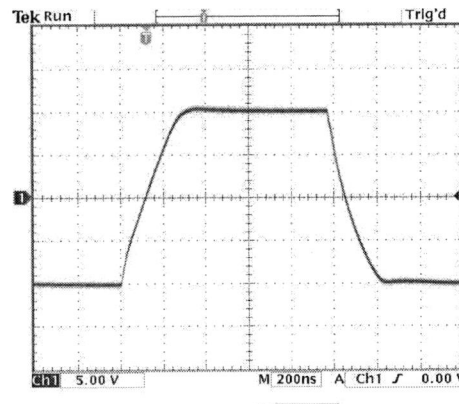

Fig.6. Gain of Ten Response to a 1V Square Wave at 500kHz

V. PERFORMANCE

The performance of the amplifier is summarized in table 1.

It can be seen that the only major parametric sacrifice over a conventional precision op-amp is the bias current of the inverting input. With the (recommended) 10kΩ feedback resistor this causes output referred errors to become dominant as the gain is reduced below ten or so, but this is an application area where conventional op-amps are probably a better choice. Fig.5 & fig.6 signal show the small and large signal transient responses respectively.

Parameter	This Work (Rfb=10kΩ)	OP-07
V_{os}	50μV	<150μV
I_b+	2nA	<7nA
I_b-	40nA	<7nA
CMRR V_{cm}=±10V	126dB	120dB
PSRR ±5V to ±15V	120dB	103dB
Voltage Gain	130dB	112dB
-3dB B.W. G=1	>10MHz	1.2MHz
-3dB B.W. G=10	4MHz	70kHz
-3dB B.W. G=100	1.8MHz	6kHz
-3dB B.W. G=1000	165kHz	600Hz
Slew Rate	50V/μS	0.2V/μS
Input Refd. Noise	8nV/√Hz	10.5nV/√Hz
Output Refd. Noise	57nV/√Hz	N/A
Supply Current	0.8mA	2.5mA

Table 1. Summary of the overall amplifier performance compared to a typical OP-07.

Fig.7 shows the amplitude response at gain settings of 1 to 1000. The feedback resistor was 10kΩ in all cases. The peaking at low gains is caused by stray capacitance at the inverting input rather than any deficiency in the phase characteristics. This could probably be improved with a better test jig or different package, but this is where a conventional op-amp is at a distinct advantage, since a small capacitor can be placed across the feedback resistor to neutralize this effect. Of course, at unity gain, a conventional op amp would not need the feedback resistor at all.

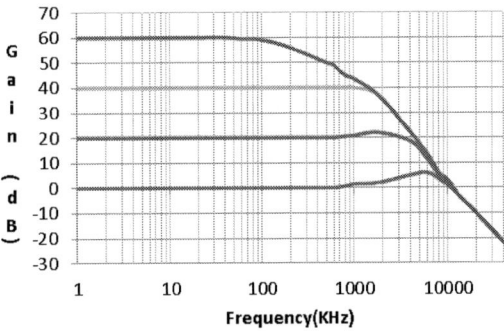

Fig.7. Small-Signal Amplitude Response at Gains of 1, 10, 100 & 1000

ACKNOWLEDGEMENT

The author wishes to thank Peter Ohlon and Lien Quach for device characterization and Wendy Huang for the mask layout. Special thanks are also due to Paul Henneuse for his enthusiasm concerning the project.

REFERENCES

[1] Donn Soderquist & George Erdi, "The OP-07 Ultra-Low Offset Voltage Op Amp" Precision Monolithics Application Note AN-13, December 1975.

[2] "A new approach to op amp design", Comlinear Corporation Application Note 300-1, March 1985.

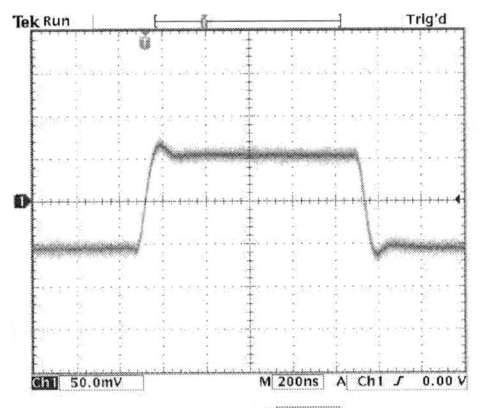

Fig.5. Unity-Gain Response to a 100mV Square Wave at 500kHz

A BiCMOS 50 MHz Input Bandwidth, 1-to-16 Channelizer Optimized for Low Power Analog Signal Classification

Hao Li[1,2], Chris M. Thomas[3], Gert Cauwenberghs[3], and Lawrence E. Larson[1]

[1]Brown University, Providence, RI
[2]Marvell Semiconductor Inc, Santa Clara, CA
[3]University of California, San Diego La Jolla, CA

Abstract- **A base-band channelizer, fabricated in a 0.18um BiCMOS process, capable of separating an input 50 MHz bandwidth signal into 16 output 3.125 MHz bandwidth signals intended for multi-channel independent component analysis (ICA) for signal classification is presented. The channelizer operates in a purely analog fashion to interface with an analog ICA system to optimize energy efficiency. Each channel within the 50 MHz bandwidth is up/down converted using active mixers and is low-pass filtered. Harmonic rejection mixers and clock predistortion techniques are employed in order to eliminate clock harmonic folding. A two-stage, quarter-band separation architecture is employed to split the 50 MHz baseband signal into 16 consecutive channels simultaneously. The measured input signal HD3 and LO 3rd-order harmonic mixing is below -40 dBc. The measured power consumption is 71.5mW from a 2.5V power supply.**

I. INTRODUCTION

Future cognitive radio systems are faced with the challenge of sensing a wide bandwidth spectrum (up to 50 MHz) over a broad frequency range for dynamic spectrum access and spectrum signal classification. Radios with dynamic spectrum access capabilities are able to quickly identify spectrum usage and efficiently access unused frequencies for transmission. Beyond smart frequency allocation, signal classification may be of interest for cases like spectrum surveillance.

One approach toward a signal classification receiver is shown in Fig. 1. An RF front-end with tunable band-select filtering can be employed to select the signal band of interest and down-convert the signal to baseband by a quadrature RF mixer. The down-converted baseband signal is further split into 16 consecutive channels, each centered at DC by the proposed channelizer. Finally, all channels are fed to a multi-channel independent component analysis (ICA) processing circuit, which senses and recognizes if there are certain modulated signal patterns transmitting in this specific band [1].

Instead of processing a digitized signal in the discrete-time domain, the ICA processor conducts all of the correlation computing in the continuous-time domain. Therefore, the channelizer has to perform the channel separation in a purely analog fashion in order to achieve better energy efficiency, compared to alternative digital approaches that would be composed of two high-speed analog-to-digital converters (ADC), banks of digital signal processing units, and dedicated digital-to-analog converters (DAC) for each channel.

In this paper, an optimized analog channelizer architecture based on a quarter-band separation scheme is proposed for better linearity and signal-to-noise ratio (SNR) performance. The two-stage quarter-band separation architecture splits a 50 MHz bandwidth baseband signal (I/Q) into 16 channels with 3.125 MHz bandwidth (I/Q) centered at DC. A prototype of the channelizer was fabricated in 0.18-μm SiGe BiCMOS process.

II. SYSTEM ARCHITECTURE

The proposed quarter-band separation scheme is illustrated in Fig. 2. The complex baseband signal with a bandwidth of $2f_c$ is divided into four consecutive segments in the frequency domain as illustrated in Fig 2. In order to split the four

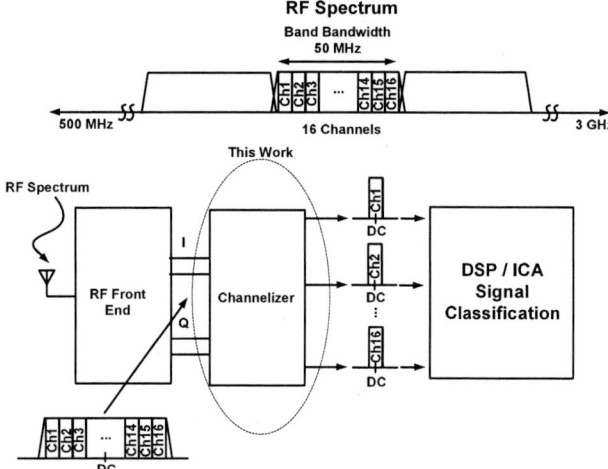

Figure 1. System architecture of a low-energy signal recognition sensor for wideband signal classification.

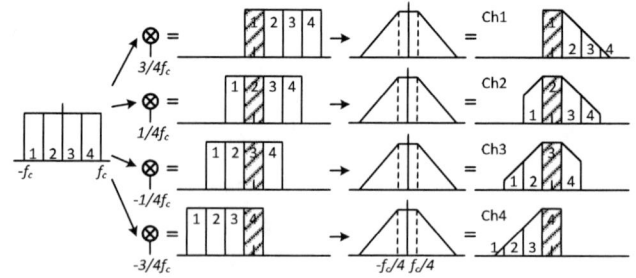

Figure 2. Illustration of quarter-band separation scheme.

978-1-4799-7231-9/14 $31.00 © 2014 IEEE

segments into four DC-centered channels, quadrature up-conversion and down-conversion mixers are adopted to move the segment of interest to DC, and low pass filters (LPF) with a cut-off frequency of $\frac{1}{4}f_c$ are used to reject the signals that are out-of-band. For segment 2 and segment 3, the local oscillator (LO) frequency of the up-conversion and down-conversion mixer is $\frac{1}{4}f_c$. While for segment 1 and segment 4, the LO frequency is $\frac{3}{4}f_c$. Additional mixing stages can be cascaded to further split each segment into more channels. Given the stage number n, the channelizer can generate 4^n consecutive channels in total.

The choice of the separation scheme and the number of stages allows a trade-off between LO harmonic folding frequencies needing to be appropriately filtered and the total number of output channels. For example, half-band separation, i.e. each stage splitting the input into 2 segments, requires 4 stages to produce 16 output channels, while quarter-band separation requires fewer stages, leading to less power consumption, less additive noise, and less in-band fluctuation due to non-ideal frequency response in the LPFs. Compared to 1/8-band separation, in which more LO frequency references ($1/8f_c$, $3/8f_c$, $5/8f_c$ and $7/8f_c$) are required and more LO harmonics appear in-band causing LO harmonic folding, quarter-band separation has a simpler LO generation and higher immunity to LO harmonic folding (3rd-order harmonic response only).

The proposed channelizer splits a 50 MHz-wide baseband signal into 16 channels based on a two-stage, quarter-band separation architecture. The fundamental circuit building block of the channelizer, the SCU (Signal Conditioning Unit), translates a complex baseband input signal up (for an input on the left side of DC) and down (for an input on the right side of DC) together with low-pass filtering and is shown in Fig. 3. Depending on the clock LO frequency, SCUs are categorized into two types: SCUa's LO reference is $1/4f_c$ (6.25 MHz for the 1st-stage and 1.5625 MHz for the 2nd-stage) while SCUb's LO reference is $3/4f_c$ (18.75 MHz for the 1st-stage and 4.6875 MHz for the 2nd-stage). The detailed diagram of the channelizer is shown in Fig. 4. The first stage, comprising of one SCUa block and one SCUb block, splits the 50 MHz band into four 12.5 MHz-wide segments, while the second stage further splits each of the segments into four 3.125 MHz-wide channels, i.e. 16 channels in total.

III. CIRCUIT IMPLEMENTATION

The channelizer is designed to receive an input signal up to 1 volt peak-to-peak fully differential and to have a programmable gain range from 0 dB to 40 dB. Moreover, the signal distortion introduced by the channelizer, including input signal harmonics and LO harmonic folding, should be below -40dBc. The circuit implementation of each of the key building blocks will be discussed in detail in the following sections.

A. Harmonic Rejection Mixer

The mixer in the SCU has stringent linearity requirements for both signal harmonic distortion and LO harmonic folding. Additionally, due to the large quantity of mixers (8 for the 1st stage, and 32 for the 2nd stage), the power consumption budget is critical. Analog multipliers, such as that in [2], usually have good linearity, but consume larger power. The Gilbert cell mixer is usually linear for input signals, but suffers from LO harmonic mixing. Therefore, an active mixer with harmonic rejection (HRM) is employed [3], which inherently eliminates the 3rd-order and 5th-order harmonic folding.

The simplified schematic of the HRM is shown in Fig. 5. Three sets of LO with 45 degree phase offset are employed for proper LO harmonic folding rejection. Input transconductance

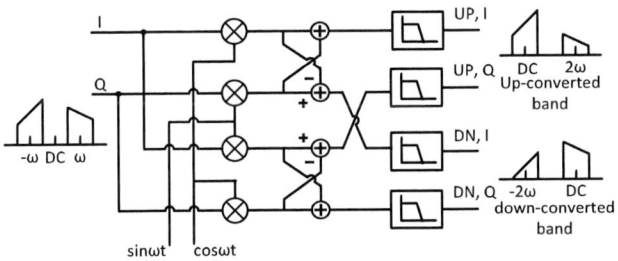

Figure 3. Diagram of signal conditioning unit (SCU).

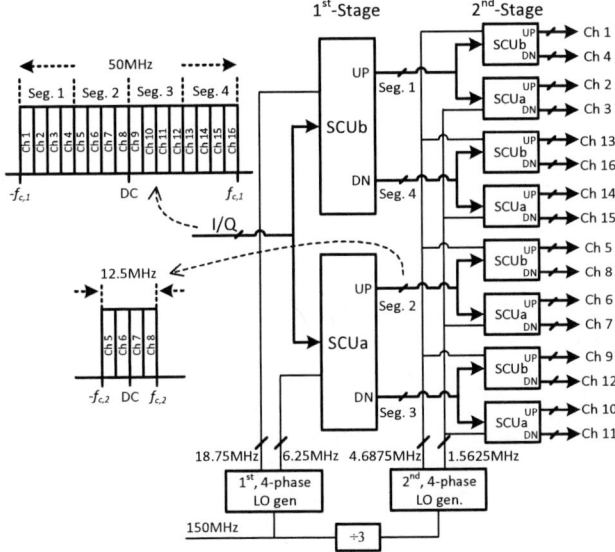

Figure 4. Diagram of the channelizer with 16 channel outputs based on 2-stage, quarter-band separation scheme.

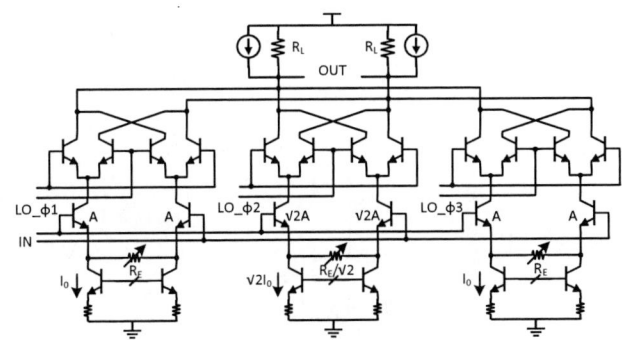

Figure 5. Schematic of active mixer with harmonic rejection.

978-1-4799-7231-9/14 $31.00 © 2014 IEEE

linearity is achieved by resistive degeneration. In addition, the degeneration resistance is tunable with digitally controlled switches (4 control bits), providing variable gain from 0 dB to 20 dB in each SCU, therefore 40 dB variable gain in total for two stages with 5 dB per step. The summation or subtraction in the SCU is realized by connecting the collectors of the quad-transistors from two quadrature mixers to the same resistive load in-phase or out-of-phase. PMOS current sources in parallel with the load resistor are employed as bleeding current to help improve the voltage headroom at the collector node of quad-transistors. The simulated power consumption is 1.5mW for the 1st stage mixer and 0.75mW for the 2nd stage mixer, both from a 2.5V supply.

B. Low-Pass Filter

Since constant group delay is a desired feature for ICA, a Bessel filter is adopted to form the LPF shown in Fig. 6. A 3rd-order Bessel low-pass filter based on a Sallen-Key structure [4] is implemented, in which a source follower is used as a unity-gain amplifier, as well as a DC-level shifter to match the input common-mode voltage of the following circuits. Resistor R_1 forms the first pole of this filter and is merged with the mixer loading resistor R_L in Fig. 5 to minimize the noise budget. The cut-off frequencies for the 1st and the 2nd stage are 6.25 MHz and 1.5625 MHz, respectively. In order to overcome process variation, switchable capacitors are employed to tune the filter cut-off frequency. The capacitance tuning range, controlled by 3 digital bits, is from 80% to 115% of default setting with 5% per LSB.

C. LO Generation and Predistortion

Although the HRM can ideally eliminate the 3rd-order and 5th-order harmonic folding, the higher 7th-order and 9th-order harmonic folding may still be significant if an LO signal with a square waveform is used. Theoretically, for a square wave LO, the HD7 and HD9 are -17dBc and -19.3dBc, respectively. To achieve better high-order LO harmonic rejection, the mixer quad-transistor operates in forward-active region instead of the on/off switching mode. Since the V-I curve for differential bipolar transistors follows a hyperbolic tangent function for large input swings, an LO predistortion circuit performing inverse hyperbolic tangent (IHT) I-V conversion will preserve the linear operation (I-I) when these two are cascaded [5]. The LO predistortion circuit shown in Fig. 7 is based on a triangle wave to sine wave conversion circuit [6], with the original

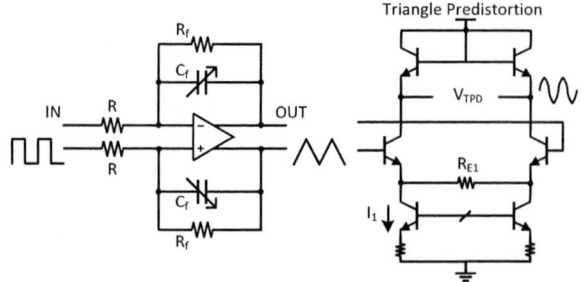

Figure 7. Schematic of LO generation and predistortion circuits.

resistive load replaced by bipolar transistor for IHT conversion. The input triangle amplitude and degeneration resistor R_{E1} are selected to minimize the HD7 and HD9 instead of the HD3 and HD5 in the sine wave conversion case, since HD3 and HD5 will be eliminated by the harmonic rejection mixer. An LO with a strong HD3 and HD5 component also provides higher conversion gain as well as less noise from the quad-transistors in the mixer, compared to a pure sine wave LO.

The triangle waveform is generated by an integrator with rail-to-rail square wave inputs. The integrator employs a switchable capacitor array for cut-off frequency tuning to overcome process variation and to adjust the triangle wave amplitude for better higher order harmonic rejection.

D. System Integration

The entire channelizer comprises 8 active mixers, 8 LPFs, and 8 LO predistorters for the first stage and 32 active mixers, 32 LPFs, and 8 LO predistorters for the second stage. The power budget for each SCU in the second stages is scaled down by a factor of 2, compared to that of the first stage, to optimize the total power consumption. A 150 MHz clock is required off-chip to generate the 4-phase LO signals of 6.25MHz/18.75MHz for the first stage and 1.5625MHz/4.6875MHz for the second stage by a series of dividers. Output buffers with a separate power supply are implemented to drive off-chip testing equipment.

IV. MEASUREMENT RESULTS

The prototype chip was fabricated in an IBM 0.18-um SiGe BiCMOS process and wire-bonded in a QFN package for PCB testing. The die micrograph is shown in Fig. 8 and the measured die size is 4.5mm x 3.7mm, including bonding pads. The measured power consumption is 71.5 mW from a 2.5 V power supply, excluding the output buffers for testing purposes.

A baseband I/Q vector signal generator was used to generate a complex input signal and a multi-channel oscilloscope FFT was used to acquire the down-converted channel. The measured input to output gain transfer function of all 16 channels is shown in Fig. 9.

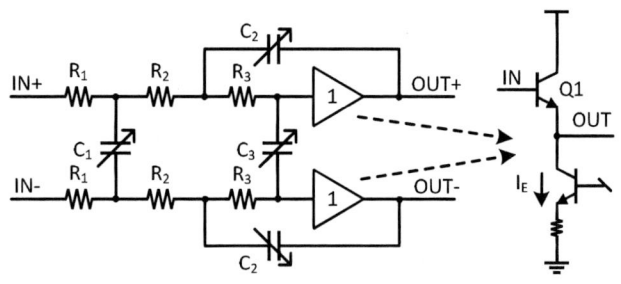

Figure 6. Schematic of 3rd-order Bessel low-pass filter.

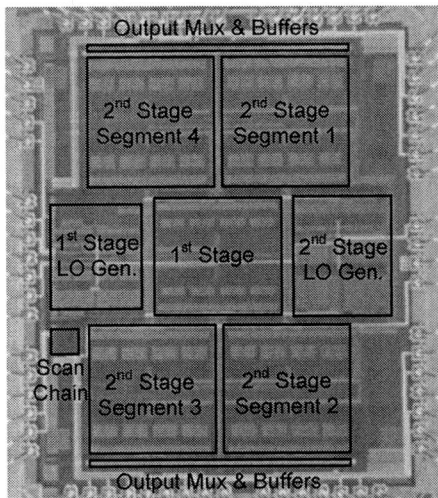

Figure 8. Die micrograph (4.5mm x 3.7mm).

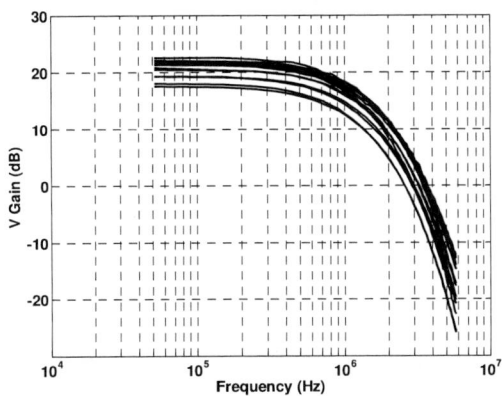

Figure 9. Channelizer gain transfer function of all 16 channels at 20 dB gain setting.

The third-order distortion measurement was performed by inputting a two-tone signal with 50 kHz spacing in the frequency region of channel 6 with a gain setting of 20 dB. The measured magnitude of the fundamental and the third order intermodulation tones at the output is shown in Fig. 10.

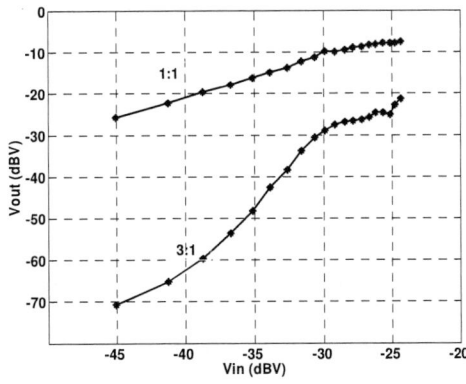

Figure 10. Measured 3rd-order inter-modulation with two-tone input and 20 dB gain setting.

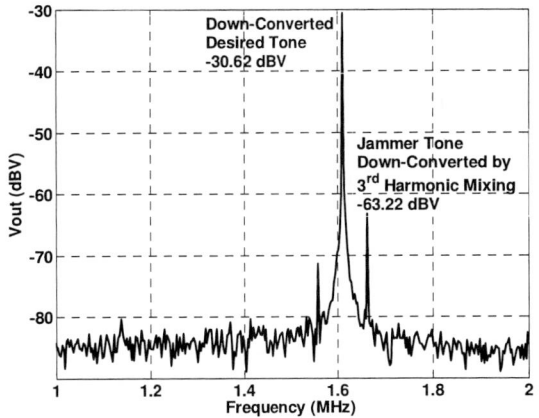

Figure 11. Measured 3rd-order harmonic mixing. Input jammer tone is 14 dB higher than the input in-band tone.

The IMD3 was better than -40dBc at -20dBV output.

The 3rd-order harmonic folding of the first stage was measured by a two-tone test with one tone placed in-band in channel 6 and the second jammer tone placed at the third harmonic of the clock frequency and 14 dB higher than the in-band tone. The measured down-converted jammer tone was 32.6 dB lower than the in-band tone, giving the total rejection of 3rd-order harmonic folding to be better than 46 dBc, as shown in Fig. 11.

V. CONCLUSION

In this paper, a 50 MHz input bandwidth, 1-to-16 BiCMOS channelizer designed for low power analog signal classification is presented. The optimized quarter-band separation architecture and the harmonic reject mixer with LO predistortion circuits achieve 3rd harmonic distortion of better than -40dBc with a low power consumption of 71.5 mW.

ACKNOWLEDGMENTS

This work was supported by the DARPA CLASIC program under Dr. William Chappel and Leidos under Dr. Dennis Braunreiter.

REFERENCES

[1] H. Agirman-Tosun, et al., "Modulation classification of MIMO-OFDM signals by independent component analysis and support vector machines," *IEEE 45th Asilomar Conference on Signals, Systems and Computers (ASILOMAR),* Nov, 2011, Pacific Grove, CA

[2] J. Weldon, et al., "A 1.75GHz highly-integrated narrow-band CMOS transmitter with harmonic-rejection mixers," *IEEE ISSCC Digest of Technical Papers,* pp 160-161, February, 2001.

[3] M. Franciotta, G. Colli and R. Castello, "A 100-MHz 4-mW four-quadrant BiCMOS analog multiplier," *IEEE Journal of Solid-State Circuits,* vol. 32, no. 10, pp 1568-1572, October, 1997.

[4] R. P. Sallen and E. L. Key, "A practical method for designing RC active filters," *IRE Transactions Circuit Theory,* vol. CT-2 , pp. 74–85, Mar.1955.

[5] P. R. Gray, P. J. Hurst, S. H. Lewis, and R. G. Meyer, *Analysis and Design of Analog Integrated Circuits,* 4th edition, New York: Wiley, page 714, 2001.

[6] R. Meyer, W. M. C. Sansen, S. Lui and S. Peeters, "The differential pair as a triangle-sine wave converter," *IEEE Journal of Solid-State Circuits,* vol. 11, issue 3, pp 418-420, June, 1976.

Linear low-power 13 GHz SiGe-Bipolar Modulator Driver with 7 V$_{pp}$ differential Output Voltage Swing and on-Chip Bias Tee

Horst Hettrich[*], Michael Möller[*†]

[*]Chair of Electronics and Circuits Saarland University, Saarbrücken, Germany
[†]MICRAM Microelectronic GmbH, Bochum, Germany

Abstract—**A linear driver amplifier IC with 7 V$_{pp}$ differential output voltage swing in SiGe Bipolar Technology is presented. The driver includes a novel on-chip bias tee, which allows for energy savings in applications with AC coupled loads compared to a reference implementation without bias tee. In addition to a low-power design, the driver was optimized regarding mutual thermal coupling between the transistors, stability and avalanche breakdown. The linearity of the driver is well suited for the intended application in Ultra Dense Wavelength Division Multiplex (UDWDM) systems, where the third harmonic of the differential output must be attenuated by at least 25 dB (relative to the fundamental amplitude) within the complete operating frequency range from 1.3 GHz to 13 GHz. The complete driver chip draws 250 mA from a 6 V power supply.**

I. Introduction

Driver designs which consider a DC-coupling to a resistive load exhibit a bias level at the output node as high as the average of min. and max. output voltage level (cf. Fig. 1a). In some applications the adjustment of the load biasing requires a higher degree of flexibility. In the target application this demand is related to the modulation scheme dependent biasing of a Mach Zehnder Modulator (MZM). This can be achieved e.g. by AC-coupling of the load to the driver and applying an external bias voltage to the load (cf. Fig. 1b). In this case, the termination resistor of the driver needs to be designed for the full (instead of the half of the) DC current I_0, which increases the parasitic capacitance and therefore decreases the bandwidth. Furthermore, as the voltage drop across the output resistor has doubled (e.g. from 1.75 V to 3.5 V) the supply voltage has to be raised by the same amount, which leads to higher power consumption. By using a bias tee configuration (cf. Fig. 1c) instead of the blocking capacitor, the output bias level is raised by half of the output voltage swing (i.e. no DC voltage drop across the output resistor) and the supply voltage can be lowered. Therefore, using a bias tee configuration is the most energy efficient way to connect an AC-coupled load to the driver circuit. In order to minimize cost and to improve the reliability, the bias tee is realized fully integrated on-chip for the present approach.

The paper is organized as follows: In Sec. II the basic structure of the driver as well as general design considerations are presented. Sec. III introduces the novel on-chip bias tee concept. Sec. IV presents a novel method to optimize the output stage of the driver regarding mutual thermal coupling

of the transistors in order to avoid current hogging. Stability considerations concerning this output stage are presented in Sec. V. Measurement results are given in Sec. VI and conclusions in Sec. VII.

II. General Circuit Design Considerations

The principal schematic of the driver, intended to be used in an Ultra Dense Wavelength Division Multiplex (UDWDM) system with up to 1000 channels [1], is shown in Fig. 2. In this system for each wavelength up to 10 DQPSK modulated data channels are modulated onto different RF carriers, which results in an analog signal with a total spectrum ranging from 1.3 GHz to 13 GHz per wavelength. This signal is created by a digital-to-analog-converter with a maximum differential output voltage swing of 800 mV$_{pp}$ [2]. Since the Mach-Zehnder-Modulator (MZM) which creates the optical signal requires an input voltage swing of 7 V$_{pp}$, the driver amplifier presented here is necessary. The required gain of ≈ 19 dB can be realized by using one single amplifier cell. Hence, no additional input stage is required, which keeps the circuit simple and saves power. For this application, the driver amplifier transfer function has to be linear in order to suppress inter-modulation between the data channels.

As already the output load is AC-coupled, also the driver input is AC-coupled by the on-chip capacitor $C_k = 1$ pF, cf. Fig. 2. The driver inputs are terminated by the resistors $R_1 = 50 \Omega$. The differential input voltage divider (R_2, R_3) is used for adjustment of the bias voltage level of the emitter followers (EF), Q_{EF1}, Q_{EF2}, and the transadmittance stage Q_{TAS}. This kind of bias adjustment is preferred to level shift by additional diodes in the signal path, which would increase the power consumption and cause an additional drop of 3 dB at 13 GHz in the small signal voltage gain.

The driver chip is operated with a negative supply voltage of -6 V. The actual amplifier stage consists of a termination resistor $R_1 = 50 \Omega$ followed by two pairs of EF Q_{EF1}, Q_{EF2}, at the input. Diodes in series to the collectors lower V_{CE} and thereby reduce the avalanche effect of the EF transistors. The second EF drives a linearized (R_4) transadmittance stage (TAS), Q_{TAS}, followed by a common base stage (CBS), Q_{CBS}, together forming a cascode, which drastically reduces Miller effect on the TAS-transistor, prevents it from breakdown and lowers the power dissipation and thus the degradation of speed

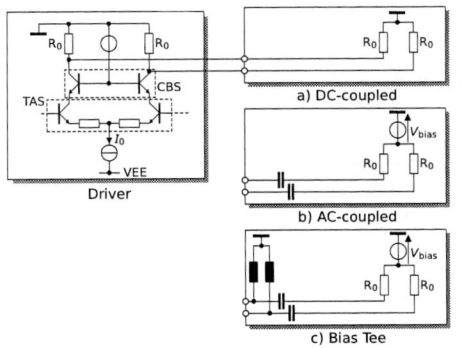

Fig. 1: State of the Art Biasing Options of the Output Stage

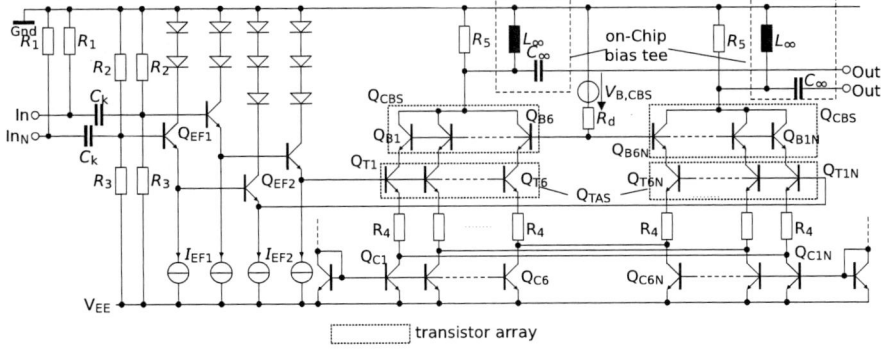

Fig. 2: Principal schematic of the driver.

by temperature in the TAS-transistors. The output is terminated by the resistor R_5, which is shunted for DC and AC-coupled to the external load by an on-chip bias tee. The latter is represented by the ideal elements L_∞, C_∞ in Fig. 2 and will be explained in detail in Sec. III.

Due to the high current ($> 140\,\mathrm{mA}$), which is necessary to generate the voltage drop of 3.5 V at the output resistors R_5, the cascode transistors are required to have a large area in order to prevent them from entering into high current region. In order to comply with the technological maximum transistor size limitation as well as to minimize the avalanche induced pinch-in effect (cf. [3]), six transistors with two emitter stripes each are operated in parallel to reach the required emitter area of each cascode transistor. As detailed in Sec. IV current hogging between the individual transistors of these array transistors is prevented by the special parallelized topology shown in Fig. 2.

The circuit was simulated based on SPICE Gummel-Poon (SGP) models, which do not properly account for high current effects. Therefore, the design considers worst-case scenarios with sufficient safety margins to stay out of the high current region. For simulation of the avalanche induced pinch-in effect, which is the main breakdown phenomenon in a CBS operating above $\mathrm{BV_{CEO}}$, a distributed six-transistor-model [3] was applied. Only negligible pinch-in was observed due to the fact that maximum collector-base-voltage and high collector current never occur at the same time.

III. ON-CHIP BIAS TEE CONCEPT

The functional principle of the bias tee (cf. Fig. 2 and 3a) is to shunt the load resistor at DC. In the operating frequency range the shunt disappears and leaves the load resistor as the output termination. This behavior can be achieved by using an inductive shunt (Fig. 3a), but the required high inductances ($> 10\,\mathrm{nH}$) can not be well realized on-chip. Instead, the inductive behaviour is realized by the symmetrical circuit shown in Fig. 3b. Under the intended differential operation (odd mode) the node connecting both load resistors R_5 lies on the symmetry axis and therefore represents ideal virtual ground. For the even mode a signal return path via $C_5 = 5\,\mathrm{pF}$ to supply voltage ground is realized. In order to keep the DC output voltage as small as possible, the base of the bias tee

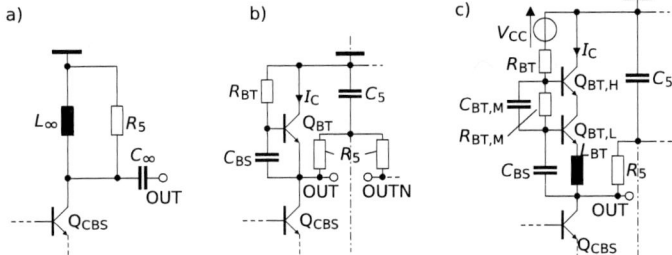

Fig. 3: On-chip bias-tee concept. a) Principle. b) Basic Concept. c) Realization.

transistor Q_{BT} is biased by the upper ground potential via R_{BT}. The bootstrapping capacitor C_{BS} together with R_{BT} are forming a voltage divider with a cut-off frequency below the operating frequency range. In the operating frequency range $\omega C_{BS} > R_{BT}^{-1}$ the voltage swing at the base of Q_{BT} equals the voltage swing at the emitter (bootstrapping), which leads to a negligible base-emitter voltage swing and therefore to a negligible modulation of the collector current I_C of Q_{BT}. However, in this basic concept already a small V_{BE}-modulation caused by the parasitic capacitance C_{CB} of Q_{BT} at its inner base leads due to the high transconductance g_m to a notable modulation of I_C. This degrades the bias tee performance at high frequencies. Simulations show, that this effect can be compensated for by introducing an inductance L_{BT} (cf. Fig. 3c) in the range of approximately 50 to 150 pH, which can be easily realized on chip. As a further modification of the basic concept in Fig. 3b, Q_{BT} is split into two transistors $Q_{BT,H}, Q_{BT,L}$, which share the output voltage swing (similar to [4]) in order to prevent Q_{BT} from breakdown (max. V_{CB} reduced from 3.1 V to 1.6 V). Careful simulations with the distributed avalanche breakdown transistor model [3], show that without this measure Q_{BT} suffers from severe current pinch-in as described in [3]. To prevent $Q_{BT,L}, Q_{BT,H}$ from entering to far into the saturation region at full swing and to keep the DC output voltage as small as possible (low supply voltage) an auxiliary external voltage source $V_{CC} > 0\,\mathrm{V}$ in series to R_{BT} is required (negligible power consumption). The degrees of freedom of the double voltage divider, $R_{BT}, C_{BS}, R_{BT,M}, C_{BT,M}$, are used to adjust the DC operating points of $Q_{BT,L}, Q_{BT,H}$ and to distribute the voltage swing equally between both transistors. The DC level of the output node is $-2.4\,\mathrm{V}$, which leads to power savings of $15\,\%$ in comparison to the solution shown in Fig. 1 b) (output node at $-3.5\,\mathrm{V}$).

IV. TEMPERATURE-OPTIMIZED LAYOUT OF THE OUTPUT STAGE

In the current design an array of six transistors is operated in parallel to reach the required emitter area for each of the four cascode transistors (cf. Fig. 2). Usually these transistors are connected in parallel using emitter degeneration resistors with small values to minimize the undesired current hogging effect. This method depends on an adequate transistor model, which accounts for thermal effects, and complex (e.g. electro-thermal) simulations. Furthermore, a tradeoff is required between the risk of current hogging and higher power consumption due to the voltage drop at the emitter degeneration resistors. Therefore, a novel concept (c.f. Fig. 2) to avoid current hogging in the related transistor arrays is applied. Instead of connecting the array transistors in parallel, the whole cascode is realized six times and connected in parallel only at the input (i.e. TAS base nodes), the output (i.e. CBS collector nodes) and the base terminals of the CBS. In this configuration the individual CBS and TAS transistors do not suffer from current hogging because their operating currents are separated from each other and are only determined by the output current of the current sources $Q_{C1} \dots Q_{C6}$ and $Q_{C1N} \dots Q_{C6N}$.

To keep the temperature and thereby the source current across the current mirror transistors constant, they are placed as far as possible from the hot spot caused by the CBS at the output (cf. Fig. 4). The maximum distance is limited due to parasitic capacitance at the collector node of the current mirror, which would decrease the common mode rejection ratio (CMRR) and increase the risk of instability due to even mode oscillation.

Fig. 4 shows the results of a static thermal 3D simulation. The temperature difference between the center and the outer CBS transistors is $4.4\,\mathrm{K}$, however, no current hogging can occur, because TAS and CBS transistor operating currents are determined solely by the output current of the related current source transistors. At the current source transistors this difference is only as small as $2\,\mathrm{K}$, which causes only small differences ($< 5\,\%$) between their respective output currents.

V. STABILITY CONSIDERATIONS

Besides commonly applied measures to mitigate the risk of instabilities in EF-chains and cascode configurations (cf. e.g. [5, 6]), the special temperature optimized topology of the cascode stage requires special attention.

As the emitter nodes of the CBS transistors of the six cascodes (cf. Sec. IV) are separated from each other, parasitic oscillation modes between the cascodes can occur. In the particular case the three transistors in each half of a

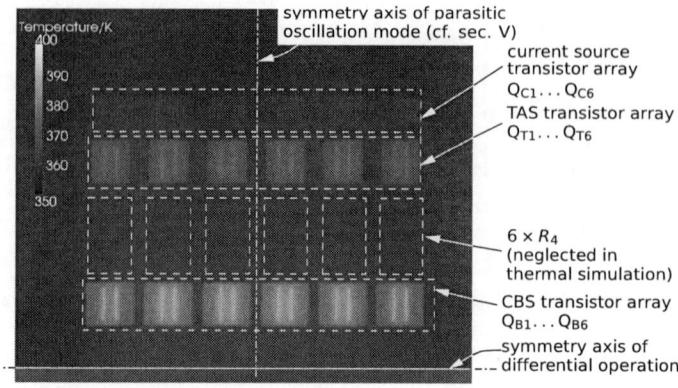

Fig. 4: Temperature Distribution in one half of the cascode. The chip backside is assumed to be at $353\,\mathrm{K}$, which is the maximum specified temperature of the target application.

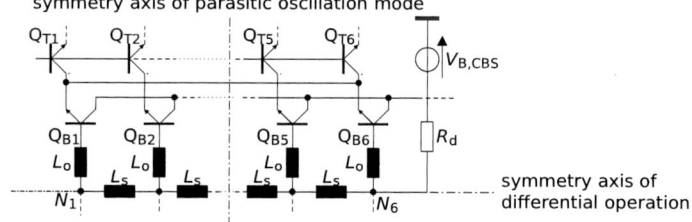

Fig. 5: Schematic of the output cascode including the parasitic wiring inductances in the CBS base branches, which severely decrease the stability of the driver. Circuit elements are located corresponding to the layout (cf. Fig. 4).

transistor array group together and oscillate in anti-phase to the transistors of the opposite half (i.e. odd mode oscillation). In contrast to the symmetry axis intended for differential operation this oscillation mode creates a new one (cf. Fig. 4 and 5). The distance between the outer transistors of both halves is quite large and contributes thereby to significant parasitic inductances between the CBS transistors (cf. Fig. 5). This configuration together with the parasitic capacitance of the TAS is forming a typical RF oscillator topology (cf. [7]). In such a structure the nodes N_1 and N_6 can oscillate in anti-phase. This parasitic oscillation can be suppressed by shorting only the emitter nodes of the corresponding CBS transistors in both halves (i.e. e.g. Q_{B1} and Q_{B6} or Q_{B2} and Q_{B5}). Since the transistors of these corresponding pairs are at the same temperature level, this measure does not induce current hogging.

VI. MEASUREMENT RESULTS

The driver chip has been fabricated in Infineons high-speed bipolar production process B7HF200 ($f_t = 200\,\mathrm{GHz}$, $f_{max} = 275\,\mathrm{GHz}$), which is based on the process described in [8]. The chip size and active area are $860 \times 500\,\mu\mathrm{m}^2$ and $380 \times 300\,\mu\mathrm{m}^2$, respectively (cf. Fig. 6a). For all measurements the chip was mounted into an RF-module similar to [9] (cf. Fig. 6b).

Fig. 7 shows the magnitude of the differential small signal gain $M_{21}^- = \frac{1}{2}(S_{31} + S_{42} - S_{32} - S_{41})$ and output return loss

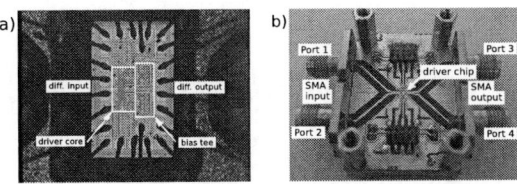

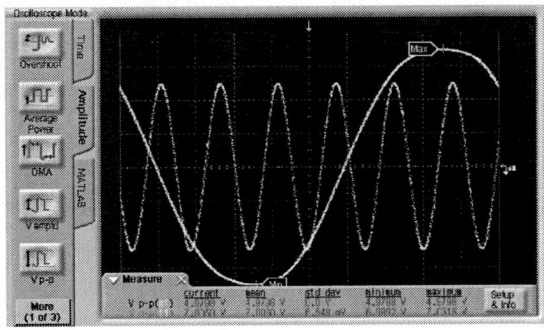

Fig. 6: Photographs of the Driver. a) Chip micrograph. b) Chip mounted in RF module.

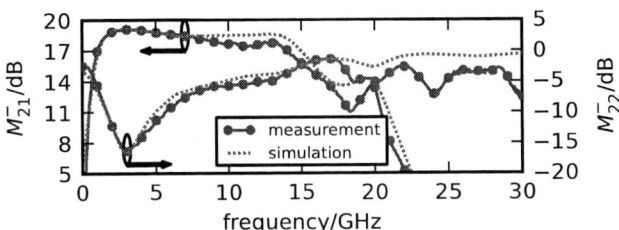

Fig. 7: Odd mode small signal parameters of the driver chip.

Fig. 8: Transient output signal at sinusoidal input drive at 2 GHz and 13 GHz of the on-chip bias tee variant.

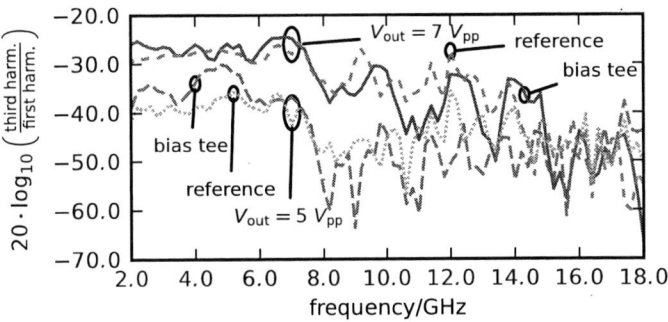

$M_{22}^{-} = \frac{1}{2}\left(S_{33} + S_{44} - S_{34} - S_{43}\right)$ of the driver with on-chip bias tee, where S_{ij} is the S-parameter of the related ports (cf. Fig. 6b). The targeted lower corner frequency of 1.3 GHz was achieved by a $C_{BS} = 8\,\mathrm{pF}$ (size: approx. $120 \times 80\,\mu\mathrm{m}^2$). Due to the high parasitic capacitance at the output node, the gain shows a slight drop over frequency. The output return loss in Fig. 7 shows a degradation at frequencies above 5 GHz, caused by the high parasitic capacitance at the output node of the driver. For comparison also the simulation results including a model of the RF module interconnects are shown in Fig. 7. A good agreement within 1 dB-variation can be observed.

The large signal characterization by eye diagrams would suffer from the high lower cutoff frequency of 1.3 GHz. Therefore, the large signal performance is demonstrated by applying a differential sinusoidal input signal at a constant input amplitude at 2 GHz and 13 GHz showing a differential output voltage swing of 7 V_{pp} up to a 3 dB-cut-off frequency of 13 GHz (cf. Fig. 8). Fig. 9 characterizes the linearity by the measurement of the ratio between third harmonic and fundamental amplitude at the driver output for maximum as well as far reduced output amplitude of 7 V_{pp} and 4.6 V_{pp}, respectively. As a reference design, in addition to the driver with on-chip bias tee a version with plain load resistor R_5 was realized. A comparison between both versions in Fig. 9 shows a slight degradation of the driver linearity (2 dB at full swing).

Fig. 9: Ratio of the third harmonic to the fundamental amplitude at constant input amplitude of 800 mV and 500 mV respectively (measured at 50 Ω at the input of the RF module) over frequency.

VII. CONCLUSION

A linear driver amplifier in SiGe bipolar technology with 7 V_{pp} differential output voltage swing and 13 GHz bandwidth intended to be used in UDWDM applications with a novel on-chip bias tee concept that improves the driver energy efficiency by 15 % was presented. In the driver design a novel concept that prevents current hogging in multi-transistor arrays of cascodes with high output currents was successfully applied. Furthermore, this new concept saves power compared to the conventional emitter degeneration concepts.

REFERENCES

[1] S. Smolorz et al., "Demonstration of a Coherent UDWDM-PON with Real-Time Processing," in *Optical Fiber Communication Conference and Exposition (OFC/NFOEC), 2011*, pp. 1–3.

[2] S. Randel et al., "Generation of a Digitally Shaped 55-GBd 64-QAM Single-Carrier Signal Using Novel High-Speed DACs," in *Optical Fiber Communication Conference, 2014*, M2A.3.

[3] M. Rickelt et al., "A novel Transistor Model for Simulating Avalanche-Breakdown Effects in Si Bipolar Circuits," *IEEE Journal of Solid-State Circuits*, vol. 37, pp. 1184–1197, 2002.

[4] C. Knochenhauer et al., "A Compact, Low-Power 40-GBit/s Modulator Driver With 6-V Differential Output Swing in 0.25-μm SiGe BiCMOS," *IEEE Journal of Solid-State Circuits*, vol. 46, pp. 1137–1146, 2011.

[5] R. Schmid et al., "SiGe Driver Circuit with High Output Amplitude Operating up to 23 Gb/s," *IEEE Journal of Solid-State Circuits*, vol. 34, pp. 886–891, 1999.

[6] R. L. Schmid et al., "Best Practices to Ensure the Stability of SiGe HBT Cascode low Noise Amplifiers," in *Bipolar/BiCMOS Circuits and Technology Meeting (BCTM), 2012 IEEE*, pp. 1–4.

[7] H. Li et al., "Millimeter-wave VCOs with Wide Tuning Range and Low Phase Noise, Fully Integrated in a SiGe Bipolar Production Technology," *IEEE Journal of Solid-State Circuits*, vol. 38, pp. 184–191, 2003.

[8] J. Bock et al., "SiGe Bipolar Technology for Automotive Radar Applications," in *Proc. Bipolar/BiCMOS Circuits and Technology Meeting, Sept. 2004*, pp. 84–87.

[9] H.-M. Rein et al., "Design Considerations for Very-High-Speed Si-Bipolar IC's Operating up to 50 Gb/s," *IEEE Journal of Solid-State Circuits*, vol. 31, pp. 1076–1090, 1996.

An Integrated Transmitter for LED-Based Visible Light Communication and Positioning System in A 180nm BCD Technology

Zongyu Dong, Fei Lu, Rui Ma, Li Wang, Chen Zhang,
Gang Chen, Albert Wang
Dept. of Electrical Engineering, University of California,
Riverside, USA, aw@ee.ucr.edu

Bin Zhao
Fairchild Semiconductor, Irvine, CA, USA

Abstract — this paper reports the first fully integrated transmitter designed in an 180nm BCD process for light-emitting diode (LED) based visible light communication (VLC) and positioning (VLP) systems. The transmitter consists of Manchester coder, precision voltage and current reference, multi-stage Cherry–Hooper amplifier, PLL, filter and LED driver. A feed-forward equalizer is used to boost the LED bandwidth for high data rate wireless streaming. Measurement shows that, driving commercial LED lighting devices, the transmitter can stream data over light at a speed of at least 12MHz.

Keywords—VLC; VLP; Visible Light; LED; PLL; Transmitter

I. INTRODUCTION

LEDs, as energy-efficient solid-state lighting devices, will replace conventional incandescent and fluorescent light bulbs in the next few years, resulting in tremendous energy savings up to 70% [1]. In addition to high lighting efficiency, LED bulbs have other advantages over traditional light sources including long life expectancy, no out-of-visible-band optical spectrum (incandescent lamps emit huge infrared light and fluorescent lamps produce additional ultraviolet spectrum), easy maintenance and environmental friendly. Uniquely, LEDs can be switched on/off at very high speed without flickering to human eyes, which means the light can be modulated to realize visible light communications while lighting. Additionally, LED lighting has a beaming feature that can be used for positioning and navigation. Therefore, the revolutionary LED-based VLC and VLP technologies open a huge opportunity for broadband wireless streaming over visible light at very high speed potentially up to multiple giga bits per second (Gbps), as well as accurate indoor and outdoor positioning and navigation applications. LED-based VLC/VLP systems will eventually realize the long-dreamed "communicate as you see" reality. Building into the existing LED lighting infrastructures, the novel LED-based VLC and VLP technologies will find countless applications in hospitals (where RF is prohibited), airports, shopping malls, warehouses, smart traffic controls, advertisements, etc. Recently, LED-based VLC and VLP technologies have gained global research interests with many testbed system demos reported [2-6]. However, almost all reported VLC systems are based on discrete PCB board electronics that are needed to drive the LEDs and process the signals. While discrete and PCB electronics based VLC

systems demonstrated the feasibility and capability, the fundamental problems arise in terms of the system size, performance, reliability and costs. It is apparent that, in order for LED-based VLC and VLP applications to become a true reality, integrated circuit based SoC and SiP (system on a chip or in a package) shall be the only solution in real world. A transceiver IC for LED-based VLC system shall ideally integrate all functions into one chip, including opto-electronic signal conversion, filtering, bandwidth enhancement, low-noise pre-amplification, power amplification, analog-to-digital conversion and digital signal processing (DSP). A SoC chip also makes it easier to adopt complex modulation methods, e.g., orthogonal frequency division multiplexing (OFDM), to boost the wireless throughput of an LED-based VLC system [4].

II. INTEGRATED TRANSCEIVER DESIGN FOR VLC SYSTEM

A typical LED-based VLC system transceiver consists of a transmitter and a receiver. Transmitter drives and modulates the LED lighting using a LED driving circuit and digital control blocks to achieve both LED lighting and light communications. At the receiver side, single photodiode (PD), PD array and imagers may be used to receive the incoming light signal and process it. Single photodiode with much higher modulation bandwidth (hundreds of MHz) is generally used for high data transmission. The photodiode converts a light signal into electrical signal. The post signal processing circuitry includes trans-impedance amplifier (TIA), limiting amplifier (LA), comparator, clock and data recovery (CDR) circuit. A DSP block will process the receiving signals and recovers the data sent by LEDs.

Since visible light communication originates from lighting, e.g., indoor residential lighting or outdoor signage lighting, certain lighting constraints have to be imposed on the modulation to make the transmitted light comfortable for human to perceive. The primary concern is the non-flickering requirement. Flickering is due to unexpected and unpredictable light intensity changes that may be recognized by human eyes. The flickering is caused by light being switched on and off, or, slow change of brightness of LED bulbs in a time period. A solution to flickering is to ensure the ratio of positive and negative driving level to be constant. Manchester encoding is a

978-1-4799-7231-9/14 $31.00 © 2014 IEEE

suitable solution to the LED VLC transmitter modulation. In addition, due to the self-clocking nature of Manchester encoding, the clock and data recovery of the Manchester encoded data at the receiver side is made much easier. In addition, low frequency part of the transmitted signal can be avoided by Manchester encoding, which makes filter design at the receiver end more flexible.

Fig. 1 shows the whole LED VLC transceiver IC diagram. A Manchester encoding circuit is used at the transmitter side and a Manchester decoding circuit in the receiver channel is used to extract the data and clock from the incoming light signals. The Manchester modulated data will be delivered to the LED driving circuit. An I²C interface is used to realize smart trimming control of LEDs for the whole chip.

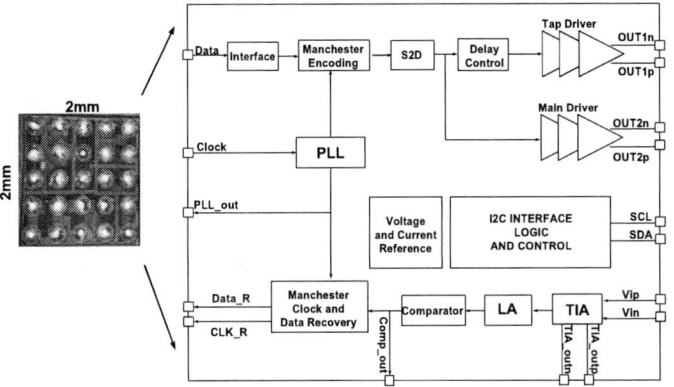

Fig. 1. Illustration of the fully integrated transceiver IC for LED-based VLC and VLP system. Left is the die photo featuring BGA bonding.

III. PRECISION VOLTAGE AND CURRENT REFERENCE

Voltage and current reference is a critical block in the LED VLC transmitter IC. Since LEDs are current driven devices, accurate control of the current or voltage level becomes very important for accurate current matching and dimming control of the LEDs. Fig. 2 shows the high-accuracy Bandgap circuit in this design featuring a 6-bit resistor trim scheme. Stacked diode-connected NPN transistors (Q1&Q3, Q2&Q4) are used in each path featuring a large area ratio (i.e., Q1/Q2=24) to reduce the op-amp offset effect. A base current type (Q6) curvature correction circuit is used to compensate Bandgap

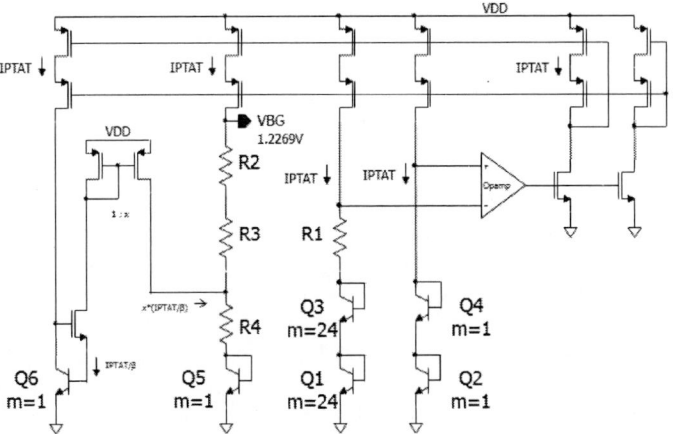

Fig. 2. A schematic for the precision Bandgap circuit with curvature correction.

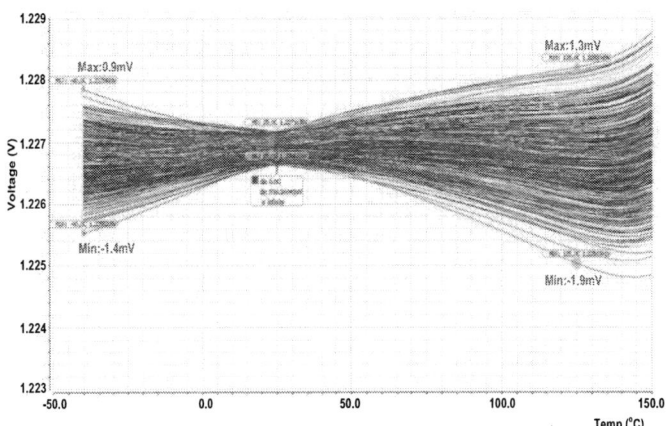

Fig. 3. 500 runs of Monte Carlo simulation of the bandgap output voltage over process corners and temperature (-40°C - 125°C) shows 0.15% variation at 3V supply.

output voltage V_{bg} across the large temperature range from -40°C to 125°C. Monte Carlo (MC) simulations are conducted to verify the V_{bg} variations across the process corners and temperature. The typical value of this Bandgap circuit is 1.2269V and each trim step is around 500µV. Based on 500 runs of MC simulation, a inaccuracy level of 0.15% can be achieved across PVT as depicted in Fig. 3.

IV. LED DRIVING CIRCUIT AND EQUALIZATION

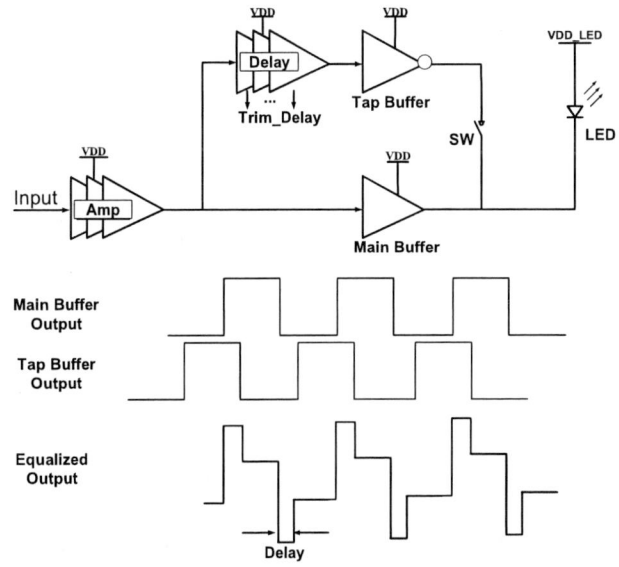

Fig. 4. A diagram for the LED driving circuit with a switchable and delay trimmable equalizer to enlarge LED bandwidth for VLC throughput.

A driving circuit is required for all LED lighting devices where power emission is a critical property. For VLC systems, LED bandwidth is a unique and important parameter that determines its communication capability because the LED, as a transmitter, relies on its frequency response to the input electrical signals, equivalent to the modulation bandwidth of the LED. The modulation bandwidth determines how fast an LED can be modulated, hence its data rates. If the input electrical signals have a much wider signal bandwidth than that of the LED modulation bandwidth, significant large signal distortion will occur. Two types of white-light LEDs are popular in the market. The first one is single blue light emitting chip

978-1-4799-7231-9/14 $31.00 © 2014 IEEE   85

combined with yellow YAG phosphor. The second one consists of three light emitting chips, namely red, green and blue. Blue light with yellow phosphor or three primary colors of red, green, and blue are mixed to generate the white light. For the phosphor based white LEDs, the typical measured modulation bandwidth is usually around 2 MHz, as it is limited by the long response time of the yellow phosphor. Methods such as using a blue filter at the receiver side to remove the slow yellow light, or using spectral efficient advanced modulation formats [3] have been reported to enhance the frequency response. However, using a blue filter could attenuate the signal and the advanced modulation formats needed, e.g., discrete multi-tone (DMT), make the transceiver more complex and expensive, hence, it is less attractive to LED VLC systems.

Fig. 4 shows the topology of and the waveforms generated by the LED driver in featuring pre-equalization to enlarge LED bandwidth in this design. The preamplifier (Amp) consists of a several-stage Cherry–Hooper amplifier, which is widely used in high-gain and broadband circuits. The output stage is implemented with current-mode logic (CML), which is suitable for driving LEDs at various conditions. Fig. 5 illustrates the driver schematic. The output stage has both main buffer and tap buffer units. The tap buffer block serves as the feed-forward equalizer and the width of the equalization pulse can be controlled by the delay trimming logic. The equalizer path can be completely shut down by turning off the equalization switch (SW). For the commercial lighting LEDs used in our VLC systems, the typical driving current is around hundreds of mA, which is a big challenge to logic CMOS technology because the MOSFET has to be huge to drive such a large current at a modulation speed of tens of MHz. For this concern, a BCD process was used in this design where the power MOSFET has much higher current conductivity, i.e., I_{dsat} around 600 μA/μm, which saves the area for the LED driver. Meanwhile, the high breakdown voltage (~22V) of the power MOSFET allows the LED driver to work under various power supply conditions.

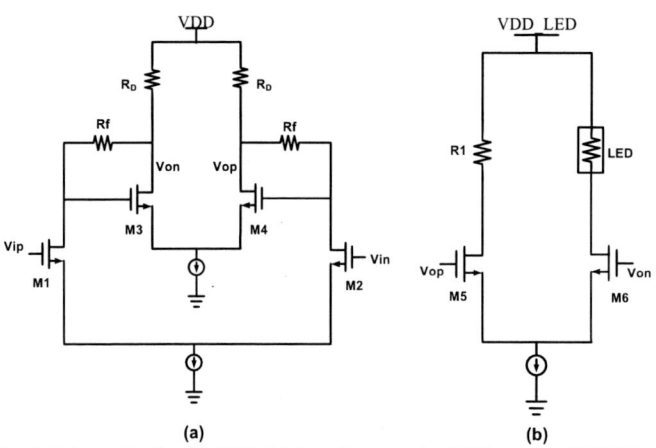

Fig. 5. Schematics for the LED driving circuits using BCD power MOSFETs: (a) Cherry-Hooper amplifier, (b) CML output stage with a tail current source.

IV. ALL-DIGITAL MANCHESTER CODING

As mentioned above, Manchester encoding is a suitable solution to modulation for LED VLC system due to its flicker removal and self-clocking nature. For Manchester encoding,

each zero is transmitted as a one followed by a zero and each one is transmitted as a zero followed by a one. Thus, a transition is introduced at the middle of each data bit, which can be used to recover the clock at the receiver. But the price paid is the doubling of the effective bit rate in the communication system. More importantly, both Manchester encoder (Fig. 6) and decoder (Fig. 7) can be realized in all digital format, which is perfect for IC design. The Manchester encoder circuit mixes the non-return-to-zero (NRZ) bit stream with the transmitting clock. Due to delay requirement for clock and data recovery, 5 times reference clock should be provided to the Manchester decoding circuit [7]. Fig. 8 depicts the simulated Manchester encoded data and its recovered clock and data after Manchester decoding circuit. It is clear that the recovered data and clock match precisely with the input data and clock.

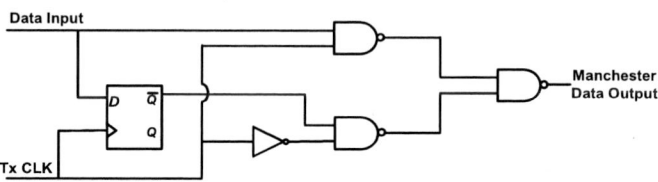

Fig. 6. A diagram for the Manchester encoder circuit in the VLC transmitter.

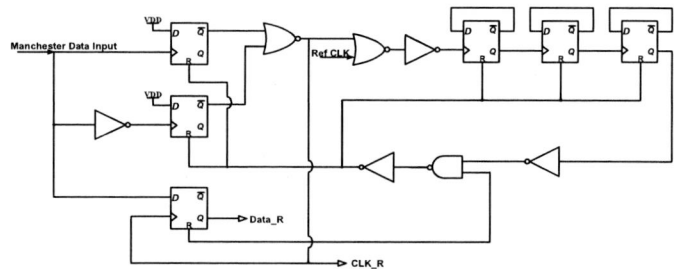

Fig. 7. A diagram for the all-digital Manchester data and clock recovery circuits in the receiver.

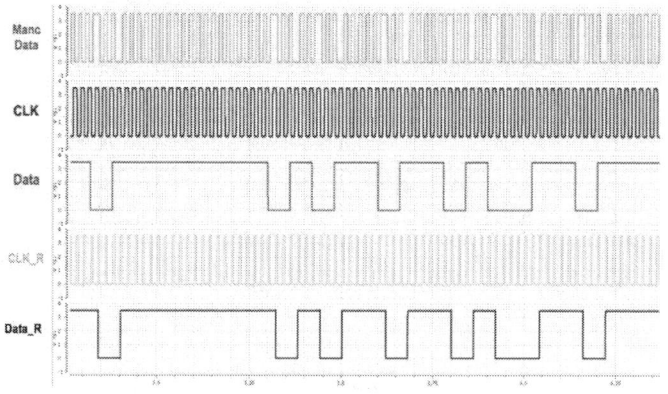

Fig. 8. Simulated Manchester encoded data at transmitter side and the recovered clock and data at receiver side match well.

Fig. 9 shows the circuit blocks of the clock generation PLL circuit, which employs a voltage-controlled ring oscillator to produce the multiple clock phases used in data multiplexing. A linear regulator supplies power to the VCO and also sets the oscillation frequency by buffering the control voltage, produced by the interaction between the phase-frequency detector (PFD), charge-pump (CP), and loop filter. As required by Manchester decoder, 5 times reference clock is generated at the divider output. As the PLL output clock is only used for generating sufficient delay, the phase noise requirement is not as strict as in conventional PLL designs. From measurement, the PLL can work well for input clock frequency from 5MHz to 25MHz.

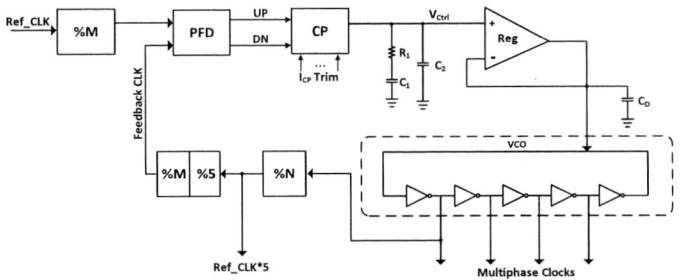

Fig. 9. A diagram for the charge pump based PLL to generate 5 times of clock for the Manchester data and clock circuits.

V. MEASUREMENT AND DISCUSSIONS

This integrated transceiver for LED-based VLC systems was designed and implemented in a foundry 180nm BCD process. The die photo is shown in Fig. 1 that occupies an area of 2mm × 2mm. Fig. 10 presents the measurement result for the reference circuit showing that it achieves V_{bg} = 1.2269V and

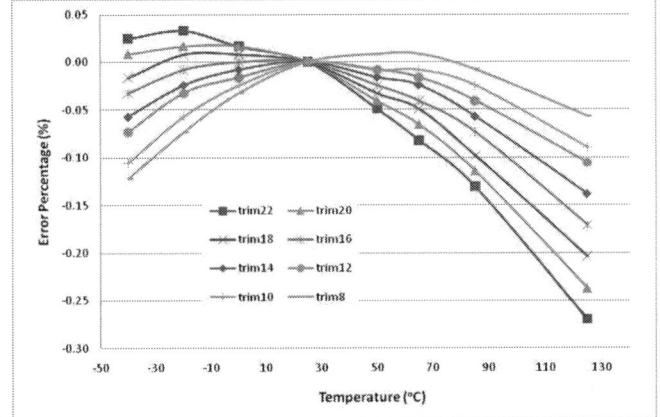

Fig. 10. Measured bandgap voltage reference achieves 0.10% variation level for a typical V_{bg} =1.2269V between -40°C to 125°C after trimming.

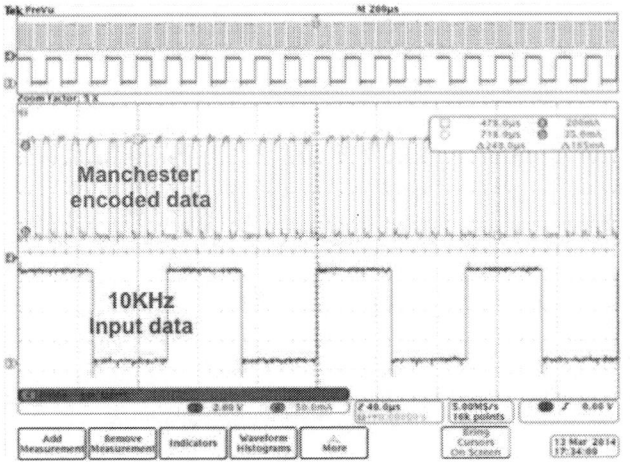

Fig. 11. Measured Manchester encoded current to drive white LEDs at 10KHz input data and 50KHz tranmitting clock.

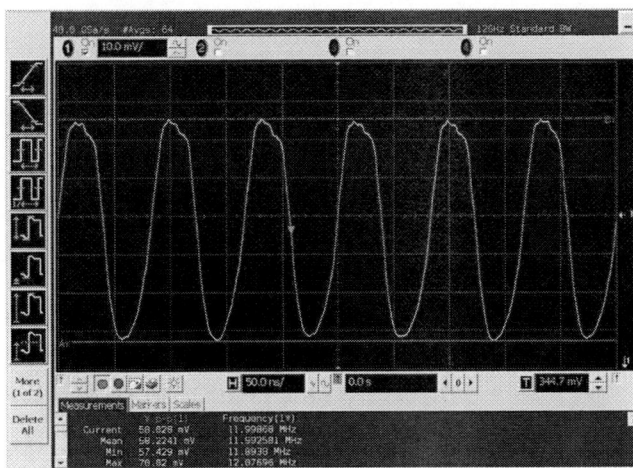

Fig. 12. Measurement of the LED VLC system, fully controlled by the transceiver IC designed shows the signal waveform received through the visible light transmitted from the LED bulb at 12MHz.

0.10% inaccuracy across temperature with best trim code 12. Fig. 11 shows the measured Manchester encoding result at the best trimming.

The transceiver IC was mounted on a PCB board to control the LED lighting source. Fig. 12 depicts the system measurement result for the VLC system controlled by the transceiver IC, which shows that it correctly transmits and receives the signal waveforms at 12MHz of the input data streams using commercial LEDs. A much high data rate is expected if customer-designed low-capacitance LEDs would be available.

VI. CONCLUSION

We report the first fully integrated transmitter designed in an 180nm BCD process for LED-based VLC/VLP systems. Measurement shows signals were wireless streamed over the visible light emitted from commercial LED lighting bulbs. It proves the feasibility to design LED VLC/VLP SoC or SiP systems for visible light communications and navigation.

REFERENCES

[1] J. K. Kim and E. F. Schubert, "Transcending the replacement paradigm of solid-state lighting," *Optics Express*, pp. 21835-21842, Dec. 2008.

[2] T. Komine and M. Nakagawa, "Fundamental analysis for visible light communication system using LED lights," *IEEE Trans. Consumer Electronics*, pp. 100-107, 2004.

[3] D. O'Brien, et al, "Indoor visible light communications: challenges and prospects," *Proc. of SPIE,* vol. 7091, 709106, 2008.

[4] Center for Ubiquitous Communication by Light (UC-Light), University of California, http://www.uclight.ucr.edu.

[5] K. Cui, et al, "Indoor optical wireless communication by ultraviolet and visible light," *Proc. of SPIE*, vol. 7464, pp. 74640D, 2009.

[6] Z. Dong, et al, "Non-line-of-sight link performance study for indoor visible light communication systems," *Proc. of SPIE Photonics and Optics*, pp. 781404-1-781404-10, 2010.

[7] F. Asgarain and A.M. Sodagar, "A High-Data-Rate Low-Power BPSK Demodulator and Clock Recovery Circuit for Implantable Biomedical Devices", *Proc. IEEE EMBS Conf. Neural Engineering*, pp. 407-410, 2009.

An Enhanced 180nm Millimeter-Wave SiGe BiCMOS Technology with f_T/f_{MAX} of 260/350GHz for Reduced Power Consumption Automotive Radar IC's

J. P. John, V. P. Trivedi, J. Kirchgessner, D. Morgan, I. To, P. Welch

Freescale Semiconductor Inc., 2100 E. Elliot Road, MD: EL319, Tempe, AZ, 85284.

Abstract — Several performance improvements on a 180nm SiGe:C BiCMOS technology targeted for improved millimeter-wave performance are described. SiGe HBT performance metrics, including f_T, f_{MAX}, and CML gate delay are improved 20-30%. f_T/f_{MAX} of 260/350GHz are achieved with a minimum gate delay of 3.2ps, without impacting the thermal budget of the technology. BEOL and ground plane optimization reduced transmission line loss to <1dB/mm at ~80GHz.

Index Terms — Silicon bipolar/BiCMOS process technology, Bipolar transistors, Millimeter wave technology, SiGe, SiGe:C.

I. INTRODUCTION

SiGe BiCMOS has proven to be ideally suited for millimeter-wave IC's, such as 77GHz automotive radar transmitters and receivers [1]-[3]. Today, mature production technologies exist with SiGe HBT peak f_T exceeding 200GHz [4]-[6], with 400GHz f_{MAX} technologies starting to come on line [7]-[11]. Continued integration and functionality demands placed on these products inevitably lead to IC power consumption concerns. A key benefit of these higher performance devices is the ability to operate at lower currents with a similar gain at a given frequency. This drives product improvements at a given application frequency, rather than a push to higher application frequencies. In addition to the SiGe HBT, transmission lines play an important role in power consumption, as lower loss leads to a lower power generation requirement.

In this paper, we describe several key performance improvements in a 180nm SiGe:C BiCMOS manufacturing technology. More aggressive layout rules, aided by advanced photolithography has enabled a shrink of the SiGe HBT emitter/base structure to improve layout parasitics for higher f_{MAX}, while optimization of the SiGe:C base epitaxy and collector doping profiles was undertaken in order to increase intrinsic device performance for higher f_T. Peak f_T and f_{MAX} of 260/350GHz have been achieved without modifying the thermal budget of the existing 180nm BiCMOS technology. Transmission line (TL) loss was reduced by 0.5dB/mm at 80GHz, through optimization of the dielectric loss and ground plane structure.

II. TECHNOLOGY

The technology baseline is an established 180nm RF BiCMOS platform [12] updated with a 200GHz f_T SiGe:C HBT device module [13], [14], high tuning ratio varactor [15], and 0.8μm thick top metal for TL. The HBT structure combines a self-aligned, selective-epi base emitter/base top structure with a novel sub-isolation buried layer (SIBL) collector structure with shallow trench isolation. The SIBL module enables reduced device collector resistance while avoiding the more expensive buried layer/epitaxy plus deep trench approach [14]. Table I summarizes some of the key SiGe HBT device parameters of the optimization results from this work, in comparison to previous SiGe HBT generations fabricated on this technology node.

TABLE I
SiGe:C BiCMOS Parameter Comparison

SiGe HBT	Units	Prior Generation SiGe HBT's			This Work
		HBT [12]	eHBT	xHBT [14]	xHBT2
W_E	nm	250	250	150	125
Beta	--	120	360	550	1500
R_{BI}	kΩ/sq	1.8	4.8	3.1	3.4
BV_{CEO}	V	3.3	1.8	1.9	1.5
BV_{CBO}	V	12.3	6.2	6.1	5.2
f_T	GHz	50	120	200	260
f_{MAX}	GHz	110	120	280	350
Gate Delay	ps	-	-	4.2	3.2

Constraints for this optimization work relate to minimizing the impact on the existing BiCMOS library and device models, restricting performance optimization thru new self-aligned architectures [8] or thermal budget scaling [9]. However, SiGe HBT layout scaling, such as in [10] remains a viable option, given the capabilities of the fabrication line.

III. SiGe:C HBT Scaling and Optimization

With the constraint to maintain both the thermal budget and the SiGe HBT device architecture,

978-1-4799-7231-9/14 $31.00 © 2014 IEEE

performance optimization of the SiGe HBT proceeded along two paths. The first path was a layout shrink of key performance limiting design rules. These consist of the emitter window width, collector active, and emitter poly width, consistent with this double-poly device architecture. More advanced photolithography was utilized to achieve the desired critical dimensions (CD) and process control. For example, the emitter poly layer was switched from an i-line to a deep ultraviolet (DUV) photoresist layer. Fig. 1 provides a TEM comparison of the scaled device along with the key scaled design rules and targeted parasitic for reduction. Beyond the CD reductions, the emitter base spacer width was reduced both to improve base resistance and to improve emitter contact formation for the scaled emitter width.

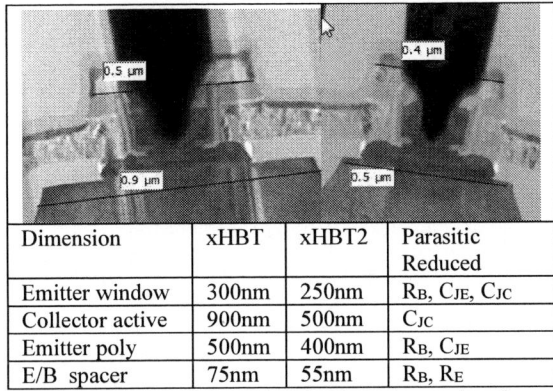

Dimension	xHBT	xHBT2	Parasitic Reduced
Emitter window	300nm	250nm	R_B, C_{JE}, C_{JC}
Collector active	900nm	500nm	C_{JC}
Emitter poly	500nm	400nm	R_B, C_{JE}
E/B spacer	75nm	55nm	R_B, R_E

Fig. 1. TEM micrograph showing the emitter-base structure of the prior generation (xHBT) and scaled device (xHBT2). A comparison of the key scaled design rules is shown.

Without thermal budget reductions, the base epitaxy as-grown profile along with the collector implant profile provides the strongest leverage for performance improvement. Increased germanium content and grading across the base, higher collector doping, and optimized overall selective epitaxy thickness were all employed to drive improved f_T.

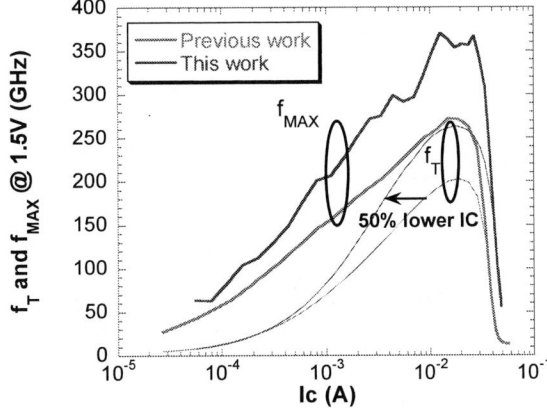

Fig. 2. f_T and f_{MAX} comparing the prior work (W_E ~150nm) and the new device (W_E ~125nm). L_E=10um. V_{CE}=1.5V. Open-short de-embedding is used.

Fig. 2 shows the results of the optimization on the f_T and f_{MAX} vs. I_C characteristic, comparing the previous generation device with the work presented here. Peak f_T/f_{MAX} of 260/350GHz is achieved, an improvement of 20-30% over the prior generation. Additionally, the current at peak performance is not appreciably increased. The current required to operate this device at a f_T of 180GHz has been reduced by approximately 50%. This metric is derived from a fairly common bias point used for these devices in our automotive radar IC's. These improvements highlight the combined improvements of both layout scaling and intrinsic profile optimization.

In order to highlight how the layout (parasitic) and intrinsic components factor into overall device performance, Fig. 3 shows CML ring oscillator gate delay vs. IDD for 4 different cases. These include the old process and old layout, the new process (epitaxy and collector) with the old layout, the old process with the scaled layout, and finally the new process and the scaled layout. Overall, gate delay has been reduced to 3.2ps with the combination of intrinsic improvements and layout scaling. The intrinsic profile improvements (peak f_T) on the old layout result in only a modest ~5% improvement (despite the ~30% higher f_T), due to the higher C_{JC} of the optimized intrinsic profile. Layout scaling is required to significantly drive ring delay improvements. Another interesting observation is the lack of improvement in ring speed with the existing process and scaled layout. The reason here is that the intrinsic epitaxy profile is not optimized for the scaled layout (emitter window width), resulting in some peak f_T degradation. In addition, the wider emitter/base spacer, contributes to degraded emitter resistance and peak f_T.

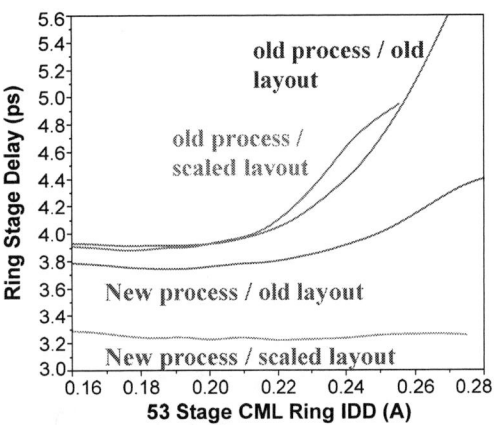

Fig. 3. Gate delay vs. IDD for a 53 stage CML ring oscillator. L_E=2um SiGe HBTs are used.

A typical Gummel plot for the new device is shown in Fig. 4. Despite the more aggressive base profile, good base current characteristics have been maintained, due to careful carbon doping profile design.

978-1-4799-7231-9/14 $31.00 © 2014 IEEE

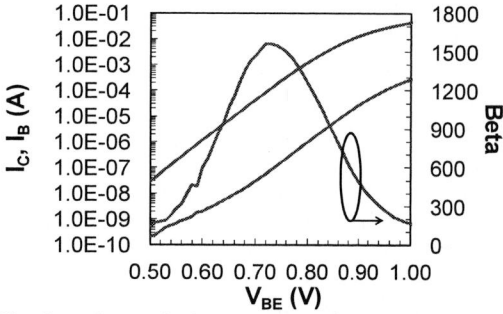

Fig. 4. Gummel characteristics (V_{CE}=1.2V) for the new device (W_E ~125nm, L_E=10um).

Additional lateral scaling of this device is possible with reduced emitter window dimension and additional structural changes. The main challenge to this scaling is managing the emitter resistance (R_E) increase (and impact to peak f_T and therefore peak f_{MAX}) due to the plugging of the narrower emitter window with emitter poly. Fig. 5 highlights the f_T degradation for further emitter window width scaling and the corresponding R_E increase. Peak f_T roll-off is mitigated significantly with an improved aspect ratio emitter/base process, indicating the importance of managing emitter resistance for these highly scaled devices.

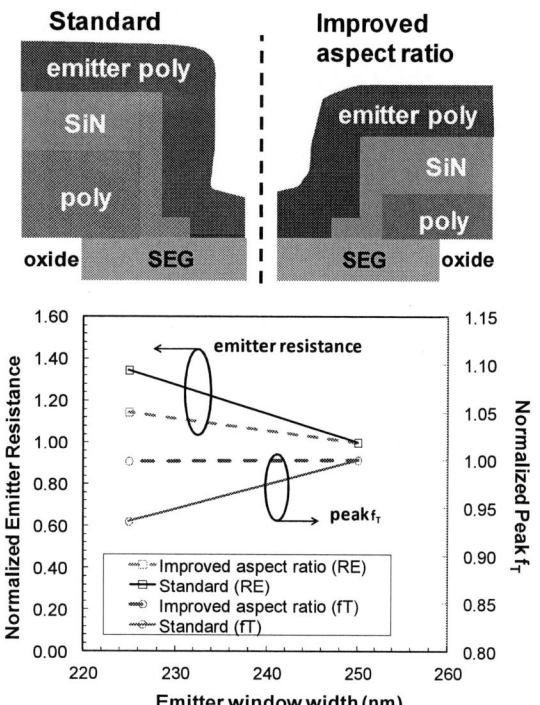

Fig. 5. Normalized emitter resistance and peak f_T vs. emitter window width for the standard and improved aspect ratio emitter structure. The top drawing compares the emitter base structure of the improved aspect ratio process to the standard process.

The influence of the R_E on performance of these devices is investigated further through various emitter contact designs. Fig. 6 plots peak f_T and R_E for four devices with identical design rules, differing only by the emitter contact design layout. These devices were fabricated with the new process on the same wafer and did not utilize the improved aspect ratio emitter structure. Emitter contact size and pitch was varied from a near ideal case (i.e. a full emitter stripe with close to 100% coverage) to a more relaxed pitch with <40% contact coverage. Peak f_T varies by 10%, while extracted R_E varies by a factor of 2.

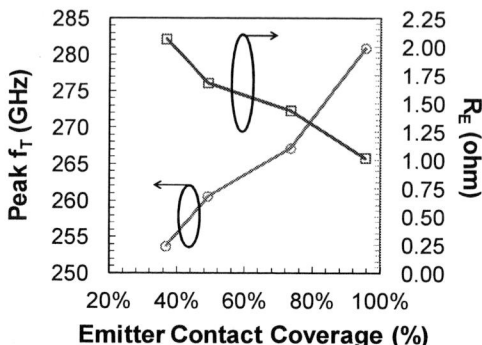

Fig. 6. Peak f_T and R_E vs. the effective contact coverage of the emitter (W_E ~125nm, L_E=10um).

IV. TRANSMISSION LINE DESIGN AND INTEGRATION

Microstrip transmission lines (TLs) are used extensively in millimeter-wave IC designs [1]-[3]. The baseline TL consists of a 0.8um thick top (Cu) metal for the signal and the lowest metal (m1) for the ground plane with ~3.3um of SiO2-equivalent inter-layer dielectric (ILD). On average, the baseline TL loss/attenuation factor (α) is ~1.5dB/mm at 80GHz. In order to achieve lower loss, ground plane optimizations, thicker ILD, and thicker top metal are explored. TLs are characterized up to 110GHz using Agilent PNA and using "different-length" de-embedding [16].

The baseline ground plane is formed using an m1 mesh in order to meet metal density rules. As shown in Fig. 7, these openings increase the loss by ~0.2dB/mm compared to a solid ground (which is not design-rule compliant). As demonstrated in Fig. 7, the ~0.2dB/mm of loss can be recovered with layout optimizations. These layout optimizations neither use any additional metals nor violate any design rules. Although not shown, the characteristic impedances of all the ground plane optimizations of Fig. 7 are comparable.

The ILD thickness is another key parameter to lower the TL loss. Fig. 8 shows that the TL loss can be lowered by 0.3-0.4dB/mm with 1-2um thicker ILD (with baseline at ~3.3um); ILD thickness is varied by repeated processing of metal-3 and/or metal-4 layers. In contrast to ILD thickness, we find (not shown here) that a thicker top metal does not provide any substantive benefit because the skin depth in Cu is

<0.5um (~0.25um) for frequencies above 30GHz (at ~77GHz).

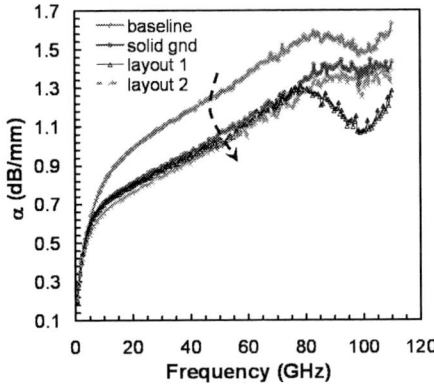

Fig. 7. Measured TL loss versus frequency for layout-based ground plane optimizations.

Based on the above findings, we integrated an enhanced-TL module with combinations of ground plane optimizations and an additional layer of thick metal identical to the thick top metal of the baseline process. The TL loss, shown in Fig. 9, is 0.9-1.0dB/mm at 80GHz, which is >0.5dB/mm lower than baseline.

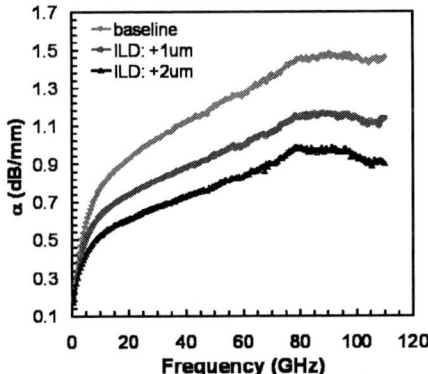

Fig. 8. Measured TL loss for ILD splits.

V. CONCLUSION

Enhancements to a 180nm millimeter wave SiGe:C BiCMOS technology have been described. Device performance improvements of 20-30% have been achieved without impacting thermal budget of the technology through a combination of vertical and lateral scaling. Peak f_T/f_{MAX} of 260/350GHz has been achieved for the SiGe HBT. Transmission line loss was reduced to <1dB/mm at 80GHz through ground plane and dielectric thickness modifications.

ACKNOWLEDGEMENT

The authors wish to acknowledge the Austin Technology and Manufacturing Center, Oak Hill Fab, and the Texas analytical labs. In particular, the authors would also like to acknowledge the special

contributions of Donna Hammock, Susan Stewart, Marc Rossow, Blaine Woodbury, and Mark Caldwell.

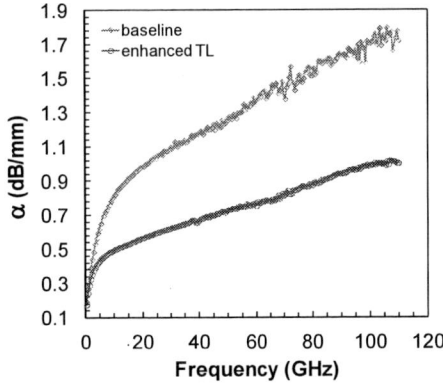

Fig. 9. Loss versus frequency for new and baseline TL.

REFERENCES

[1] D. Salle, et al, "A Fully Integrated 77GHz FMCW Radar Transmitter Using a Fractional-N Frequency Synthesizer," *EuRAD*, pp. 149-152, 2009.

[2] S. Pacheco, et al, "SiGe technology and circuits for automotive radar applications," *SiRF11*, pp. 141–144, Jan. 2011.

[3] S. Trotta, et al, "A transceiver chipset for automotive LRR and SRR systems in the 76–77 and 77–81 GHz bands in SiGe BiCMOS technology," *RFIC Symp. Dig.*, pp. 1–4, Jun. 2011.

[4] G. Avenier, et al, "0.13um SiGe BiCMOS Technology Fully Dedicated to mm-Wave Applications," *IEEE J. Solid-State Circuits*, Vol. 44, No. 9, , pp. 2312-2321, Sept. 2009.

[5] E. Preisler, et al, "A Millimeter-Wave Capable SiGe BiCMOS Process with 270GHz FMAX HBTs Designed for High Volume Manufacturing," *Proc. BCTM*, pp. 74-77, 2011.

[6] B. Orner et al., "A 0.13 µm BiCMOS technology featuring a 200/280 GHz (fT /fmax) SiGe HBT," *Proc. BCTM*, pp. 203-206, 2003.

[7] P. Chevalier, et al, "A Conventional Double-Polysilicon FSA-SEG Si/SiGe:C HBT Reaching 400 GHz fMAX," *Proc. BCTM*, pp. 1-4, 2009.

[8] B. Heinemann et al., "SiGe HBT technology with fT/fmax of 300GHz/500GHz and 2.0 ps CML gate delay," *IEDM* 2010.

[9] B. Geynet, et al, "SiGe HBTs Featuring fT >400GHz at Room Temperature," *Proc. BCTM*, pp. 121-124, 2008.

[10] P. Chevalier, et al, "Towards THz SiGe HBTs," *Proc. BCTM*, pp. 57-65, 2011.

[11] R.A. Camillo-Castillo, et al, "SiGe HBTs in 90nm BiCMOS technology demonstrating 300GHz/420GHz fT/fMAX through reduced Rb and Ccb parasitic," *Proc. BCTM*, pp. 227-230, 2013.

[12] J. Kirchgessner, et al, "A 0.18 µm SiGe:C RFBiCMOS technology for wireless and gigabit optical communication applications," *Proc. BCTM*, pp. 151-154, 2001.

[13] J. P. John, et al, "Development of a Cost-Effective, Selective-Epi, SiGe:C HBT Module for 77GHz Automotive Radar," *Proc. BCTM*, pp. 247-250, 2006.

[14] J. P. John, et al, "Novel Collector Structure Enabling Low-Cost Millimeter-Wave SiGe:C BiCMOS Technology," *IEEE RFIC*, pp. 559-562, 2007.

[15] V. P. Trivedi, et al., "Hyperabrupt-junction varactor for mmWave SiGe:C BiCMOS, enabling 77GHz VCO/TX with 13-15GHz tuning range," *Proc. BCTM*, pp. 82-85, 2010.

[16] A. M. Mangan, et al., "De-embedding transmission line measurements for accurate modeling of IC designs", *IEEE Trans. Elec. Dev.*, Vol. 53, No. 2, 235-241, Feb. 2006.

A 90nm SiGe BiCMOS Technology for mm-wave and high-performance analog applications

John J. Pekarik, J. Adkisson, P. Gray, Q. Liu, R. Camillo-Castillo, M. Khater[1], V. Jain, B. Zetterlund, A. DiVergilio, X. Tian, A. Vallett, J. Ellis-Monaghan, B. J. Gross, P. Cheng[*], V. Kaushal[*], Z. He, J. Lukaitis, K. Newton, M. Kerbaugh, N. Cahoon, L. Vera[2], Y. Zhao[2], J. R. Long[2], A. Valdes-Garcia[1], S. Reynolds[1*], W. Lee[1*], B. Sadhu[1], D. Harame

IBM Corporation, Essex Junction, VT 05452, [1] IBM T. J. Watson Research Center Yorktown Heights, NY 105, [2] Delft University of Technology Mekelweg 4, 2628CD Delft, The Netherlands, [*] formerly with IBM

Abstract— **We present the electrical characteristics of the first 90nm SiGe BiCMOS technology developed for production in IBM's large volume 200mm fabrication line. The technology features 300 GHz f_T and 360 GHz f_{MAX} high performance SiGe HBTs, 135 GHz f_T and 2.5V BV_{CEO} medium breakdown SiGe HBTs, 90nm Low Power RF CMOS, and a full suite of passive devices. A design kit supports custom and analog designs and a library of digital functions aids logic and memory design. The technology supports mm-wave and high-performance RF/Analog applications.**

Keywords—Silicon bipolar process technology; Silicon germanium; Heterojunction bipolar transistors; Millimeter wave technology.

I. INTRODUCTION

IBM's 90nm SiGe BiCMOS (9HP) is the world's first 90nm SiGe BiCMOS technology in production. The technology is optimized to support mm-wave and high-performance analog applications including RF transceivers, high bandwidth analog to digital converters, optical networks, terahertz imaging and sensing, automotive radars and instrumentation [1-4]. These applications incorporate increasing amounts of control circuitry, digital signal processing, and built-in test functions, which require increased levels of CMOS density to remain affordable, as well as CMOS-based analog functions. IBM's 90nm low-power CMOS technology [5] was chosen as the base for this BiCMOS offering because of its CMOS density, digital performance/power design point, and RF features. For example digital logic cores are ¼ the size of equivalent 180nm CMOS nodes and ½ the size of 130nm CMOS nodes. The RF-centric features and 200mm wafers of IBM's 90nm CMOS node make it more attractive compared to advanced "digital" CMOS processes on 300mm wafers. Advantages include a rich menu of RF passives with an optional thick dielectic BEOL module and a lower mask set cost.

9HP features a high-performance npn HBT with f_T = 300GHz, f_{MAX} = 360GHz and BV_{CEO} = 1.7V. The progression of f_T and f_{MAX} through multiple generations of IBM SiGe HBTs is illustrated in Fig. 1. The current density, J_C, at peak f_T increases almost proportionally with the increase in f_T across these technologies.

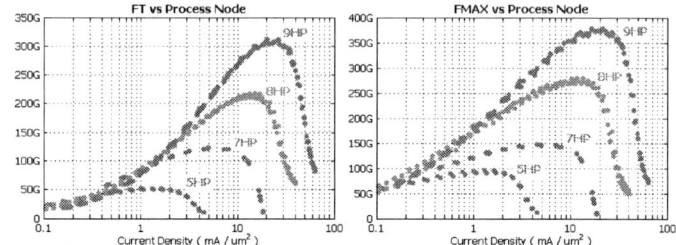

Fig. 1. f_T and f_{MAX} progression through multiple generations of IBM SiGe.

The transistor is derived from the HBT in IBM's 130nm SiGe BiCMOS8HP (8HP) technology [6] and employs vertical and selective lateral scaling along with a self-aligned emitter-base junction. A medium-breakdown npn HBT having f_T = 135GHz, f_{MAX} = 350GHz and BV_{CEO} = 2.5V, shares the same transistor structure with modified collector implants [7]. Five n/p MOS transistor pairs from IBM's 90nm low-power CMOS offer choices to circuit designers ranging from n-channel FETs with f_T = 145GHz to 3.3V-tolerant FETs. A broad selection of passive devices formed from elements of the CMOS, the bipolar and the BEOL processes support flexibility of circuit topology and physical layout. A Process Design Kit (PDK) supports custom design flows with schematic symbols, parametric layout cells (pcells), design rule checking, layout vs. schematic checking, and parasitic extraction.

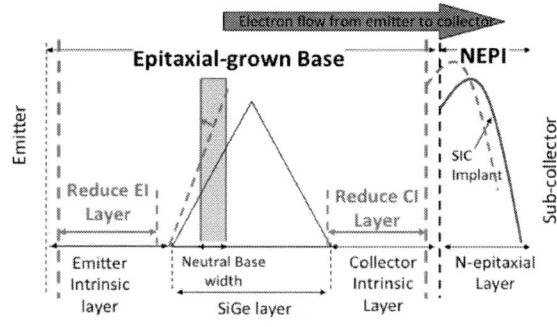

Fig. 2. Schematic illustration of the vertical scaling elements employed in 9HP.

978-1-4799-7231-9/14 $31.00 © 2014 IEEE

II. TECHNOLOGY OVERVIEW

A. HBT Architecture

A combination of vertical and lateral scaling from the previous 130nm SiGe BiCMOS technology (8HP) was used to achieve 300 GHz peak f_T and 360 GHz f_{MAX} at V_{CB}=0.3V. In order to increase f_T, the vertical profile was modified as illustrated in Fig. 2. The intrinsic silicon layers on both the emitter- and collector-side of the SiGe epitaxial growth sequence were shrunk accomodating the lower thermal cycle of the final anneal in the 90nm CMOS which is also used to set the position of the emitter-base and collector-base junctions. The grading of germanium alloy composition was made steeper and the peak germanium concentration was increased. The boron concentration in the base was increased providing lower base resistance with the narrow base width afforded by the reduced post-epitaxy thermal cycle. And finally, the dose of the collector implant was increased.

The final HBT structure used in 9HP is similar to that used in 8HP but selected elements of the transistor structure were modified to reduce base resistance and collector-base capacitance in order to increase f_{MAX}. Several approaches were considered including scaling the previous generation (8HP) and a self-aligned pedestal alignment approach [8]. The elements of lateral scaling used in 9HP include a lateral shrink of the emitter window and collector implant, tighter lithographic alignment of the emitter to the collector – taking advantage of the finer resolution and improved overlay of more-advanced lithography tools used in the 90nm CMOS – and a self-aligned emitter-base junction. Process-window experiments were conducted to empirically determine the optimal alignment overlay and mask size dimensions for the collector implant. The results are illustrated in Fig. 3.

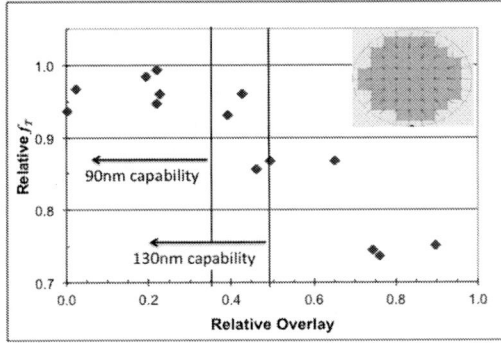

Fig. 3. Relative f_T vs. relative overlay error which was deliberately varied across a wafer as illustrated by the error-vectors in the inset. The vertical lines indicate the improved overlay tolerance afforded by the use of more-advanced lithography.

Fig. 4 is a cross section of the 9HP HBT. After non-selective epitaxial growth of the intrinsic base, an oxide/polysilicon/dielectric stack is formed. The emitter opening is patterned and following formation of an inner sidewall the base is undercut and an epitaxial growth forms the link between the intrinsic and extrinsic base. The process continues with the emitter polysilicon layer definition. As a benefit of this emitter-base self-alignment scheme, the diffusion of boron from the extrinsic-base polysilicon is reduced resulting in lower C_{CB}.

Fig. 4. TEM micrograph of self-aligned emitter-base junction.

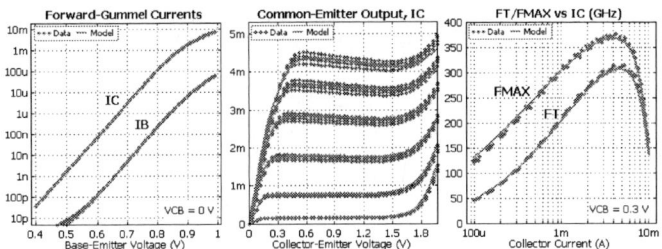

Fig. 5. Characteristics of the high-performance HBT showing measured data and SPICE model simulations for a 0.1x2.0μm² drawn emitter.

B. HBT characteristics

The key characteristics of the high-performace npn are shown in Fig. 5. Both collector and base current exhibit ideal exponential behavior to the limit of the instrumentation, β at V_{BE}=0.72V and V_{CB}=0 is ~470 and emitter resistance is 8Ω. BV_{CEO} is 1.7V and BV_{CBO} is 5.2V. High-frequency characteristics were extracted using two-port S-parameter measurements with Load-Reflect-Reflect-Match (LRRM) calibration and open and short de-embedding to remove parasitic impedances associated with measurement cables, probes, wiring and pads. Good agreement with the HICUM L2 (v2.31) compact model is shown for both DC and AC characteristics.

C. 90nm CMOS

This 90nm SiGe BiCMOS work is distinguished from other published and commercial SiGe BiCMOS offerings based on 180nm and 130nm nodes. Electrical specifications of 180nm, 130nm and 90nm IBM RF CMOS nodes are summarized in Table I. The gate density and SRAM cell area show the significant improvement in integration density. The improved high-frequency performance of the FETs better supports high-performance analog circuits for the 90nm lithography node.

TABLE I. SCALING OF CMOS TECHNOLOGY ELEMENTS WITH NODE

Technology elements	180nm	130nm	90nm
L_{GATE} / contacted gate pitch (μm)	0.145 / 0.70	0.092 / 0.48	0.067 / 0.42
V_{DD} for thin oxide FET	1.8V	1.2V	1.2V
V_{DD} for thick oxide FET	3.3V	2.5V / 3.3V	1.8V /2.5V / 3.3V

978-1-4799-7231-9/14 $31.00 © 2014 IEEE

Technology elements	180nm	130nm	90nm
SRAM cell (μm^2)	4.84	2.48	1.18
Density (Kgates/mm^2)	100	200	400
Ion ($\mu A/\mu m$) / Ioff (nA/μm) (Reg. Vt NFET)	600 / 0.02	530 / 0.3	550 / 0.2
Ion ($\mu A/\mu m$) / Ioff (nA/μm) (Reg. Vt PFET)	260 / 0.03	190 / 0.25	215 / 0.35
NFET Peak f_T (GHz)	55	95	145
NFET Peak f_{MAX} (GHz)	145	200	225

D. Passive Devices Offerred

Table II summarizes the SiGe 9HP passive devices and their key figures of merit. Each device is fully supported in the design kit with linked schematic symbols, layout cells and models which are coordinated with design rule checking, layout versus schematic and parasitic extraction design verification tools.

TABLE II.   PASSVE DEVICES AND THEIR KEY HARACTERISTICS

Device	Key Characteristics
Resistors	
N+ diffusion	$R_{SHEET}=100\Omega$/square $\pm12\%$
P+ polysilicon	$R_{SHEET}=380\Omega$/square $\pm16\%$
High-value polysilicon	$R_{SHEET}=2080\Omega$/square $\pm24\%$
Precision polysilicon	$R_{SHEET}=287\Omega$/square $\pm8\%$
Silicided polysilicon	$R_{SHEET}=7.6\Omega$/square $\pm30\%$
Sub-collector	$R_{SHEET}=8.8\Omega$/square $\pm20\%$
Thin-film	$R_{SHEET}=62\Omega$/square $\pm8\%$
Varactors	
nMOS	$C(1.2, -0.5V)=11.7, 3.4fF/\mu m^2$
pMOS	$C(-1.2, 0.5V)=11.4, 1.5fF/\mu m^2$
Thick nMOS	$C(2.5, -0.5V)=6.65, 3.0fF/\mu m^2$
Hyper-abrupt junction	$C(0, 1.5, 3V)= 2.7, 1.5, 0.97fF/\mu m^2$
Capacitors	
MIM(single/dual)	$2.7/5.4fF/\mu m^2$
High-Q MIM	$0.22fF/\mu m^2$
VNCAP[b]	$1.44fF/\mu m^2$

a. 1.8V and 3.3V under- / over-drive variations of these devices are available

b. 4 levels minimum pitch + 2 levels double pitch stacked wires.

Other devices supported in the design kit but not listed above include a one-time programmable e-fuse, inductors, bondpads, microstrip and coplanar transmission lines, and distributed passive devices formed from transmission-line discontinuities. A full ESD protection toolkit is also provided.

E. PIN and Schottky diodes

Diodes are useful in millimeter-wave applications serving as high-frequency small-signal switches or as non-linear elements in passive mixers. The SiGe BiCMOS 9HP PIN diode model will be described by DiVergilio in a separate paper at this symposium. Fig. 6 (a) shows insertion loss at 1V forward bias and isolation at 4V reverse bias, as a function of frequency, for a typical anode dimension. The SiGe BiCMOS 9HP Schottky barrier diode was described by Jain at the 2013

BCTM [9]. A cutoff frequency can be defined as the product of the on-state resistance and off-state capacitance. The cutoff frequency for the SBD as a function of various anode geometries is shown in Fig. 6(b).

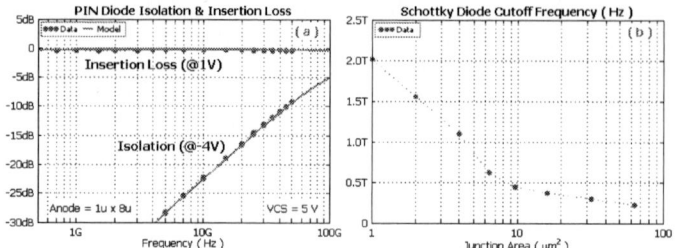

Fig. 6.   Typical diode characteristics: (a) PIN diode insertion loss @1V and isolation @-4V for a 1μm x 8 μm anode, and (b) Schottky diode cutoff frequency vs junction area.

F. mm-Wave Metal Stack

One SiGe BiCMOS 9HP metal stack that emphasizes all of the 200mm wafer 90nm node wiring options is the 10 level stack which features low-K dielectrics with 1X-, 2X-, and 4X-pitch Cu wiring and an optional thick dielectric add-on module which includes a 3.0μm Copper layer, starting at about 10μm above the silicon, as well as a 4.0 μm Aluminum layer. Fig. 7 shows a cross-sectional SEM of the wiring stack.

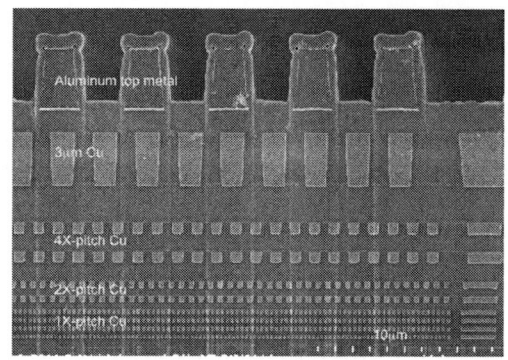

Fig. 7.   Cross section showing 10-level metal stack.

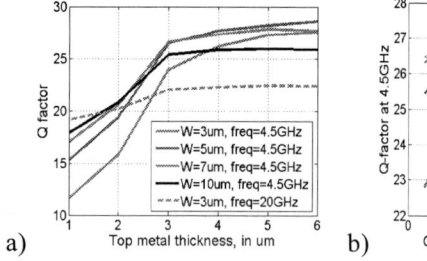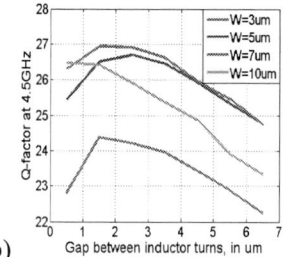

Fig. 8. a) Inductor Q-factor vs. metal thickness for various widths (spacing=3.5μm). b) Inductor Q factor versus metal spacing for various widths.

A study of inductors was performed to quantify how parameters of the technology backend, such as metal thickness, metal pitch, and insulator thickness between conductors and substrate affect Q. As shown in Fig.8, an insulator thickness, t_i of 10μm between the silicon and top copper metal was selected to control loss to the substrate. Skin and proximity effects at RF diminish the benefit on inductor Q of increasing metal thickness (t_m) beyond approximately 3μm, as shown in Fig. 8a. Metal pitch is important even at 3.0μm Cu layers. Fig. 8b shows the layout optimization around the width and space for a 3.0μm copper layer.

III. BENCHMARK CIRCUIT DEMONSTRATIONS

Benchmark circuits were fabricated with 9HP to validate the performance of the technology, the device models and the design flow. A few examples are presented here. Others have been published elsewhere or are in draft.

A. Broadband amplifier

A broadband amplifier serves as an effective technology benchmark because bandwidth is directly related to transistor f_{MAX}. The simplified circuit block (Fig. 9a) and frequency response of a multi-transistor feedback amplifier circuit (Fig. 9b) across multiple SiGe BiCMOS generations demonstrates this concept. The Darlington amplifier is configured with input and output impedances of 50Ω and low-frequency gain of 12dB. The -3dB frequencies across 4 BiCMOS technology generations are: 13.5GHz (0.5μm 5HP), 27.5GHz (180nm 7HP), 54GHz (130nm 8HP), and 79GHz (90nm 9HP). Excellent agreement is shown between simulation and fabricated hardware for the 90nm technology.

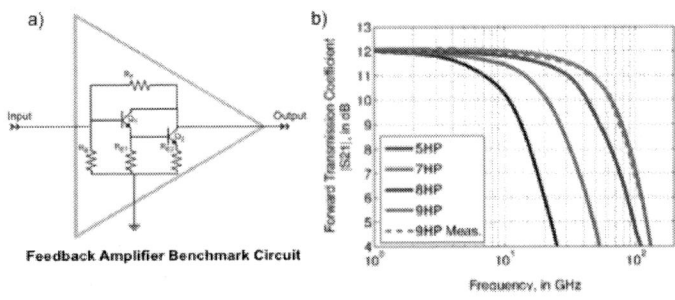

Fig. 9. (a) circuit block. (b) Comparison of forward gain (S_{21}) of a multi-transistor SiGe-HBT feedback amplifier across 4 technology generations.

B. Frequency doubler and LNA

Narrow-band circuits can also serve as suitable benchmarks. The HBT performance of 9HP provides the ability to create mm-wave circuits with very low power consumption. To demonstrate this, a W-band frequency doubler and a three stage 50–90 GHz low noise amplifier (LNA) have been designed and measured. The single-stage frequency doubler provides +0dBm output power at 84GHz to a single-ended 50Ω load with a power consumption of only 12mW and 0dB conversion loss. The measured output power at 84GHz and the conversion gain vs. applied differential input power at 42GHz are shown in Fig. 10(a). The LNA employs three common-emitter stages biased from a 1.25V supply. S-parameter and noise figure measurements are presented in Fig. 10(b). Leveraging the high f_T/f_{MAX} in 9HP, the three-stage LNA attains a small signal gain > 16dB from 50 to 90GHz with just 10mW of power consumption. A noise figure of 5 +/-1dB was measured in the 50-65GHz frequency range.

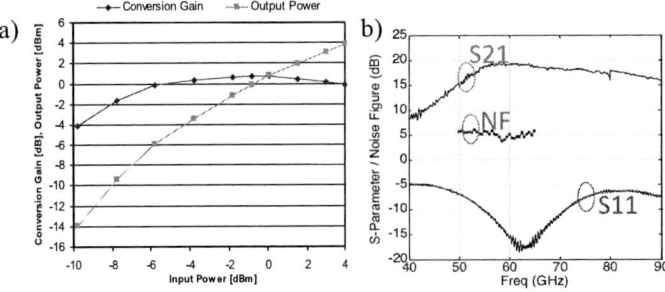

Fig. 10. Exemplary results on low-power mmWave circuits in SiGe 9HP (a) Measured conversion gain and output power for a frequency doubler with an output frquency of 84GHz. (b) Measured s-parameters and noise figure for a 3-stage LNA.

IV. CONCLUSION

The first production-ready 90nm SiGe BiCMOS technology has been presented. It offers a diverse menu of transistors and passive devices to support high frequency applications such as mm-wave wireless communication, serial links above 100Gbps, radar and imaging. Early circuit efforts illustrate its potential capability in these and other high-performance applications.

ACKNOWLEDGMENT

The authors acknowledge the technical support of numerous colleagues at IBM's Essex Junction facility and The Trusted Access Program Office (TAPO), ITT Exelis, Semtech and Tektronix for their support.

REFERENCES

[1] P. Candra et al., "A 130nm SiGe BiCMOS technology for mm-Wave applications featuring HBT with fT/fMAX of 260/320 GHz", IEEE Radio Freq Integrated Circuits Symp (RFIC), pp. 381-384, 2-4 Jun 2013

[2] J. W. May, and G. M. Rebeiz, "Design and Characterization of W - Band SiGe RFICs for Passive Millimeter-Wave Imaging", IEEE Trans on Microwave Theory and Techniques, vol.58, no.5, p.1420, May 2010

[3] M. Chu et al., "A 40 Gs/s Time Interleaved ADC Using SiGe BiCMOS Technology", IEEE J of Solid-State Circuits, vol.45, no.2, pp. 380 - 390, Feb. 2010

[4] D. L. Harame et al., "Current Status and Future Trends of SiGe BiCMOS Technology", IEEE Trans on Electron Devices, Vol. 48, No. 11, pp. 2575, Nov 2001

[5] http://www.mosis.com/vendors/view/ibm/9lprf

[6] Orner, B.A, et al. , "A 0.13 μm BiCMOS technology featuring a 200/280 GHz (fT/fmax) SiGe HBT," Bipolar/BiCMOS Circuits and Technology Meeting (BCTM), 2003 IEEE , pp. 203- 206, 28-30 Sept. 2003.

[7] J. Pekarik, et al., "Co-integration of high-performance and high-breakdown SiGe HBTs in a BiCMOS technology", (BCTM), 2012 IEEE. pp145-148, 30 Sept.-3 Oct. 2012.

[8] Liu, Q. Z., et al. "A Self-Aligned Sacrificial Emitter Process for High Performance SiGe HBT in BiCMOS." ECS Transactions 50.9 (2013): 121-127

[9] V. Jain, et al. ""Schottky Barrier Diodes in 90nm SiGe BiCMOS process operating near 2.0 THz cut-off frequency," (BCTM), 2013 IEEE , pp.73,76, Sept. 30 2013-Oct. 3 2013.

Device and circuit performance of SiGe HBTs in 130nm BiCMOS process with f_T/f_{MAX} of 250/330GHz

Vibhor Jain*[1], T. Kessler[2], B. J. Gross[1], J. J. Pekarik[1], P. Candra[1], P. B. Gray[1], B. Sadhu[3], A. Valdes-Garcia[3],
P. Cheng[2], R. A. Camillo-Castillo[1], K. Newton[1], A. Natarajan[3,4], S. K. Reynolds[3], D. L. Harame[1]

[1]IBM, 1000 River Road, Essex Junction, VT 05452 USA
[2]Formerly at IBM, Essex Junction, VT 05452 USA
[3]IBM T.J. Watson Research Center, 1101 Kitchawan Road, Yorktown Heights, NY 10598
[4]Now with Oregon State University, Corvallis, OR 97330
*vibhorj@us.ibm.com

Abstract— A high performance (HP) SiGe HBT in IBM's 130nm SiGe BiCMOS8XP technology demonstrating peak f_T/f_{MAX} of 250/330GHz is reported for mm-wave and high performance RF/analog applications. The HBT has been developed as an optional device within the existing IBM 130nm SiGe BiCMOS8HP technology which includes a full suite of 130nm RFCMOS FETs, passives and mm-wave distributive passive devices. CML ring oscillators fabricated using HP HBTs in 8XP demonstrate 16% lower delay per stage than 8HP. A VCO design in 8XP has 40% lower phase noise at 36GHz compared to an identical design in 8HP. LNA and PA designs in 8XP show 6dB and 3dB higher gain respectively at 94GHz.

Keywords— SiGe, BiCMOS, HBT, ring oscillator, mm-wave, LNA, PA, VCO

I. INTRODUCTION

A high performance SiGe BiCMOS technology enables potential high speed, mixed signal and analog applications including wireless telecommunication networks, terahertz imaging and sensing, high bandwidth analog to digital converters, automotive radars and instrumentation [1-6].

SiGe BiCMOS8XP technology (8XP) has been developed based on IBM's current SiGe BiCMOS8HP (8HP) technology [7-8]. The HBT integration flow is essentially unchanged with lateral and vertical scaling implemented to meet bipolar performance requirements while preserving manufacturing compatibility with 130nm RF CMOS. 8XP features a high-performance (HP) npn HBT with f_T = 250GHz, f_{MAX} = 330GHz and BV_{CEO} = 1.65V. Specifically the transistor is derived from 8HP technology employing vertical and selective lateral scaling along with a self-aligned emitter-base junction. A high breakdown (HB) npn HBT having f_T = 67GHz, f_{MAX} = 225GHz and BV_{CEO} = 3.3V, shares the same transistor integration with modified sub-collector implant. Improvement in HBT performance for the current 130nm BiCMOS process allows

circuit designers to re-utilize the existing circuit blocks for improved circuit performance. Several identical benchmark circuit designs - CML ring oscillator, VCO, mm-wave LNA and mm-wave PA were measured in both 8HP and 8XP to investigate the improvement in circuit performance with the improved HBT in 8XP.

An overview of the HBT architecture and results is presented in Section II. Section III discusses the benchmark circuit results followed by conclusions in Section IV.

II. HBT OVERVIEW

A. HBT Architecture

A combination of vertical and lateral scaling from the 8HP technology was used to improve HBT performance [7]. In order to increase f_T, the vertical profile was modified as illustrated in Fig. 1. The thickness of intrinsic silicon layer on the collector side was reduced and doping in the selectively implanted collector (SIC) was increased to reduce collector transit delay (τ_C) and for higher J_{KIRK}. The peak germanium concentration in the intrinsic base was increased, the intrinsic base thickness was reduced and the quasi-electric field in the neutral base was increased through steeper Ge grade, all contributing to a lower base transit time (τ_B). The boron concentration in the base was increased providing lower base pinch and link resistance with the narrower base width.

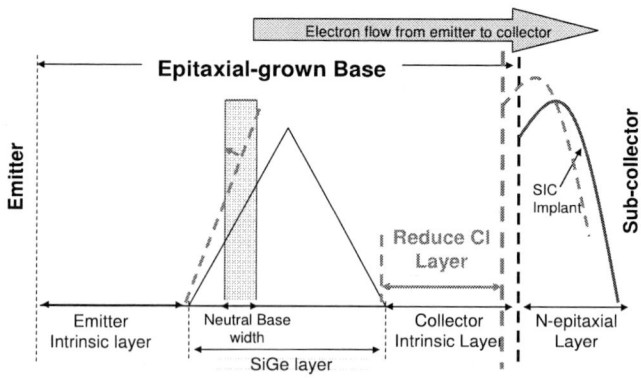

Fig. 1. Schematic illustration of the vertical scaling elements employed in 8XP

The design of the presented SiGe8HP circuits was funded by DARPA Strategic Technology Office (STO) under contract # HR0011-11-C-0136 (Si-Based Phased-Array Tiles for Multifunction RF Sensors, DARPA Order No. 8320/00, Program Code 1P30). The views, opinions, and/or findings contained in this presentation are those of the author and should not be interpreted as representing the official views or policies, either expressed or implied, of the DARPA or the Department of Defense

f_{MAX} was improved through reduction in parasitic base resistance by employing a self-aligned emitter-base contact scheme [9] for lower link resistance and a higher base doping for lower pinch resistance. Width of the SIC implant was scaled for lower collector base capacitance (C_{CB}). The emitter-base self-alignment scheme also reduces the diffusion of boron from the extrinsic-base polysilicon to intrinsic device region, resulting in lower C_{CB} [10]. Details of the BiCMOS process flow are given in [3][7] and the self-aligned HBT integration is described in [9].

B. HBT Results

Fig. 2 shows measured DC gummel curves for a single HBT having emitter area, A_E, of $0.12 \times 2.5 \mu m^2$ in Collector-Base-Emitter-Base-Collector (CBEBC) configuration at

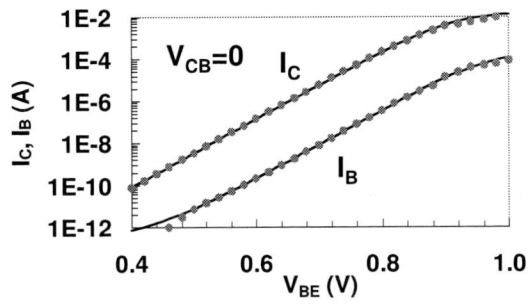

Fig 2. DC Gummel curves for HP SiGe HBT having $A_E = 0.12 \times 2.5 \mu m^2$ at $V_{CB} = 0V$ (red dots) and HiCUM v2.31 model simulations (Black line)

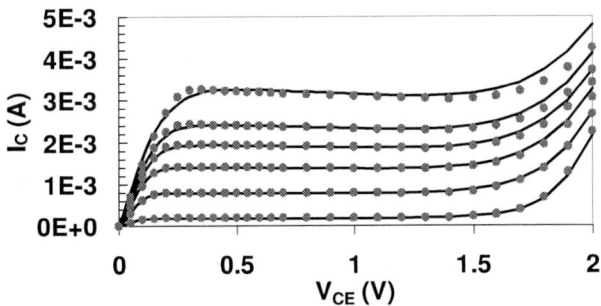

Fig 3. Forced I_B Output curves (I_C-V_{CE}) for the HP SiGe HBT having $A_E = 0.12 \times 2.5 \mu m^2$ at I_B = 0.3, 1.5, 3, 4.5, 6 and 9μA. Red dots represent measured data and black lines represent HiCUM v2.31 model simulations

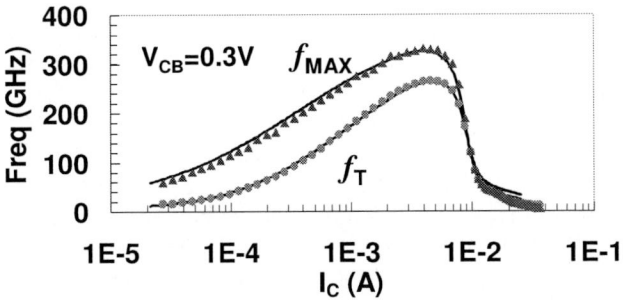

Fig 4. Extracted f_T (●) and f_{MAX} (▲) vs I_C curves for HP HBT at V_{CB}=0.3V. Solid black lines represent HiCUM v2.31 model simulations

V_{CB}=0V along with the HiCUM v2.31 compact model simulations. β for the single device is 650 at $V_{CB} = 0V$ and $V_{BE} = 0.72V$. Fig. 3 shows measured forced I_B output curves (I_C-V_{CE}) for the HP SiGe HBT which has a BV_{CEO} of 1.65V, BV_{CBO} of 5.65V and an emitter resistance of 4Ω.

AC parameters for the HBT were extracted using two port S-parameter measurements with Load-Reflect-Reflect-Match (LRRM) calibration and open and short de-embedding to remove parasitic impedances associated with measurement cables, probes, wiring and pads. Fig. 4 shows extracted f_T and f_{MAX} vs I_C curves for HP HBT at V_{CB}=0.3V. f_T and f_{MAX} were extracted using extrapolations from -20dB/decade fit to the measured current gain H_{21} and Mason's Unilateral gain U. The HBT demonstrates a peak f_T of 250 GHz ($A_E = 0.12 \times 2.5 \mu m^2$) at

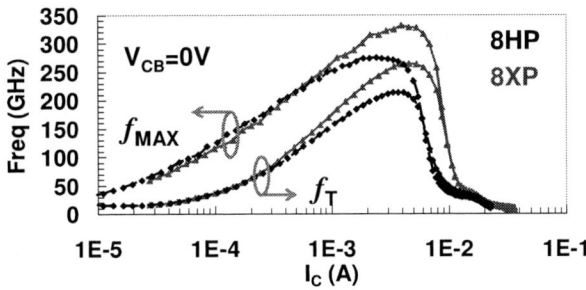

Fig. 5. f_T and f_{MAX} vs I_C curves for HP HBT in 8HP and 8XP at V_{CB}=0V demostrating the improvement in peak f_T/f_{MAX} and I_C at peak performance for $A_E = 0.12 \times 2.5 \mu m^2$ device

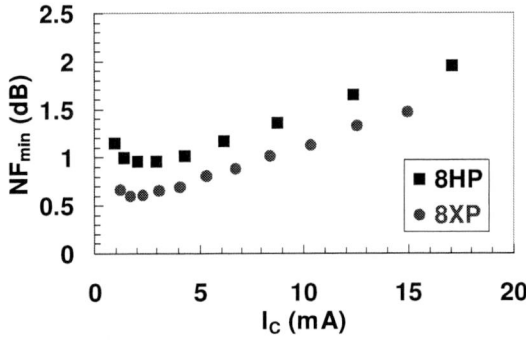

Fig. 6. Measured NF for a $0.12 \times 18 \mu m^2$ HP HBT in 8XP and 8HP at freq = 12GHz and V_{CE} = 1.35V showing a 0.3dB reduction in NF

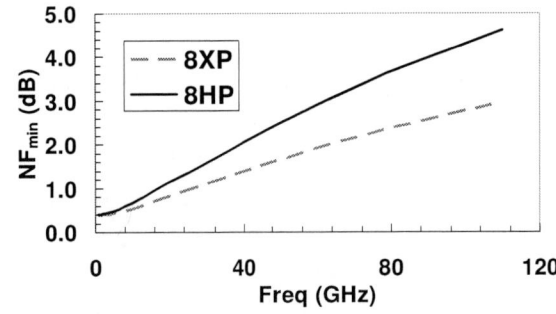

Fig. 7. Simulated NF_{min} as a function of frequency for a $0.12 \times 18 \mu m^2$ HP HBT in 8XP and 8HP at V_{CB} = 0.3V and I_C = 2mA

$I_C = 4.4\text{mA}$ ($J_{KIRK} = 14.67\text{mA}/\mu\text{m}^2$) and $V_{CE} = 1.2\text{V}$. Peak f_{MAX} at the same bias condition is extracted as 330GHz. Good agreement with the HICUM v2.31 compact model is shown for both the DC and AC characteristics.

Compared to the high performance HBT in 8HP, peak f_T has improved by 40GHz from 210GHz to 250GHz and peak f_{MAX} has improved from 265GHZ to 330GHz (at $V_{CB} = 0.3\text{V}$). Fig. 5 shows improvement in peak f_T and f_{MAX} performance and J_{KIRK} for the HP device in 8XP compared to 8HP at $V_{CB} = 0\text{V}$. Base resistance R_B has decreased by ~30% due to lower link and pinch resistance components. An increase in SIC doping and thinner collector intrinsic layer has increased the intrinsic C_{CB} of the HBT by 20%. The reduction in R_B and improvement in f_T compared to 8HP improves the NF of 8XP HP HBT by ~0.3dB at 12GHz (Fig. 6). This improvement in NF is more prominent at higher frequency as shown in Fig. 7. At 100GHz, for identical I_C bias of 2mA, NF of 8XP HBT is lower by ~1.6dB.

The 8XP technology also co-integrates a high breakdown (HB) NPN HBT demonstrating simultaneous peak f_T and f_{MAX} of 67GHz and 225GHz at $V_{CB}=1\text{V}$ and $I_C = 0.4\text{mA}$ (Fig. 8). These HBTs have $BV_{CEO} = 3.3\text{V}$ and $BV_{CBO} = 11.7\text{V}$. Peak f_T has improved from 60GHz to 67GHz compared to HB HBTs in 8HP.

III. CIRCUIT RESULTS

A. Ring Oscillators

Bipolar CML ring-oscillators were designed using HBTs in CBEBC configuration with $A_E = 0.12\text{x}2.5\mu\text{m}^2$. The CML delay-stages were designed with load-resistance = 100Ω, using p+ polysilicon resistors. The ring-oscillator has 24 delay-stages with fan-out = 1 and a hard-wired inversion to produce a negative loop gain. The ring-oscillator drives a buffer amplifier, which in turn drives a 17-stage CML divider to produce an output signal in the 40 KHz range.

Unloaded-gate-delay vs bias-current-per-stage (I_{TAIL}) for the CML ring-oscillator is shown in Fig. 9. Results from the ring-oscillator circuits using 8XP HP HBTs are compared to circuits with identical layout processed in 8HP. The ring-oscillator results demonstrate that 8XP has faster switching delays, and can also operate at higher bias currents before self-saturation limits switching performance. Improvement in delay for 8XP is primarily due to lower R_B [11]. For 8XP design,

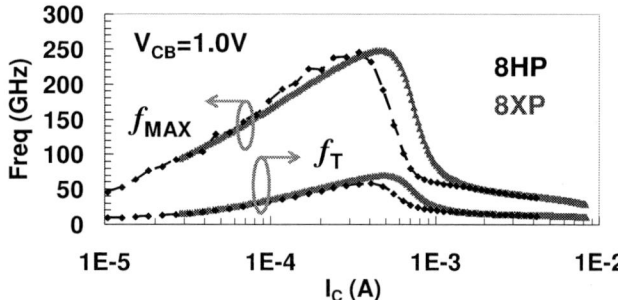

Fig 8. f_T and f_{MAX} vs I_C curves for High Breakdown HBT in 8HP and 8XP at $V_{CB} = 1.0\text{V}$ for $A_E = 0.12\text{x}2.5\mu\text{m}^2$ device

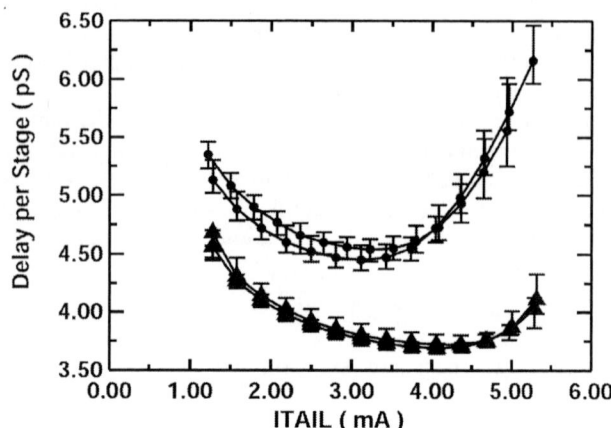

Fig. 9. Ring-Oscillator delay-per-stage vs bias-current-per-stage for 8XP (▲) vs 8HP (●) at room temperature. 2 wafers for each technology. The symbols are wafer mean values, and the error-bars represent the range of measured performance from 15 sites per wafer.

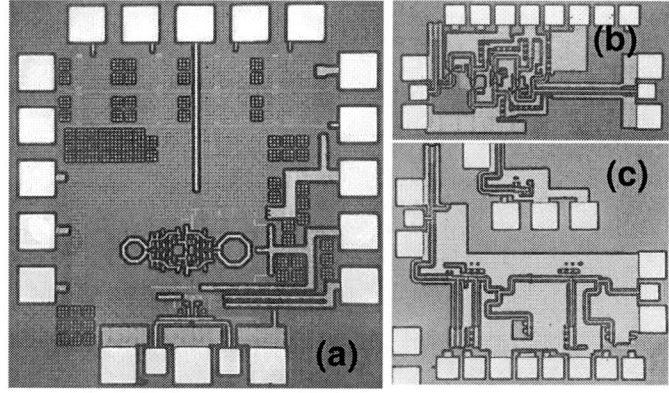

Fig. 10. Die photos of (a) VCO (b) mm-wave LNA and (c) mm-wave PA

lowest delay per stage (τ_{PD}) is measured as 3.72ps at $I_{TAIL} = 4.06\text{mA}$ while for 8HP it is 4.45ps at $I_{TAIL} = 3.12\text{mA}$, an improvement of 16%. For the same I_{TAIL} current of 3.12mA, 8XP design has a τ_{PD} of 3.80 ps. These ring oscillator designs have not been optimized for lower τ_{PD} and are being used as a process monitor in the technologies. Optimization of RO design can reduce τ_{PD} by another 30-40% for the same bias conditions.

B. Low noise mm-wave LC VCO

Low noise frequency synthesizers at mm-wave frequencies are critical for achieving higher data rate communications [5-6]. The lower base resistance and higher f_T/f_{MAX} of the SiGe8XP technology enables low noise mm-wave VCO designs. To compare the performance improvement, identical state-of-the-art VCOs were implemented in the 8HP and 8XP technologies (Fig. 10(a)). The VCOs are bipolar adaptations of the LiTVCO principle presented in [12]. They were measured under identical bias conditions using the same test setup. The phase noise performance across different offset frequencies from a 36GHz carrier is shown in Fig. 11. As seen in the figure, the phase noise in the 8XP design improves by about

978-1-4799-7231-9/14 $31.00 © 2014 IEEE

1.4dB compared to the 8HP design, signifying a 40% reduction in phase noise.

C. mm-wave LNA and PA

The higher f_T/f_{MAX} of the 8XP technology helps to achieve higher amplifier gain at mm-wave frequencies. In order to compare and quantify the improvement in gain, identical 4-stage low noise amplifier (LNA) and 2-stage power amplifier (PA) breakout designs were fabricated in 8HP and 8XP processes (Figs 10(b) and 10(c)). These breakout designs, designed to operate at 94GHz, are similar to those integrated in [6] as part of a scalable mm-wave phased array. The S-parameters of these designs were measured with identical supply voltages and DC current biasing. The same setup comprising a vector network analyzer and a broadband test set was used for all measurements. The LNA consumes 18mA of bias current while the PA consumes 22mA from a 2.7V supply. As seen from the S_{21} in Fig. 12, the LNA gain improves by ~7dB and the PA gain improves by ~3dB, both at 94GHz. This demonstrates a significant gain and power efficiency improvement over the SiGe8HP designs at mm-wave frequencies. The performance benefits can also be leveraged as a reduction in power consumption for these circuits.

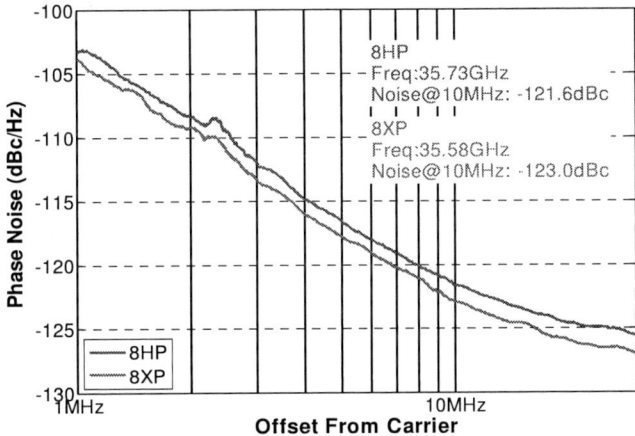

Fig. 11. Phase noise comparison of state-of-the-art VCO in 8HP and 8XP

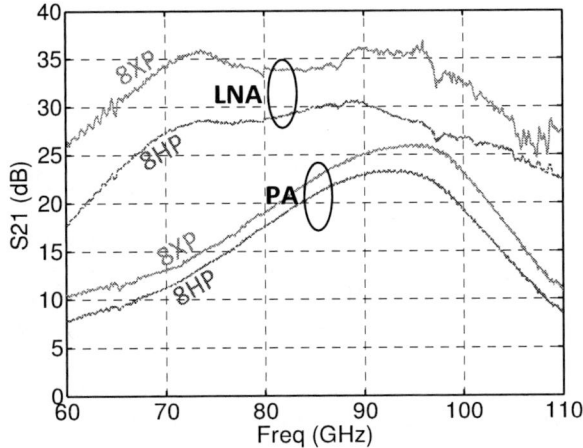

Fig 12. Gain comparison of mm-wave LNA and PA in 8HP and 8XP

IV. CONCLUSIONS

A high performance SiGe HBT demonstrating f_T/f_{MAX} ~ 250/330GHz has been successfully integrated into IBM's 130nm SiGe BiCMOS8XP technology to enable high performance mm-wave designs. In order to demonstrate the improvement in circuit performance due to the technology enhancements, identical CML ring oscillators, mm-wave LNAs, PAs and VCOs were implemented in both SiGe8HP and SiGe8XP technologies. Compared to the 8HP circuits, the LNA gain in 8XP improved by ~7dB at 94GHz, while the PA gain improved by ~3dB at the same frequency. Moreover, a 40% phase noise reduction was achieved in the 8XP 36GHz LC VCO compared to the 8HP VCO. CML ring oscillator delay also improved by 16%. These circuit results illustrate the potential capability of the new technology for mm-wave and high performance RF/analog applications.

ACKNOWLEDGMENT

The authors would like to thank C. Lamothe, A. Beganovic, C. Hedges, J. Rascoe and D. Cook for AC and DC measurements.

REFERENCES

[1] J. W. May, and G. M. Rebeiz, "Design and Characterization of W-Band SiGe RFICs for Passive Millimeter-Wave Imaging", IEEE Trans on Microwave Theory and Techniques, vol.58, no.5, pp.1420-1430, May 2010

[2] M. Chu et al., "A 40 Gs/s Time Interleaved ADC Using SiGe BiCMOS Technology", IEEE J of Solid-State Circuits, vol.45, no.2, pp. 380 - 390, Feb. 2010

[3] D. L. Harame et al., "Current Status and Future Trends of SiGe BiCMOS Technology", IEEE Trans on Electron Devices, Vol. 48, No. 11, pp. 2575, Nov 2001

[4] H. B. Wallace, "Advanced Millimeter-Wave Multifunction Systems and the Implications for Semiconductor Technology", IEEE Compound Semiconductor Integ Circuit Symp (CSICS), pp.1-3, 16-19 Oct. 2011

[5] Natarajan, et al., "A Fully-Integrated 16-Element Phased-Array Receiver in SiGe BiCMOS for 60-GHz Communications", IEEE J of Solid State Circuit, vol.46, no.5, pp.1059-1075, May 2011

[6] A. Valdes-Garcia, et al., "A fully-integrated dual-polarization 16-element W-band phased-array transceiver in SiGe BiCMOS", IEEE Radio Frequency Integrated Circuits Symposium (RFIC), pp.375-378, 2-4 Jun 2013

[7] P. Candra et al., "A 130nm SiGe BiCMOS technology for mm-Wave applications featuring HBT with fT/fMAX of 260/320 GHz", IEEE Radio Freq Integrated Circuits Symp (RFIC), pp. 381-384, 2-4 Jun 2013

[8] B. A. Orner, et al., "A 0.13 μm BiCMOS technology featuring a 200/280 GHz (fT/fMAX) SiGe HBT", Proc of the Bipolar/BiCMOS Circuits and Technology Meeting, pp. 203- 206, 28-30 Sept. 2003

[9] J. Pekarik et al., "A 90nm SiGe BiCMOS Technology for mm-wave and high-performance analog applications", submitted Proc BCTM, 2014

[10] P. Cheng et al., "A novel Ccb and Rb reduction technique for high-speed SiGe HBTs", Proc. IEEE BCTM, 2012

[11] M. Alioto, G. Palumbo, "Oscillation frequency in CML and ESCL ring oscillators," IEEE Transactions on Circuits and Systems I: Fundamental Theory and Applications, vol.48, no.2, pp. 210-214, Feb 2001

[12] B. Sadhu, et al., "A linearized, lowphase-noise VCO-based 25-GHz PLL with autonomic biasing," IEEE Journal of Solid-State Circuits, vol.48, no.5, pp.1138–1150, May 2013

978-1-4799-7231-9/14 $31.00 © 2014 IEEE

Low-Voltage Organic Field-Effect Transistors for Flexible Electronics

Ute Zschieschang [a], Reinhold Rödel [a], Ulrike Kraft [a], Kazuo Takimiya [b],
Tarek Zaki [c], Florian Letzkus [d], Jörg Butschke [d], Harald Richter [d], Joachim N. Burghartz [d],
Wei Xiong [e], Boris Murmann [e], Hagen Klauk [a]

[a] Max Planck Institute for Solid State Research, Heisenbergstr. 1, 70569 Stuttgart, Germany
[b] Emergent Molecular Function Research Team, RIKEN Advanced Science Institute, Wako, Saitama, Japan
[c] Institute for Nano- and Microelectronic Systems (INES), University of Stuttgart, Germany
[d] Institut für Mikroelektronik/IMS CHIPS, Stuttgart, Germany
[e] Center for Integrated Systems, Stanford University, Stanford, USA

Abstract - **A process for the fabrication of bottom-gate, top-contact (inverted staggered) organic thin-film transistors (TFTs) with channel lengths as short as 1 µm on flexible plastic substrates has been developed. The TFTs employ vacuum-deposited small-molecule semiconductors and a low-temperature-processed gate dielectric that is sufficiently thin to allow the TFTs to operate with voltages of about 3 V. The p-channel TFTs have an effective field-effect mobility of about 1 cm^2/Vs, an on/off ratio of 10^7, and a signal propagation delay (measured in 11-stage ring oscillators) of 300 ns per stage. For the n-channel TFTs, an effective field-effect mobility of about 0.06 cm^2/Vs, an on/off ratio of 10^6, and a signal propagation delay of 17 µs per stage have been obtained.**

I. Introduction

Organic thin-film transistors (TFTs) can typically be fabricated at temperatures below about 100 °C and thus not only on glass substrates, but also on a variety of unconventional substrates, such as plastics and paper. This makes organic TFTs potentially useful for the realization of flexible, large-area electronics applications, such as rollable or foldable information displays [1], conformable sensor arrays [2], and plastic circuits [3]. In some of the more advanced applications envisioned for organic TFTs, such as the integrated row and column drivers of flexible active-matrix organic light-emitting diode (AMOLED) displays [4], the TFTs will have to be able to control electrical signals of a few volts at frequencies of several megahertz. For portable applications powered by small batteries, an additional requirement is a low power consumption, which implies a complementary circuit technology and thus the availability of both p-channel and n-channel TFTs with sufficient static and dynamic performance on plastic substrates.

II. Materials and Design Considerations

The first requirement for achieving high switching frequencies in organic TFTs is efficient charge transport in the organic semiconductor layer. This requirement can be met by choosing small-molecule semiconductors that provide good molecular ordering and usefully large field-effect mobilities even when processed at temperatures below 100 °C. In the case of organic p-channel TFTs, for example, the alkylated thienoacene C_{10}-DNTT that was recently developed in the group of Kazuo Takimiya at Hiroshima University has shown very promising field-effect mobilities in the range of 10 cm^2/Vs [5,6]. For organic n-channel TFTs, a number of naphthalene and perylene tetracarboxylic diimides equipped with strongly electron-withdrawing core substituents and fluoroalkyl chains at both imide positions have recently been developed, such as NTCDI-Cl$_2$-(CH$_2$C$_3$F$_7$)$_2$ (which was recently developed in the group of Frank Würthner at the University of Würzburg [7]) and PTCDI-(CN$_2$)-(CH$_2$C$_3$F$_7$)$_2$ (recently developed in the group of Antonio Facchetti and Tobin Marks at Northwestern University [8]), both of which have shown electron mobilities exceeding 1 cm^2/Vs along with excellent air stability.

The second requirement for achieving high switching frequencies in organic TFTs is a small channel length [9-12]. To meet this requirement, a TFT process in which high-resolution silicon stencil masks are employed for the patterning of the source and drain contacts and all other components of the transistors, including the gate electrodes and the organic semiconductor layer, has recently been developed [11,12]. With this process, bottom-gate, top-contact organic TFTs with a channel length of 1 µm can be fabricated on plastic substrates without exposing the organic semiconductors to potentially harmful organic solvents and photoresists.

978-1-4799-7231-9/14 $31.00 © 2014 IEEE

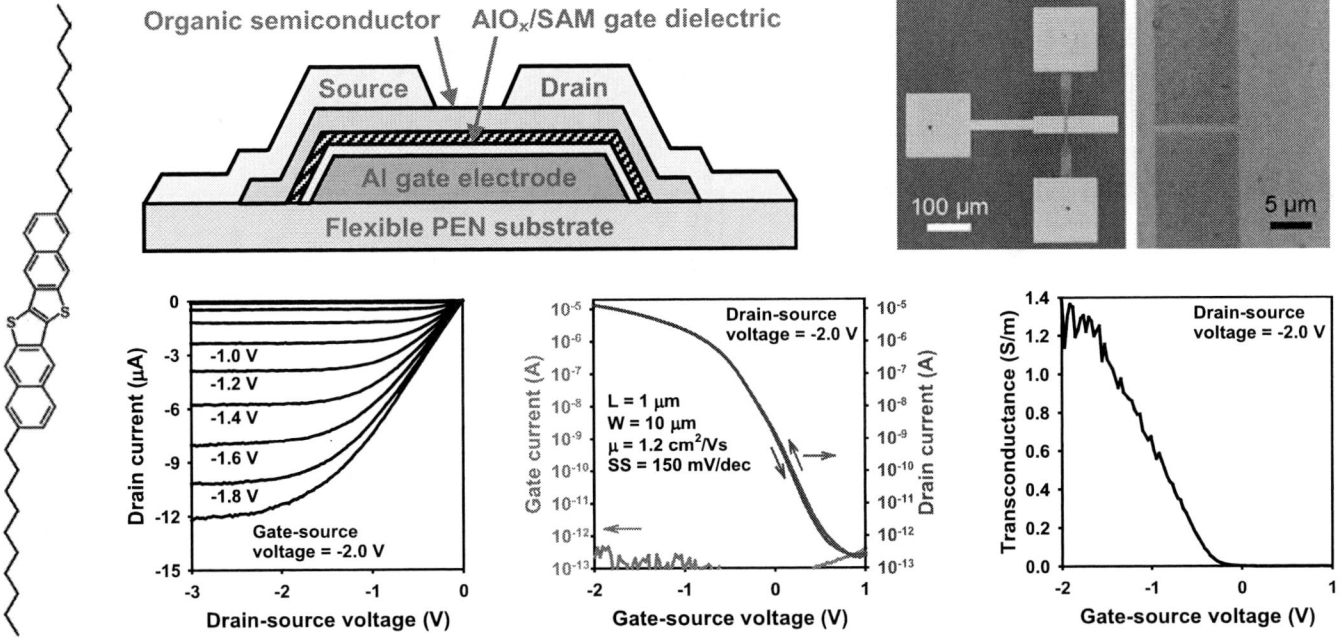

Fig. 1. Schematic cross-section, photographs, and measured current-voltage characteristics of a C_{10}-DNTT p-channel TFT with a channel length of 1 µm fabricated on a flexible polyethylene (PEN) substrate. The TFT has an effective field-effect mobility of 1.2 cm^2/Vs, an on/off current ratio of 10^7, a subthreshold swing of 150 mV/decade, and a width-normalized transconductance of 1.2 S/m. Also shown is the chemical structure of the organic semiconductor (C_{10}-DNTT) employed for these TFTs [12].

III. TRANSISTOR FABRICATION PROCESS

A set of four stencil masks is required to fabricate either p-channel or n-channel TFTs: one mask each to pattern the gate electrodes, the gate vias, the organic semiconductor layer, and the source/drain contacts. To fabricate both p-channel and n-channel TFTs on the same substrate for the realization of complementary circuits, a total of five masks are required, since in this case two different semiconductors need to be individually deposited and patterned.

In the first process step, a thin layer of aluminum is deposited directly onto the substrate by thermal evaporation in vacuum through the first stencil mask. In order to define the locations for the gate vias, a thin layer of gold is then deposited by thermal evaporation in vacuum through the second stencil mask onto specific locations on the aluminum outside of the active TFT areas. In the third step, a hybrid gate dielectric composed of a 3.6-nm-thick layer of aluminum oxide (obtained by briefly exposing the surface of the aluminum gate electrodes to an oxygen plasma) and a 1.7-nm-thick self-assembled monolayer (SAM) of *n*-tetradecylphosphonic acid (obtained by briefly immersing the substrate into a 2-propanol solution of the phosphonic acid) is then produced [11,12]. The exact thickness of this gate dielectric can be easily controlled by the plasma conditions and by the choice of the alkylphosphonic acid, and it forms only on the surface of the aluminum gate electrodes, but not in the gold-covered via locations. In the fourth and fifth steps, thin layers of the organic semiconductors (one each for the p-channel and n-channel TFTs) are deposited onto the AlO$_x$/SAM gate dielectric by sublimation in vacuum through the third and fourth stencil masks. Finally, a thin layer of gold is deposited by thermal evaporation in vacuum through the fifth stencil mask to define the source and drain contacts of the TFTs as well as the interconnects for the integrated circuits. The highest temperature during the fabrication process is 100 ºC.

IV. PERFORMANCE OF p-CHANNEL TFTs

Figure 1 shows two photographs and the measured current-voltage characteristics of a C_{10}-DNTT p-channel TFT with a channel length of 1 µm and a channel width of 10 µm fabricated on a flexible, 125-µm-thick polyethylene naphthalate (Teonex® Q65 PEN; kindly provided by William A. MacDonald, Du-Pont Teijin Films, Wilton, UK) substrate. The small thickness (5.3 nm) and large capacitance per unit area (800 nF/cm^2) of the AlO$_x$/SAM gate dielectric allow the TFTs to operate with low voltages of about 3 V. The TFTs have an effective hole mobility in the saturation regime of 1.2 cm^2/Vs (limited by the contact resistance [11]), a subthreshold swing of 150 mV/decade, and a width-normalized transconductance of 1.2 S/m. While the transconductance is smaller by several orders of magnitude than that of state-of-the-art silicon MOSFETs [13], it is believed to be the largest transconductance reported thus far for organic TFTs on plastic substrates.

978-1-4799-7231-9/14 $31.00 © 2014 IEEE

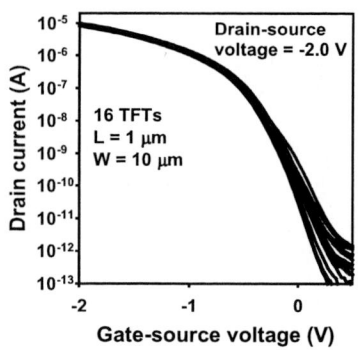 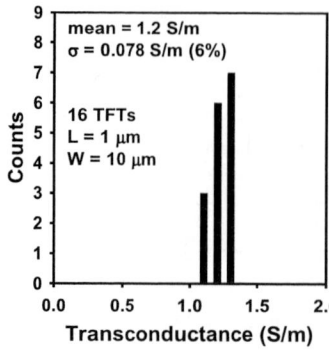 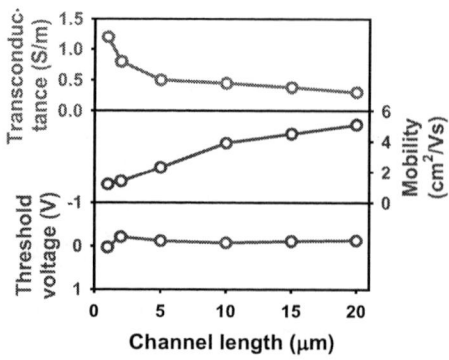

Fig. 2. Distribution of the measured transfer characteristics and distribution of the width-normalized transconductance in an array of sixteen C_{10}-DNTT TFTs with a channel length of 1 μm fabricated on a flexible PEN substrate [12].

Figure 2 shows the measured transfer characteristics and the distribution of the extracted width-normalized transconductance in an array of sixteen C_{10}-DNTT p-channel TFTs with a nominal channel length of 1 μm and a nominal channel width of 10 μm fabricated on a flexible PEN substrate. Across the array of 16 TFTs, the width-normalized transconductance varies between a minimum of 1.16 S/m and a maximum of 1.40 S/m, with a standard deviation of 6%. According to the standard FET equations, the transconductance is inversely proportional to the channel length, so assuming that the other transistor parameters, including the effective field-effect mobility and the gate-dielectric capacitance, are constant across the substrate, a standard deviation in the transconductance of 6% may simply reflect a standard deviation in the channel length of 6%, which in the case of an intended channel length of 1 μm corresponds to a standard deviation of 60 nm. This is substantially larger than the standard deviation of the gate length of silicon MOSFETs [14], but in contrast to silicon MOSFETs, the organic TFTs were fabricated without lithographic pattern reduction, which may explain the larger standard deviation in the feature size of the TFTs. Figure 2 also shows how the width-normalized transconductance, the effective mobility and the threshold voltage depend on the channel length. As can

be seen, the effective mobility decreases significantly with decreasing channel length, due to the fact that the relative influence of the contact resistance on the total device resistance increases with decreasing channel length [11]. Note that the contact resistance of organic TFTs is larger than in silicon MOSFETs due to the lack of contact doping.

The third requirement for achieving high switching frequencies (in addition to a large field-effect mobility and a small channel length) is a small gate capacitance. One component of the gate capacitance is the parasitic capacitance formed by the geometric overlaps between the gate electrode and the source/drain contacts, so reducing not only the channel length, but also the gate overlap can be useful in view of high-frequency TFT operation [11]. The gate-channel, gate-source and gate-drain capacitances and their contributions to the device characteristics have been analyzed by admittance measurements [15] and compact modeling [16]. Figure 3 shows the schematic and the photograph of an 11-stage ring oscillator with output buffer comprised of unipolar inverters with saturated load based on C_{10}-DNTT p-channel TFTs fabricated on a flexible PEN substrate. In the most aggressive design, the organic TFTs have a channel length (L) of 1 μm, a gate overlap (L_C) of 5 μm, and channel widths (W) of 24 μm

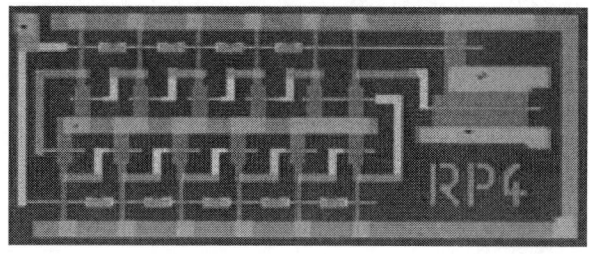

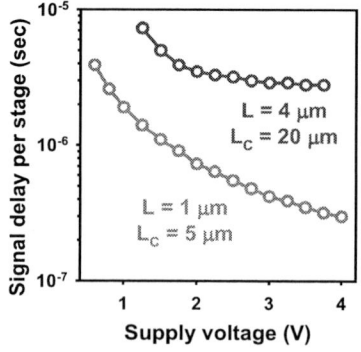

Fig. 3. Circuit schematic, photograph and measured signal propagation delay per stage as a function of the supply voltage of 11-stage unipolar ring oscillators fabricated with C_{10}-DNTT TFTs with two different channel lengths (1 μm and 4 μm) on a flexible PEN substrate. For a channel length of 1 μm, the measured stage delay is 1.9 μs at a supply voltage of 1 V, 730 ns at 2 V, 420 ns at 3 V, and 300 ns at 4 V [12].

978-1-4799-7231-9/14 $31.00 © 2014 IEEE

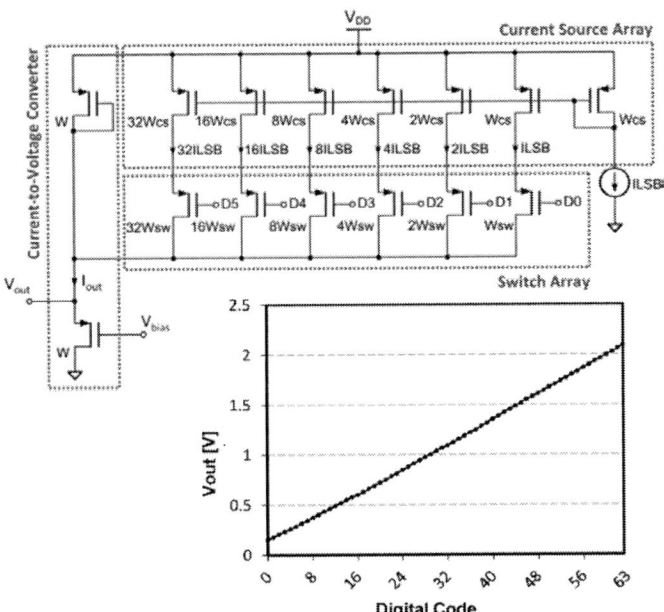

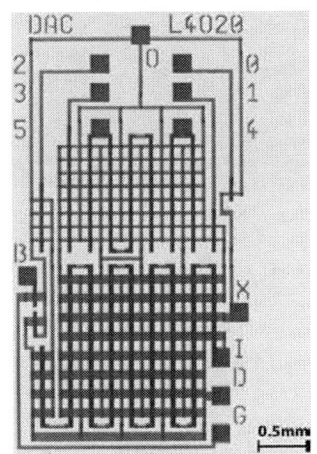

Fig. 4. Circuit schematic, photograph, and measured transfer function of a 6-bit digital-to-analog converter based on low-voltage organic p-channel TFTs with a channel length of 4 μm fabricated on a glass substrate. The DAC has a supply voltage of 3.3 V and a maximum sampling rate of 100 kS/s [19].

(for the drive TFTs) or 72 μm (for the load TFTs). In a more relaxed design, the channel length is 4 μm and the gate overlap is 20 μm. Also shown in Figure 3 are the signal propagation delays per stage measured in these ring oscillators and plotted as a function of the supply voltage. For the more aggressive dimensions ($L = 1$ μm, $L_C = 5$ μm), the measured stage delay is 1.9 μs at a supply voltage of 1 V, 730 ns at 2 V, 420 ns at 3 V, and 300 ns at 4 V. These are believed to be the first organic TFTs fabricated on flexible plastic substrates demonstrating cutoff frequencies above 1 MHz at supply voltages below 10 V. For larger supply voltages, cutoff frequencies up to 30 MHz have been reported [17,18].

V. D/A CONVERTER BASED ON ORGANIC p-CHANNEL TFTs

Figure 4 shows the schematic, a photograph, and the measured transfer function of a 6-bit binary-weighted current-steering digital-to-analog converter (DAC) based on low-voltage organic p-channel TFTs with a channel length of 4 μm fabricated on a glass substrate [19]. The DAC has a supply voltage of 3.3 V, a circuit area of 2.6×4.6 mm^2, and a maximum sampling rate of 100 kS/s. Owing to the high precision and small line-edge roughness (<50 nm) of the silicon stencil masks, the relative current mismatch of the TFTs is 4% [19], providing maximum differential and integral nonlinearities (DNL, INL) of 0.69 and 1.16 LSB at 1 kS/s after calibration [19].

VI. PERFORMANCE OF n-CHANNEL TFTs

By choosing an organic semiconductor with a large electron affinity, i.e., with a large energy difference between the lowest unoccupied molecular orbital (LUMO) and the vacuum level, it is possible to realize organic n-channel TFTs. An example is the recently developed core-chlorinated and fluoroalkyl-substituted naphthalene tetracarboxylic diimide NTCDI-Cl$_2$-(CH$_2$C$_3$F$_7$)$_2$ [7].

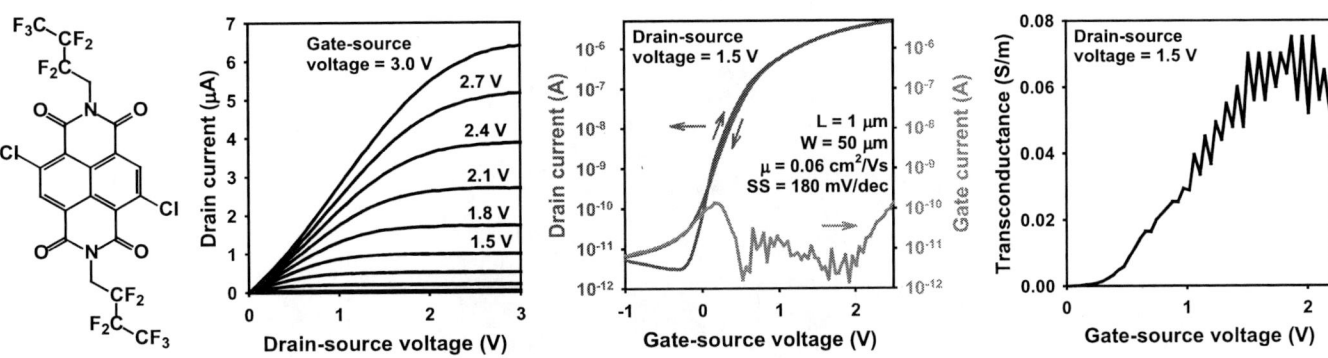

Fig. 5. Electrical characteristics of an NTCDI-Cl$_2$-(CH$_2$C$_3$F$_7$)$_2$ n-channel TFT with a channel length of 1 μm. The TFT has an effective field-effect mobility of 0.06 cm^2/Vs, an on/off current ratio of 10^6, a subthreshold swing of 180 mV/decade, and a width-normalized transconductance of 0.06 S/m. Also shown is the chemical structure of the organic semiconductor NTCDI-Cl$_2$-(CH$_2$C$_3$F$_7$)$_2$ employed for these n-channel TFTs [20].

978-1-4799-7231-9/14 $31.00 © 2014 IEEE   103

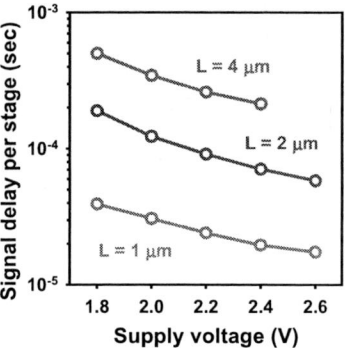

Fig. 6. Photograph and measured signal propagation delay per stage as a function of the supply voltage of 11-stage complementary ring oscillators fabricated with and NTCDI-Cl$_2$-(CH$_2$C$_3$F$_7$)$_2$ n-channel TFTs and DNTT p-channel TFTs on a flexible PEN substrate [20].

Figure 5 shows the chemical structure of this semiconductor, along with the measured current-voltage characteristics of an NTCDI-Cl$_2$-(CH$_2$C$_3$F$_7$)$_2$ n-channel TFT with a channel length of 1 µm and a channel width of 50 µm [20]. The TFTs have an effective electron mobility in the saturation regime of 0.06 cm^2/Vs, an on/off ratio of 10^6, a subthreshold swing of 180 mV/decade, and a width-normalized transconductance of 0.06 S/m. While these parameters are notably inferior to those of the C$_{10}$-DNTT p-channel TFTs shown in Figure 2, they represent the best performance currently achievable in air-stable, low-voltage organic n-channel TFTs with such a small channel length.

Although organic n-channel TFTs with significantly larger electron mobilities have been reported, these TFTs either had much larger channel lengths (thus suppressing the detrimental influence of the contact resistance on the total device resistance, leading to effective electron mobilities up to 3.5 cm^2/Vs [21]) or they could only be operated in an inert gas or in vacuum, due to a lack of air stability of the semiconductor (an example is the fullerene C$_{60}$) [22].

The fact that the mobility of the n-channel TFTs is significantly smaller than that of the p-channel TFTs implies that the dynamic performance of complementary circuits based on organic n-channel and p-channel TFTs will be limited by the longer signal propagation delay of the n-channel devices. Figure 6 shows the photograph of an 11-stage complementary ring oscillator based on NTCDI-Cl$_2$-(CH$_2$C$_3$F$_7$)$_2$ n-channel TFTs and DNTT p-channel TFTs fabricated on a flexible PEN substrate, along with the signal delays measured for ring oscillators with minimum feature sizes of 1 µm, 2 µm and 4 µm. For the most aggressive design (L = 1 µm), the measured stage delay is 17 µs at a supply voltage of 2.6 V, which is indeed slower by more than an order of magnitude compared with the signal delay of the unipolar all-p-channel ring oscillators shown in Figure 3. However, regardless of the lower dynamic performance, these results demonstrate the feasibility of realizing low-voltage, low-power complementary circuits on flexible plastic substrates.

VII. ORGANIC COMPLEMENTARY A/D CONVERTER

As an early demonstration of low-voltage, low-power mixed-signal organic complementary circuits, a 6-bit switched-capacitor (C-2C) analog-to-digital converter (ADC) based on 19 thin-film capacitors, 27 organic p-channel TFTs and 26 organic n-channel TFTs (all having a channel length of 20 µm) was fabricated on a glass substrate [23]. The circuit schematic, a photograph, and the measured transfer function are shown in Figure 7. The ADC has a supply voltage of 3 V, a circuit area of 28 × 22 mm^2, and a maximum sampling rate of 100 S/s.

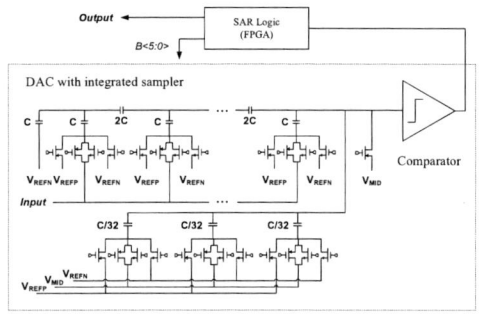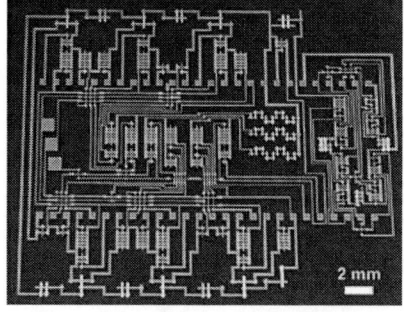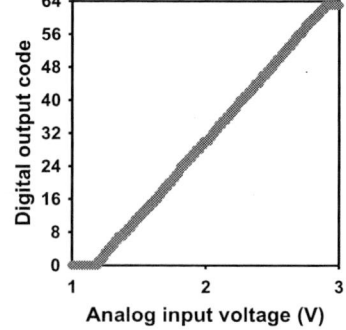

Fig. 7. Circuit schematic, photograph, and measured transfer function of a 6-bit complementary analog-to-digital converter fabricated on a glass substrate. The ADC has a supply voltage of 3 V and a maximum sampling rate of 100 S/s [23].

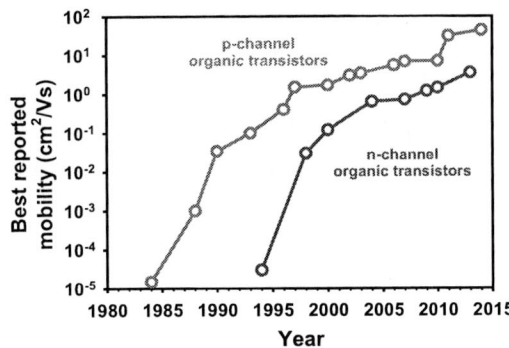

Fig. 8. Historic development of the field-effect mobilities of organic p-channel and n-channel TFTs reported in each calendar year, showing that the gap between the performance of p-channel and n-channel organic TFTs is slowly closing.

VIII. CONCLUSIONS

Organic thin-film transistors can be fabricated at low process temperatures of typically about 100 °C, which makes them useful for flexible electronics applications on unconventional substrates, such as plastics or paper. In this work, low-voltage p-channel and n-channel organic TFTs with a channel length of 1 μm as well as unipolar and complementary mixed-signal circuits with promising static and dynamic performance were demonstrated.

A number of challenges remain. One is the problematic compromise between the performance and the long-term stability of organic transistors [24], although at least in the case of p-channel TFTs, some progress has recently been made through the development of organic semiconductors with large ionization energies and tight molecular packing that provide both large carrier field-effect mobilities and good air stability [25]. Another challenge is the significant gap between the carrier mobilities of organic p-channel and organic n-channel TFTs (>10 cm²/Vs in p-channel, <5 cm²/Vs in n-channel TFTs). However, it appears that this gap is slowly closing, thanks to increased research in the area of organic n-channel TFTs (see Figure 8 and [26]).

ACKNOWLEDGEMENTS

This work was partially funded by BASF SE, by the German Ministry of Education and Research under the Grants 13N10205, 13N12084, and 1612000463, and by the German Research Foundation (DFG) under Grants KL 2223/5-1 and KL 2223/5-2.

REFERENCES

[1] M. Noda, N. Kobayashi, M. Katsuhara, A. Yumoto, S. Ushikura, R. Yasuda, N. Hirai, G. Yukawa, I. Yagi, K. Nomoto, and T. Urabe, "An OTFT-driven rollable OLED display," *J. Soc. Inf. Display*, vol. 19, p. 316 (2011).

[2] T. Sekitani, T. Yokota, U. Zschieschang, H. Klauk, S. Bauer, K. Takeuchi, M. Takamiya, T. Sakurai, and T. Someya, "Organic nonvolatile memory transistors for flexible sensor arrays," *Science*, vol. 326, p. 1516 (2009).

[3] K. Myny, E. van Veenendaal, G. H. Gelinck, J. Genoe, W. Dehaene, and P. Heremans, "An 8-bit, 40-instructions-per-second organic microprocessor on plastic foil," *IEEE J. Solid-State Circ.*, vol. 47, p. 284 (2012).

[4] P. Schalberger, M. Herrmann, S. Hoehla, and N. Fruehauf, "A fully integrated 1-in. AMOLED display using current feedback based on a five-mask LTPS CMOS process," *J. Soc. Inf. Display*, vol. 19, p. 496 (2011).

[5] M. J. Kang, I. Doi, H. Mori, E. Miyazaki, K. Takimiya, M. Ikeda, and H. Kuwabara, "Alkylated dinaphtho[2,3-b:2',3'-f]thieno[3,2-b]thiophenes (C_n-DNTTs): Organic semiconductors for high-performance thin-film transistors," *Adv. Mater.*, vol. 23, p. 1222 (2011).

[6] R. Hofmockel, U. Zschieschang, U. Kraft, R. Rödel, N. H. Hansen, M. Stolte, F. Würthner, K. Takimiya, K. Kern, J. Pflaum, and H. Klauk, "High-mobility organic thin-film transistors based on a small-molecule semiconductor deposited in vacuum and by solution shearing," *Org. Electronics*, vol. 14, p. 3213 (2013).

[7] J. H. Oh, S. L. Suraru, W. Y. Lee, M. Könemann, H. W. Höffken, C. Röger, R. Schmidt, Y. Chung, W. C. Chen, F. Würthner, and Z. Bao, "High-performance air-stable n-type organic transistors based on core-chlorinated naphthalene tetracarboxylic diimides," *Adv. Funct. Mater.*, vol. 20, p. 2148 (2010).

[8] N. A. Minder, S. Ono, Z. Chen, A. Facchetti, and A. F. Morpurgo, "Band-like electron transport in organic transistors and implication of the molecular structure for performance optimization," *Adv. Mater.*, vol. 24, p. 503 (2012).

[9] Y. Y. Noh, N. Zhao, M. Caironi, and H. Sirringhaus, "Downscaling of self-aligned, all-printed polymer thin-film transistors," *Nature Nanotechnology*, vol. 2, p. 784 (2007).

[10] M. Kitamura, Y. Kuzumoto, S. Aomori, and Y. Arakawa, "High-frequency organic complementary ring oscillator operating up to 200 kHz," *Appl. Phys. Express*, vol. 4, p. 051601 (2011).

978-1-4799-7231-9/14 $31.00 © 2014 IEEE

[11] F. Ante, D. Kälblein, T. Zaki, U. Zschieschang, K. Takimiya, M. Ikeda, T. Sekitani, T. Someya, J. N. Burghartz, K. Kern, and H. Klauk, "Contact resistance and megahertz operation of aggressively scaled organic transistors," *Small*, vol. 8, p. 73 (2012).

[12] U. Zschieschang, R. Hofmockel, R. Rödel, U. Kraft, M. J. Kang, K. Takimiya, T. Zaki, F. Letzkus, J. Butschke, H. Richter, J. N. Burghartz, and H. Klauk, "Megahertz operation of flexible low-voltage organic thin-film transistors," *Org. Electronics*, vol. 14, p. 1516 (2013).

[13] C.-H. Jan, "A 22nm SoC platform technology featuring 3-D tri-gate and high-k/metal gate, optimized for ultra low power, high performance and high density SoC applications," *2012 IEEE International Electron Device Meeting (IEDM) Technical Digest*, p. 44 (2012).

[14] A. Asenov, S. Kaya, and A. R. Brown, "Intrinsic parameter fluctuations in decananometer MOSFETs introduced by gate line edge roughness," *IEEE Trans. Electr. Dev.*, vol. 50, p. 1254 (2003).

[15] T. Zaki, S. Scheinert, I. Hörselmann, R. Rödel, F. Letzkus, H. Richter, U. Zschieschang, H. Klauk, and J. N. Burghartz, "Accurate capacitance modeling and characterization of organic thin-film transistors," *IEEE Trans. Electr. Dev.*, vol. 61, p. 98 (2014).

[16] S. Scheinert, T. Zaki, R. Rödel, I. Hörselmann, H. Klauk, and J. N. Burghartz, "Numerical analysis of capacitance compact models for organic thin-film transistors," *Org. Electronics*, vol. 15, p. 1503 (2014).

[17] K. Myny, S. Steudel, S. Smout, P. Vicca, F. Furthner, B. van der Putten, A. K. Tripathi, G. H. Gelinck, J. Genoe, W. Dehaene, and P. Heremans, "Organic RFID transponder chip with data rate compatible with electronic product coding," *Org. Electronics*, vol. 11, p. 1176 (2010).

[18] M. Kitamura and Y. Arakawa, "High current-gain cutoff frequencies above 10 MHz in n-channel C_{60} and p-channel pentacene thin-film transistors," *Jpn. J. Appl. Phys.*, vol. 50, p. 01BC01 (2011).

[19] T. Zaki, F. Ante, U. Zschieschang, J. Butschke, F. Letzkus, H. Richter, H. Klauk, and J. N. Burghartz, "A 3.3 V 6-bit 100 kS/s current-steering digital-to-analog converter using organic p-type thin-film transistors on glass," *IEEE J. Solid-State Circuits*, vol. 47, p. 292 (2012).

[20] R. Rödel, F. Letzkus, T. Zaki, J. N. Burghartz, U. Kraft, U. Zschieschang, K. Kern, and H. Klauk, "Contact properties of high-mobility, air-stable, low-voltage organic n-channel thin-film transistors based on a naphthalene tetracarboxylic diimide," *Appl. Phys. Lett.*, vol. 102, p. 233303 (2013).

[21] F. Zhang, Y. Hu, T. Schuettfort, C. A. Di, X. Gao, C. R. McNeill, L. Thomsen, S. C. B. Mannsfeld, W. Yuan, H. Sirringhaus, and D. Zhu, "Critical role of alkyl chain branching of organic semiconductors in enabling solution-processed n-channel organic thin-film transistors with mobility of up to $3.50 \text{ cm}^2\text{V}^{-1}\text{s}^{-1}$," *J. Am. Chem. Soc.*, vol. 135, p. 2338 (2013).

[22] H. Li, B. C. K. Tee, J. J. Cha, Y. Cui, J. W. Chung, S. Y. Lee, and Z. Bao, "High-mobility field-effect transistors from large-area solution- grown aligned C_{60} single crystals," *J. Am. Chem. Soc.*, vol. 134, p. 2760 (2012).

[23] W. Xiong, U. Zschieschang, H. Klauk, and B. Murmann, "A 3V, 6b Successive Approximation ADC using Complementary Organic Thin-Film Transistors on Glass," *2010 IEEE International Solid-State Circuits Conference (ISSCC) Technical Digest*, p. 134 (2010).

[24] E. D. Glowacki, M. Irimia-Vladu, M. Kaltenbrunner, J. Gasiorowski, M. S. White, U. Monkowius, G. Romanazzi, G. P. Suranna, P. Mastrorilli, T. Sekitani, S. Bauer, T. Someya, L. Torsi, and N. S. Sariciftci, "Hydrogen-bonded semiconducting pigments for air-stable field-effect transistors," *Adv. Mater.*, vol. 25, p. 1563 (2013).

[25] U. Zschieschang, F. Ante, D. Kälblein, T. Yamamoto, K. Takimiya, H. Kuwabara, M. Ikeda, T. Sekitani, T. Someya, J. Blochwitz-Nimoth, and H. Klauk, "Dinaphtho[2,3-b:2',3'-f]thieno[3,2-b]thiophene (DNTT) thin-film transistors with improved performance and stability," *Org. Electronics*, vol. 12, p. 1370 (2011).

[26] Y. Zhao, Y. Guo, and Y. Liu, "Recent advances in n-type and ambipolar organic field-effect transistors," *Adv. Mater.*, vol. 25, p. 5372 (2013).

978-1-4799-7231-9/14 $31.00 © 2014 IEEE

Radio-Frequency Flexible Electronics: Transistors and Passives

Jung-Hun Seo, Zhenqiang Ma[*]

Department of Electrical and Computer Engineering
University of Wisconsin-Madison
Madison, WI 53706 USA
*mazq@engr.wisc.edu

Weidong Zhou

Department of Electrical Engineering
University of Texas-Arlington
Arlington, TX 76019 USA

Abstract— Flexible electronics have customarily addressed low-frequency applications because the traditional materials for flexible electronics, such as polymer and non-crystalline inorganic semiconductors, have poor electronic properties. Fast flexible electronics that operate at radio frequencies (RF), particularly at microwave frequencies, could lead to a number of novel RF applications that rigid chip based solid-state electronics cannot easily fulfill. Single-crystal semiconductor nanomembranes that can be released from a number of wafer sources are mechanically very flexible *and* exhibit outstanding electronic properties that are equivalent to those of their bulk counterparts. These thin, flexible single-crystal materials can furthermore be placed, via transfer printing techniques, onto nearly any substrate, including flexible polymers, thus creating the opportunity to realize RF flexible electronics. In this paper, we will present various RF transistors made of semiconductor nanomembranes on plastic substrates, along with RF passives fabricated on the same flexible substrates. We will also elaborate on the difference between such flexible electronics and those made from thinned rigid wafers.

Keywords—Si Nanomembrane, Flexible transistors, Flexible Inductors, Flexible Capacitors

I. INTRODUCTION

Over the past decade, flexible electronics have been extensively investigated for their unique characteristics, such as light weight, robustness, bendability, stretch and/or fold, and mountability on uneven surfaces [1-2]. These days, flexible electronics have already been implemented in primarily low-speed applications, such as flexible displays [3-4], simple electronic tags [5-6], and bio-sensors [7]. These applications have used amorphous/ polycrystalline silicon or organic polymers which only have low electron mobility. On the other hand, though, there has been a desire for years for high-speed flexible electronics that can operate beyond 1 GHz operation frequency. Once they have high speed capability, flexible electronics can be used in a much wider range of applications than their low-speed counterparts, and such fast flexible electronics provide superior performance and application advantages. It is well known that high-speed devices consume much less power if they operate at a reduced speed, and this dramatically enhances operation time in battery powered devices. Wirelessly connected devices enabled only by high operation speed/frequency are more convenient to use than wired devices and a high-frequency wireless system is also generally more compact than a low-frequency one [8]. Nonetheless, current high-speed devices have predominantly been made of rigid chips because none of the traditional flexible active semiconductors can satisfy the requirements of these high-speed applications.

Such technical barriers have recently been broken by the introduction of high-mobility transferable single-crystal Si nanomembranes (NMs) from silicon-on-insulator (SOI) [9]. With the development of effective doping processing techniques [10], and advanced active device structures [8, 11-13], truly high-speed flexible electronic devices with good mechanical flexibility could be realized. Some examples of this new development are thin-film transistors (TFTs), PIN-diodes and single-pole single-throw (SPST) switches. Beyond these applications, however, the development of high-frequency microwave passive components, such as inductors, capacitors, and switches, on the same low-temperature substrates remain very important to implement high-speed microwave flexible electronic devices in functional circuits and systems [14-15].

In this paper, we report the recent development and progress in Si NMs based active devices such as TFTs, PIN-diodes and SPST switches as well as passive components, such as flexible high-frequency inductors and/or capacitors, which can be monolithically integrated onto a flexible substrate to build a flexible microwave system/ circuit.

II. MICRO WAVE THIN-FILM TRANSISTORS

A. Evolution of RF flexible Si NM TFTs

The high-speed flexible electronics, also referred as fast flexible electronics, are able to replicate the high performance applications made on rigid substrate in flexible forms.

978-1-4799-7231-9/14 $31.00 © 2014 IEEE

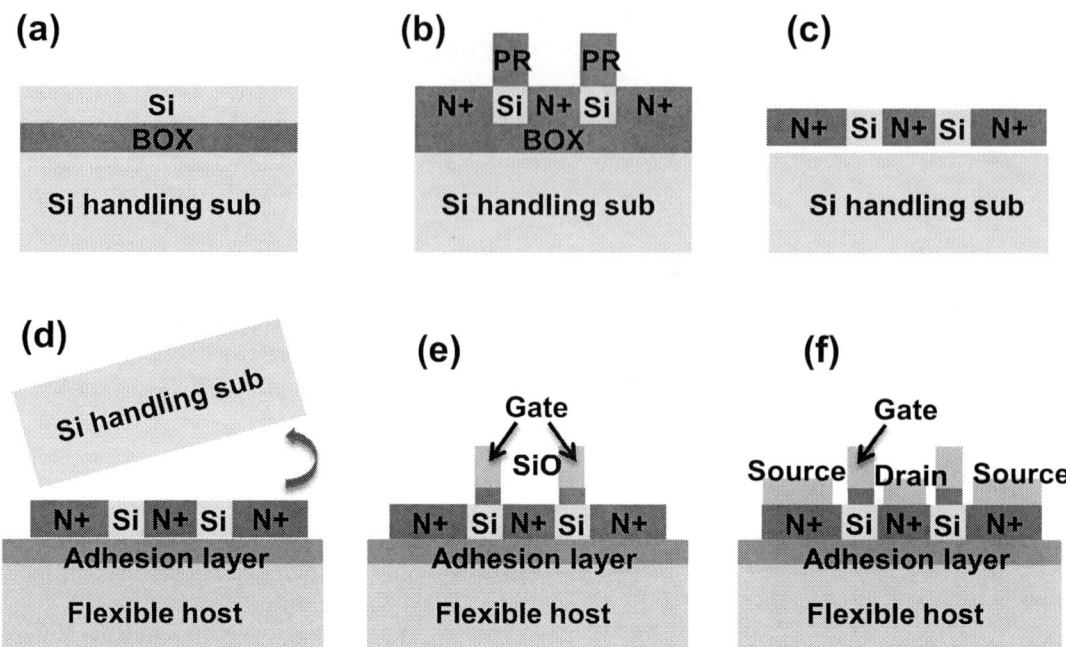

Figure 1. Illustrates the gate-after-source/drain process flow on selectively doped transferred Si NMs on a flexible substrate.

As we briefly mentioned above, the realization of fast flexible electronics requires delicate fabrication techniques and high carrier mobility materials. Based on these criteria, single crystal Si NMs are the ideal option due to their releasable and transfer printable processability. In order to manipulate single crystal Si NMs as active materials for fast flexible electronics, several key processes, including selective etching and transfer-printing, need to be addressed. The detailed process can be found in elsewhere [8, 16]. In short, the process starts with etching the top Si layer of the SOI wafer using the lithography process and reactive ion etching (RIE) into the desired shapes, such as ribbons or membranes with mesh holes. The exposed buried oxide (BOX) layer is then selectively etched in a concentrated hydro fluoride (HF) solution. After etching, the Si NMs can be transferred onto an adhesive coated flexible substrate and cured so that the transfer-printed Si NMs permanently bond to the flexible substrate.

The conventional CMOS process often requires high temperatures for dopant diffusion or recrystallization to repair ion-implantation induced damages and activate dopants. However, such high temperature processes are not compatible with flexible substrates. For this reason, alternative sophisticated processes for fast flexible electronics are necessary to circumvent the high temperature processes and potential damage by chemicals. As an example, typically, heavy doping is needed to realize low-resistance contacts on the source and drain electrodes in TFTs, for which ion implantation can be effectively used before the NMs are released and transferred. By performing high temperature processing procedures on Si NMs before their release, high temperature processing on the low temperature flexible substrate can be avoided. Figure 1 illustrates the gate-after-

source/drain process using selectively doped transferred Si NMs on a flexible substrate. The detailed processing conditions for TFTs fabrication can be found in elsewhere [8, 16]. It should be noted that PMOS TFTs were fabricated due to the high electron mobility in Si and particularly for the better microwave performance, however, NMOS TFTs are also possible, if the fabrication goes with N-type SOI wafer. Thus, various CMOS flexible applications are also possible to realize [22].

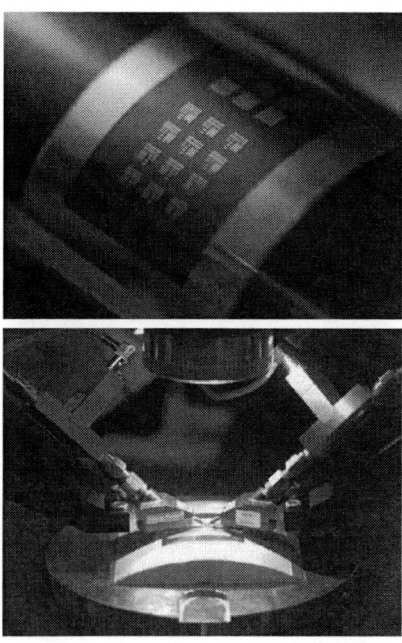

Figure 2. (top) A bent device array on a bending fixture. (bottom) Bending setup for RF measurements.

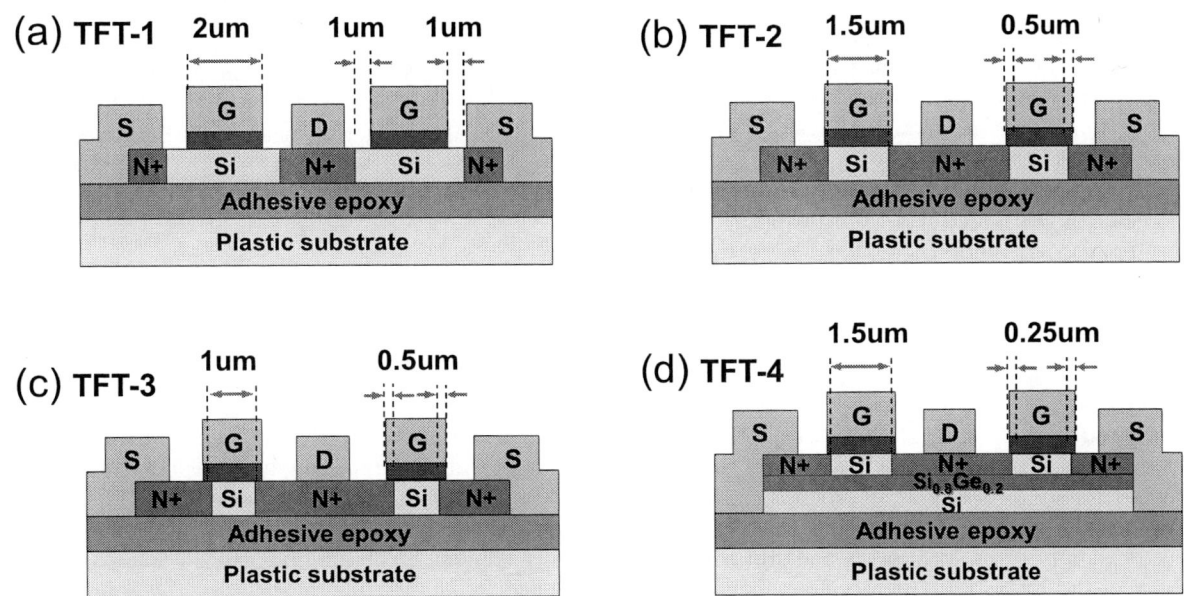

Figure 3. (a)-(c) The cross sections of three TFTs (TFT-1 ~ TFT-3) showing the variations of the critical dimensions. (d) The cross section of Si/SiGe/Si strained TFTs (TFT-4) to enhance the NM mobility.

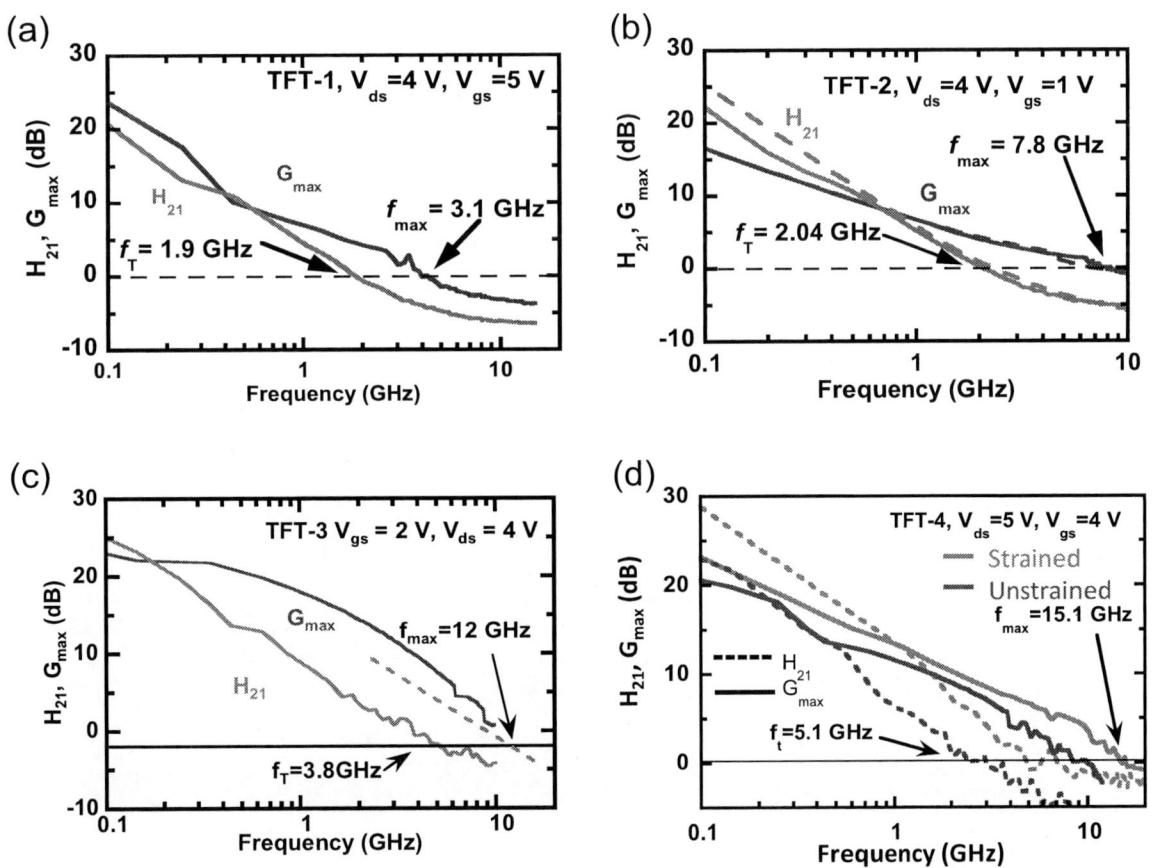

Figure 4. (a)-(d) The RF characteristics of the TFT-1 ~ TFT-4, illustrated in Figure 3. The f_T and f_{max} have increased from 1.9 GHz/ 3.1 GHz to 5.1 GHz/15.1 GHz, respectively. Note that TFT-4 has larger features than TFT-3. Bias conditions for each TFT are shown on the top right of each figures.

978-1-4799-7231-9/14 $31.00 © 2014 IEEE

Figure 2(a) and (b) shows bent device array on a bending fixture and the bending setup for RF measurement. All SiNM flexible TFTs remain intact and operational under high-bending conditions; a convex radius of curvature of 15.5 mm translates into an external strain of ~1.1%. Figure 3(a)-(d) shows the evolution of TFTs with downscaled gate and active materials. Device speed enhancement can be achieved by shrinking the device's critical dimensions and/or by enhancing the carrier mobility of the active material. In comparison to rigid chip CMOS, the critical dimensions of these devices once they have been downscaled are rather modest. Generally large-area applications rather than compact electronic applications are addressed using flexible electronics, for which devices with large features can be used. In addition, flexible substrates are generally unacceptable for the alignment of small features. Figure 3(d) shows the mobility enhancement strategy for TFTs using strained Si NMs [8]. A strain-sharing technique using symmetrical tri-layer structure is employed to apply strain to Si NMs and still maintain the flat topology of the NMs. To further realize effective doping and maintain the flatness in the trilayer structure, we use doped and strained NMs, which can be achieved by growing a stressor layer (SiGe) on doped and thinned Si NMs platforms. Figure 4(a)-(d) shows the RF characteristics of the devices. As shown in figure 4(a) and (c), f_{max} increases ~4times when the gate length is reduced from 2 μm to 1 μm. Figure 4(d) shows that the RF characteristics of the devices (strained TFTs) with the highest speed reached 15.1 GHz of f_{max} when

we used a 1.5 μm gate length. This, in turn, means that about 18 GHz of f_{max} is achievable at a gate length of 1 μm, which is more than 50% improvement compared to the unstrained Si NM TFTs. The results indicate that Si NM has tremendous potential as an active material for high speed flexible TFTs.

III. MICROWAVE FLEXIBLE PASSIVE DEVICES

A. Flexible Microwave Inductors and Capacitors

In conventional rigid chip based CMOS technology, on-chip MIM capacitors are generally implemented between interconnected metal layers using plasma-enhanced chemical vapor deposited (PECVD) dielectrics. The typical PECVD deposition temperature is about 250–350 °C. Such a temperature is not suitable for plastic substrates because the *Vicat softening temperature* (VST) of most of plastic substrates are below 200 °C. In this study, to form the MIM capacitors, a 200 nm thick silicon monoxide (SiO) (ε_r = 5.8–10, depending on stoichiometry) layer was deposited by an e-beam evaporation system in a vacuum on top of the bottom electrode of the capacitor at room temperature [15]. Following the SiO deposition, another metal layer (M2), consisting of Ti/Au, was then evaporated on top of the patterned SiO dielectric layer. This then became the top electrode and formed the MIM structure of the capacitors [Figure. 5(b)].

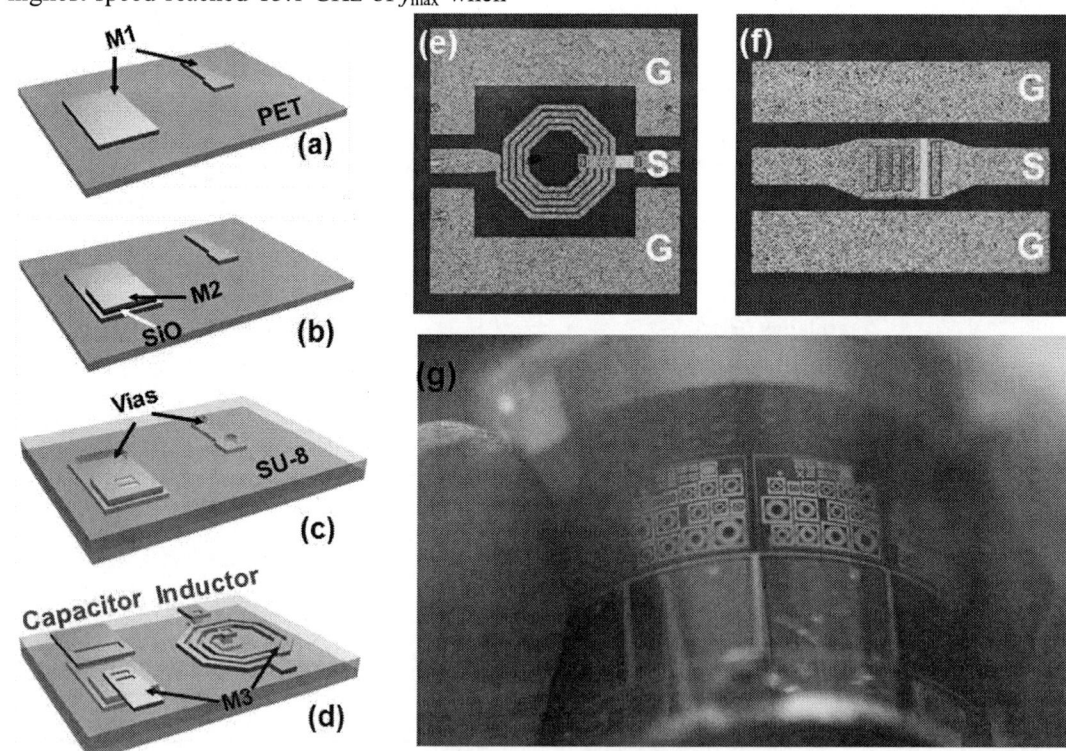

Figure 5. (a)-(d) Illustration of fabrication process for integrated flexible spiral inductors and MIM capacitors. (e) a 4.5-turn spiral inductor, (f) 88×88 μm² MIM capacitor, and (g) finished inductor and capacitor arrays on a bent PET substrate.

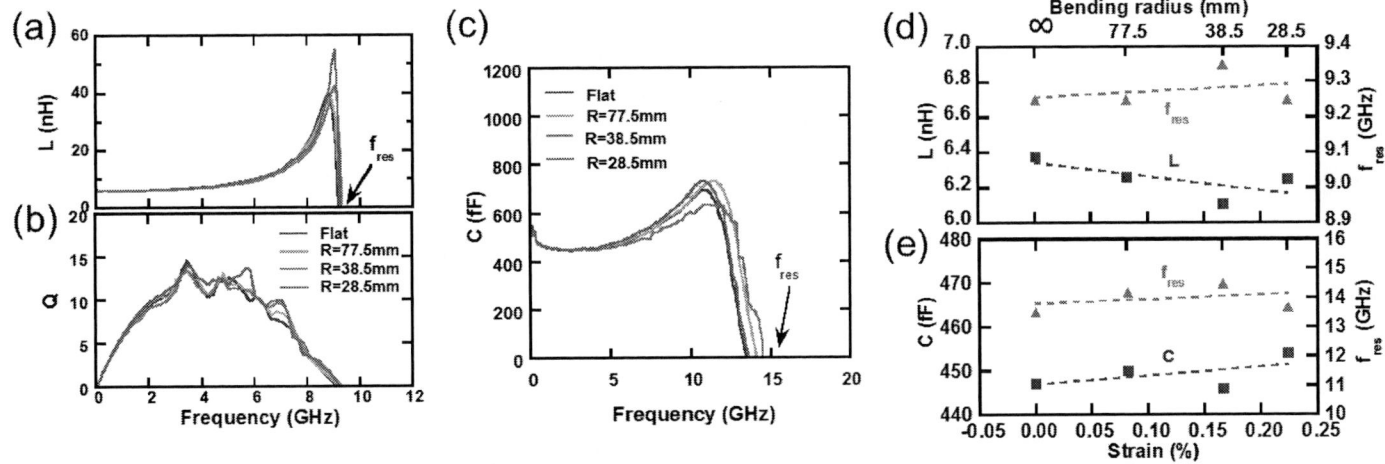

Figure 6. Measured (a) L values and (b) Q values of a 4.5-turn spiral inductor as a function of frequency under flat and bending conditions. (c) Measured capacitance values of a 40x40 um² MIM capacitor as a function of frequency under flat and bending states. The f_{res} is indicated by the zero capacitance values. (d) Measured L (values take at 45 MHz) and f_{res} of a 4.5-turn spiral inductor, and (e) measured C (values take at 4 GHz) and f_{res} of a 40x40 um² MIM capacitor under different bending radii.

For the MIM capacitors, a higher dielectric constant (high-k) and a thinner structure are necessary for high-density capacitors and thus capacitor size must be reduced. However, such fabrication steps are contradictory to that of spiral inductors. A smaller intermetal dielectric constant (low-k) will result in a smaller parasitic capacitance and thus higher resonance/operation frequencies for inductors. To solve this dilemma, we used relatively thick (~1 μm) SU-8 (MicroChem) (dielectric constant $\varepsilon_r = 3$) as the intermetal dielectric but a very flexible polymer material for the spiral inductors. The SU-8 was spun on the sample surface, followed

by a photolithography patterning step to open up via holes which allow access to the bottom and top electrodes of the capacitors and the central lead of the inductors. For the MIM capacitors, a higher dielectric constant (high-k) and a thinner structure are desired for high-density capacitors and reduced capacitor size. On the other hand, it is expected that a smaller inter-metal dielectric constant (low-k) in inductors will result in a smaller parasitic capacitance and therefore higher resonance/operation frequencies for inductors. For the inductors, a Ti/Au top interconnect metal (M3 in figure 5 (d)) was evaporated to form the spiral metal of the inductors and to make interconnects (e.g., ground-signal-ground).

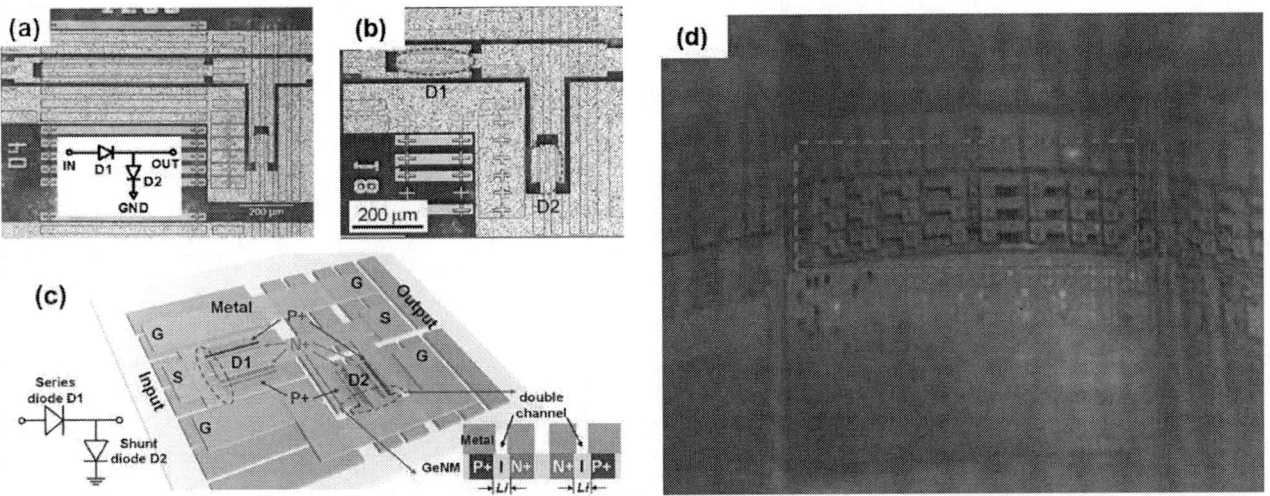

Figure 7. Microscope image of (a) a Si NM and (b) PIN diode SPST switch on a plastic substrate and its schematic circuit diagram. (c) Structure schematics of the flexible Ge switch as an example. Si NM based switch also has same layout. (d) Optical image of flexible SPST switches on a plastic substrate [Ref. 16].

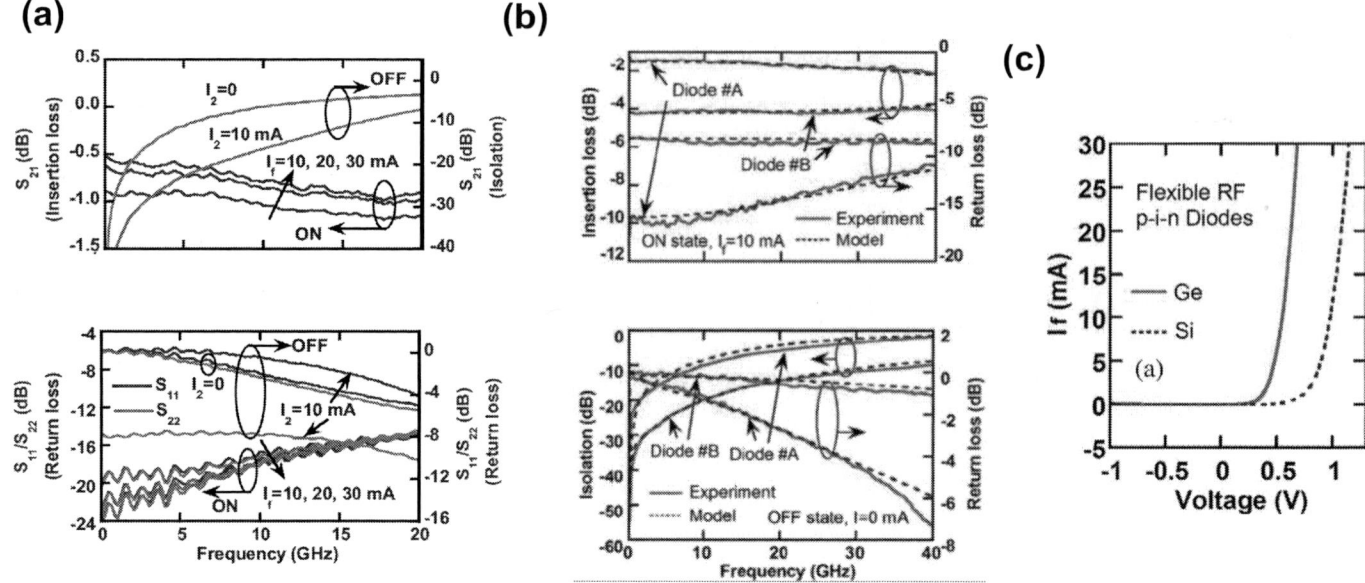

Figure 8. (a) (top) Measured insertion loss (ON state) and isolation (OFF state), and (bottom) return loss of a shunt-series Si NM PIN SPST switch. (b) RF (top) On state (bottom) OFF state of a shunt-series Ge NM PIN SPST switch, (c) DC characteristics between flexible single-crystal Si and Ge diodes on plastic substrates, with similar device structure and equivalent bias.

The reason for using a thicker metal for M3 was to reduce the parasitic resistance of the devices (particularly for multiple turn inductors) in order to allow high-frequency operations. The use of the octagonal shape for the inductors was intended to reduce the sharp-bending microwave loss at high frequencies. Overall, three metal layers and two dielectric layers were used to form the integrated passive components on a single PET substrate. The thin (maximum thickness: ~3.5 μm) and planar structures of the devices make them highly robust to mechanical bending and are also suitable for operating at high-frequencies. The highest process temperature during fabrication was under 115 °C (the highest photoresist baking temperature), which easily satisfies the tolerance of most low-temperature plastic substrates. More importantly, this fabrication process is completely compatible with that used to fabricate microwave flexible TFTs, as shown in a previous section. Figures 5(e) and 5(f) show microscopic images of the 4.5-turn inductor and the 88×88 μm² MIM capacitor and inductor on a plastic substrate, respectively. The spiral metal lines of the inductors have a width of 15 μm and spacing of 4 μm. Figure 5(g) shows the inductor and capacitor arrays on a bent PET substrate. The critical passive component parameters, such as the inductance/capacitance (L/C), the quality factor (Q-factor), and the resonant frequency (f_{res}), were extracted by measured scattering- (S-) parameters followed by the "thru", "open-short" and "short-open" de-embedding procedure to extract the internal device parameters. Due to the carefully designed layer thickness and material selection, both the inductors and the capacitors showed relatively stable inductance and capacitance under the bending condition. The stable f_{res} values under the bending condition suggest that the microwave characteristics using these components will be minimally affected by bending, which is beneficial for the circuit/system design. Moreover, increased f_{res} value can be possible by applying optimal device design, such as thickness and shape of metal lines.

B. Flexible Microwave Diodes and Switches

In addition to flexible inductors and capacitors, flexible switches are also necessary for a functional flexible RF system. In this section, we will discuss the demonstrations of flexible RF switches and diodes using Si NM and Ge NM. The fabrication process for the RF switches is almost identical to that of TFTs, except for an additional step of ion implantation of P-type doping and the fact that there is no gate stack needed, which actually simplifies the process for flexible substrates. It should also be noted that the fabrication of these components is completely compatible with the flexible Si NM TFT process that we demonstrated in the section above. Figure 7(a) shows microscope images of the flexible RF Si single-pole single-throw (SPST) switch, which consists of shunt and series P–intrinsic (I, unintentionally lightly p-type doped)–N (PIN) diodes (area of diodes: D1 = 240 μm² and D2 = 40 μm²). Figure 7(b) shows a microscopic image of a flexible RF Ge NM SPST switch with same device geometries. For the fabrication of Ge NM based RF switches, the Ge annealing condition after ion implantation for recrystallization should be different from that of Si NM. The detailed implantation and annealing conditions can be found elsewhere [17-19]. The I-V characteristic of the Ge NM PIN diodes showed ~1.2 of the ideality factor at a turn-on voltage of 0.45V, while the Si NM PIN diodes showed a similar 1.05 of the ideality factor but at a slightly higher turn-on voltage of 0.75V, due to the fact that Si has a higher bandgap than Ge. Both diodes had a very low reverse current, about few nA at -5V.

RF Si NM switches connecting 2 individual diodes showed a low insertion loss of 0.93 dB and the isolation is 6.9 dB at 20 GHz in the ON state, as shown in Figure 8(a). A comparison of the performances (I-V characteristics) of the individual flexible Ge NM and Si NM diodes with similar structures is shown in Figure 8(c). The Ge NM and Si NM diodes have an identical PIN channel width of 800 μm (the GeNM is 250 nm thick, and the SiNM is 200 nm thick, due to the limitation of GeOI and SOI wafers). "L_i" which is the length of intrinsic area, was set at 5 μm for the Ge NM diode and 2 μm for the Si NM diode to maintain similar power handling capability. The same forward current (I_f) of 10 mA is used. The flexible Ge NM diode showed a much lower turn-on voltage than the Si diode, validating the advantage of the Ge material [20-21]. The I-V characteristic of the Ge NM PIN diodes showed ~1.2 of the ideality factor with the a turn-on voltage of 0.45V while the Si NM PIN diodes showed a similar 1.05 of the ideality factor but at a slightly higher turn-on voltage of 0.75V, due to its higher bandgap than Ge. Both diodes have a very low reverse current, about few nA at -5V. The Ge NM diodes show a better frequency response at the RF regime compared to the Si NM diodes, which indicates comparable forward-bias performance and better reverse-bias performance (or comparable reverse-bias performance and better forward bias performance if the Ge diode has a similar L_i to the Si NM diode). In addition, our results indicate that the flexible Ge diodes show less performance variation under mechanical bending strain compared to the flexible Si diodes. However, Si NM RF switches have the advantage of being compatible with other CMOS active components while maintaining reasonably high performance as well as mechanical robustness.

IV. SUMMARY

In this report, high-frequency flexible Si NM TFTs, and microwave flexible inductors, capacitors, and switches have been demonstrated on low-temperature plastic substrates. We successfully demonstrated RF flexible TFTs with an f_{max} of 15 GHz with good mechanical flexibility. In the inductors and capacitors, both high Q and high f_{res}, around 10 GHz, were achieved. Robust mechanical characteristics were also exhibited in these high-frequency passives. RF switches made with both Si NM and Ge NM demonstrated very low isolation and insertion loss at 20 GHz. As a result, all passive components can fulfill the frequency requirements of flexible RF circuits up to 10~20 GHz. With the demonstration of such flexible passive components, not only can more advanced flexible RF systems be realized but we can expect lowered costs and enlarged form factors in the fabrication process.

ACKNOWLEDGMENT

The work was supported by AFOSR under grant FA9550-091-0482. The program manager is Dr. Gernot Pomrenke.

REFERENCES

[1] S. R. Forrest, "The path to ubiquitous and low-cost organic electronic appliances on plastic," Nature, vol. 428, pp. 911-918, 2004.

[2] Y. Sun, and J. A. Rogers, "Inorganic semiconductors for flexible electronics," Adv. Mater., vol. 19, pp. 1897-1916, 2007.

[3] G. Crawford, ed. "Flexible flat panel displays," John Wiley & Sons, 2005.

[4] L. Zhou, A. Wanga, S.-C. Wu, J. Sun, S. Park, and T. N. Jackson, "All-organic active matrix flexible display," Appl. Phys. Lett., vol. 88, pp. 3502, 2006.

[5] P. F. Baude, D. A. Ender, M. A. Haase, T. W. Kelley, D. V. Muyres, and S. D. Theiss, "Pentacene-based radio-frequency identification circuitry," Appl. Phys. Lett., vol. 82, pp. 3964-3966, 2003.

[6] B. Crone, A. Dodabalapur, Y-Y. Lin, R. W. Filas, Z. Bao, A. LaDuca, R. Sarpeshkar, H. E. Katz, and W. Li, "Large-scale complementary integrated circuits based on organic transistors," Nature, vol. 403, pp. 521-523, 2000.

[7] S. CB Mannsfeld, T. CK Benjamin, R. M. Stoltenberg, C. V. HH Chen, S. Barman, B. VO Muir, A. N. Sokolov, C. Reese, and Z. Bao, "Highly sensitive flexible pressure sensors with microstructured rubber dielectric layers," Nat. Mater., vol. 9, pp. 859-864, 2010.

[8] H. Zhou, J.-H. Seo, D. M. Paskiewicz, Y. Zhu, G. K. Celler, P. M. Voyles, W. Zhou, M. G. Lagally, and Z. Ma, "Fast flexible electronics with strained silicon nanomembranes," Sci. Rep., vol. 3, pp. 1291, 2013.

[9] E. Menard, K. J. Lee, D-Y. Khang, R. G. Nuzzo, and J. A. Rogers, "A printable form of silicon for high performance thin film transistors on plastic substrates," Appl. Phys. Lett., vol. 84, pp. 5398-5400, 2004.

[10] H.-C. Yuan, Z. Ma, M. M. Roberts, D. E. Savage, and M. G. Lagally, "High-speed strained-single-crystal-silicon thin-film transistors on flexible polymers," J. Appl. Phys., vol. 100, pp. 013708, 2006.

[11] H.-C. Yuan, and Z. Ma, "Microwave thin-film transistors using Si nanomembranes on flexible polymer substrate," Appl. Phys. Lett., vol. 89, pp. 212105, 2006.

[12] H.-C. Yuan, G. K. Celler, and Z. Ma, "7.8-GHz flexible thin-film transistors on a low-temperature plastic substrate," J. Appl. Phys., vol. 102, pp. 034501, 2007.

[13] G. Qin, J.-H. Seo, Y. Zhang, H. Zhou, W. Zhou, Y. Wang, J. Ma, and Z. Ma, "RF Characterization of Gigahertz Flexible Silicon Thin-Film Transistor on Plastic Substrates Under Bending Conditions," IEEE Electron Device Lett., vol. 34, pp. 262-264, 2013.

[14] G. Qin, T. Cai, H.-C. Yuan, J.-H. Seo, J. Ma, and Z. Ma, "Flexible radio-frequency single-crystal germanium switch on plastic substrates," Appl. Phys. Lett., vol. 104, pp. 163501, 2014.

[15] L. Sun, G. Qin, H. Huang, H. Zhou, N. Behdad, W. Zhou, and Z. Ma, "Flexible high-frequency microwave inductors and capacitors integrated on a polyethylene terephthalate substrate," Appl. Phys. Lett., vol. 96, pp. 013509, 2010.

[16] K. Zhang, J.-H. Seo, W. Zhou, and Z. Ma, "Fast flexible electronics using transferrable silicon nanomembranes," J. Phys. D: Appl. Phys., vol. 45, pp. 143001, 2012.

[17] J.-H. Seo, Y. Zhang, H.-C. Yuan, Y. Wang, W. Zhou, J. Ma, Z. Ma, and G. Qin, "Investigation of various mechanical bending strains on characteristics of flexible monocrystalline silicon nanomembrane diodes on a plastic substrate," Microelectron. Eng., vol. 110, pp. 40-43, 2013.

[18] G. Qin, H.-C. Yuan, G. K. Celler, W. Zhou, and Z. Ma, "Flexible microwave PIN diodes and switches employing transferrable single-crystal Si nanomembranes on plastic substrates," J. Phys. D: Appl. Phys., vol. 42, pp. 234006, 2009.

[19] H.-C. Yuan, G. Qin, G. K. Celler, and Z. Ma, "Bendable high-frequency microwave switches formed with single-crystal silicon nanomembranes on plastic substrates," Appl. Phys. Lett., vol. 95, pp. 043109, 2009.

[20] G. Qin, H.-C. Yuan, Y. Qin, J.-H. Seo, Y. Wang, J. Ma, and Z. Ma, "Fabrication and Characterization of Flexible Microwave Single-Crystal Germanium Nanomembrane Diodes on a Plastic Substrate," IEEE Electron Device Lett., vol. 34, pp. 160-162, 2013.

[21] H.-C. Yuan, J. Shin, G. Qin, L. Sun, P. Bhattacharya, M. G. Lagally, G. K. Celler, and Z. Ma, "Flexible photodetectors on plastic substrates by use of printing transferred single-crystal germanium membranes," Appl. Phys. Lett., vol. 94, pp. 013102, 2009.

[22] S.-W. Hwang, X. Huang, J.-H. Seo, J.-K. Song, S. Kim, S. Hage-Ali, H.-J. Chung, H. Tao, F. G Omenetto, Z. Ma, and J. A Rogers, "Materials for bioresorbable radio frequency electronics," Adv. Mater., vol. 25, no. 26, pp. 3526-3531, 2013.

Quantum-Well Transistor Laser for Optical Interconnect and Photonic Integrated Circuits

M. Feng, *Fellow, IEEE*, H. W. Then, *Member, IEEE,* F. Tan, M. K. Wu, R. Bambery, and Nick Holonyak, Jr., *Life Fellow, IEEE*

Department of Electrical and Computer Engineering and Micro and Nanotechnology Laboratory, University of Illinois, 1406 West Green Street, Urbana, Illinois 61801 USA

Email: mfeng@illinois.edu

ABSTRACT — The heterojunction bipolar transistor laser, inherently a fast switching device, operates by transporting small minority base charge densities $\sim 10^{16}$ cm^{-3} over nano-scale base thickness (< 900A) in picoseconds. The insertion of quantum-wells and tilted charge in the short base of a transistor reduces recombination lifetime below 30 ps which is critical for extending the direct modulation bandwidth of the semiconductor laser towards 200 GHz. Three-port operation expands the use of the transistor laser (TL) to optical interconnect and photonic integrated circuits.

Index Terms — Heterojunction bipolar transistor, quantum well laser, three-port laser, transistor laser (TL).

I. INTRODUCTION

The transistor invention by Bardeen and Brattain in 1947 lead us to realize that the 'magic" of the transistor is intrinsically in the base. And, it is the base that potentially offers more, particularly when we arrive at the direct-gap, high speed, high current density heterojunction bipolar transistor (HBT) and realize the base although thin (10-100nm), has room for more layering (bandgap and doping) and can be modified. Employing quantum-wells (QWs) [1] and cavity reflection, we can re-invent the base region and its mechanics, reduce the current gain and achieve stimulated recombination, i.e., realize a quantum-well transistor laser (QWTL) – a novel device with an electrical input, an electrical output and an optical output. The result is unique transistor in form and operation, as well as a unique three-terminal laser [2-5].

We note that quantum-well base region and stimulated recombination (stimulated emission), besides yielding a transistor laser, changes the transistor into an active element that can used for nonlinear and switching applications (both electrical and optical) as we recently demonstrated [6-10]. Transistor laser with a 200μm length demonstrated an error-free data transmission @ 13.5 Gb/s [8]. With two electrical inputs through the base, we demonstrated new mixer with both electrical and optical

outputs in nonlinear operation region and signals addition to produce new waveform in linear operation region [9]. Recently, vertical cavity transistor laser with microcavity and ultralow power operation was reported [11, 12, 13].

II. TRANSISTOR LASER OPERATION

The p-i-n diode laser (DL) is a charge-storage recombination device, the carriers piling up in a reservoir-like double heterostructure before recombining with holes in the quantum well (QW) see Fig. 1(a). The saturation-mode dynamics give rise to an average recombination lifetime of the order of 1 ns. In contrast, the transistor laser is a dynamic charge-slope device with the base recombination rate determined by the carrier gradient (shown in Fig. 1(b)) rather than by the carrier density as in the diode laser. By diffusing the injected carriers across a thin oppositely doped QW base active region in an n-p-n bipolar heterostructure, slowly recombining carriers can be removed by the collector. Hence, the intrinsic spontaneous carrier recombination lifetime in the base of the transistor can be "clamped" at the same magnitude as the base transit time (as low as 2 ps). Thus, the transistor laser can achieve 30 x faster recombination lifetime compared to the diode laser. Only "fast" recombination can compete with a "fast" collector.

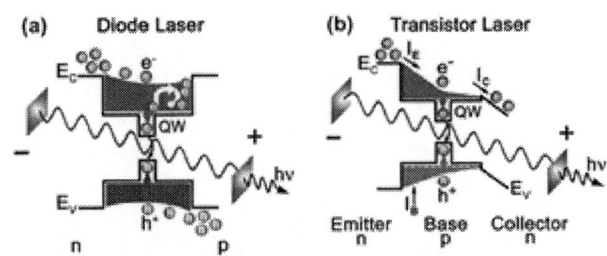

Fig. 1. Schematic energy band diagram and charge distribution in (a) a p-i-n double heterojunction diode laser (DL), and (b) a n-p-n heterojunction bipolar quantum-well transistor laser (QWTL).

978-1-4799-7231-9/14 $31.00 © 2014 IEEE

III. EDGE-EMITTING TRANSISTOR LASER

A. Collector I_C-V_{CE} & Optical L-V_{CE} Outputs

The TL shifts from spontaneous to stimulated recombination process in the base which results in a unique beta compression ($\beta \sim \Delta Ic/\Delta Ib$) in the collector I-V characteristics, mapping out differential optical gain and the photon generation process in the base quantum-well [6]. Figure 2 illustrates a transistor laser (L = 400 μm) with I_B = I_{TH} = 22mA at 12°C with (a) Collector I-V characteristics and (b) Light L-V characteristics.

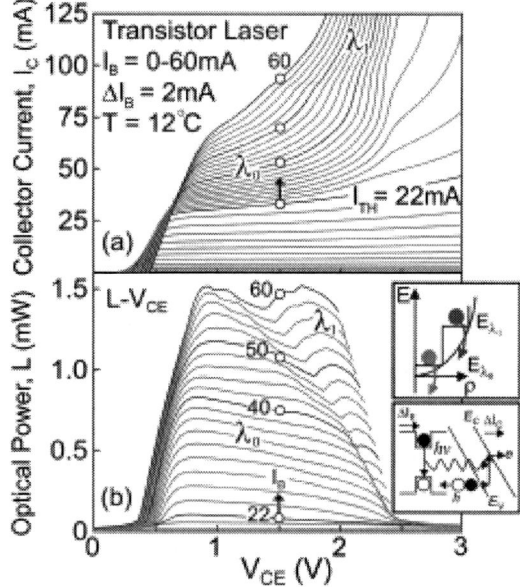

Fig. 2. A QWTL (L = 400 μm): (a) Collector I_C-V_{CE} characteristics exhibiting (gain) compression owing to the shift from spontaneous (I_B = I_{TH}) to stimulated recombination (coherent) in the base, and (b) single-facet laser optical L-V_{CE} characteristics ($I_B > I_{TH}$) with the optical power shifting in energy, and sharply in amplitude, from Eλ_0 ($\lambda_0 \sim$ 1000 nm), to Eλ_1 ($\lambda_1 \sim$ 970 nm).

B. Modulation Bandwidth of Transistor Laser

The intrinsic spontaneous recombination speed in the base of the transistor can compete with and can be "clamped" at the same order of magnitude as the QW base region transit time. The small-signal linear optical response, $H(\omega)$ can be written as

$$H(\omega) = \frac{A_o}{1 - \omega^2/\omega_n^2 + j2(\omega/\omega_n)\xi}, \qquad (1)$$

Where the natural frequency $\omega_n^2 = (\eta/\tau_{ph}\tau_{B,spon})(I_B/I_{TH} - 1)$, $\xi = 1/(2\omega_n\tau_{B,spon}) + \tau_{ph}\omega_n/2$, and A_o is a normalization

factor. A damping factor is defined as $\gamma = 2\omega_n\xi = 1/\tau_{B,spon} + \Gamma vg'N_{ph}$. The resonance frequency, f_R, is given by $f_R = \omega_n/2\pi (1-2\xi^2)^{1/2}$, and the magnitude of the resonance peak is given by $|H(\omega_R)|^2 = A_o^2/[4(1-\xi^2)\xi^2]$. The resonance peak can be reduced by increasing ξ. Since $\xi = 1/(2\omega_n\tau_{B,spon}) + \tau_{ph}\omega_n/2$, for a given τ_{ph} and g', the damping of the resonance can be obtained by both a faster $\tau_{B,spon}$, and by increasing ω_n^2 or I_B/I_{TH} [7]. In Figure 3, we measured and fitted optical frequency response of a transistor laser showing absence of carrier-photon resonance owing to "fast" $\tau_{B,spon}$ = 29 ps. Slight resonance (< 5 dB) is observed only at higher bias consistent with $\xi \sim 1/(2\omega_n\tau_{B,spon})$. Resonance-free response of the QW transistor laser [7] at all biases ranging from (a) I_B = 30, (b) 40, (c) 60 and (d) 100 mA.

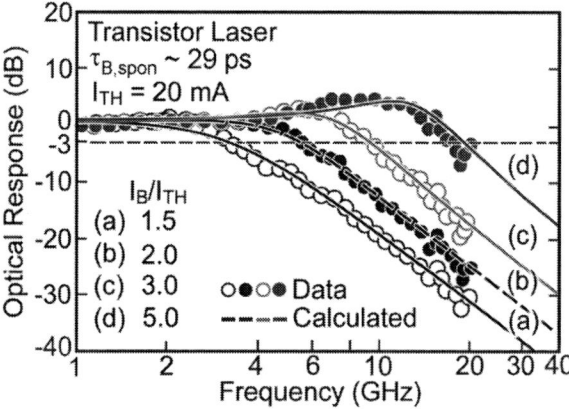

Fig. 3. Resonance-free response of transistor laser at biases ranging from (a) I_B = 30, (b) 40, (c) 60 and (d) 100 mA.

C. Voltage Modulation of Transistor Laser via Photon Assisted Tunneling at Collector

Fig. 4 shows the schematic band diagram of the tunnel junction transistor laser with all the key physical processes labeled [9]. I_E is the emitter current (minority current in the base) with the junction in forward bias; I_B is the re-supply of holes by the usual base ohmic contact; I_{jkT} is the re-supply of holes by the Franz-Keldysh (FK) photon-assisted tunneling; I_{rT} represents the re-supply of holes via the direct tunneling of electrons; and I_t is the usual minority carrier current of injected electrons that do not recombine in the base and are collected. The collector current I_C consists of the usual transport component across the base I_t, the Franz-Keldysh portion I_{jkT}, and the direct tunnel junction current I_{rT}, or

$$I_C = I_t + I_{rT} + I_{fkT}. \tag{2}$$

The base recombination current, I_{Br}, is expressed as the sum of the hole components, or

$$I_{Br} = I_B + I_{fkT} + I_{rT}. \tag{3}$$

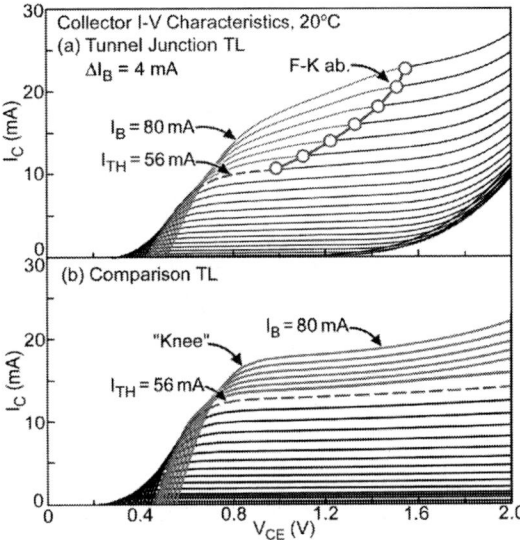

Fig. 4 Schematic band diagram of a tunnel junction transistor laser shown with a generic resonator cavity.

Fig. 5 shows the collector *I-V* characteristics of (a) the tunnel junction transistor laser, and (b) the comparison transistor laser of lesser collector doping and no tunnel junction. The forward-active mode of the tunnel junction transistor laser operation (i.e., the base-collector junction in reverse bias) is indicated by the collector current, I_C, being nearly constant (i.e., flat) despite further increase in V_{CE} above the "knee" voltages of 0.4 V ($I_B = 56$ mA) to 0.8 V ($I_B = 80$ mA). The effects of collector tunneling (Fig. 5a) are evident from the upward slope in the collector current, I_C vs. V_{CE}, that otherwise would be relatively flat (see Fig. 5b for comparison). In the comparison transistor laser, collector tunneling is negligible; hence, $I_C \approx I_t$ (usual collector current). However, in the tunnel junction transistor laser, I_C increases as a function of V_{CE} owing to the various tunneling components (Eq. 3). In the presence of a stimulated-emission optical field and laser operation of the tunnel junction transistor laser, the tunneling process occurs predominantly via Franz-Keldysh (photon-assisted) absorption ($I_C \approx I_t + I_{fkT}$). Direct tunneling (not photon-assisted) can be observed at higher V_{CE} biases ($I_C = I_t + I_{fkT} + I_{rT}$).

The collector *I-V* characteristics of the tunnel junction transistor laser agree in form well with its optical output, with the *LI-V* characteristics shown in Fig. 6. In the operation of the tunnel junction transistor laser under weak collector junction field (left region 1 of Fig. 6), collector tunneling (photon-assisted, $I_{fkT} > 0$) enables the efficient supply of holes to the QW active region, and thus

improves the laser optical output to twice that of the comparison transistor laser. The holes supplied by collector tunneling need only relax a distance of ~30 nm (from collector to the base QW), as opposed to the lateral distance of several μm traversed by holes supplied (from collector to the base QW), as opposed to the lateral distance of several μm traversed by holes supplied by the

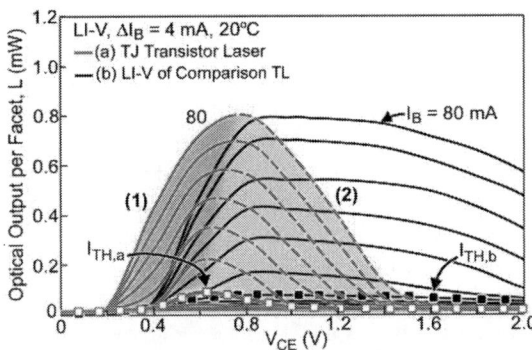

Fig. 5 The collector *I-V* characteristics of (a) a tunnel junction transistor laser (TL) and (b) a comparison transistor laser without a collector tunnel junction. Below the knee voltages, the transistors are biased in saturation. The tunneling process is evident from the slope of the tunnel junction transistor laser collector current, I_C, vs. emitter-collector voltage bias, V_{CE} (0.4 – 1.6 V), which otherwise would be flat as for the collector *I-V* of the (b) comparison transistor laser.

Fig. 6 The dependence of optical output of the tunnel junction (TJ) transistor laser on V_{CE}, indicating the enhancement ($V_{CE} < 0.8$ V) and quenching ($V_{CE} \geq 0.8$ V) of the laser output by FK photon-assisted tunneling (photon absorption). The *LI-V* of the comparison TL (b) shows similar behavior occurring gradually except at higher $V_{CE} \geq 1.6$ V.

D. Error-Free Data Transmission @ 13.5 Gb/s

In order to investigate the feasibility of TLs in high speed data transmission, it is important to characterize the signal integrity. For a given bias condition, we measure the eye diagram and the bit error rate (BER) of the same TL device. A non-return-to-zero (NRZ) 2^7-1 bit length pseudorandom binary series pattern (PRBS7) with peak-to-peak ac voltage swing V_{pp} = 1.5 V is generated by the signal generator module of an Agilent N4901B 13.5 Gb/s serial BERT. The data sequence is combined with the DC bias via an Agilent 11612B bias network and transmitted into the BE junction of the TL device via a 40 GHz GSG probe. The fiber coupled laser output is converted to an electrical signal by a New Focus 1414-50 25 GHz photodetector and amplified by a Picosecond Pulse Labs 5882 amplifier with 40 GHz bandwidth and 16 dB power gains. The amplified signal is then fed into an Agilent 86100C oscilloscope with an Agilent 86117A module for capturing the eye diagram.

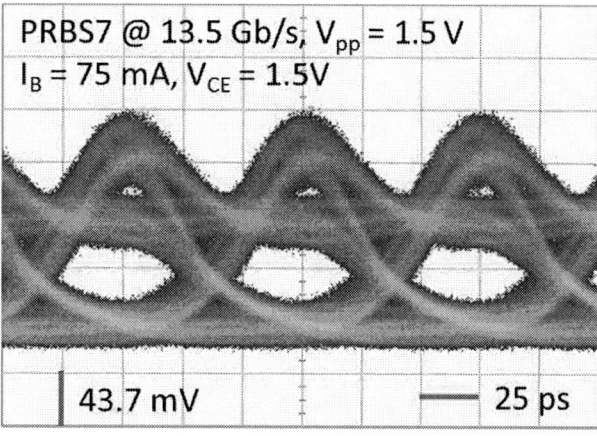

Fig. 7 Measured un-averaged optical eye diagram of the QWTL (L = 200 μm) operating at 13.5 Gb/s with PRBS7 data sequence modulation. The bias conditions are I_B = 75 mA and V_{CE} = 1.5 V, and the corresponding ac voltage swing amplitude is V_{pp} = 1.5 V.

Figure 7 shows an un-averaged optical eye diagram of a 200 mm cavity length QWTL at 13.5 Gb/s biased at I_B = 75 mA (I_{TH} = 36 mA) and V_{CE} = 1.5 V at T = 20°C, where a clean and open "eye" for the optical laser modulation signal is shown [8]. The overshoot feature is due to the relatively high relaxation oscillation amplitude ($\sim$ 4.2 dB) and frequency response flatness of the post amplifier as shown in the optical microwave response. To test the bit error rate (BER) performance, we employ a continuously

variable neutral density filter in the free space laser beam path to attenuate the fiber coupled optical power. The photodetector and the amplifier are connected to the error analyzer module of the Agilent N4901B serial BERT.

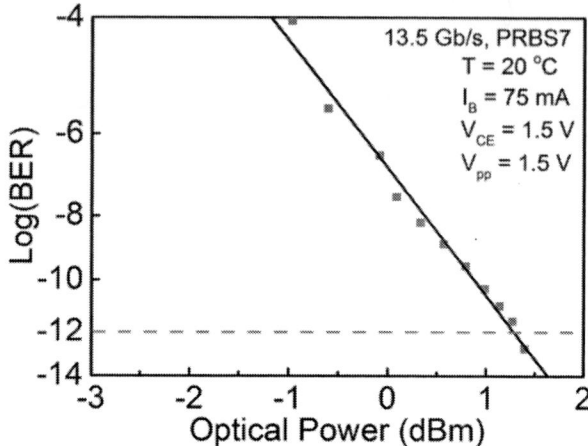

Fig. 8 Measured bit error rate (BER) vs. received optical power of the QWTL operating at T = 20 °C with 13.5 Gb/s PRBS7 data sequence modulation. The DC bias and modulation condition are the same as the eye diagram measurement.

Figure 8 shows the measured BER of the device as a function of received optical power at 13.5 Gb/s with a PRBS7 data sequence. The device bias condition is the same as the eye diagram measurement (I_B = 75 mA and V_{CE} = 1.5 V). For error-free data transmission, the BER needs to be less than 10^{-12}. The measured device shows error-free 13.5 Gb/s data transmission for received optical power above 1.4 mW [8]. At an optical power P_0 = 1.95 mW, the device shows no error for continuous bit acquisition $> 2 \times 10^{12}$. Note that the bandwidth of this device at I_B = 75 mA is f_{-3dB} = 8.1 GHz, and previously we reported a different and better-performing TL with bandwidth f_{-3dB} = 13.7 GHz and resonance amplitude < 3 dB at 20 °C. This device is expected to achieve > 20 Gb/s error-free data transmission (currently limited by our BER testing capability).

E. Transistor Laser as a Signal Mixer

A three-terminal tunnel junction transistor laser, employed as a nonlinear microwave device, is much more convenient for circuit matching. It utilizes the nonlinearity that results from the strong coupling between minority carrier injection (I_E), electron-hole recombination (I_B) and the collector junction field (V_{CB}), mediated by photon-assisted tunneling (I_{fkT}) at the collector junction. The tunnel junction transistor laser, therefore, enables a new nonlinear signal processing (adding and mixing) device operating above laser threshold for improved optical output power. It can serve as a powerful nonlinear optoelectronic component for applications such as frequency multiplication, frequency synthesis, and signal-processing.

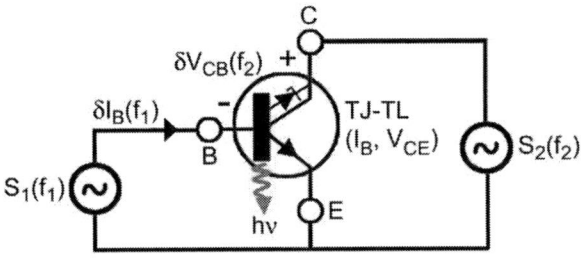

Fig. 9 Schematic circuit showing a TJ-TL in a common-emitter configuration for frequency multiplication and signal mixing of input signals $S_1(f_1)$ and $S_2(f_2)$.

For signal mixing, the tunnel junction transistor laser can be readily arranged in a two-port microwave electrical configuration (common-emitter, common-base as well as common-collector), and can be conveniently accessed with two separate GSG (ground-signal-ground) probes. Because of the tunnel junction design, the collector terminal can now be used as an additional input port. For convenience and simplicity in fabrication, the tunnel junction transistor laser is fabricated in the common-emitter configuration (TJ-TL CE) shown in Fig. 9. The tunnel junction transistor laser is biased above threshold (I_B = 80 mA, V_{CE} = 0.8 V), and provides a single tone input at frequency f_1 = 2.0 GHz at the BE-port of the tunnel junction transistor laser, and a separate tone at frequency f_2 = 2.1 GHz at the CE-port. The choice of f_1 and f_2 is made for convenience in displaying and subsequently identifying the various harmonics in the output optical signal [9]. In Fig. 10 the signal mixing produces harmonics ($mf_1 \pm nf_2$) as high as the ($m+n$) = 11th order, with the highest harmonic being $4f_1+7f_2$ = 22.7 GHz despite being bandwidth-limited to 20 GHz by the detector amplifier.

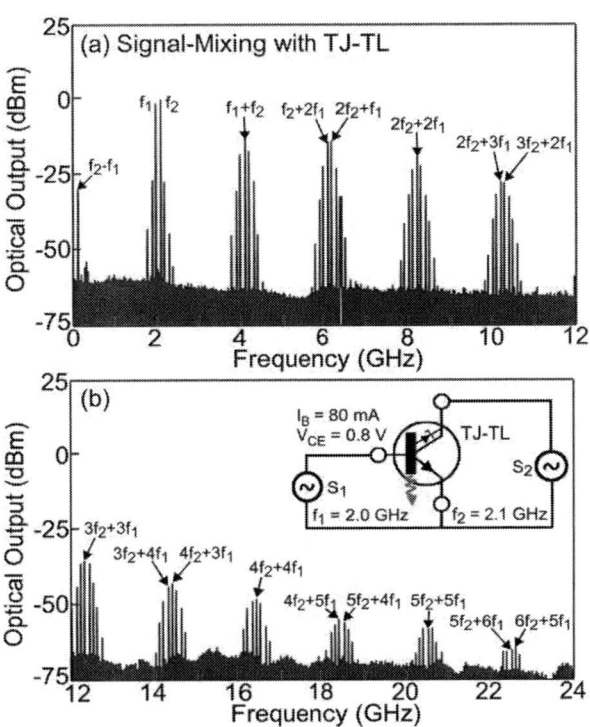

Fig. 10 Signal mixing with a TJ-TL in the common-emitter configuration of Fig. 9, producing optical output with harmonics $mf_1 \pm nf_2$. The input tones are of equal RF power (0 dBm).

F. Physical Model of Transistor Laser

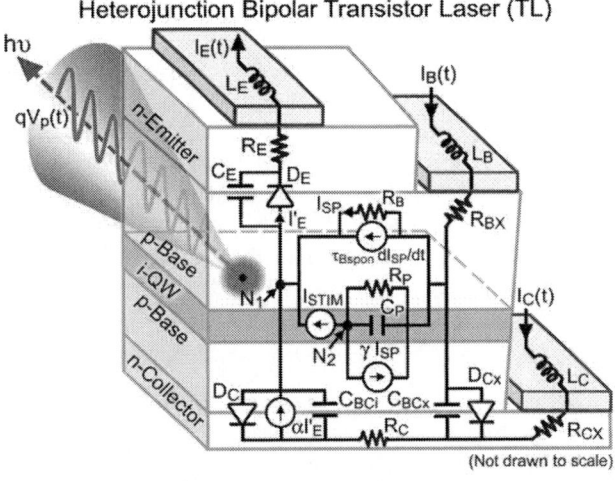

Fig. 11 Physical model of a three-port heterojunction bipolar transistor laser formulated in equivalent electrical and optical elements.

For use in extracting physical transistor laser parameters using microwave measurements, we derive a model of the transistor laser by extending Kirchhoff's law to include consistently both electrical and optical equivalent circuit elements [10] as shown in Fig. 11. We are primarily interested in operating the transistor with forward-biased emitter junction and reverse-biased collector junction.

IV. VERTICAL CAVITY TRANSISTOR LASER

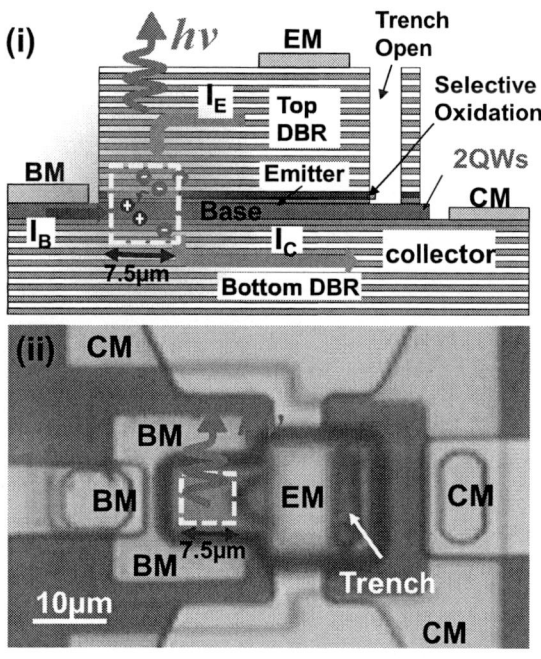

Fig. 12. (i) The schematic of selective oxidation confined high "Q" vertical cavity transistor laser (VCTL). The emitter metal (EM) is evaporated on top of the DBRs, and the trench is open behind EM for lateral oxidation underneath the EM for current and optical mode confinement. The base metal (BM) surrounds the cavity with BM-to-cavity distance less than 2 μm. The optical microscopic image (ii) shows the VCTL device with the oxidation confined aperture of 6.5 x 7.5 μm² (yellow box).

To pursue the high speed directly modulated laser for energy efficient data transmission, vertical cavity configuration TL with a high "Q" distributed Bragg reflector (DBR) offers several merits over an edge-emitting TL in terms of smaller optical cavity volume, lower mirror loss, lower parasitics, and lower threshold base current. Data are presented for low threshold n-p-n

vertical cavity transistor laser (VCTL) with improved cavity confinement by trench opening and selective oxidation. The oxide-confined VCTL with 6.5 x 7.5 μm² oxide aperture dimension as shown in Fig 12 demonstrates a threshold base current at 1.6 mA and an optical power of 150 μW at I_B = 3mA operating at -80 °C due to the mismatch between the quantum well emission peak and the resonant cavity optical mode [11, 12]. The VCTL operation switching from spontaneous to coherent stimulated emission is clearly observed in optical output power L-V_{CE} characteristics as shown in Fig. 13. The collector output I_C − V_{CE} characteristics demonstrate the VCTL can lase at transistor's forward active mode with a collector current gain β = 0.48 as shown in Figure 13 insert.

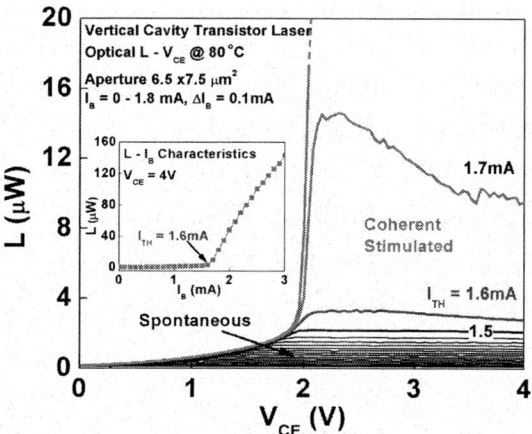

Fig. 13. The optical output (L − V_{CE}) characteristics near the threshold current reveal detailed recombination process. A great enhancement of light intensity is demonstrated as the high-Q vertical cavity transistor switching from spontaneous to coherent stimulated operation. The inset shows the optical output (L- I_B) characteristics at V_{CE} = 4V. The emission intensity takes off at I_{TH} = 1.6mA. For I_B < I_{TH}, the spontaneous emission intensity is less than 2 μW. At I_B = 3 mA (> I_{TH}), the coherent stimulated emission intensity exceeds above 150 μW.

Figure 14 shows the emission spectra of an RCLET and a deposited-mirror VCTL with base current I_B = 2.5mA and V_{CE} = 2V. The measurement is taken at -75°C since the QW emission is detuned from bottom DBR mirror at room temperature. The emission from the RCLET (without deposited mirror) has a broad peak at 971.3nm with a full wave half maximum (FWHM) of 6nm. After depositing the 11 pairs of the SiO_2/TiO_2 DBR the device shows stimulated emission at 969.68nm with a FWHM smaller than 0.3 Å. The inset shows the stimulated emission spectra on a logarithmic scale. It shows three distinct modes. The mode spacing between the fundamental mode and first excited mode is 1.91 nm with a side mode suppression ratio (SMSR) of 33.25 dB. Using this mode spacing, we estimate the aperture of the device to be 4 x 3.3 μm^2, making it the smallest reported

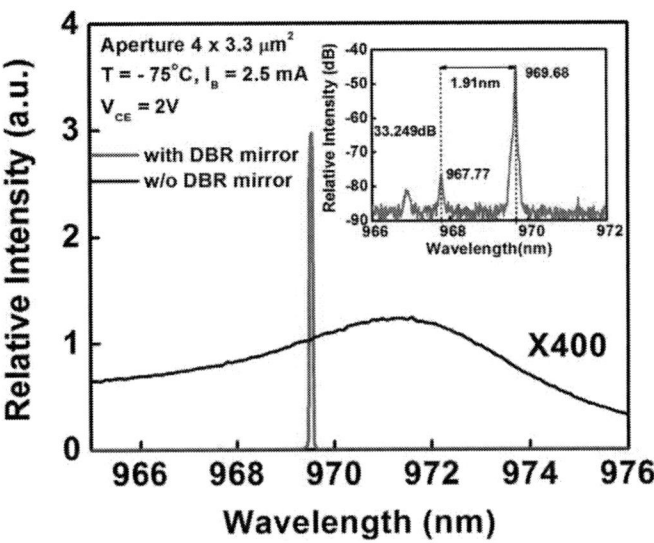

Fig. 14. The emission spectra of an resonant cavity light emitting transistor (RCLET) in black and a deposited mirror microcavity VCTL in red at $-75^\circ C$ with $I_B = 2.5$mA and $V_{CE} = 2$V. The inset is a log plot of the stimulated emission spectra of 4 x 3.3 μm^2 microcavity VCTL.

transistor laser for low power operation [13].

V. ENERGY EFFICIENCY DATA TRANSFER

We performed calculations on data/energy efficiency for the electrical and optical interconnects based on 0.5 meter copper line, VCSEL, oxide-VCSEL and Microcavity oxide-VCSELs as well as transistor lasers [14]. Fig. 7 displays a plot of data/energy efficiency (Tb/J) as a function of data rate (Gb/s) for electrical and optical interconnects with different threshold current for VCSELs and TLs. The green circles (VCSELs) and blue circles (Oxide-VCSELs) represent published experimental results. The red circles (Microcavity-oxide VCSELs) are UIUC published results. In this plot, the electrical interconnect (copper line) is limited to 1Tb/J at 10 Gb/s. The VCSEL ($I_{TH} = 1$ mA) based optical interconnect is limited to 2Tb/J @ 10 Gb/s and 0.7Tb/J @ 20 Gb/s. It shows 2x data/energy advantages over copper interconnect. The oxide-VCSELs with $I_{TH} = 0.4$ mA based optical interconnect shows 7Tb/J @ 10 Gb/s which is 7x better than copper interconnect. The Microcavity-VCSELs with $I_{TH} = 0.2$ mA demonstrate 20Tb/J @ 10 Gb/s which is 20x better than copper interconnects. However, the Microcavity VCSEL (diode laser) based optical interconnect is limited to lower efficiency of 1Tb/J @ 45 Gb/s data rate. Hence, there is a great need to innovate a coherent source to transmit data at a rate > 40Gb/s with an energy efficiency > 20Tb/J [14].

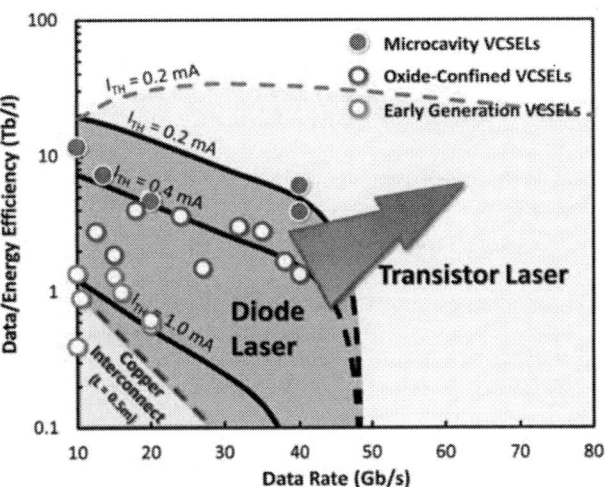

Fig. 15. Calculated data/energy efficiency (Tb/J) vs data rate (Gb/s) for electrical and optical interconnect via copper interconnect and optical interconnect by diode lasers (VCSELs, oxide-VCSELs, Microcavity VCSELs) and transistor laser.

The innovation of the transistor laser with pico-second recombination lifetime and $I_{TH} = 0.2$ mA has potential for data/energy efficiency >20 Tb/J @ 100 Gb/s. In addition, the output collector current can be directly connected back to the base input as a feedback loop to improve laser temperature stability and optical and electrical dynamic ranges. Without any doubt, the three-port transistor lasers will play important role in enabling photonic integrated circuits.

VI. Conclusion

The ability of the quantum-well transistor laser to reduce radiative-recombination lifetime < 30 ps is a fundamental "breakthrough" in semiconductor LED and Laser technology, offering the possibility of extending the direct-modulation bandwidth of the light emitting diode and transistor to >10 GHz and the semiconductor laser to >100 GHz. Three-port operation of transistor laser provides feedback control and high impedance modulation capability for integration, hence, it expands its use to optoelectronic interconnects and enables a new generation of high-performance photonic integrated circuits.

ACKNOWLEDGMENT

N. Holonyak, Jr. is grateful for the support of the Bardeen Chair (Sony) of Electrical and Computer Engineering and Physics, and M. Feng for the support of the Holonyak Chair of Electrical and Computer Engineering. We wish to thank Dr. Michael Gerhold and acknowledge the support of the Army Research Office, Grant No. W911NF-12-1-0394.

REFERENCES

[1] M. Feng, N. Holonyak, Jr., and R. Chan, "Quantum-Well-Base Heterojunction Bipolar Light Emitting Transistor," *Appl. Phys. Lett.*, vol. 84, pp. 1952, 2004.

[2] G. Walter, N. Holonyak, Jr., M. Feng, and R. Chan, "Laser Operation of a Heterojunction Bipolar Light-Emitting Transistor," *Appl. Phys. Lett.*, 85, 4768 (15 November 2004).

[3] M. Feng, N. Holonyak, Jr., G. Walter, and R. Chan, "Room Temperature Continuous Wave Operation of a Heterojunction Bipolar Transistor Laser," *Appl. Phys. Lett.*, 87, 131103-1/131103-3 (26 Sept. 2005).

[4] M. Feng and N. Holonyak, Jr. "The Metamorphosis of the transistor into a laser", *Optics & Photonics News(OPN)*, Optical Society of America (OSA) p. 44-49, March (2011)

[5] H.W. Then, M. Feng, and N. Holonyak, Jr., "Physics of base charge dynamics of transistor laser," *Appl. Phys. Lett.*, 96, 113509 March 17th (2010)

[6] M. Feng, H. W. Then, N. Holonyak, Jr., and G. Walter, "Resonance-Free Frequency Response of a Semiconductor Laser," *Appl. Phys. Lett.*, 95, 033509 (2009).

[7] F. Tan, R. Bambery, M. Feng, and N. Holonyak, Jr., "Transistor laser with simultaneous electrical and optical output at 20 and 40 Gb/s data rate modulation," *Appl. Phys. Lett.*, 99, 061105 (2011)-Aug 10.

[8] F. Tan, R. Bambery, M. Feng and N. Holonyak, Jr., "Transistor Laser with a 13.5 Gb/s Error Free Data Transmission," *IEEE Photonics Technology Letters,* TBD, (2014) – Accepted for Publication-online version in June, 2014.

[9] H. W. Then, C. H. Wu, G. Walter, M. Feng, and N. Holonyak, Jr., "Electrical-optical signal mixing and multiplication (2→22 GHz) with a tunnel junction transistor laser," *Appl. Phys. Lett.,* **94**, 101114 (2009).

[10] H. W. Then, M. Feng, and N. Holonyak, Jr., "Microwave circuit model of the three-port transistor laser," *J. Appl. Phys.,* 107, 094509 (2010).

[11] M. K. Wu, M. Feng and N. Holonyak, Jr.,"Voltage Modulation of the Vertical Cavity Transistor Laser via Intra-Cavity Photon Assisted Tunneling*," Appl. Phys. Lett.,* 101, 081102 (2012)- August 20.

[12] M. K. Wu, M. Liu, F. Tan, M. Feng and N. Holonyak, Jr., "Selective oxidation cavity confinement for low threshold VCTL," *Appl. Phys. Lett.*, 103, 011104 (2013).

[13] M.K. Wu, M. Liu, R. Bambery, M. Feng and N. Holonyak, Jr., " Low Power Operation of a Vertical Cavity Transistor Laser via the Reduction of Collector Offset Voltage," *IEEE Photonics Technology Letters* , **26**, 1003-6 (2014)-May 15

[14] M. Feng, F. Tan, M.K. Wu, R. Bambery and N. Holonyak, Jr., "Transistor Laser: Three Port Operation for Optoelectronics Interconnects and Electronic-Photonic Integrated Circuits," *GOMATECH-2013* Session 10: Photonic Interconnects and Microphotonics (Paper No. 10.1) March 13, 2013.

Integration Challenges for High-Performance Carbon Nanotube Logic

James B. Hannon, Hongsik Park, George S. Tulevski, and Wilfried Haensch

Abstract—As the scaling of silicon-based devices becomes more challenging, alternative channel materials are being actively explored. One approach is to replace the silicon channel with nanoparticles – for example, carbon nanotubes – that offer higher performance and better scaling potential. However, the incorporation of nanoparticles requires the development of new "bottom up" fabrication techniques to grow or place particles at precise locations on a substrate. The inherent randomness of these assembly processes has an obvious impact on device yield, which must be taken into account in optimizing the layout of a device. Here we describe a simple statistical analysis of device yield that can give insight into the self-assembly process, and is particularly useful for characterizing nanoparticle self-assembly from solution.

Keywords—*nanoparticles, carbon nanotubes, placement, self assembly.*

I. INTRODUCTION

IN conventional semiconductor fabrication, devices are fabricated from layers of materials deposited onto a substrate. The process of depositing and patterning layers to form a device is known as "top down" manufacturing. Integration of devices built from nanoparticles, such as semiconductor nanowires (NWs) or carbon nanotubes (CNTs), requires new methods for device fabrication. These "bottom up" approaches involve guiding or growing nanoparticles so that they are positioned at precise locations of a substrate. The assembly processes involved can be difficult to characterize, are often inherently random, and for the most part, are poorly understood. Here we describe a statistical analysis of nanoparticle placement that can be used to characterize the basic features of the placement process.

II. CARBON NANOTUBE DEVICES

With the anticipated end of conventional scaling [1], continued performance gains in high-performance logic will come from the introduction of new materials and new device geometries. CNTs are one candidate material for replacing silicon as the channel material in logic devices. CNTs combine a small body with ballistic transport, making them inherently fast and scalable [2]. Logic devices fabricated from CNTs have shown the potential to out-perform conventional silicon devices [3]. However, before CNT electronics can become viable, several integration challenges must be overcome. Processor-level simulations of system performance suggest that CNT

The authors are with the IBM Research Division, IBM T.J. Watson Research Center, 1101 Kitchawan Road, Yorktown Heights, NY, 10598 USA (email: jbhannon@us.ibm.com)

devices must be small in order to achieve significant power and performance gains [4], [5]. Specifically, the optimized device in these simulations contains 5 or 6 CNTs in the channel, with a spacing between CNTs of about 10 nm, and a channel length of 10 nm [5].

There are two main approaches to fabricating this type of device. A common method is based on growth and transfer [6]–[9]. The growth is carried out via chemical vapor deposition (CVD) on a substrate, such as sapphire [6] or quartz [7], that induces alignment of the CNTs. CNT densities as high at 60 CNTs/μm have been achieved using this growth method [8]. After growth, the CNTs are then transferred from the growth substrate to a CMOS substrate for device fabrication. One disadvantage of the growth-and-transfer approach is that there is no method to completely eliminate metallic CNTs during growth. These metallic CNTs must be removed after transfer. One approach is to electrically 'burn out' the metallic CNTs [10]. Shulaker *et al.* have used this method to reduce the metallic fraction of transferred CNTs to below 0.01% [11]. However, even if the metallic tubes can be removed with 100% efficiency, they still increase device variability. On average, devices that contained a metallic CNT will have fewer semiconducting CNTs, and hence, lower on-state current than those containing all semiconducting CNTs.

An alternate approach to growth-and-transfer is CNT placement from solution. For example, a substrate is chemically modified to guide CNTs in solution to specific areas on a surface. The solution can be purified to remove metallic CNTs before deposition, thus isolating the process of purification from growth [5]. Metallic fractions below 0.1% have been demonstrated using solution-based purification methods [5], [12]. In one of the first demonstrations of CNT self-assembly, a combination of hydrophobic and hydrophilic monolayers was used to position CNTs at predefined locations on a surface [13]. Recently Park *et al.* demonstrated extremely high placement density using a combination of surface chemistry and patterned substrates [14]. In this approach, a patterned oxide substrate is used to define the target areas for CNT placement. A thin SiO_2 layer is grown on top of a HfO_2 starting surface. The SiO_2 is etched to form HfO_2 'trenches' in the surrounding SiO_2 field oxide. The surface is then exposed to a positively-charged monolayer (NMPI) which binds to HfO_2 but not to SiO_2. CNTs are dispersed in solution using a negatively-charged surfactant. The surfactant binds to the NMPI monolayer via an ion exchange mechanism. An SEM image of CNTs placed using this method is shown in Figure 1. Very little is known about how CNTs in solution diffuse to the surface and bind in the trenches. Clearly, there is high degree of randomness: some trenches have no CNTs while others have

978-1-4799-7231-9/14 $31.00 © 2014 IEEE

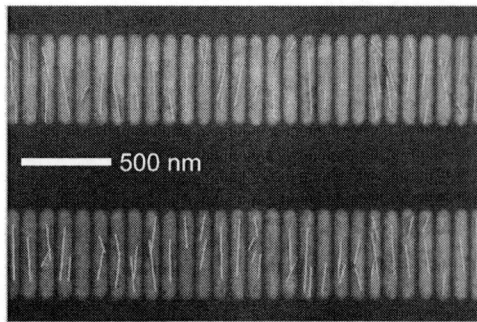

Fig. 1. SEM image of HfO$_2$ trenches surrounded by an SiO$_2$ field oxide. The trenches are 80 nm × 500 nm, and the trench pitch is 100 nm. The positions are of individual CNTs are indicated by orange lines.

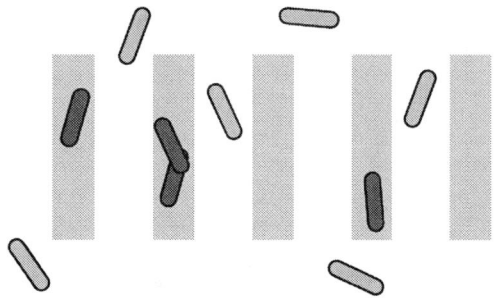

Fig. 2. The N target areas for deposition on a substrate are shown in yellow. The M nanoparticles located in the target areas are shown in red. Nanoparticles that are located outside the target areas (gray) are ignored

two or more. In the following, we describe a simple statistical model that can give insight into the deposition process. We developed the model to describe CNT placement from solution, but it can be applied to the generic problem of nanoparticle placement, or even nanoparticle growth.

III. RANDOM DEPOSITION MODEL

We assume there are N well-defined regions on a surface at which nanoparticle placement is desired. These target areas would correspond to the HfO$_2$ trenches in the CNT deposition case described above. The bins are shown schematically in Figure 2. There are M_0 nanoparticles in solution, some fraction of which are eventually deposited onto the surface. Of these, there are M which are deposited in the target regions (shown in red in Figure 2). Particles that are deposited on the surface outside the target areas (shown in gray) are ignored. Our analysis is concerned with how the M nanoparticles are distributed in the N target areas. Focusing on M simplifies the analysis considerably since the details of the adsorption process are ignored. Furthermore, M can, in principle, be measured by imaging the N target areas, whereas quantities such as M_0 are difficult to quantify. If the target areas are large compared to the particle size, or the interaction between particles is weak, a target area can contain more than one particle. In general, the probability of a particle landing in a particular target area will depend on the occupation. For example, adsorption in an occupied target area might be far less

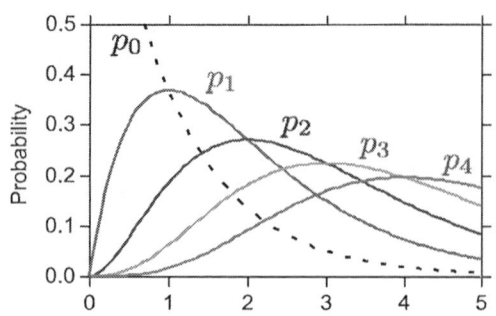

Fig. 3. Probabilities p_n for having target areas with n nanoparticles, as a function of the average density, M/N. Note the largest value of p_1 is 0.37, which occurs when $M/N = 1$.

probable than adsorption in an empty one. The blocking effect will be reflected in the probability, p_n, that a particular target area will contain n particles. The simplest case corresponds to no blocking effect. For this case, the probability, p, of an adsorbed particle being located in a particular trench is simply $1/N$. That is, all target areas are equally likely, independent of occupation. The probability of finding n particles in a particular target area is given by the binomial theorem:

$$p_n = \binom{M}{n} p^n (1-p)^{M-n} = \quad (1)$$

$$\frac{M!}{n!\,(M-n)!}\, p^n (1-p)^{M-n} \quad (2)$$

p_n is proportional to the probability of having n particles in a particular target area, p^n, and to the probability of the other $M-n$ particles adsorbing in different target area, $(1-p)^{M-n}$. The combinatorial factor is simply the number of ways of choosing n particles from a population of M. The values of p_n as a function of average occupation, M/N, are shown in Figure 3. We note that, to some degree, M/N reflects the probability that a particle in solution is deposited into a target area. That is, if M/N is 'large' deposition is 'easy.'

If the goal of the deposition process is to place a precise number of nanoparticles (e.g., one) in each target area, then it is clear from Figure 3 that random deposition makes this goal unattainable. Even if the deposition is tuned so that the average occupation is one ($M/N = 1$), the yield of target areas with exactly one particle (p_1) is only 37%. Furthermore, the fraction of empty target areas (p_0) is also 37%, and 26% of target areas have more than one particle. Increasing the average occupation does not increase p_1, in fact, p_1 decreases with increasing M/N, while p_2 increases. A blocking effect is needed to reduce the inherent randomness of deposition from solution. One way of enhancing the blocking effect is to make the size of the target area comparable to the size of the nanoparticle. In Section V we discuss this limit for CNT deposition. But first, we consider an important case for which the random deposition model is accurate and useful.

IV. CARBON NANOTUBE PURITY

As noted above, an advantage of placing CNTs from solution is that the population can be purified before placement.

978-1-4799-7231-9/14 $31.00 © 2014 IEEE

N	N_o	N_s	$N - N_o$	M	purity (%)
10,590	6,248	75	4,342	5,588	98.7
4,425	1,593	83	2,832	4,521	98.2

TABLE I.   STATISTICS FROM TWO EXPERIMENTS TO DETERMINE THE METALLIC FRACTION OF A CNT POPULATION FROM DEVICE MEASUREMENT. SYMBOLS ARE DEFINED IN THE TEXT.

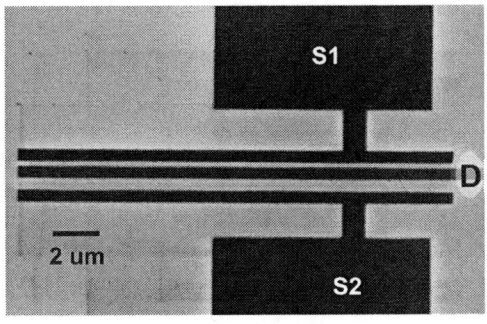

Fig. 4.   Device structure used to measure CNT purity. The image shows two source electrodes ($S1$, $S2$) that share a common drain (D). The channel length of the upper device is 250 nm, while that of the lower device is 500 nm. The channel width is 20 μm.

A key issue in CNT purity is assessing the purity of a given population [5]. One method is to use absorption spectroscopy to detect the presence of metallic CNTs in solution. However, metallic fractions less than 1% are difficult to measure. For these pure populations, direct electrical measurement appears to be the only viable method for determining the purity [15]. The procedure is to build thousands of devices from a low density of randomly-dispersed CNTs on a substrate. The purity is determined by counting the number of metallic devices. An image of a device structure we have used to measure purity is shown in Figure 4. The devices are large, with channel lengths of 250 and 500 nm, and a channel width of 20 μm. The target areas for CNT deposition are consequently quite large. If the density of deposited CNTs is low, then the probability of one CNT blocking the adsorption of another is vanishing small. For large devices and low CNT density, the random deposition model described in Section III should be accurate.

It might be assumed that, if the CNT density is low, the metallic fraction is simply given by the ratio of the number of shorted devices, N_s, to the number of conducting (i.e., not open) devices. The implicit assumption is that each non-open device contains only one CNT. However, for random CNT deposition, this need not be the case. In fact, the metallic fraction is actually equal to M_s/M, where M is the number of CNTs on the surface that span the source and drain of a device, and M_s is the number of these that are metallic. In general $M_s > N_s$ and $M > N - N_o$. The difficulty is that routinely imaging thousands of devices to determine M is impractical.

While measuring M directly is time consuming, performing electrical measurements on thousands of devices is not. We can use the yield of open devices to estimate M for cases where the CNT deposition is truly random. Specifically, the number of open devices, N_o, can be expressed as $N_o = N p_0$, where p_0 is given by Eq. 2:

$$N_o = N p_0 = N \left(1 - \frac{1}{N}\right)^M \tag{3}$$

solving for M:

$$M \approx -N \ln\left(\frac{N_o}{N}\right) \tag{4}$$

which is valid for random placement and $N \gg 1$. To illustrate the consequences, consider a situation in which 1000 devices are fabricated in a deposition experiment and 800 are found to be open. Even for this low device yield, one would significantly underestimate the number of placed particles by assuming $M \approx N - N_o = 200$. The actual value of M would be closer to 223.

We have used this analysis to estimate the metallic fraction of our purified CNT solutions. A summary of two separate experiments is given in Table I. In both cases, several thousand devices were fabricated on an SiO_2 surface covered with a low-density randomly-dispersed CNTs. In the first experiment, analysis of the fraction of open devices (59%) suggest that 5,588 CNTs were placed. 75 shorted devices were measured. If each of the shorted devices contained only one metallic CNT, then the metallic fraction of the population is 75/5,588 or 1.3%. Similar analysis yields a metallic fraction of 1.8% for the second experiment.

In principle, the shorted devices could contain more than one metallic CNT. In fact, if the number of shorted devices is large, this is likely. An analog of Eq. 4 can be used to estimate the number of metallic CNTs placed (M_s) from the number of non-shorted devices, $N - N_s$:

$$M_s \approx -N \ln\left(\frac{N - N_s}{N}\right) \tag{5}$$

For large number of devices, and low metallic fractions, $M_s \approx N_s$.

V.   CARBON NANOTUBE PLACEMENT

Thus far the assumption is that nanoparticles that land on the surface will be found in any of the target areas with equal probability, independent of the occupation. For the appropriate device geometry, this assumption is valid, and quantities that are tedious to measure, such as M, can be accurately estimated from electrical measurements. However, for large-scale device fabrication for logic applications, random deposition leads to unacceptable variation in the device yield.

One way to reduce the randomness in nanoparticle deposition is to combine a highly-selective deposition scheme with small target areas. Small target areas will enhance any blocking effect associated with the deposition. We model this effect with a simple blocking parameter, q. We assume that the probability of a nanoparticle being placed in a target area is proportional to q^n, where n is the number of nanoparticles in the target area. The random model corresponds to $q = 1$. Using this model, we simulate the deposition by sequentially placing each of M particles into N target areas. Before each particle is placed, the deposition probability for each target area is recomputed based on the current occupation.

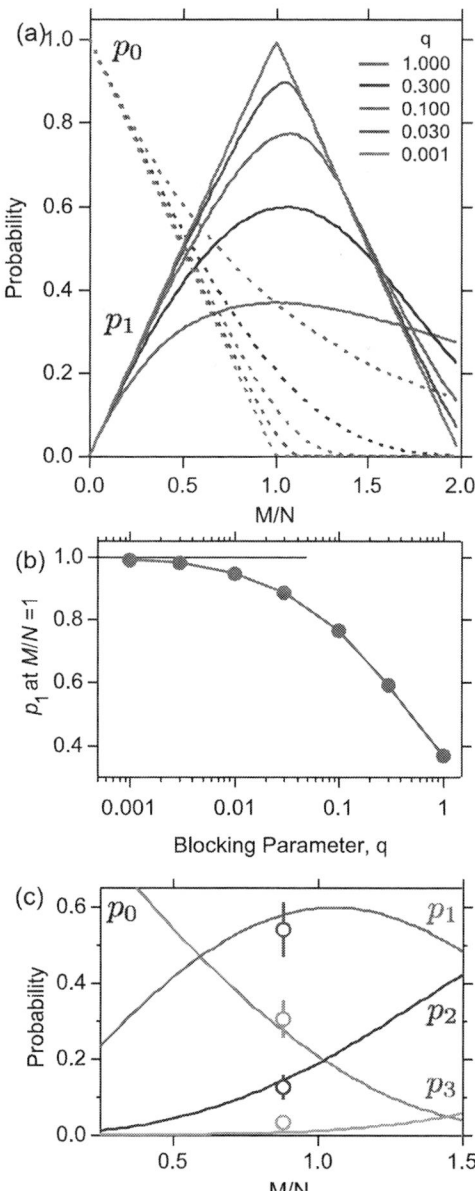

Fig. 5. (a) Computed values of p_0 (dashed) and p_1 (solid) as a function of M/N for selected values of the blocking parameter, q. (b) p_1 at $M/N = 1$ as a function of q. (c) Comparison of measured placement yields for 244 devices with those calculated for $q = 0.3$. The target areas for these devices were HfO$_2$ trenches 150 nm wide and 2 μm long.

The results of a simulation for $N = 1000$ is shown in Figure 5(a). The curves were obtained by averaging the results of 2000 simulations. The figure shows p_0 and p_1 as a function of M/N for five different values of the blocking parameter, q. The red curve ($q = 1$) corresponds to the random model shown in Figure 3. As the blocking parameter is increased, the yield of target areas with only one particle, p_1, increases, while p_0, p_2, and all other p_n's decrease. In the limit of small q, the distributions become essentially linear. That is, all of the empty target areas fill up before any double occupation occurs.

The simulations suggest that the desired deposition outcome – one nanoparticle in each target area – requires a very small blocking parameter. To illustrate this, in Figure 5(b) the value of p_1 at $M/N = 1$ is shows as a function q.

The analysis of CNT purity using very large devices could be accurately modeled with a random deposition model, that is, with no blocking effect. We now show that the blocking effect can be increased if smaller target areas are used. Devices similar to those shown in Figure 1 were fabricated using the approach described in [14]. The HfO$_2$ target areas (trenches) were 150 nm wide and 2 μm long. Following CNT deposition from solution, the substrate was imaged using SEM. $M = 113$ CNTs were observed in a total of $N = 128$ devices, corresponding to an average occupation of $M/N = 0.88$. The measured distribution of empty, single-, double-, and triple-CNT devices is shown in Figure 5(c). The measured yields are inconsistent with completely random deposition. For example, the yield of single-CNT trenches is 54%, while random deposition would suggest a yield of only 35%. Direct simulation shows that a blocking parameter of $q = 0.3$ gives a good description of the yields, as shown in Figure 5(b). Note that only a modest data set, 128 devices, is needed to get a good estimate of the blocking parameter. Roughly speaking, the analysis shows that CNTs are three times more likely to land in an empty trench compared to an occupied trench. This analysis shows that reducing the size of the target area can significantly reduce the inherent randomness of nanoparticle placement.

VI. CONCLUSION

We have applied a simple statistical model to the analysis of nanoparticle placement at predefined target areas of a substrate. The deposition is characterized by p_n, the fraction of target areas that contain n nanoparticles. For many applications, ideal deposition corresponds to $p_1 = 1$ with all other $p_n = 0$. In other words, each target area contains exactly one nanoparticle. We show that completely random deposition leads to significant variation in the occupation, and that p_1 never exceeds 0.37. We apply the random model to the problem of determining the fraction of metallic CNTs in solution. We show that reducing the size of the target area can lead to deposition that is inconsistent with completely random deposition. In this case, the presence of a nanoparticle in a particular target area significantly reduces the probability of additional deposition. We show how a simple blocking parameter can be used to characterize this effect.

REFERENCES

[1] W. Haensch, E. Nowak, R. Dennard, P. Solomon, A. Bryant, O. Doku-maci, A. Kumar, X. Wang, J. Johnson, and M. Fischetti, "Silicon CMOS devices beyond scaling," *IBM Journal of Research and Development*, vol. 50, no. 4.5, pp. 339–361, July 2006.

[2] A. Javey, J. Guo, Q. Wang, M. Lundstrom, and H. Dai, "Ballistic carbon nanotube field-effect transistors," *Nature*, vol. 424, no. 6949, pp. 654–657, 08 2003. [Online]. Available: http://dx.doi.org/10.1038/nature01797

[3] F. Kreupl, "Electronics: Carbon nanotubes finally deliver," *Nature*, vol. 484, no. 7394, pp. 321–322, 04 2012. [Online]. Available: http://dx.doi.org/10.1038/484321a

[4] L. Wei, D. Frank, L. Chang, and H. S. P. Wong, "A non-iterative compact model for carbon nanotube FETs incorporating source exhaustion effects," in *Electron Devices Meeting (IEDM), 2009 IEEE International*, Dec 2009, pp. 1–4.

[5] G. S. Tulevski, A. Franklin, D. Frank, J. Lobez, Q. Cao, H. Park, A. Afzali, S. Han, J. B. Hannon, and W. Haensch, "Towards high-performance digital logic technology with carbon nanotubes," 2014, to appear in *ACS Nano*.

[6] A. Ismach, D. Kantorovich, and E. Joselevich, "Carbon nanotube graphoepitaxy: Highly oriented growth by faceted nanosteps," *Journal of the American Chemical Society*, vol. 127, no. 33, pp. 11 554–11 555, 2005, pMID: 16104703.

[7] S. J. Kang, C. Kocabas, T. Ozel, M. Shim, N. Pimparkar, M. A. Alam, S. V. Rotkin, and J. A. Rogers, "High-performance electronics using dense, perfectly aligned arrays of single-walled carbon nanotubes," *Nat Nano*, vol. 2, no. 4, pp. 230–236, 04 2007. [Online]. Available: http://dx.doi.org/10.1038/nnano.2007.77

[8] L. Ding, D. Yuan, and J. Liu, "Growth of high-density parallel arrays of long single-walled carbon nanotubes on quartz substrates," *Journal of the American Chemical Society*, vol. 130, no. 16, pp. 5428–5429, 2014/07/23 2008. [Online]. Available: http://dx.doi.org/10.1021/ja8006947

[9] K. Ryu, A. Badmaev, C. Wang, A. Lin, N. Patil, L. Gomez, A. Kumar, S. Mitra, H. S. P. Wong, and C. Zhou, "CMOS-analogous wafer-scale nanotube-on-insulator approach for submicrometer devices and integrated circuits using aligned nanotubes," *Nano Letters*, vol. 9, no. 1, pp. 189–197, 2014/07/23 2008. [Online]. Available: http://dx.doi.org/10.1021/nl802756u

[10] P. Avouris, "Carbon nanotube electronics," *Chem. Phys.*, vol. 281, no. 2-3, pp. 429–445, 2002.

[11] M. M. Shulaker, G. Hills, N. Patil, H. Wei, H.-Y. Chen, H. S. P. Wong, and S. Mitra, "Carbon nanotube computer," *Nature*, vol. 501, no. 7468, pp. 526–530, 09 2013. [Online]. Available: http://dx.doi.org/10.1038/nature12502

[12] M. S. Arnold, A. A. Green, J. F. Hulvat, S. I. Stupp, and M. C. Hersam, "Sorting carbon nanotubes by electronic structure using density differentiation," *Nat Nano*, vol. 1, no. 1, pp. 60–65, 10 2006. [Online]. Available: http://dx.doi.org/10.1038/nnano.2006.52

[13] Y. Wang, D. Maspoch, S. Zou, G. C. Schatz, R. E. Smalley, and C. A. Mirkin, "Controlling the shape, orientation, and linkage of carbon nanotube features with nano affinity templates," *Proc. Nat. Acad. Sci.*, vol. 103, no. 7, pp. 2026–2031, 2006.

[14] H. Park, A. Afzali, S.-J. Han, G. S. Tulevski, A. D. Franklin, J. Tersoff, J. B. Hannon, and W. Haensch, "High-density integration of carbon nanotubes via chemical self-assembly," *Nat. Nano.*, vol. 7, no. 12, pp. 787–791, 12 2012. [Online]. Available: http://dx.doi.org/10.1038/nnano.2012.189

[15] G. S. Tulevski, A. D. Franklin, and A. Afzali, "High purity isolation and quantification of semiconducting carbon nanotubes via column chromatography," *ACS Nano*, vol. 7, no. 4, pp. 2971–2976, 2014/07/24 2013. [Online]. Available: http://dx.doi.org/10.1021/nn400053k

Technologies for Very High Bandwidth Real-time Oscilloscopes

Peter J. Pupalaikis, *Fellow, IEEE,* Brian Yamrone, *Life Member, IEEE,* Roger Delbue, *Member, IEEE,*
Amarpal S. Khanna, *Fellow, IEEE,* Kaviyesh Doshi, *Member, IEEE,* Balamurali Bhat, *Member, IEEE,*
and Anirudh Sureka

Invited Paper

Abstract—Technologies and design considerations are presented for the design of very high bandwidth oscilloscopes. These include chip, DSP and microwave technologies employed in some of the fastest waveform digitizers in the world.

Index Terms—Analog-digital conversion, Signal processing, Signal sampling, Signal restoration, Microwave frequency conversion, Oscilloscopes, Oscillography, Indium phosphide.

I. INTRODUCTION

THIS PAPER will present various topics regarding IC design and system architecture that are considerations for the design and production of very high-speed real-time oscilloscopes. Briefly, a real-time oscilloscope is, at the heart, a waveform digitizing instrument that acquires waveforms in a single trigger event. This means that, once armed, it must digitize every waveform point at a given sample-rate, one point after another. It cannot benefit from assumptions of repetitiveness of a signal. For the purpose of this paper, we consider very high-speed real-time oscilloscopes as those occupying bandwidths in the range of tens of Gigahertz to around one hundred Gigahertz with commensurate sample-rates that satisfy the Nyquist criteria of acquiring the entire band from DC to the bandwidth with enough sample rate for there to be no frequency aliasing of the signal. Generally, this means sample rates of around three times the bandwidth. Finally, the record length required for such oscilloscopes depends on the application, but a general goal is to acquire tens of milliseconds of data per acquisition.

II. THE MOORE'S LAW OF SCOPE BANDWIDTH

In Figure 1 we see the progression of oscilloscope bandwidth over time for real-time scopes. This plot shows only scope introductions that represent the highest bandwidth scope from each the three high-end vendors. Here we see a steady log-linear progression that is shown as the line marked *LeCroy Trend* which is the trend considering only scopes manufactured by Teledyne LeCroy - but which is fairly representative of the entire data set. The trend indicates an average yearly increase in bandwidth of approximately 28 percent which amounts

Peter J. Pupalaikis, Brian Yamrone, Roger Delbue, Balamurali Bhat, Kaviyesh Doshi, and Anirudh Sureka are with Teledyne LeCroy, Chestnut Ridge, NY USA, e-mail: peterp@lecroy.com.
Amarpal (Paul) Khanna is with National Instruments, San Jose, CA USA, e-mail: pkhanna@PhaseMatrix.com.

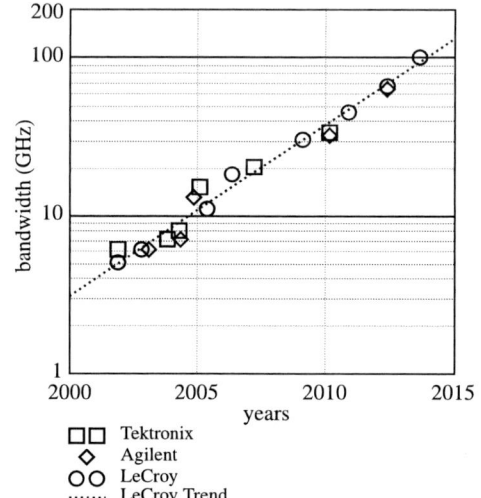

Figure 1. Oscilloscope Bandwidth Versus Time During the Last Decade

to a doubling of bandwidth every 2.8 years. While the plot covers only the first decade of this century, this trend has been fairly constant for the last thirty five years. This leads to the conclusion:

The oscilloscope bandwidth progression is simply another expression of Moore's Law.

While often misstated, Moore's law [1] states that transistor density doubles every 18 months. It says nothing about speed and the implication is that this doubling of transistor density is economical. However, so far, this doubling has been economical and has mostly come with higher speed. Transistor speed is roughly inversely proportional to minimum line thickness and using this loose rule-of-thumb we would predict transistor speed to double every 36 months, or three years. Cause and effect, however, is difficult to interpret. Transistor speeds have certainly benefited over time, but lately, there is little connection between the number of transistors packed into a CMOS microprocessor chip and the speed of bipolar transistor technology which would be used for a scope front-end. Our theory is that the bandwidth progression is *driven* by testing needs as opposed to being enabled by chip technology.

As a final note on this topic, it would be wrong to connect the real concept behind Moore's Law to oscilloscopes because of the economics involved. The cost of an oscilloscope has increased linearly with bandwidth while the cost of computers

978-1-4799-7231-9/14 $31.00 © 2014 IEEE

(and related products) have decreased while growing in performance. This is partly because the volume for oscilloscopes is comparatively very small.

III. THE MAIN TECHNOLOGY VECTORS FOR DIGITAL OSCILLOSCOPES

Oscilloscope performance is pushed to higher speeds along three key technology vectors. These are:

1) Chip Technology - chips are custom designed in high-speed processes for increasing speed.
2) DSP and Corrections Technology - as speeds are pushed, signal fidelity degrades near the upper limits. Signal fidelity is restored through the use of digital signal processing (DSP) and other correction algorithms. Sometimes, this allows the chips to be pushed higher in speed.
3) DBI Technology - digital bandwidth interleaving (DBI) is a technology that allows for the doubling and tripling of the speeds limited by chip technology. This is achieved through the combination of channel resources.

Oscilloscope synchronization technology is also utilized to connect multiple acquisition systems to counteract loss of channels through combination of resources.

IV. CHIP TECHNOLOGY

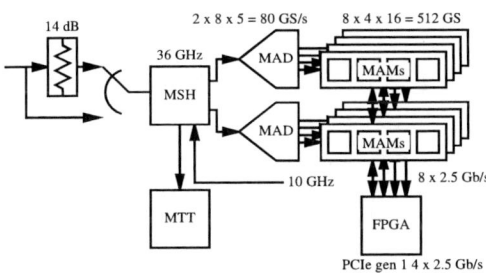

Figure 2. High End Scope Channel Block Diagram

The main technology responsible for oscilloscope performance is chip technology. The underlying performance of a the oscilloscope is determined in the front-end, analog-to-digital converter (ADC) and memory. The front-end is usually a dual-purpose variable gain amplifier and track & hold. Unlike most systems that the reader might be familiar with whereby an analog buffer feeds the ADC containing the track & hold and digitizing elements, in oscilloscopes, because of the speed, the track & hold is generally placed with the amplifier. This is to reduce downstream bandwidth, signal fidelity, and importantly, signal path matching requirements. This is because the digitizer itself might consist of multiple ADC chips and each ADC chip internally consists of multiple ADCs that are time-interleaved. Providing multiple held signals in the amplifier section reduces the bandwidth requirements at the input of the ADC to approximately twice the sample rate of the track & hold which is easier to generate and transmit and is easier to match to each other. The ADC itself has another track & hold whose purpose is to hold this lower speed signal for conversion.

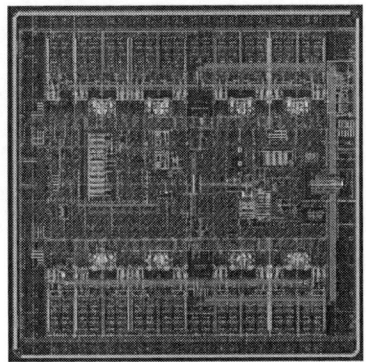

Figure 3. Teledyne LeCroy MSH722B

A high-end LeCroy scope architecture is shown in Figure 2. This is the design of a 36 GHz, 80 GS/s scope channel. It shows an input section which is basically a switched attenuator which provides for 14 dB of attenuation. This feeds an MSH[1] which serves the dual purpose of providing 18 dB of continuous gain in multiple combinations of fixed and variable gain amplifiers and providing sixteen 5 GS/s tracked and held outputs. The MSH is implemented in IBM's 8HP Silicon Germanium (SiGe) process. Each of these outputs has a sample phase that provides for interleaving of the final result. Some interesting features of the MSH is that it consists of sets of gain stages and driver stages that successively fan out the analog signal. The first driver stage drives five outputs, one to the trigger system and four to four drivers which each drive four track & holds. The chip is designed to take its input directly from the center of the chip and to provide its tracked and held outputs on the periphery. The package for the MSH has a built in coaxial connector so that coaxial cables from the outside of the scope are connected directly to the chip package, maintaining a high bandwidth coaxial environment for as long as possible.

Eight of the tracked and held MSH outputs go to one MAD and eight go to another. The MSH also supplies independent sample clocks to each MAD input. Each MAD provides for 40 GS/s sample rate and contains eight virtually independent 5 GS/s internal ADCs. The MAD is implemented in IBM's 7HP SiGe process. Finally, each 5 GS/s data stream is stored in MAMs which are custom designed memories. Each MAM chip takes data at 10 GS/s and can store up to 16 million samples. Note that the MAD provides a data muxing arrangement internally so that even at lower sample rate configurations that might employ less ADCs, these lower number of ADCs can still access the full memory provided by the MAMs.

The readout system consists of an FPGA exposing, on one end, four lanes of PCI Express gen 1 for up to 10 Gb/s (8 Gb/s payload) and on the other end, eight 2.5 Gb/s links that go directly to the MAMs and acquisition system controller in a daisy-chained arrangement. The readout system can sustain approximately 400 MS/s readout rates.

---

[1]There is some special LeCroy nomenclature employed here. All LeCroy chips are referred to as three letters beginning with an 'M' which stands for monolithic. The complete three letter acronym tries to be descriptive. Therefore, an MSH is a monolithic sample/hold, an MAD is a monolithic ADC, and an MAM is a monolithic acquisition memory.

978-1-4799-7231-9/14 $31.00 © 2014 IEEE

Finally, in Figure 2 we also see one other custom chip, the MTT. This is a combination trigger and timebase chip. It is implemented in IBM 5HP SiGe process and enables smart triggers and is the main acquisition controller.

A. *Indium Phosphide Front-end Amplifier*

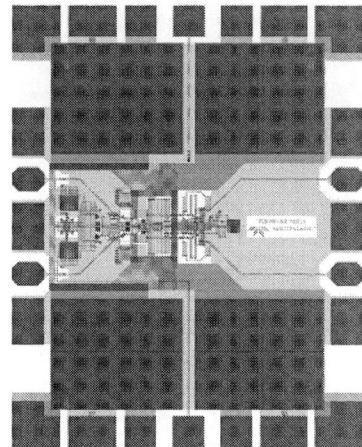

Figure 4.   DC-65 GHz Variable Gain Amplifier

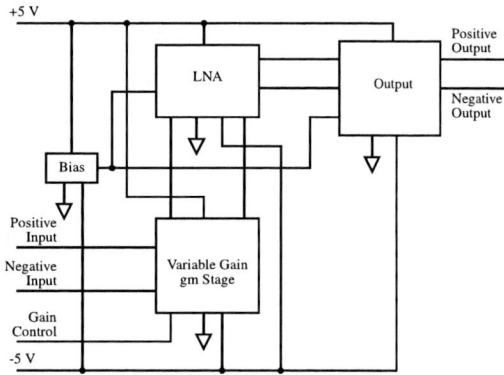

Figure 5.   InP Amplifier Block Diagram

After LeCroy joined Teledyne in 2012, Teledyne LeCroy began designing into Teledyne Scientific's Indium Phosphide (InP) technology [2]. This allows for an increase in front-end performance beyond the current 36 GHz. A goal would be to develop a multi-staged track & hold topology where the InP chip sets the bandwidth.

Most are familiar with III-V processes such as Gallium Arsenide in RF IC design. Despite their high frequency capabilities, the bandwidth of chips developed in these processes tend to be narrow band. In oscilloscope designs, the requirements are for very broad band performance including DC, which is also surprisingly problematic. Another characteristic important for oscilloscope front-ends is thermal tails caused by self-heating of the bipolar devices. Thermal tails cause the step response to slowly reach the final state over a very long (like 1 μs) time period.

In order to gain familiarity with Teledyne Scientific's process, an evaluation front-end amplifier was developed. The chip was fabricated using Teledyne Scientific's 500 nm InP HBT process. The process offers devices with 350 GHz cut-off frequencies, a BVcbo of 5 V, low-loss BCB inter-layer dielectric and four levels of metal interconnect.

The design requirements for an evaluation amplifier were, bandwidth in excess of 65 GHz, signal-to-noise ratio (SNR) of greater than 40 dB, thermal tails less than 1 %, total harmonic distortion (THD) better than -40 dB, and a continuous variable gain range of 10 dB. An amplifier like this would be cascaded with another such amplifier followed by a track & hold in an actual oscilloscope front-end.

A block diagram for the proposed amplifier is shown in Figure 5. Here we have three main sections: a variable gain gm stage, a low noise amplifier (LNA) stage and an output stage. Each stage presents certain challenges which will now be discussed.

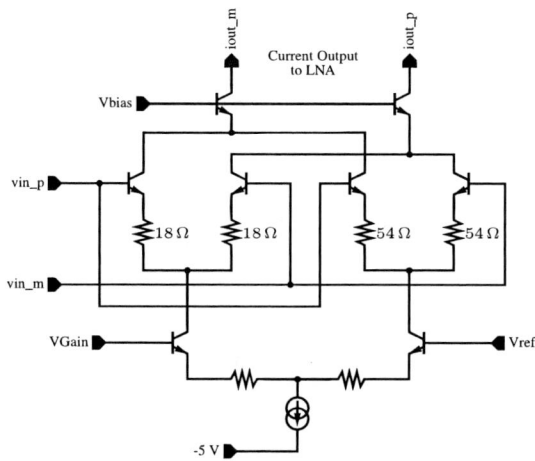

Figure 6.   Variable Gain gm Stage (Simplified)

1) Variable Gain gm Stage: The variable gain gm stage is shown schematically in Figure 6. The function of this stage is to convert a single ended input voltage to a differential output current. The amplitude of this current is dependent on the gain setting of the stage. In addition, this stage provides the variable gain function. An ideal solution to provide this function would be to use FETs, but these are not available at present, so the solution is entirely bipolar. Our solution to achieve low noise, high bandwidth good linearity and low thermal tails is to connect two long tail pairs in parallel each with a different emitter degenerating resistor and each with a variable tail current. The variable tail current satisfies the continuously variable gain requirement. The tail current is set by the gain control voltage which has a range from -3.5 to -1.5 V which provides continuous variable gain from between 0.98 and 3.35. In order to provide the 40 dB signal to noise ratio large devices were required, which tended to lower the stages bandwidth. By carefully sizing the devices a good compromise for both SNR and bandwidth was achieved.

2) LNA Stage: The LNA stage is shown schematically in Figure 7. In order to keep rbb noise to a minimum this stage

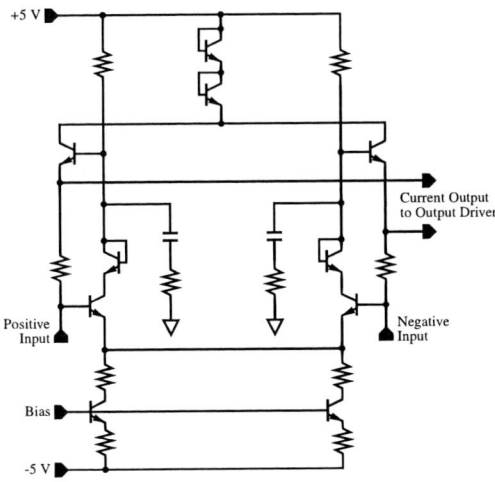

Figure 7. LNA Schematic (Simplified)

requires relatively large devices. By using large devices, the thermally generated rbb noise is kept low, but due to the large capacitance the bandwidth suffered. In order to satisfy both noise and bandwidth, a feedback scheme was used. The large output impedance of the gm stage drives into the low input impedance of this stage's shunt feedback amplifier. The stability of the amplifier is controlled by an RC network and adjusted for a phase margin of about 55 degrees. This allowed for decent bandwidth but with a slight bit of peaking (about 1.5 dB in simulation before parasitic extraction).

3) Output Stage: The low output impedance of the LNA drives into the high input impedance input of the output stage. This stage provides two emitter followers in series to provide the necessary current required to drive the 50 ohm loads to 500 mVpp differential and not present a load to the preceding stage. Two series connected emitter followers could lead to instability. Due to wiring inductance in the base this stage had at least 30 dB of peaking after parasitic extraction. A series resistor added to the base reduced the peaking to around 5 dB.

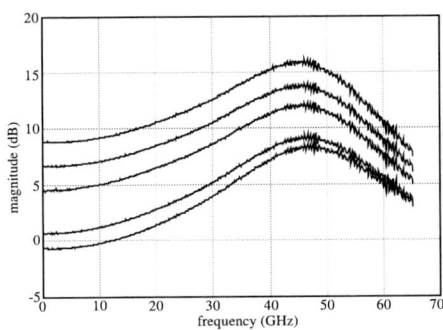

Figure 8. InP Amplifier Frequency Response

4) Conclusion: The test chip was fabricated on a multi-project wafer and the die is shown in Figure 4. Parasitic extraction was performed using HFSS® from ANSYS for all wiring and all parasitic models were s-parameters. The extracted simulations utilized very large, multi-port s-parameter blocks with the transistor models reinserted. We found that the use

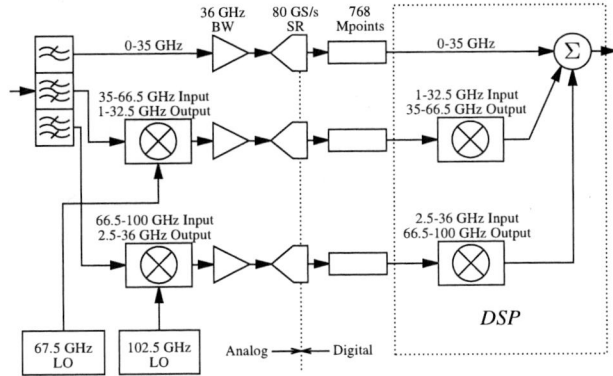

Figure 9. DBI System Block Diagram

of s-parameters does not work well for transient simulations and we are working on a better design/simulation flow.

Comparing the measured results with the simulated extracted results showed a good match. The frequency response is shown with various gain control settings in Figure 8 where we see the expected 5 dB of peaking and in excess of 65 GHz bandwidth.

The chip achieves all of the design goals, but does exhibit about 1.5 dB of compression which will be improved in future versions. Our evaluation shows viability for future oscilloscope designs.

V. Digital Bandwidth Interleaving (DBI)

Digital Bandwidth Interleaving, or DBI [3][4] is a Teledyne LeCroy patented technique in which multiple channel resources are combined to an oscilloscope channel that is virtually a combination of the bandwidths of the combined channels. This technique is analogous to the well-known technique of time-interleaving, which combines channel resources to obtain higher channel sample rates. DBI is the only technique that increases both bandwidth and sample rate. This technique has been employed in the design of several generations of high-end oscilloscopes [5][6].

A DBI architecture that utilizes this traditional hardware is shown in Figure 9. In this design, three 36 GHz, 80 GS/s scope channels are utilized. The scope channels are combined with a microwave front-end and a DSP back-end. The combination of channel resources is common in oscilloscope designs and is often selectable by the user.

The microwave front-end shown stylistically in Figure 9 consists of a multiplexer [7][8] that separates the incoming signal into multiple frequency bands feeding multiple down-converters to frequency translate the separate bands down to a frequency range suitable for acquisition by digitizing channels. Note that for the lowest frequency band, no down-converter is employed and the signal is fed directly to the oscilloscope front-end. For the high-frequency bands, the down-converter modules have digitally controllable gain adjust and the usable band does not extend to DC.

Each down-converter consists of variable attenuation and gain elements, a mixer, a local oscillator (LO), image reject filters and fixed gain amplifiers. There are also mechanisms

978-1-4799-7231-9/14 $31.00 © 2014 IEEE   131

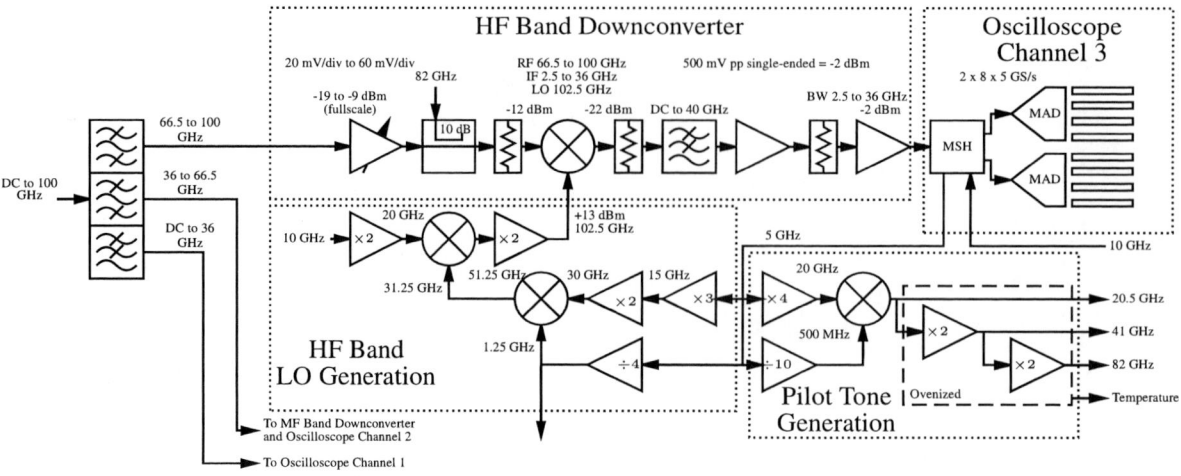

Figure 10.   HF Band Down-converter Block Diagram

for locking the reference clock of the LO to the oscilloscope timebase and sample clock generation along with means for inserting similarly locked calibration tones into the mixer [9] radio frequency (RF) inputs in each band for LO phase calibration. These methods are used for subsequent digital LO regeneration.

Each down-converter is driven by a LO that is set higher than the edge of the frequency band. In this way, the input to the down-converter is translated down and flipped over in frequency (i.e. high-side down-conversion is employed).

In this discussion, the following nomenclature is used to refer to the bands:

- LF - low frequency band - DC to 35 GHz
- MF - medium frequency band - 35 to 66.5 GHz
- HF - high frequency band - 66.5 to 100 GHz

On the DSP side, processing is performed to regenerate the input signal from the three 80 GS/s acquisitions. The recombination involves upsampling from 80 to 240 GS/s, digitally synthesizing a phase-locked rendition of the LO, up-converting the signal back to its original frequency band, rejecting unwanted images, and combining the results. The final result is an 100 GHz, 240 GS/s waveform acquisition [10].

A. Downconverter

A detail of the HF downconverter is shown in Figure 10 and is illustrative of all of the downconverters (with appropriate changes of frequency).

In oscilloscope design, one tends to think of signal strength in terms of a full-scale signal at a given volt/division setting. There are eight divisions vertically on an oscilloscope screen[2]. This oscilloscope handles ranges from 10 mV/div to 80 mV/div, thus the scope accepts full-scale signals from approximately −18 to 0 dBm. Thus, after the high frequency band has been split off by the multiplexer, the first task of

[2]Teledyne LeCroy scopes have eight major divisions vertically. Scopes generally have eight or ten

the downconverter is to provide variable attenuation and gain such that the signal provided to the high linearity mixer is at a constant maximum level of −10 dBm, set 24 dB down from the LO for maximum mixer spur levels of −41 dBc.

There are additional filters placed before and after the mixer. The initial filter is designed so that no signal in the IF band of approximately 2.5-36 GHz is present at the RF port, otherwise the poor (approximately 15 dB) port-port isolation of the mixer causes spurs in the IF. The final filter is designed to reject mainly the LO that leaks to the IF port at about 0 dBm – this would saturate any downstream amplifier.

The fixed gain IF amplifier at the final stage sets the final output power to −4 dBm, which represents almost the full-scale range of the ADC (reserving 3 dB for digital compensation).

The gain/attenuation of the downconverter is finely balanced to achieve a noise figure of 8 to 18 dB depending on the gain range with a worst case SNR of approximately 34 dB. Regarding linearity, it achieves IM3 of −50 dBm and OIP3 in excess of 19 dBm.

B. Local Oscillator Generation

In Figure 10, we see a rather complicated clocking and LO generation. Some of this complication is necessity and some is related to reuse of prior designs.

The 102.5 GHz LO is generated by doubling 51.25 GHz. 51.25 GHz is obtained by mixing 31.25 GHz and 20 GHz (10 GHz system clock multiplied by 2) and 31.25 GHz is generated by mixing 30 GHz and 1.25 GHz. 30 GHz is obtained by multiplying 5 GHz MSH clock output by 6 and 1.25 GHz is obtained by dividing 5 GHz MSH clock by 4 and a copy of the 1.25 GHz is fed back to the channel for phase locking the 102.5 GHz LO to the system phase reference. At startup, the 1.25 GHz signal comes up randomly with one of the four phases of 0°, 90°, 180° and 270° with respect to the system phase and during calibration of the system one of the phases is appropriately chosen and the absolute calibration phase value is stored. During normal startup the divider is reset repeatedly until the phase of the 1.25 GHz is the same as the stored value of the phase.

The phase of the LO drifts over temperature because of high frequency mixing and multiplying and this needs to be compensated for, otherwise this will result in the phase of the HF/MF band to be shifted with respect to the phase of the LF band. To compensate for the LO phase drift, three coherent tones of 20.5 GHz, 41 GHz (20.5 GHz times two) and 82 GHz (41 GHz times two) are injected into the LF, MF and HF bands, respectively and this coherency is maintained throughout the operating temperature by adjusting the phase of the respective MF and HF band digital LO synthesizer. During the calibration, the reference phase shift between the 41 GHz and 20.5 GHz and 82 GHz calibration tones and the 20.5 GHz reference calibration tone is measured and stored during normal operation. Whenever the system detects a change in temperature, the user input is isolated and these coherent calibration tones are switched on and the phase difference between them is re-measured and the LO phase drift is recalibrated. The multipliers that generate the calibration tones are ovenized to maintain the coherency over temperature. A temperature sensor is also installed in the oven to monitor the temperature and can also be used to apply a secondary correction to correct for the phase drift of the tones.

VI. DIGITAL SIGNAL PROCESSING

Once the down converted signals, along with the LF base-band signals are acquired by the three 80 GS/s digitizers and stored in high-speed memory, the waveforms are read out and processed through the DSP processing as shown in Figure 11. All signals undergo an interleave correction filter, an adaptor, upsampler and fractional delay filter. The interleave correction removes the spurious response due to mismatched ADCs. The adaptor handles integer sample propagation time differences and the fractional delay handles the fractional sample time differences. The path propagation time differences are calibrated and stored in the instrument during factory calibration. The upsampler is required because mixing of the signal will cause images that, if not upsampled, will alias into bands of interest. The low image filter is used to define the image of interest after upsampling and before mixing is performed. The mixing action is performed between the LO and the signal from the low image filters. First the LO phase is determined. Because the system sample clock and the LO generation are locked together, the LO phase is a function of location within the acquisition memory. Once the phase is recovered, the digital LO is generated presumably in phase with the actual LO (but at least phase locked). The high image filters keep only the desired image. Sometimes, a crossover phase correction is employed to ensure that all bands sum constructively and holes do not appear in the response [11]. Prior to summing the band signals, a gain adjustment is applied to rescale the signals onto the same vertical scale and the waveforms are summed. The resultant signal is then compensated, first in magnitude, then in phase to produce the 100 GHz, 240 GS/s waveform acquisition.

The DSP processing system utilizes server class Intel® Xeon® X5660 processors (2.8 GHz per core, six cores per

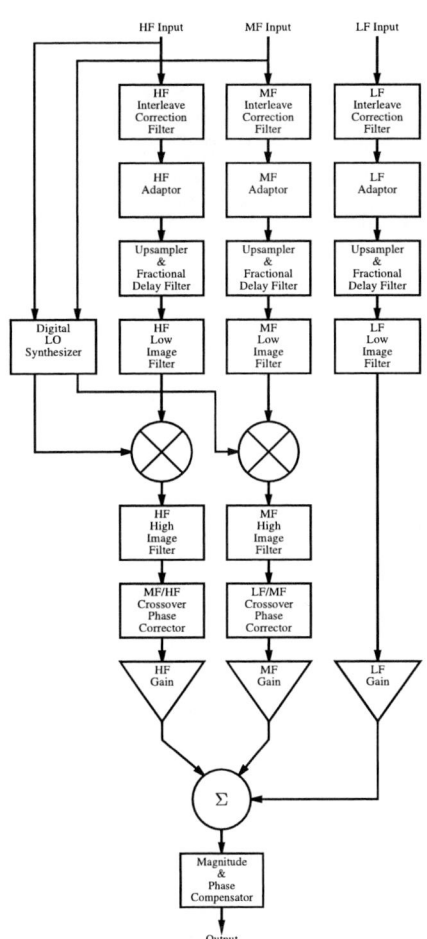

Figure 11.   DSP Block Diagram

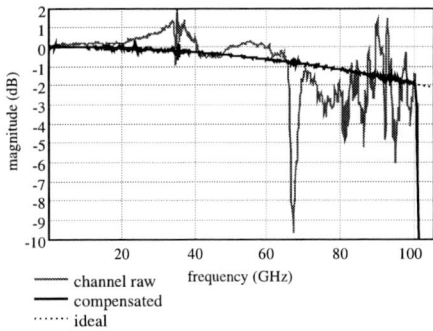

Figure 12.   Magnitude Compensation

processor, and two processors for a 33.6 GHz total effective clock speed with up to 192 GB of RAM. Using Intel performance libraries, it can theoretically sustain calculation rates of about 100 GFLOPS. At approximately 20,000 FLOPs per sample point, the system achieves approximately 5 Mpoint/s throughput.

One of the most important elements in the DSP processing system is the magnitude and phase compensation system. The uncompensated magnitude response of the complete system is shown in Figure 12. Note that there are large variations in the

978-1-4799-7231-9/14 $31.00 © 2014 IEEE

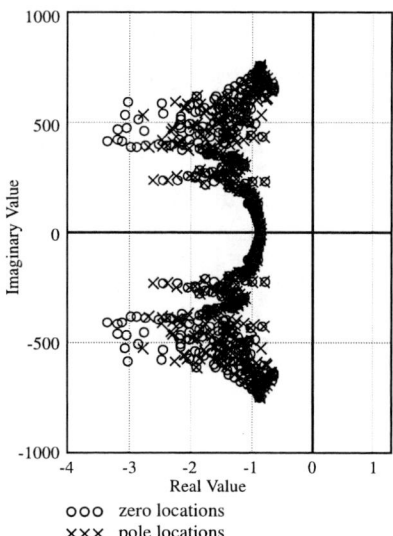

ooo zero locations
xxx pole locations

Figure 13. Pole Zero Locations of Magnitude Compensator

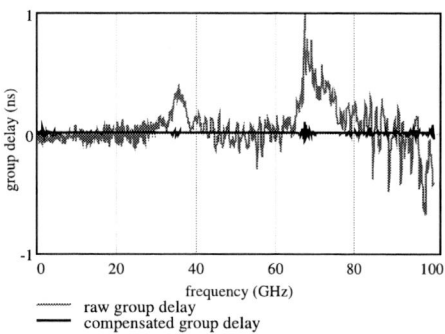

raw group delay
compensated group delay

Figure 14. Group Delay Compensation

cross-over regions - the regions where the LF and MF intersect and where MF and HF intersect. During factory calibration, the delay of each path is adjusted so that the response in the cross-over region is maximized. As seen in Figure 12, the cross-over region between LF and MF bands has response much close to 0 dB. The high and the low peaks are to be expected since the phase may not add up constructively for the entire region. The cross-over region between MF and HF band shows a response that is 10 dB down. One of the reasons for such a response is that the response of the RF deck rolls off sharply in this region and hence adjusting the delay doesn't help much. In practice we do not want to have such a drop in the response because the filter has to boost the signal by 10 dB to make the final response close to 0 dB. Such an amplification will also boost the spurs in the region leading to incorrect behavior. Along the same lines, we do not want the raw response to have a peak greater than 3 dB either. A peak in the raw response indicates an amplification of the acquired analog signal. This amplification can increase the level of the analog signal, and this can saturate the ADCs leading to clipping of the waveform. In any case, the data shown is for a prototype unit and this dip in the raw response is being fixed in a future version.

More common raw responses are within -6 dB and 2 dB, and the filters are calculated so that the final response is a smooth one as shown in Figure 12. To correct the raw response an infinite impulse response (IIR) compensation filter [12] is designed using pole/zero fitting. For the raw response shown in Figure 12, up to 500 poles and zeros are used to generate the IIR filter with desired response. These analog pole and zero locations[3] are plotted in Figure 13 where we see the large numbers of poles and zeros, many of them almost overlapping. They are fitted using techniques involving the Levenberg-Marquardt algorithm [13]. The original guess at the pole and zero locations fall onto a line of equal bandwidth close in to the $j\omega$ axis and restrictions are placed on how close they are allowed to get to the $j\omega$ axis to control inter-point behavior. The compensated response is shown overlaid with the raw response in Figure 12 where we see a nearly perfect fourth-order Bessel response with -2 dB[4] attenuation at 100 GHz.

The phase or group delay compensation is as important as the magnitude compensation as the phase mostly determines the goodness of the step response and other time-domain oscilloscope characteristics. The raw group delay of the system (after magnitude response correction) is shown in Figure 14. As can be seen, the delay in the cross-over region between MF and HF is almost 1 ns. Such a group delay would cause the step response to be completely distorted.

We always compensate the group delay after the magnitude response. This is because the magnitude response is compensated with poles and zeros that also affect the phase response. The phase response is compensated with poles and zeros in an all-pass filter arrangement. All-pass filters are combinations of left half plane poles and right half plane zeros that affect the phase response but have no effect on the magnitude response. The filters are fit with a fuzzy logic grading system based on the step response as phase is an unreliable target for a fit [14]. For DBI, direct all-zero FIR filter design techniques are employed [15].

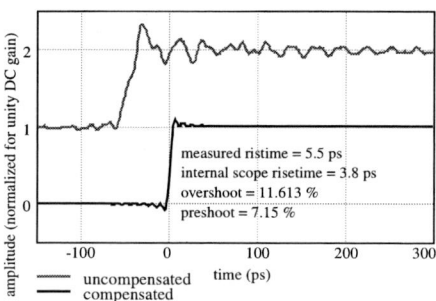

uncompensated
compensated

Figure 15. Uncompensated and Compensated Step Response

The step response with corrected and uncorrected group

---

[3]These locations are approximate as they are derived from digital filter pole and zero locations that line up on a rim on the unit circle in the z-domain. They are shown in the analog domain using the fact that $z = e^{\frac{s}{Fs}}$ which implies $s = \ln(z) \cdot Fs$. They are shown in the analog domain to gain insight into the analog behavior this filter is correcting.

[4]In oscilloscopes, we set the goal for the scope bandwidth point at -2 dB to provide 1 dB of margin to hopefully guarantee that the scope bandwidth specification is held in between calibration periods.

delay is shown in Figure 15. The group delay correction causes the system to be linear phase and provides for a step response rise time of about 4 ps.[5] Teledyne LeCroy scopes offer programmable responses in three flavors: eye-diagram mode (Bessel roll-off and linear phase), flat (flat response and linear phase) and pulse-response (Bessel roll-off and minimum phase[16]).

VII. PERFORMANCE

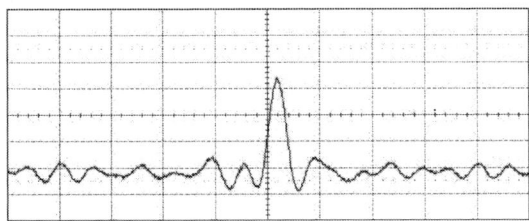

Figure 16.   Optical Impulse Response

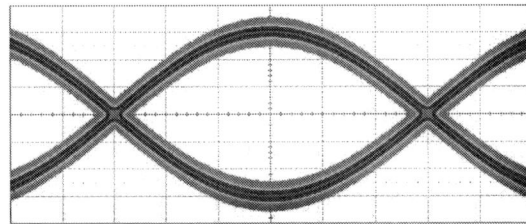

Figure 17.   Eye Diagram

To demonstrate the performance of our 100 GHz oscilloscope employing all of the technology discussed thus far, consider two oscilloscope screens showing first, an optical impulse response in Figure 16 and an eye diagram shown in Figure 17[6]. In Figure 16, the impulse is displayed at 20 ps per division, so the pulse width is about 10 ps for a 5 ps rise and fall time. Based on other measurements we've performed with sampling oscilloscopes, this is the actual rise/fall time of this pulse, so the speed limit of the oscilloscope did not actually come into play. The trace was averaged to reduce noise.

In Figure 17, we have an eye pattern formed by our serial data analysis software on an acquired sinewave of 98.96 GHz according to the oscilloscope. It is displayed at 840 fs per horizontal division. The jitter measured is 63 fs random jitter with 20.1 fs deterministic jitter. This shows not only the vertical quality of the oscilloscope but also the quality of the oscilloscope timebase as well.

---

[5]The general rule of thumb for scopes is that the risetime multiplied by the bandwidth is a constant between 0.35 and 0.45 - 0.35 is for a single-pole rolloff and 0.45 would be for a steep (multi-pole) rolloff. High-end real-time scopes generally have very steep rolloffs.

[6]Normally, for oscilloscope performance, one wants to show screenshots of pseudo-random bit sequence (PRBS) waveforms and step response, but step and optical PRBS generators of sufficient quality and speed to properly show the oscilloscope performance were not available at the time this paper was written.

ACKNOWLEGMENT

The authors wish to acknowledge the other fine engineers who design high-speed analog and digital ICs at Teledyne LeCroy. These are Joan Sokil, Keith Roberts, Philippe Convers, Jim Kelly, Andy Sexton, Mark Gorbics, Juan Mena-Gonzalez and Isiah Schwartz.

The authors would also wish to acknowledge the other fine engineers at National Instruments involved in the design of high-frequency microwave components used in Teledyne LeCroy oscilloscopes. These are Michael Neumann, Irfan Ashiq, Utkarsh Unnikrishna and the late Alex Feldman.

The authors wish to thank Dr. Laleh Najafizadeh of Rutgers University for her part in our invitation as an invited paper.

REFERENCES

[1] G. E. Moore, "Cramming more components onto integrated circuits," *Electronics*, vol. 38, no. 8, April 1965.

[2] M. Urteaga, M. Seo, J. Hacker, Z. Griffith, A. Young, R. Pierson, P. Rowell, A. Skalare, and M. Rodwell, "InP HBT integrated circuit technology for terahertz frequencies," in *Compound Semiconductor Integrated Circuit Symposium (CSICS), 2010 IEEE*, Oct 2010, pp. 1–4.

[3] P. Pupalaikis and D. Graef, "High bandwidth real-time oscilloscope," *US Patent 7,058,548*, June 2006.

[4] ——, "High bandwidth oscilloscope," *US Patent 8,566,058*, May 2007.

[5] P. Pupalaikis, "An 18 GHz bandwidth, 60 GS/s sample rate real-time waveform digitizing system," in *Microwave Symposium, 2007. IEEE/MTT-S International*, June 2007, pp. 195–198.

[6] P. J. Pupalaikis and M. Schnecker, "A 30 GHz bandwidth, 80 GS/s sample rate real-time waveform digitizing system," in *Optical Fiber Communication Conference.* Optical Society of America, 2010, p. JThA52.

[7] I. Ashiq and A. Khanna, "Ultra-broadband contiguous planar DC-35-65 GHz diplexer using softboard suspended stripline technology," in *Microwave Symposium Digest (IMS), 2013 IEEE MTT-S International*, June 2013, pp. 1–4.

[8] ——, "A novel ultra-broadband DC-36-66 GHz, hybrid diplexer using waveguide and SSL technology," in *44th European Microwave Conference*, Oct 2014.

[9] A. Feldman, U. Unnikrishna, I. Ashiq, and A. Khanna, "An ultra-broadband planar millimeter-wave mixer with IF bandwidth covering 0.5 to 34 GHz," in *44th European Microwave Conference*, Oct 2014.

[10] J. Love, "LeCroy Reignites Scope Wars With 100 GHz Oscilloscope," *EE Times*, July 2013. [Online]. Available: http://www.eetimes.com/document.asp?doc_id=1319038

[11] P. Pupalaikis, "Method of crossover region phase correction when summing signals in multiple frequency bands," *US Patent 7,711,510*, May 2010.

[12] ——, "Digital frequency response compensator and arbitrary response generator system," *US Patent 6,701,335*, March 2004.

[13] K. Levenberg, "A method for the solution of certain non-linear problems in least squares," *The Quarterly of Applied Mathematics*, vol. 2, pp. 164–168, 1944.

[14] P. Pupalaikis and F. Lamarche, "Digital group delay compensator," *US Patent 7,050,918*, May 2006.

[15] A. K. Sureka, "Group delay compensation using IFFT filters," *U.S. Patent Application 11/264,530*, August 2006.

[16] N. Damera-Venkata, B. L. Evans, and S. R. McCaslin, "Design of Optimal Minimum Phase FIR Filters Using Discrete Hilbert Transforms," *IEEE Transactions on Signal Processing*, vol. 48, no. 5, pp. 1491–1495, May 2000.

Ultra-Wide-Bandwidth Oscilloscope Architectures and Circuits

Dan Knierim   Tektronix Fellow, Beaverton, OR, USA

Abstract -- Users of test equipment such as oscilloscopes expect performance and accuracy beyond the level of their device under test in order to insure measurement results correspond to the DUT, not to limitations of the test equipment. This drives the use of bipolar circuitry at the front-end of high-bandwidth oscilloscopes, even if targeted at testing devices in a marketplace dominated by CMOS. Several circuit schematics in 130nm and 90 nm SiGe BiCMOS are presented to demonstrate how bipolar transistors are typically employed to provide broadband DC-coupled amplifiers, samplers, and oscilloscope triggers in the 30 to 70 GHz range. The performance metrics of these circuits depend on many parameters of the underlying process beyond Ft and Fmax, and the design effort and simulation accuracy on many effects beyond the core bipolar device model. Examples drawn from the schematics will illustrate many other facets of optimizing a fabrication process for such uses.

I. INTRODUCTION

Digital oscilloscopes, or more generically, digital waveform recorders, have been a staple in physics and electronics laboratories for three to four decades, providing the ability to digitize and numerically analyze behavior of a device under test (DUT). A block diagram of a typical digital oscilloscope acquisition system is shown in Fig 1.

The earliest digital oscilloscopes, built in the 1970s, employed bipolar technology in all blocks for performance reasons. Over the years, CMOS technology has been encroaching from the right side of the diagram, with the compute engine turning to CMOS in the 1980s, the waveform memory and address counter in the 1990s, and the ADC in some cases in the 2000s [1]. At lower performance levels, the input amplifier, sampler, and trigger circuits may also be

implemented in CMOS. There are two reasons for this shift toward CMOS technology: increasing performance of CMOS devices and increasing density of CMOS processes allowing for the use of parallelization to further increase net speed.

Parallelization in the compute engine takes the form of multi-core processors to improve throughput. Parallelization of the waveform memory increases available record length as well as supporting ADC sample rates in excess of the memory write cycle. Likewise, parallelization of the samplers and ADCs (more often referred to as time-interleaving) increases the overall sample rate. However, parallelization does not increase the analog bandwidth of the input amplifier, samplers, or trigger circuitry. In fact, the extra loading of multiple interleaved samplers may actually decrease the analog bandwidth of the input amplifier. It is for this reason that the input circuits have and will remain bipolar at the high-performance end of the market.

To mitigate the bandwidth loss of the input amplifier due to loading from a large number of interleaved samplers, a cascaded parallelization scheme may be employed. In this case, the input amplifier may drive a relatively small number of interleaved samplers, each of which in turn drives more interleaved samplers running at a slower rate. An example of a "2 by 4" cascaded time-interleaved acquisition system is shown in Fig. 2.

This provides the same factor of 8 increase in sample rate over a single ADC as would a simple 8-way interleave, but allows for a higher bandwidth by loading the input amplifier with only two samplers instead of eight. However, there are limits to use of the cascaded time-interleaved architecture. From Nyquist theory, we know that the output of the first

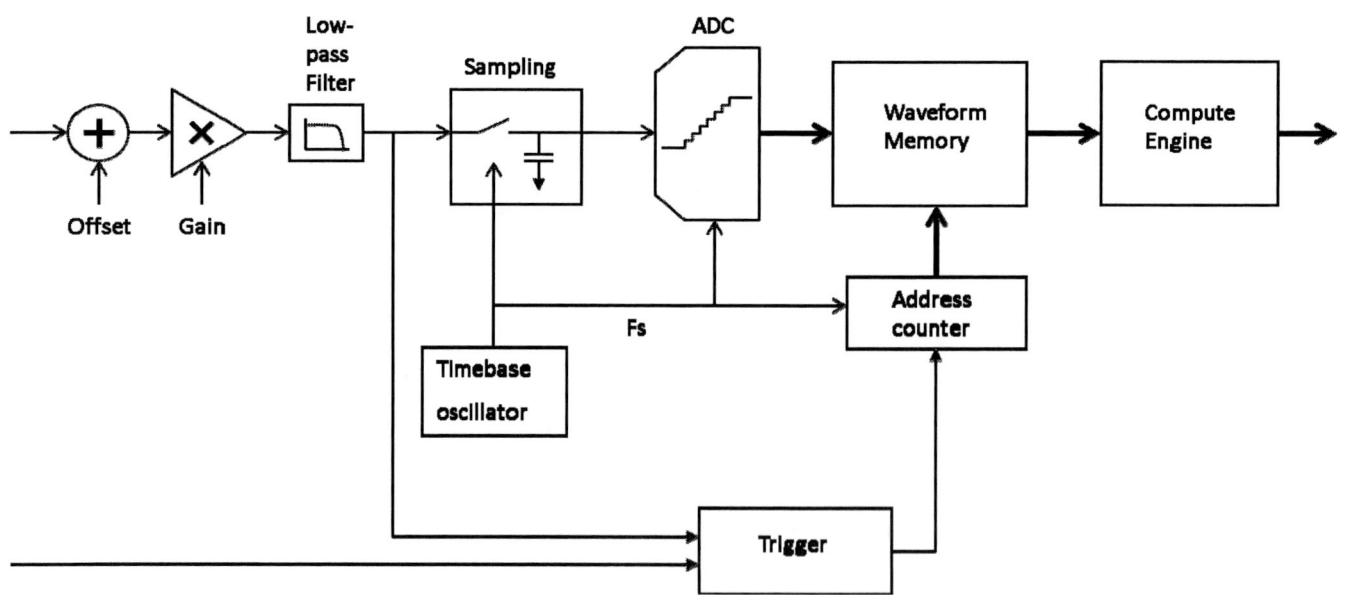

Fig. 1: Digital Oscilloscope Acquisition System

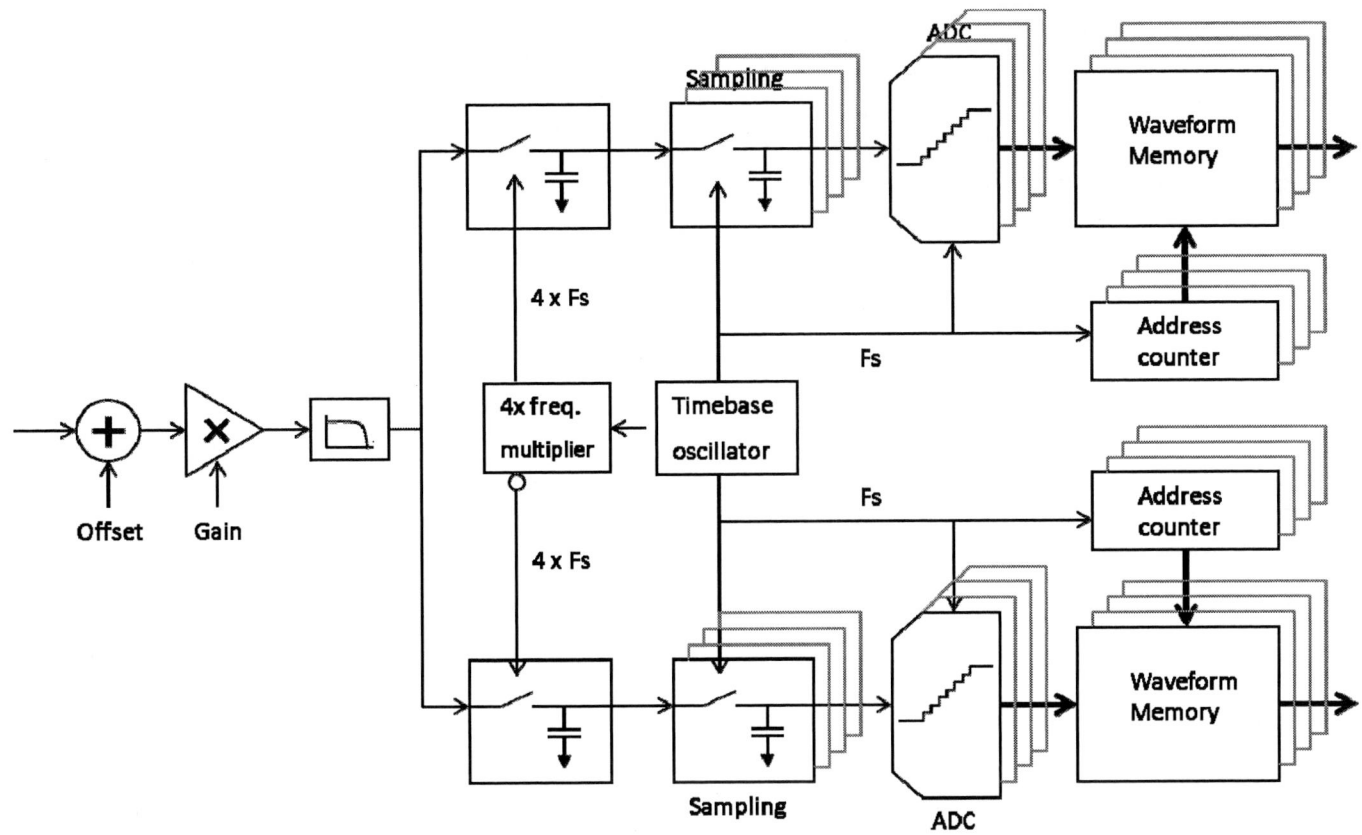

Fig. 2: : "2 by 4" Cascaded Time-Interleaved Acquisition System (trigger system not shown)

samplers, sampled at a rate of 4·Fs, contain unique spectral content from DC to half the sample rate, or 2·Fs. Thus the second set of samplers must have a minimum bandwidth of this amount, or they will not be able to capture the information content of the first samplers. An equivalent statement in the time domain is that the second samplers must be able to settle to the sampled values from the first samplers before a new sample is presented.

II. ASYNCHRONOUS TIME INTERLEAVE

This limitation is quite significant when using cascaded time-interleaving to increase the sample rate and/or bandwidth of an existing acquisition system design, as systems are normally designed with sufficient interleaving to prevent aliasing by insuring the sample rate is *above* twice the bandwidth. Such existing acquisition channels cannot then be used with pre-samplers running at the full interleaved sample rate of the downstream channels. However, we can still gain the sample rate and bandwidth benefits of cascaded interleaving if the pre-samplers are run not at the downstream sample rate, but rather at twice the downstream bandwidth [2] (thus satisfying the Nyquist criterion). For instance, if each ADC runs at 25 GS/s, and four ADCs are interleaved to produce a 100 GS/s channel with ~40 GHz bandwidth, two such 100 GS/s channels could be interleaved and pre-sampled at 75 GS/s. If the input amplifier can drive the two pre-

samplers with sufficient bandwidth, this system would support real-time capture of signals up to 70 GHz.

We refer to this as "Asynchronous Time Interleaving" or ATI, as the pre-sampling function is not operating at the same time or frequency as the downstream ADC conversion. The hardware modification to Fig. 2 to implement ATI is simply to change the 4x frequency multiplier to, in this example, a 3x multiplier. The bigger change is in interpretation of the digitized samples from the waveform memory. In traditional synchronous time interleaving, each ADC output word represents the digitized value of a single sample of the input waveform, whether sampled directly or with a cascade of two or more samplers. In the case of ATI, we know from Nyquist theory that the ADC words contain the information content of the original signal, but there is no longer a direct one-to-one correlation between ADC words and samples of the input signal. The Digital Signal Processing (DSP) needed to reconstruct a stream of digital values representing the input signal is easiest to understand in the frequency domain, as shown in Fig. 3.

The input spectrum (as drawn in the left-most graphs) is shown with a slight tilt from DC to 70 GHz, then a sharp drop to a negligible value at and above 75 GHz (insured by the input anti-aliasing filter). The sampling function of the pre-samplers is a harmonic mixing in the frequency domain, passing the original input spectrum and also shifting it by integer multiples (harmonics) of the 75 GHz sampling clock. In the case of a

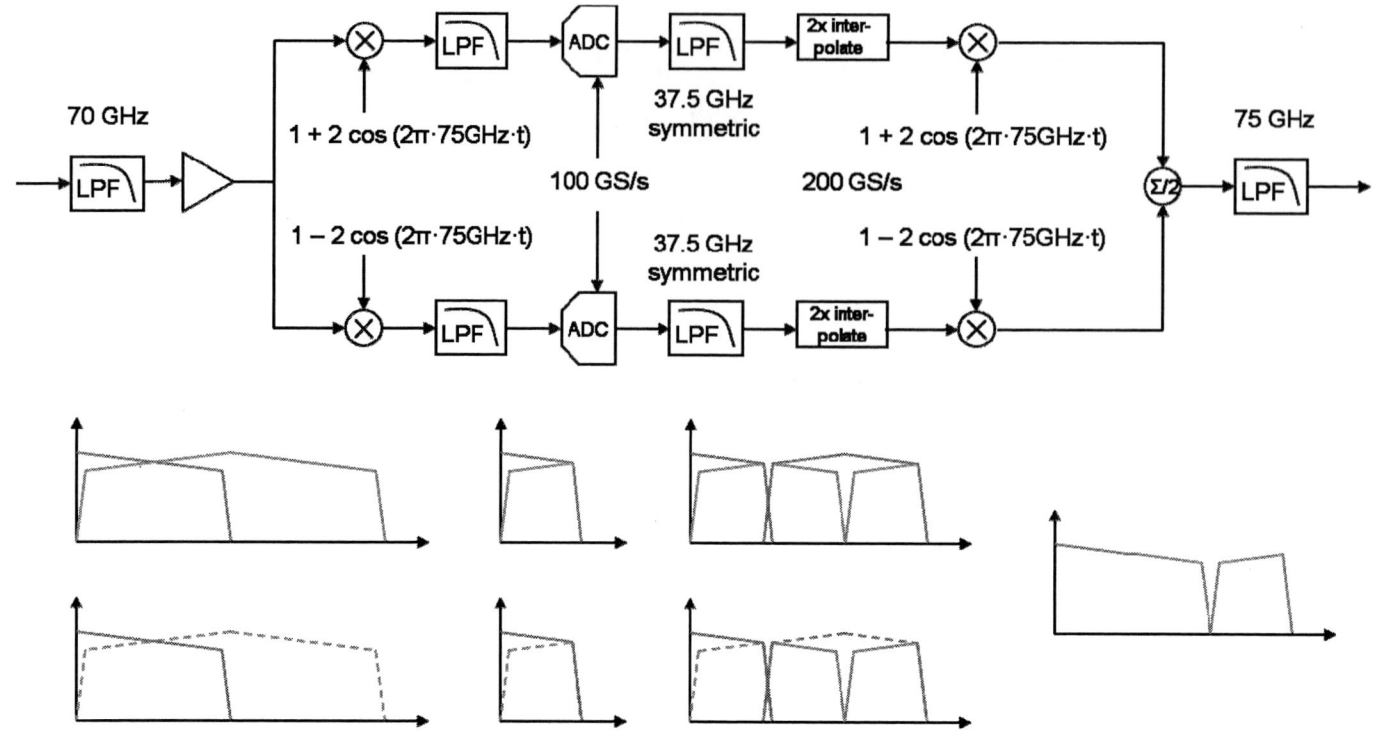

Fig. 3: Asynchronous Time Interleaving of ADC channels

two-way interleave, only the fundamental (1st harmonic) of the sampling clock is pertinent. This leaves the spectrum at the output of the pre-samplers containing both the original input spectrum and the mirrored and up-shifted spectrum. Note that the original spectra in both paths are in-phase, but the shifted spectrum in the lower path is 180° out of phase with respect to the upper path, due to the half-cycle time-shift of the interleaved pre-samplers (shown by the dotted line).

The second set of graphs show these same spectra after being low-pass filtered (by the inherent bandwidth limit of the ADC channels, as well as any additional input anti-aliasing filters and/or output DSP filters to provide matched and symmetric responses). These 100 GS/s data streams are then up-sampled (interpolated) to 200 GS/s and multiplied digitally by the same mixing functions used in the analog domain. This shifts some spectral content back to its original frequency, but also leaves many shifted spectral artifacts as well. Note that all spectral content (whether unshifted, shifted once in analog, shifted once in digital, or shifted twice) remains in-phase in the upper path. In the lower path, however, the unshifted and twice-shifted content ends-up in-phase, while the content that was shifted only once (analog or digital) and thus is at the wrong frequency also ends up out-of-phase. An average of the two data streams then reinforces the unshifted and twice-shifted content, while canceling the once-shifted artifacts. The spectrum out of the averager then contains the lower-frequency content of the original input (unshifted), the higher-frequency content of the original input (shifted twice, down and back up to its original frequency) and a final artifact of the higher-

frequency content shifted twice in the same direction. This artifact is removed in the final low-pass filter.

ATI maintains the advantages of standard synchronous time interleaving, while allowing operation with existing ADC channel designs that have sub-Nyquist bandwidth. In specific, by shifting or folding the higher-frequency input content on top of the lower-frequency content, both ADC channels are digitizing the input signal, independent of input frequency. This is in contrast to band-splitting architectures which feed the lower band to one ADC channel and a down-converted version of the higher band to another ADC channel [3]. Since only one ADC channel digitizes any given portion of the input spectrum, the final step of the reconstruction in a band-splitting approach is an addition of the signal content from the channels, rather than the average operation used with ATI. This averaging effect provides a noise benefit for ATI architectures over band-splitting.

III. CIRCUIT DESCRIPTION

To achieve 70 GHz bandwidth in the input amplifier and pre-sampling functions, a minimalist circuit approach is taken using a 90nm SiGe HBT process. A simplified schematic is shown in Fig. 4.

The input amplifier consists of an emitter follower and two differential pairs. The input capacitance of the follower is peaked with a bridged T-coil implemented using transmission-line segments and an interdigitated metal capacitor. One differential pair converts the input voltage (referenced to the offset voltage) into a differential current signal, which is then

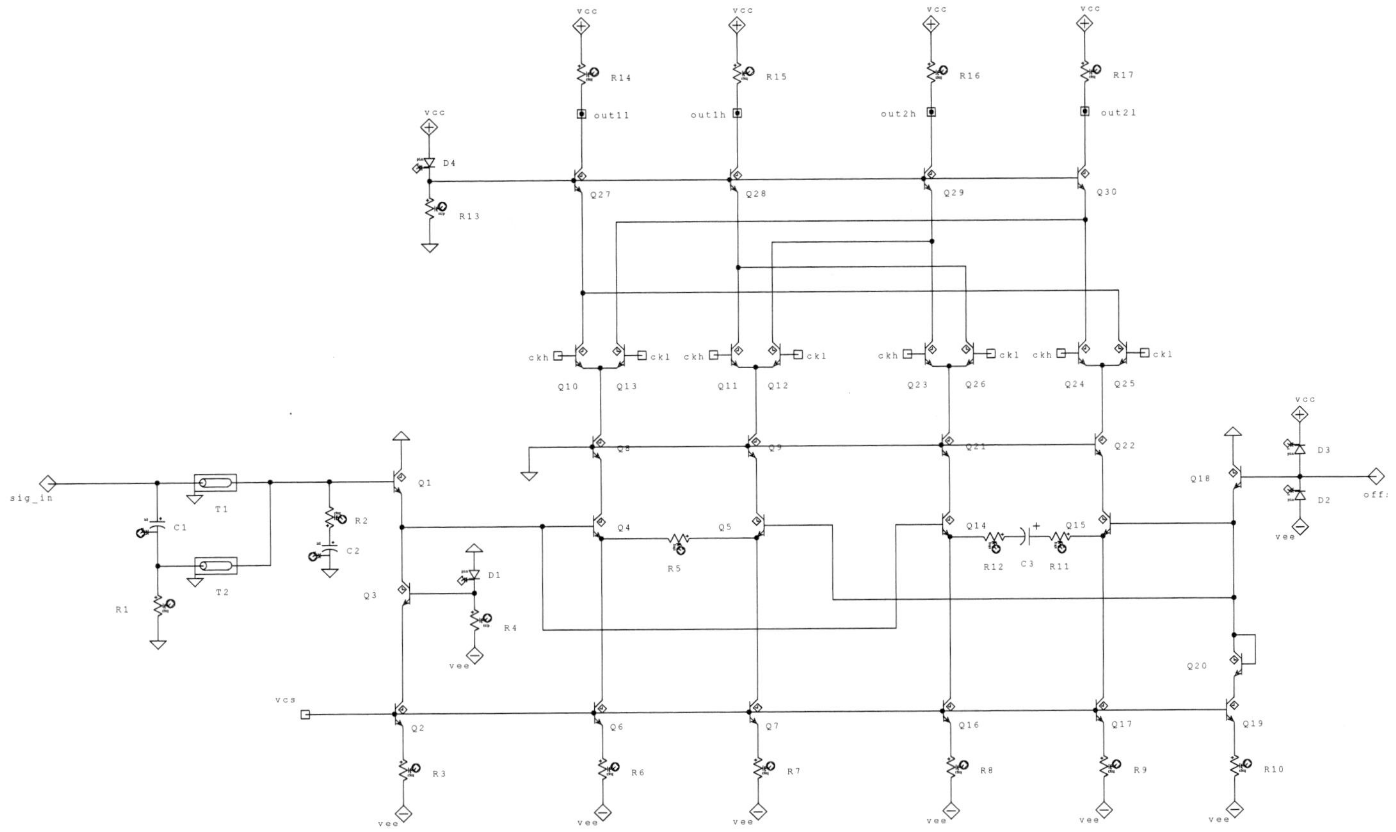

Fig. 4: Asynchronous Time Interleave input amplifier and pre-sampler schematic

switched by the 75 GHz clock to provide signal to one output for half a clock cycle and to the other output for the other half clock cycle. The second differential pair's output current is switched to the outputs on opposite phases of the clock, so as to maintain a constant output common-mode voltage. Additionally, a capacitively coupled emitter degeneration network in this second pair provides some signal boost at higher frequencies. The collector connections are swapped to account for this high-band boost appearing on the opposite half-clock-cycle.

IV. SIMULATION AND MEASUREMENT RESULTS

A prototype of this circuit was designed based on simulations with target models for a process still under development, automated extraction of parasitic capacitance, and hand extraction of series inductance on critical nodes. Measured results (as shown in Fig. 5) exhibit quite significant discrepancies from the design goal of a nominally flat response to near the 70 GHz target bandwidth.

In an attempt to understand the measured prototype results and thereby inform design of the production version, we have since simulated this circuit using many different revisions of models for the process, including both VBIC and HiCUM primitives and covering different process states. The circuit response variation over the range of models is roughly 2 dB. We have also simulated temperature and process variation

(Monte Carlo) sweeps, each adding another dB of uncertainty. But the single dominant variable in determining simulation results is the parasitic extraction method used, accounting for over 4 dB of uncertainty (about as much uncertainty as all other variables combined). Accurately modeling and extracting parasitics, especially series inductance and resistance, has always been difficult, but is becoming critical as we push to ever-higher bandwidths.

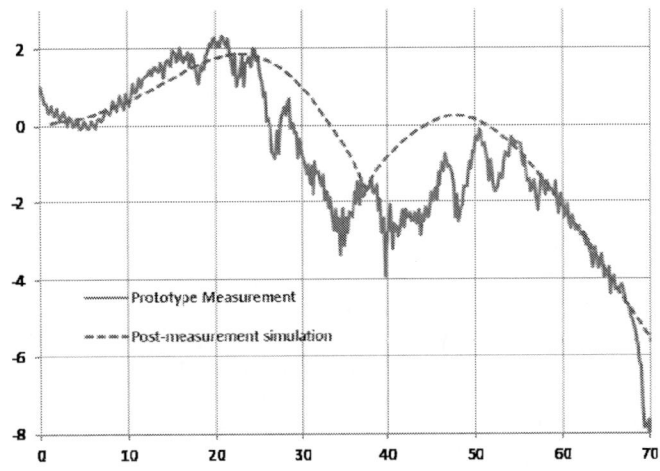

Fig. 5: Measured results of prototype system and post-measurement simulation fit (relative amplitude in dB vs frequency in GHz)

To achieve a reasonable match to the measured results, we used a combination of transmission-line primitives available in the process PDK, hand-extraction of critical nets (especially those that could be implemented as microstrip in layout), and automated extraction of more complex sections of layout. In general, we find hand-extractions tend to under-estimate series inductance, perhaps due to insufficient consideration of the ground return current path length. On the other hand, commercial extraction tools/decks tend to over-estimate series inductance, leading designers to be overly conservative in damping perceived resonances and lowering bandwidth in the process. The post-measurement simulation in Fig. 5 is based on this mix of extraction techniques, applying a multiplier of 0.8 on series inductances from the automated extraction deck, but full extracted capacitances. The remaining discrepancies between measurement and simulation are believed to be related to interconnect issues in the prototype system outside the IC.

V. PROCESS FEATURES AIDING DESIGN

A. Availability of Ground Planes

Parasitic extraction may be mostly a "tools" issue, but process design can certainly help. Suitable metal layer(s) to achieve a large-area ground plane make it much easier to design and model interconnect as strip or micro-strip lines with a known ground return path. "Suitable" here means no maximum trace width restrictions, low sheet resistance, auto-generated holes ("cheesing") if present with dimensions comparable to metal and/or dielectric thickness, and presence of one or more additional low sheet resistance metal layer(s) on which to route low-loss transmission lines.

B. Device Design for Layout

Another significant process design issue is device layout. A "CBEBC" device layout will generally have the lowest Rb and Rc for a given emitter opening due to the dual, parallel contacts. However, a quick check of the schematic in Fig. 4 will show many instances of "stacked NPNs" wherein the collector of one NPN connects to the emitter of the next in the stack, often continuing through four or five stacked devices. A "CBE" device layout allows such stacked devices to be placed very close together and wired simply at the lowest metal layer. The "CBEBC" structure requires more area and longer interconnect, adding both capacitance and inductance to the layout parasitics. We have laid out and simulated (with extracted parasitics) both structures, and find the "CBEBC" layout penalty generally more than offsets the benefits of the raw device performance advantage.

C. Bipolar Figures-of-Merit

A process' transit frequency Ft and maximum frequency of oscillation Fmax have been traditionally quoted as process figures-of-merit with regard to speed, but many other device parameters impact the bandwidth of a system designed on that process. We seldom bias any device at its Ft or Fmax peak, generally selecting devices two to four times larger for a given bias current. The use of a larger device is typically required to minimize some resistance parameter to achieve desired circuit performance.

Input devices are often oversized to minimize base resistance and the associated thermal noise. More often, a device is oversized to control emitter resistance. Resistance in the emitter of an emitter follower driving a capacitive load creates a low-pass filter. Resistance in the emitter of a cascode stage (generally placed to shield a load resistor from the drive capacitance below) also forms a low-pass filter with that drive capacitance. If emitter resistance were well controlled, these effects might be suitably peaked with an appropriate inductive element. But to the extent that emitter resistance is not well controlled, the circuit designer's only option is to oversize the device until Re is suitably small.

Another device resistance of note is thermal resistance. Unlike most RF circuits, oscilloscope acquisition systems are DC-coupled, so shifts in bias conditions over temperature are passed to the output along with the signal that may have caused the temperature shift. A typical high-Ft SiGe HBT may self-heat by 50°C or more if biased at peak Ft, leading to a roughly 50mV decrease in Vbe, quite substantial compared to the 100 to 500 mV ranges of signal amplitude typical of high-speed oscilloscope circuits. One circuit topology used to minimize these thermal errors is to put devices on a "balanced" load line, wherein bias current tends to increase in the same proportion as bias voltage decreases and vice-versa, thus minimizing change in power dissipation. Another circuit topology uses the thermal error in one device (say an emitter follower, which typically reduces power dissipation as the base is raised) to compensate the thermal error in another device (say a common-emitter gm stage, which tends to increase current and therefore power as the base is raised). But neither method is universally applicable or perfectly balanced/compensated, so larger devices with lower thermal resistance are often the solution of choice.

Many typical circuits are also impacted by collector-substrate capacitance, which is not seen in Ft at all, and impacts Fmax only to the extent the capacitance is lossy due to series substrate resistance. For these reasons, I generally use "RC" products as process figures-of-merit rather than Ft or Fmax. Depending on the application, the "R" in the product may be Rb, Re, or Θj, and the "C" is generally Cjc, Cje, Cjs, or a combination thereof.

D. Usefulness of Common-Base Curves

Another glance at the schematic of Fig. 4 shows that devices are generally biased with a relatively low impedance (compared to Vt/Ib) on the base and a relatively high impedance (compared to Vt/Ie) on the emitter. These topologies do not match the typical DC characterization methods used to plot Forward-Gummel or Common-Emitter curves. A better fit is the Common-Base characterization, wherein the base is grounded and the emitter current and collector voltage are swept. Fig. 6 shows a collection of plots from such a Common-Base characterization.

These curves provide substantial information about the use of the device at or above the collector-emitter breakdown BVceo. For instance, the value of BVceo at low current can be read by adding the Vbe at low emitter current as read off the vertical axis of the lower-left plot (~0.77V) to the Vcb at low

978-1-4799-7231-9/14 $31.00 © 2014 IEEE   140

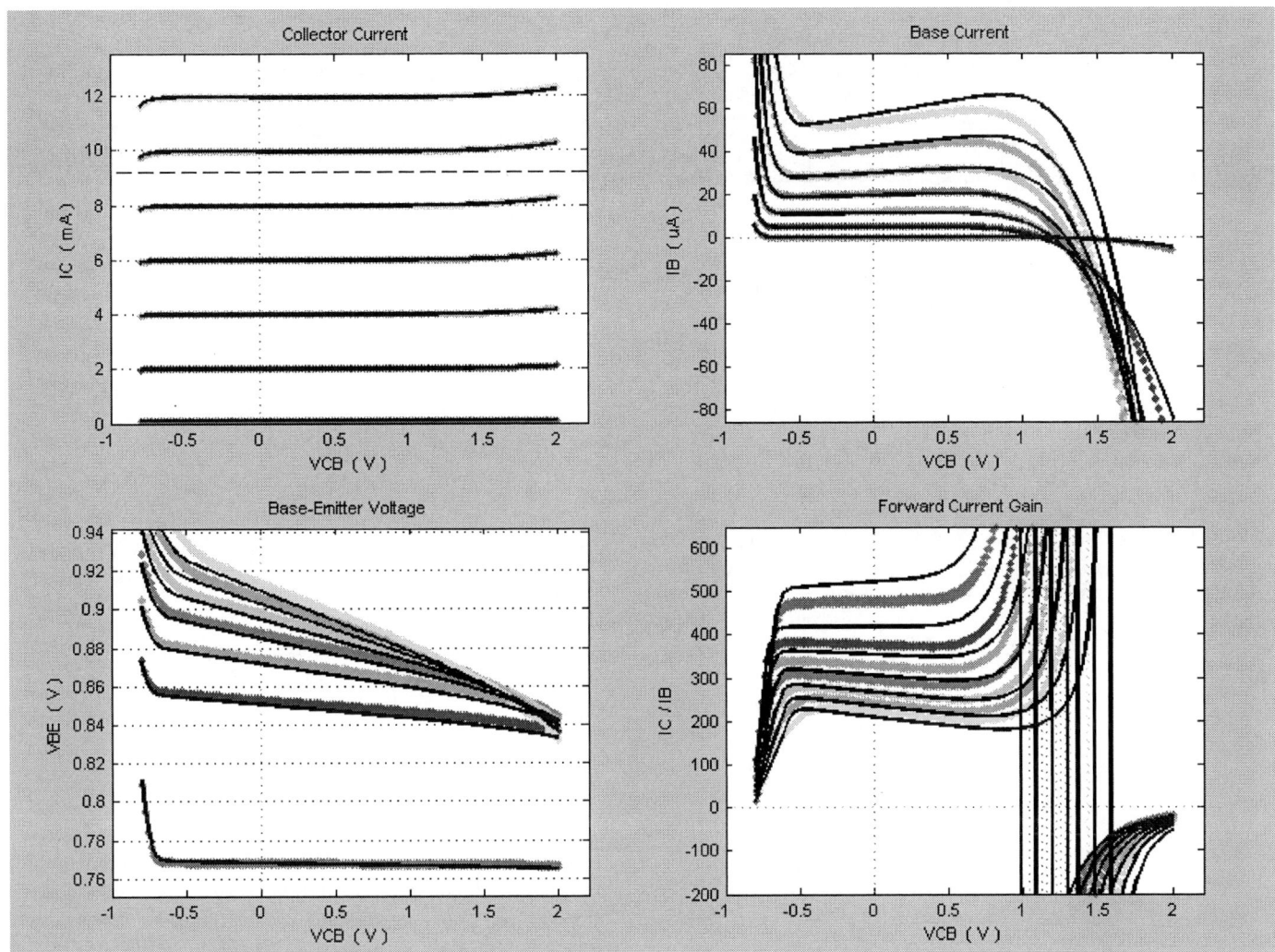

Fig. 6: Common-Base Device Characteristics

emitter current and zero base current (infinite beta) as read off the horizontal axis of the lower-right plot (~1.00V). The increase in "breakdown" collector voltage at higher currents is also readily discernable, as are the reasons for the increase (an increase in Vbe, a reduction in beta before avalanche multiplication, and an increase in drop across the collector resistance). These plots nicely demonstrate the ability to run the device well above BVceo with relatively minor impact on Ic, while also showing the extent of the negative and non-linear Ib for circuit designs sensitive to such. I highly recommend publishing these common-base curves as a standard part of DC process characterization.

E. Other Devices

Resistor parameters can be just as important as HBT parameters in determining the best process for a design. Besides the obvious desire for low parasitic capacitance, resistor tolerance plays an important role for any design that is constrained by a power dissipation budget. Given that devices are running below Ft, the circuit bandwidth is generally set by an RC product, whereas the power dissipation is set by V^2/R. For those of us who cannot count on binning chips into different speed grades for different applications, we must design to meet power specifications at the lowest process sheet resistance, but simultaneously meet bandwidth specifications at the highest process sheet resistance. Thus we gain on both ends of the tolerance. For instance, moving from a ±15% sheet resistance specification to ±10% allows us to lower nominal design values by 5% and thus lower the maximum value by 10%, providing a nominal 10% boost in minimum bandwidth. The improved resistor tolerance provides well more bandwidth benefit than a 10% increase in Ft or Fmax, while also allowing for tighter specifications on I/O impedances (e.g. for transmission line terminations or the like).

The most crucial devices we use to implement high-bandwidth oscilloscopes are NPN HBTs and passives (capacitors, resistors, inductors, transmission lines), but we have found a "PIN" diode extremely useful as well. Although not truly a PIN (no intrinsic layer), the diode is a "P+ N− N+" stack formed from the NPN base, epitaxial collector, and sub-collector, but with no emitter (to allow for a lower resistance contact to the base region) and no SIC implant (to approach the Intrinsic layer of a true PIN).

Fig. 4 shows two uses of this device. One is for Electro-Static Discharge (ESD) protection. The key figure-of-merit for an ESD protection diode is the Ron·Coff product. The lightly-doped epitaxial collector and the heavily doped sub-collector and SiGe base regions together generally lead to an RC product better than any other junction available on the process. Another use is a simple, low-impedance level-shift for biasing purposes. A diode-connected transistor (base tied to collector) can often be used in a similar fashion, but may need to be bypassed with a capacitor to keep the impedance and noise low at high frequency.

The "PIN" diode also serves well in its more typical role as a switch. PIN diode switches are common in AC-coupled RF circuits, but can easily be used for DC-coupled switching uses as well, as long as care is taken to match DC drops across PINs. Again, the Ron·Coff product is the appropriate figure-of-merit for switching applications. An electronically-switchable DC to 35 GHz attenuator using PIN diode switching is shown in Fig. 7. CMOS-based bias switches select one of the three pairs of NPN current sources to enable one signal path from input to output at a time. The bias cell for the NPN current sources (not shown) is tuned to insure their collector currents each match the long-tail pull-up resistor current, thus matching the diode currents while minimizing bias current flow through the series attenuator resistors.

F. Model Accuracy vs Efficiency

The ATI input amplifier and pre-sampler IC (the signal portion of which is shown in Fig. 4) is a relatively small IC by our standards, employing a few hundred HBTs. The 35 GHz input amplifier used in place of the ATI circuit for two-channel 100 GS/s operation (the selectable attenuator portion of which is shown in Fig. 7) contains just under 4K HBTs. The interleaved 100 Gs/s sampler ICs contain 11K HBTs each. The trigger IC contains 41K HBTs. The number of CMOS devices used for various control and DC adjustment purposes is roughly an order of magnitude higher. We make good use of the integration density allowed by SiGe, both for HBTs and CMOS.

As mentioned earlier, interconnect modeling tends to be the most problematic part of the simulations we run during the design process. But perhaps even more problematic is the simulations we <u>don't</u> run because they take too long, or don't converge, or we're still waiting for earlier simulations to complete. The device counts mentioned above demonstrate that we can always benefit from faster simulation models, in that we can simulate larger sections of the ICs at once to better look for interactions between circuits or blocks. In this author's immediate recollection, I can find two examples of IC turns necessitated by inadequate device models, five examples of IC turns caused by inadequate interconnect models, and five examples of IC turns caused by simulations not even run. Given a choice, I will take the faster and less accurate model that allows me to run more simulations.

VI. CONCLUSION

In conclusion, I would like to thank the process designers and engineers that have allowed me to do what I love to do for the past thirty years: push the state-of-the-art in bipolar circuit designs and high-performance oscilloscopes. But I would also like to remind us all that there is much, much more to a state-of-the-art process than Ft and Fmax. Attention to layout parasitics (both minimizing and accurately modeling them) and device resistances that limit use at peak-Ft bias are equally important, as are the presence and specifications of other devices necessary to complete the circuit design. Finally, with a mix of designs from a few devices to tens of thousands of devices, it is appropriate to have a selection of device models trading off between accuracy and execution speed.

REFERENCES

[1] K. Poulton, et al, "A 20GS/s 8b ADC with a 1MB Memopry in 0.18μm CMOS", in *2003 IEEE International Solid-State Circuits Conference Digest of Technical Papers*, pp. 318-319.

[2] D. Knierim, "Test and measurement instrument including asynchronous time-interleaved digitizer using harmonic mixing," US patent 8,742,749.

[3] P. Pupalaikis and D. Graef, "High bandwidth real-time oscilloscope," US patent 7,058,548.

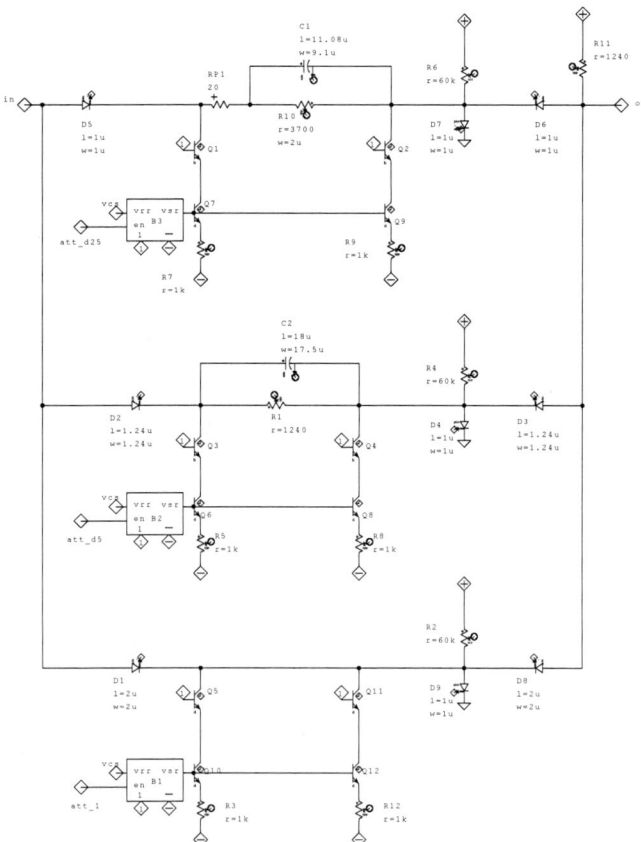

Fig. 7: PIN diode switching used in DC to 35 GHz attenuator

A 27GHz, 31dBm Power Amplifier in a 0.25μm SiGe:C BiCMOS technology

J. Essing[1], D. Leenaerts[1,2], R. Mahmoudi[1]

[1]Mixed-Signal Microelectronics Group, Eindhoven University of Technology, the Netherlands
[2]NXP Semiconductors, Eindhoven, the Netherlands

Abstract — **This paper describes an 8-way in-phase current combining power amplifier (PA) for Ka-band applications implemented in a 0.25um SiGe:C BiCMOS technology. The PA achieves a saturated output power of 29.7dBm at 27GHz with a maximum PAE of 10.5%. After applying load-pull, this output power increases further to a level of 31dBm with a maximum PAE of 13%. The small-signal gain is 24.5dB and the saturated gain is more than 14.7dB in the band of interest. The consumed area is only 2.83mm².**

Index Terms — **power amplifier, millimeter-wave, power combining, SiGe BiCMOS**

I. INTRODUCTION

The Ka-band offers a wide range of applications, like VSAT, DVB, LMDS and radar. Although compound III-V transistors can be used for implementing the power amplifier in these systems, silicon transistors are becoming more favorable due to their low cost and high integration capability. However, operating in the (near) mm-wave frequency regime complicates the realization of Watt-level output powers due to the low breakdown voltage of these silicon transistors. Recently, silicon-based power amplifiers employing transmission-line based power combiners achieve Watt-range (P>0.5W) output powers in the 45GHz band [1], [2].

The presented work achieves 29.7dBm of output power at 27GHz in a 0.25μm SiGe:C BiCMOS technology which is 1.3dB more power than the highest reported output power in in the (near) mm-wave frequency regime [1]. After applying load-pull, the PA delivers another 1.3dB more power to result in 31dBm. It employs 8-way output power combining, a 4-way power splitter at inter-stage matching and an input splitter.

II. CIRCUIT DESIGN

The circuit diagram of the PA is shown in Fig. 1. It consists of a 2-way transmission-line based input power splitter (PS), with each path followed by a driver (AS1). The inter-stage network consists of a 4-way transmission-line based power splitter, with at each splitter output an output stage (AS2). The output powers of the in total eight output stages are combined towards the 50Ω load by employing an 8-way transmission-line based power combiner (PC).

The output stage (AS2) consists of a cascoded transistor pair, with a low-voltage common-emitter (LV-CE) device and a high-voltage common-base (HV-CB) device, as shown in Fig. 2. This configuration enhances the output stages' gain and

available breakdown voltage [3]. The latter increases the required load impedance, which is beneficial for output network losses, and also reduces the required DC-current. This reduces the minimum required metal track width for preventing electromigration. The devices are scaled such that both operate near peak G_{max} current density and in their "safe operating area" (SOA) regarding electro-thermal breakdown [4]. This resulted in a HV-CB device with a 0.4x20.4x20μm emitter area, which is a factor 2.5 larger than the LV-CE emitter area (0.4x20.4x8μm). Resistive base biasing is implemented for both devices with an additional high-pass shunt RC network (41Ω, 2p) at the LV-CE device's base. This improves electrical stability at lower frequencies and also the devices' thermal stability as it functions as base-ballasting. The output stages' load impedance Zas2 its optimum is 3.9+j8.8Ω, where the inductive imaginary part resonates out the output capacitance of the output stage at the fundamental frequency.

The 8-way output power combiner (PC) uses in-phase current combining [5], [1] by employing CPW transmission-lines (TLo0-TLo4), which are kept short to minimize insertion loss. The lines' characteristic impedances and lengths are selected together with lumped series inductors (Lo2) and shunt inductors (Lo4) to perform impedance matching at the same time. Capacitance Co4 is inserted for DC-blocking purposes. Combining transmission lines and lumped components for matching reduces losses and area. Deep trench isolation (DTI) is implemented below the passives to improve their Q-factor by reducing the substrate parasistics. To obtain the combiner's optimum design parameters for minimum loss targeting the output stages' optimum load impedance, an exhaustive search optimization routine was implemented in Matlab. This routine is based on an even-mode analysis of the combiner and uses data-based models for the various matching components to improve accuracy, which are extracted from the technology library. The resulting combiner's loss is 2.4dB with the even-mode impedance transformation path from 50Ω load towards the output stages' output shown in Fig. 3. The combiner's performance is verified by employing an EM-simulator. The layout implementation of the combination of TLo4, Lo4 and Co4 is shown in Fig. 4 with the inductive loop formed by shunt inductor Lo4 visible. As inter-stage network (ISN) two times a 4-way CPW-line in-phase current splitter (TLm1-TLm2) is implemented together with lumped shunt capacitances (Cm2) for matching and series capacitances (Cm1) for DC-blocking purposes. This 4-way splitting topology reduces the number of required parallel drivers

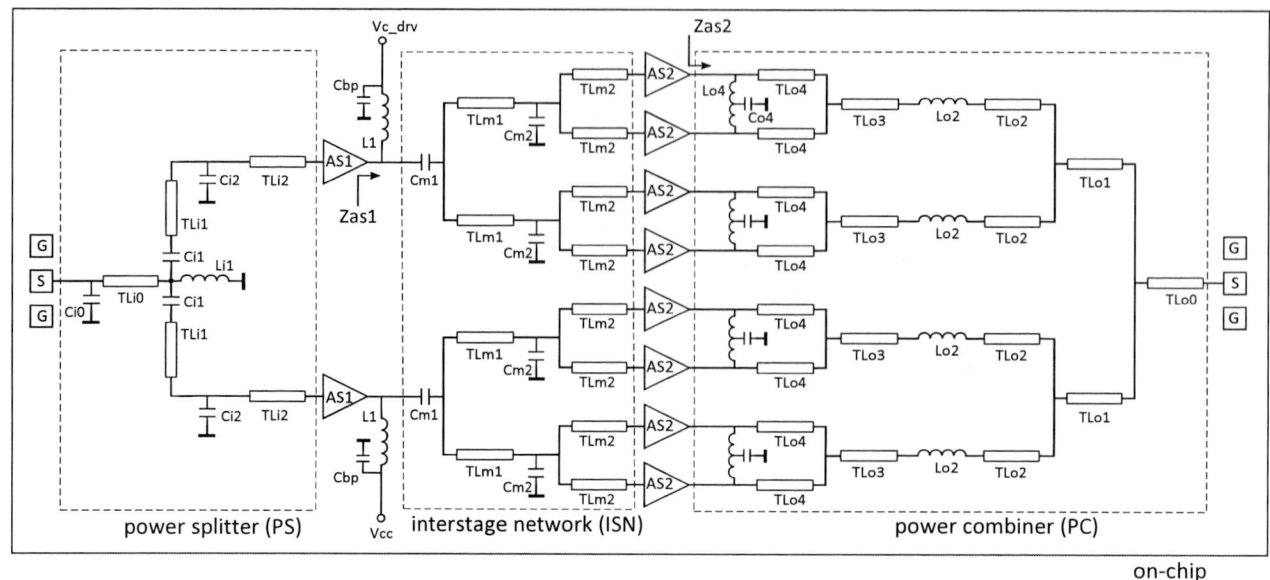

Fig. 1 Circuit diagram of the 8-way power combined PA.

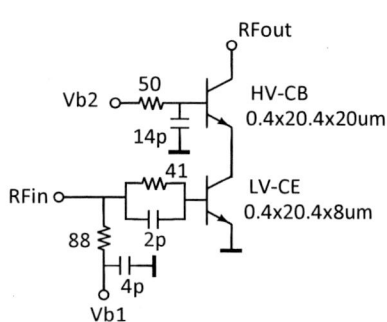

Fig. 2. Circuit diagram of both output stage AS2 and driver AS1.

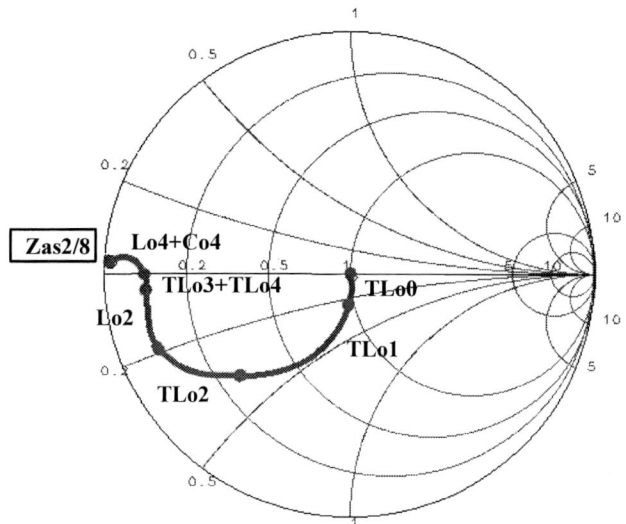

Fig. 3. Impedance transformation 8-way power combiner from 50Ω load to impedance Zas2 after even-mode analysis.

to only two, which reduces the number of required RF chokes (L_1) significantly and hence reduces area. Chokes L_1 are a combined lumped/microstrip-line inductor design and C_{bp} is a distributed bypass capacitor with a total value of 64pF to offer a broadband low impedance path, favorable for low frequency stability. Next to this, both drivers AS1 are operating now fully symmetrical. For the drivers AS1 the same cell used as output stages AS2 is implemented (see Fig. 2).

The 2-way CPW-line in-phase current splitter (PS), using additional lumped components, transforms the input impedance finally to ~50Ω. A high-pass filter is implemented with a shunt inductor (Li1) and series capacitances (Ci1) to reduce gain at lower frequencies.

A combined RC-extraction/EM-simulation is performed on the design to include all the device parasitics, layout interconnect, inductors and bond pads carefully. Stability of the design is checked at distinctive internal nodes using the method described in [6] as internal (odd-mode) oscillation loops can exist within this multi-cell topology. In that case the

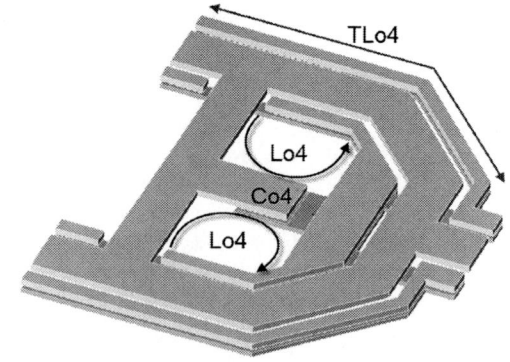

Fig. 4. Layout implementation of the combination of TLo4, Lo4 and Co4, with the inductive loop formed by shunt inductor Lo4 shown.

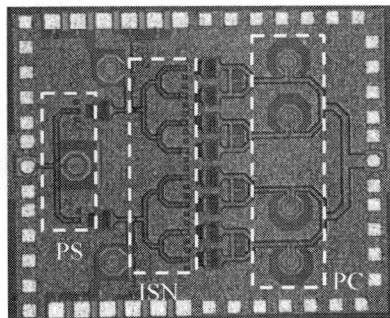

Fig. 5. Chip photograph of 8-way power combining PA. The chip size is 1.86mmx1.52mm (2.83mm²).

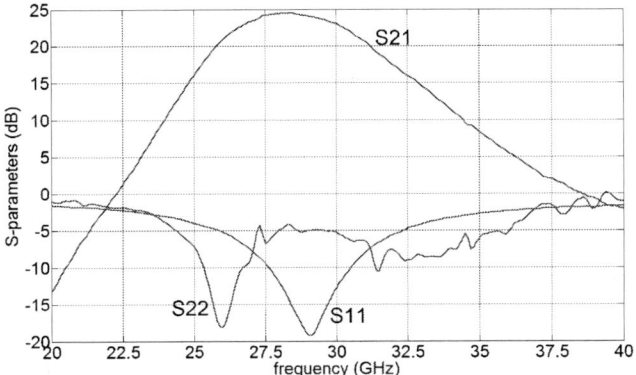

Fig. 6. Measured S_{11}, S_{22} and S_{21}. The peak S_{21} is 24.5dB at 28.3GHz.

conventional k-factor method would not be sufficient to ensure stability.

III. MEASUREMENT RESULTS

The PA is implemented in a 0.25um SiGe:C BiCMOS technology with a peak f_t/f_{max} of 216/177 GHz. The chip photograph of the realized PA with a die size of 1860x1520µm is shown in Fig. 5.

A. Small-signal Measurements

On-wafer small signal measurements are carried out using a 67GHz Agilent PNA-X and Cascade Microtech probes. Fig. 6 shows the results in which a maximum gain of 24.5dB can be observed at 28.3GHz. The results reveal a 3dB-bandwidth of 3.8GHz (from 26.5 to 30.3 GHz) and the S_{11} is below -10dB from 27.7GHz to 30.5GHz. Considering S_{22}, a mismatch was observed compared with simulation results, possibly caused by modeling inaccuracies.

B. Large-signal Measurements

Large-signal measurements are performed using an Agilent NVNA which results in accurate fully calibrated measurement results. The measured saturated output power, gain, collector efficiency and PAE versus frequency are shown in Fig. 7 at an output stage's supply voltage (V_{CC}) of 5.5V and driver's supply voltage (V_{Cdrv}) of 4V. The output stage (AS2) its CE base bias voltage ($Vb1_{AS2}$, see Fig. 2) is set to 0.7V, which

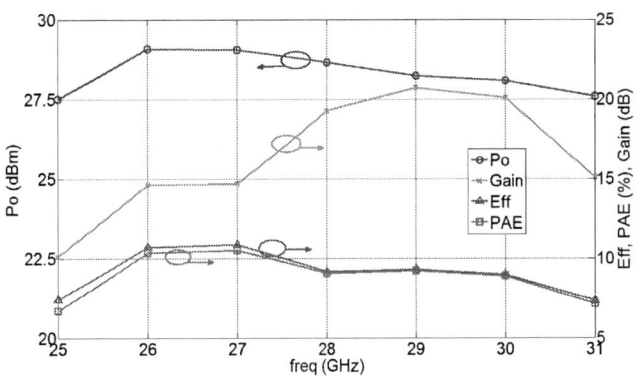

Fig. 7. Measured saturated output power, gain, collector efficiency and PAE versus frequency at V_{CC}=5.5V, $Vb1_{AS2}$=0.7V and Z_L=50Ω.

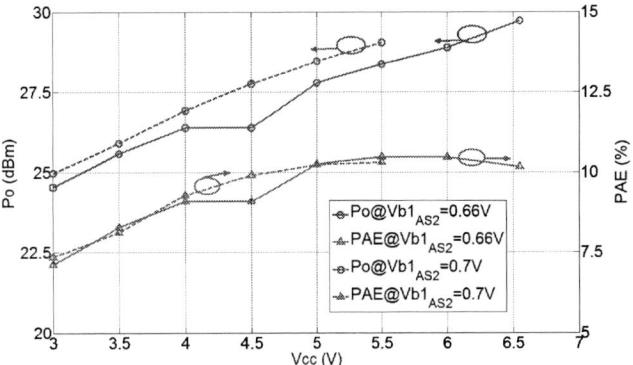

Fig. 8. Measured saturated output power and PAE versus supply voltage V_{CC} at 27GHz for $Vb1_{AS2}$ is 0.7V and 0.66V and at Z_L=50Ω.

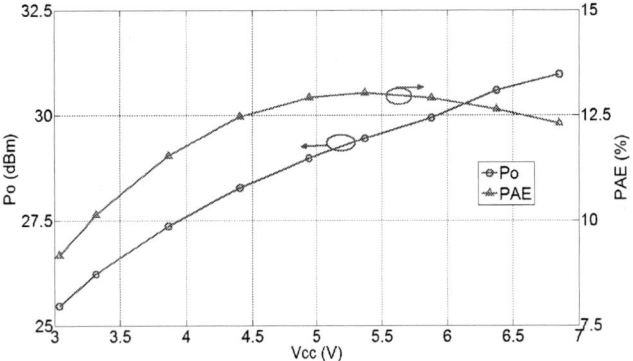

Fig. 9 Measured saturated output power and PAE versus supply voltage V_{CC} at 27GHz for $Vb1_{AS2}$ is 0.7V and at Z_L=21.2-j18Ω.

corresponds with class AB operation. The CB base bias voltage ($Vb2_{AS2}$) is set to 1.4V to optimize PAE. At 27GHz, the PA achieves a maximum saturated output power of 29.1dBm with a PAE of 10.3% and a collector efficiency of 10.7%. The saturated gain is 14.7dB and the output power 1dB bandwidth is about 4GHz, from above 25GHz to 30GHz, and corresponds with a 14.2% fractional bandwidth.

978-1-4799-7231-9/14 $31.00 © 2014 IEEE   145

Table I
Comparison with (near) mm-wave silicon PAs with $P_{sat} \geq 23$dBm

	[1]	[2]	[7]	[2]	[8]	[9]	This work	This work*
Technology	0.13um BiCMOS	45nm SOI CMOS	45nm SOI CMOS	45nm SOI CMOS	0.13um BiCMOS	0.20um BiCMOS	**0.25um BiCMOS**	**0.25um BiCMOS**
Topology	16-way in-phase current combiner	8-way lumped λ/4 combiner	4-stacked 2-bit DAC	λ/4 load modulated switched PA	4-way transformer combiner	4-way transformer combiner	**8-way in-phase current combiner**	**8-way in-phase current combiner**
freq. (GHz)	42	37	45	45	60	22	**27**	**27**
Supply (V)	4 / 2.4	4.8	5.1	2.4 / 2.6	4	1.8	**4 / 6.6**	**4 / 6.9**
$P_{sat,max}$ (dBm)	28.4	27.3	24.3	23.4	23	23	**29.7**	**31.0**
PAE_{max} (%)	10	10.7	14.6	6.7	6.3	19.7	**10.5**	**13**
$Gain_{max}$ (dB)	18.5	19.4	18	11.9	20	20	**20.2**	**20.7**
Area (mm^2)	5.55	4.16	0.77	4.16	3.42	6	**2.83**	**2.83**

*at load impedance Z_L=21.2-j18Ω

The measured saturated output power and PAE versus supply voltage V_{CC} are shown in Fig. 8 at a frequency of 27GHz for a base bias voltage $Vb1_{AS2}$ of both 0.7V and 0.66V. The reduction in bias voltage from 0.7V to 0.66V extends the maximum allowable supply voltage before breakdown as it reduces the current conduction angle and hence DC-current consumption. This extends the large-signal SOA regarding electro-thermal breakdown. The driver's supply voltage V_{Cdrv} is set fixed to 4V. At $Vb1_{AS1}$ is 0.66V and a supply voltage of 6.55V, the PA achieves a maximum saturated output power of 29.7dBm with a PAE of 10.2%. This extends the maximum saturated output power with 0.6dB compared with $Vb1_{AS2}$ set to 0.7V.

Due to the mentioned output mismatch (S_{22}), the output stages were not terminated in their optimum load impedance. Therefore load-pull was applied, which resulted in an optimum load impedance of 21.2-j18Ω instead of 50Ω. At this optimum impedance, the measured saturated output power and PAE versus supply voltage V_{CC} are shown in Fig. 9 at a frequency of 27GHz at $Vb1_{AS2}$ is 0.7V. The PA achieves a maximum saturated output power of 31dBm with a PAE of 13% at a V_{CC} of 6.9V. An additional off-chip matching network is still required to match to 50Ω.

Table I summarizes and compares the performance of (near) mm-wave silicon PAs with a saturated power of 23dBm and larger. The proposed PA delivers 1.3dB more saturated output power compared to [1] and has therefore the highest reported output power in the (near) mm-wave regime. After applying Z_L=21.2-j18Ω, the PA delivers another 1.3dB more power. The achieved PAE and gain are comparable with the referenced works. The consumed area is only 2.83mm^2.

IV. CONCLUSIONS

This paper presented an 8-way in-phase current combining power amplifier for Ka-band applications, implemented in a 0.25um SiGe:C BiCMOS technology. It achieves a saturated output power of 29.7dBm at 27GHz, which is the highest reported output power for a PA operating in the (near) mm-wave regime. After applying load-pull, the PA output power increases further to a level of 31dBm.

ACKNOWLEDGMENT

The author would like to thank M. v.d. Heijden, J. Bergervoet, Q. Ma, C. Lu, Y. Pei and Y. Chen for helpful discussion, R. van Dommele for measurement support and Focus Microwaves for load-pull tuner support.

REFERENCES

[1] Wei Tai; Carley, L.R.; Ricketts, D.S., "A 0.7W fully integrated 42GHz power amplifier with 10% PAE in 0.13μm SiGe BiCMOS," Solid-State Circuits Conference Digest of Technical Papers (ISSCC), 2013 IEEE International , vol., no., pp.142,143, 17-21 Feb. 2013.

[2] Bhat, R.; Chakrabarti, A.; Krishnaswamy, H., "Large-scale power-combining and linearization in watt-class mmWave CMOS power amplifiers," Radio Frequency Integrated Circuits Symposium (RFIC), 2013 IEEE , vol., no., pp.283,286, 2-4 June 2013.

[3] J. Andrews, J. Cressler, and M. Mitchell, "A High-Gain, Two-Stage, X-Band SiGe Power Amplifier," IEEE/MTT-S International Microwave Symposium, pp. 817–820, Jun. 2007.

[4] Vanhoucke, T.; Hurkx, G. A M, "Unified electro-thermal stability criterion for bipolar transistors," Bipolar/BiCMOS Circuits and Technology Meeting, 2005. Proceedings of the , vol., no., pp.37,40, 9-11 Oct. 2005.

[5] Bohsali, M.; Niknejad, A.M., "Current combining 60GHz CMOS power amplifiers," Radio Frequency Integrated Circuits Symposium, 2009. RFIC 2009. IEEE , vol., no., pp.31,34, 7-9 June 2009.

[6] M.Ohtomo, "Stability Analysis and Numerical Simulation of Multidevice Amplifiers, " IEEE Transactions on MTT, June/July 1993, vol. 41, pp. 983-991.

[7] A. Balteanu, et al., "A 45GHz, 2b Power DAC with 24.3dBm Output Power, >14Vpp Differential Swing, and 22% Peak PAE in 45nm SOI CMOS," IEEE Radio Frequency Integrated Circuits Symp., pp. 319 – 322, June 2012.

[8] U. R. Pfeiffer and D. Goren, "A 23-dBm 60-GHz Distributed Active Transformer in a Silicon Process Technology," IEEE T-MTT, vol. 55, no. 5, pp. 857 –865, May 2007.

[9] Cheung, T.S.D.; Long, J.R., "A 21-26-GHz SiGe bipolar power amplifier MMIC," Solid-State Circuits, IEEE Journal of , vol.40, no.12, pp.2583,2597, Dec. 2005

A Differential SiGe Power Amplifier Using Through-Silicon-Via and Envelope-Tracking for Broadband Wireless Applications

Jerry Tsay, Matthew Sapp, Michael Phamvu, Travis Hall, Ryan Geries, Yan Li*, Jerry Lopez, and Donald Y. C. Lie

Dept. of Electrical & Computer Engineering, Texas Tech University, Lubbock, TX, 79409, USA
*RF Micro Devices (RFMD) Design Center, Phoenix, AZ, USA

Abstract — In this paper, a differential SiGe power amplifier (PA) is designed using a bipolar differential pair in a 0.35-μm SiGe BiCMOS technology with through-silicon-via (TSV). Measured using continuous wave (CW) and Long Term Evolution (LTE) modulated waveforms, significant gain expansion (3-5 dB) is observed. The PA reaches power-added-efficiency (PAE) of 61.7% / 51.2% / 40.0% at P_{OUT} = 25.6 / 25.4 / 25.7 dBm at supply voltages of V_{CC} = 2.8V / 3.3V / 4.2V, respectively, with 800 MHz CW input. With the help of the envelope-tracking (ET) technique, the measured PAE improves by 7.3% / 10.4% / 15.4% compared to the fixed supply PA at power back-off regions at P_{OUT} = 19.9 / 22.1 / 22.4 dBm, achieving PAE of 38.4% / 43.4% / 38.6% at 800 MHz for LTE 16QAM 5 MHz and passing the LTE spectrum emission mask (SEM) without predistortion. This SiGe ET-PA shows promise for operation as the medium power (MP) PA for efficiency enhancement in the back-off regions.

Index Terms — envelope modulator (EM), envelope tracking (ET), ET-PA, LTE, SiGe power amplifier (PA).

I. INTRODUCTION

Silicon-based radio frequency (RF) power amplifiers (PAs) have recently become serious contenders for the handset PA market because of their higher monolithic integration and functionalities on-chip, promising potential lower cost and overall size reduction for the RF front-end module (FEM) and/or the PA module (PAM). The CMOS PA has been introduced into the low-end 2G/2.5G market for several years now, but with the slight performance inferiority and perceived concerns on yield and reliability, they have only grabbed a few percent of the handset PA market from the incumbent III-V PAs. However, for the WLAN PAM/FEM market, SiGe PA dominates here with its better performance over CMOS PA that means all the specs (especially for the 5-6 GHz bands), and with lower cost than the III-V PAs. For the 3G/4G handsets such as the Long Term Evolution (LTE), silicon-based PAs have reported comparable linear P_{OUT} with competitive power-added-efficiency (PAE) values against those of the III-V compound based PAs [1]-[2]. In particular, since SiGe devices can reach higher breakdown voltages and do not suffer from oxide reliability that can plague CMOS PAs, and with the recent addition of the through-silicon-via (TSV)

technology for much lower stray inductance to ground and better thermal release (a standard process feature for III-V PAs for decades), SiGe handset PAs have finally become very attractive [3], with a relatively large number of 3G SiGe PA already inside commercial handsets today.

3G/4G standards adopt spectral-efficient modulation schemes with inherent high peak-to-average-power ratio (PAPR), and these broadband amplitude-modulated signals require stringent PA linearity specs, meaning the RF PA often has to run in the power back-off modes to meet the tough linearity requirements at the expense of efficiency. For example, it is common to see commercial LTE PAs reach a peak PAE of 40% at around 28 dBm, but at P_{OUT} = 17 dBm back-off region the PAE becomes <20%. One major challenge is then to design its PAE to be high for *both* the peak and power back-off regions to save the overall *average* power. Therefore, multi-power modes 3G/4G PA design are preferred. Three power mode configurations (High Power/Medium Power/Low Power modes) are popular with path switching between 2 sets of PAs and/or bias adjustment with/without the additional DC-DC converter in the PAM [4]. In this work, therefore, we present a differential SiGe PA with TSV as a medium power (MP) PA for efficiency enhancement. The PA is also mated with an efficient discrete linear-assisted envelope modulator (EM) to form an ET-PA, which has shown impressive efficiency enhancement at power back-off compared with the fixed supply PA [5-11]. Circuit design will be described in Sect. II. Measurement data are presented and discussed in Sect. III, and the conclusion is given in Sect. IV.

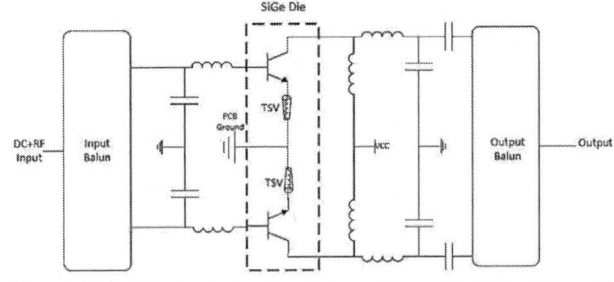

Fig. 1 Simplified schematic of the differential SiGe PA with TSV used in this work.

II. SIGE PA CIRCUIT DESIGN WITH TSV

For the SiGe PA design with TSV, the differential pair was designed in the 0.35-μm IBM 5PAe SiGe BiCMOS technology with a device emitter area of 1200 μm^2. The HP (high performance) option was adopted for the SiGe BJTs. A TSV is a vertical connection etched through a silicon wafer and filled with metal deposition, which provides very small parasitic inductance and resistance to ground. Fig. 1 shows the schematic of this design. Due to the extremely low-parasitic TSVs, this differential PA has very small ground impedance at the virtual ground point, and it is bondwired to a PCB in FR4 (see Fig. 2). Conductive silver based epoxy was used for mounting the chips to the PCB to ensure good ground conductivity. The input/output matching was designed on the PCB with lumped passive components of the same values as reported in [12].

Fig. 2 Microphotograph of the differential SiGe PA with TSV and grounded from the die onto a PCB.

III. EXPERIMENT RESULTS

We first measured the differential PA with continuous wave (CW) input at 800 MHz with V_{CC} of 2.8V, 3.3V, and 4.2V. Fig. 3 shows the measured PAE and gain vs. P_{OUT} at 800 MHz. The loss of the input/output baluns is de-embedded from the measurements. With the strong gain expansion observed, the PA is far from reaching its P_{1dB} compression point at 25 dBm (note even at 31 dBm this PA has not reached P_{1dB} yet but data not shown here as one PA was destroyed during testing; in that case measured $P_{0.6dB}$ = 31 dBm with PAE = 65.7% and V_{CC} = 3.5V). Since thermal effects start to be significant at $P_{OUT} \geq 26$ dBm, we purposely avoided these higher power regions to not destroy the devices. The PA achieves PAE of 61.7%/ 51.2%/ 40.0% and gain of 16 / 16.8 / 18.1 dB at P_{OUT} = 25.6 / 25.4 / 25.7 dBm with supply voltages of V_{CC} = 2.8V / 3.3V / 4.2V at 800 MHz.

Fig. 4 shows the measured PAE, gain vs. P_{OUT} for the SiGe PA using 5 MHz 16QAM LTE signal with PAPR ~7.5 dB. One can see the PA reaches high peak PAE values of 45.4% / 41.6% / 23.7% at 23.7 / 24.7 / 22.6 dBm for V_{CC} = 2.8V / 3.3V / 4.2V (note the PA at V_{CC} of 4.2V suffered from thermal runaway issues at higher power so P_{OUT} is limited to ~25 dBm). However, at the power back-off regions of P_{OUT} = 19.9 / 22.1 / 22.4 dBm for V_{CC} = 2.8V / 3.3V / 4.2V, and the PAE drops to

31.1% / 33.0% / 23.2%. This large PAE reduction of up to 14.3% in less than 4 dB back-off prompts us to investigate if an ET-PA can improve the power back-off efficiency while also meeting the linearity specs.

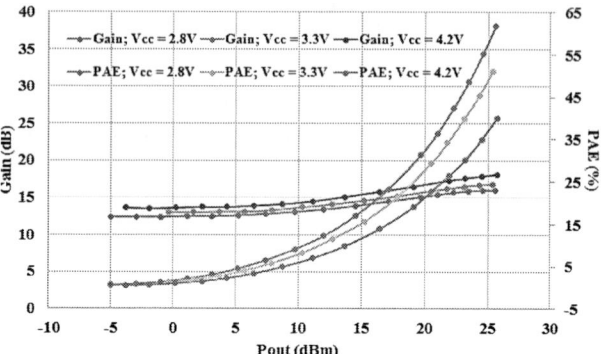

Fig. 3 Measured CW performance of the differential SiGe PA with TSV at V_{CC} = 2.8V / 3.3V / 4.2V and 800 MHz. Note the gain expansion of the PA.

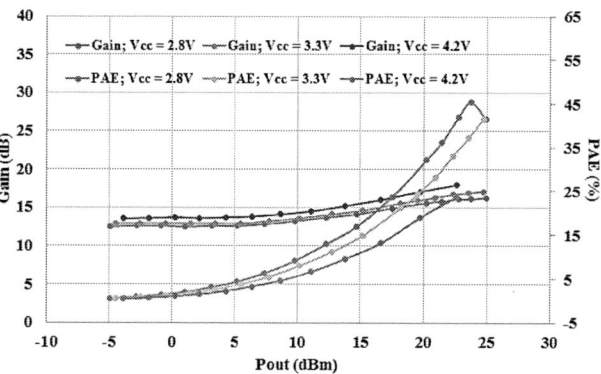

Fig. 4 Measured PAE and gain vs. P_{OUT} for the differential SiGe PA using 5 MHz LTE 16QAM signal with V_{CC} = 2.8V / 3.3V / 4.2V at 800 MHz.

Fig. 5 shows a simplified ET-PA system setup used in this work. The linear-assisted EM design used is rather similar to the double-buck EM reported in Ref. [12]. Figs. 6-8 show the measured output spectra at P_{OUT} of 19.9 / 22.1 / 22.4 dBm for the ET-PA system at V_{CC} = 2.8V / 3.3V / 4.2V, respectively, with the 5 MHz 16QAM LTE input.

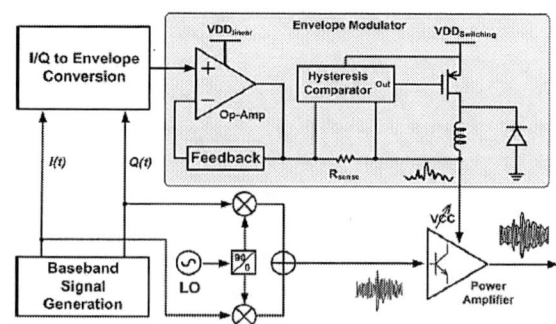

Fig. 5. Simplified transmitter block diagram with the ET-PA technique using a linear-assisted EM.

978-1-4799-7231-9/14 $31.00 © 2014 IEEE

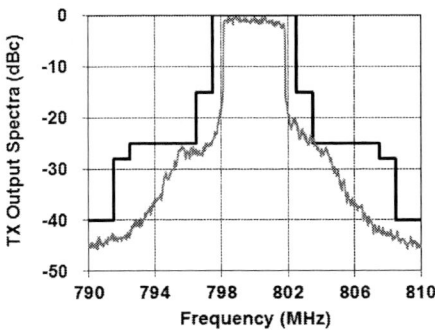

Fig. 6 Measured output spectra of the SiGe ET-PA at P_{OUT} = 19.9 dBm with LTE 16QAM 5 MHz input signal at 800 MHz with V_{CC} = 2.8V. The LTE SEM mask (black) also included.

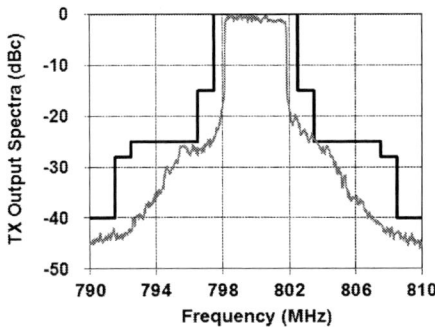

Fig. 7 Measured output spectra of the SiGe ET-PA at P_{OUT} = 22.1 dBm with LTE 16QAM 5 MHz input signal at 800 MHz with V_{CC} = 3.3V. The LTE SEM mask (black) also included.

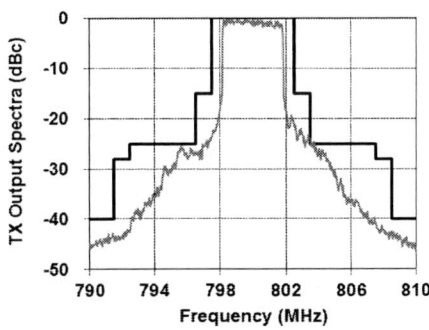

Fig. 8 Measured output spectra of the SiGe ET-PA at P_{OUT} = 22.4 dBm with LTE 16QAM 5 MHz input signal at 800 MHz with V_{CC} = 4.2V. The LTE SEM mask (black) also included.

One can see from Figs. 6-8 that the ET-PA passes the LTE SEM specs. In general, the fixed supply PA has a slightly better Adjacent Channel Leakage Ratio (ACLR) performance (spectra not shown), but the ET-PA has a much higher efficiency, as summarized in Tables 1-3 for the same P_{OUT} levels. Fig. 9 shows the detailed measured PAE and gain of the fixed supply differential PA with LTE 16QAM 5 MHz signal at 800 MHz vs. the ET-PA (V_{min} = 0.8V) at V_{CC} = 2.8V. Note the ET-PA has a simple 0.8V DC voltage shift at the collector to keep V_{CC} above 0.8V to prevent the PA from being momentarily shut down with the modulated waveform [13]. For P_{OUT} = 19.9 dBm where both cases pass the

LTE spectral mask specs, the ET-PA has a 1.8 dB lower gain of 13.8 dB compared to the fixed supply PA, but the ET-PA has a 7.3% higher PAE. This efficiency improvement is also partly enhanced by the waveform shaping with reduced PAR on the collector voltage, at a slight sacrifice of higher EVM [13]. Nevertheless, please note the SEM and ACLR linearity specs are usually much more stringent than the EVM specs for LTE applications [4], as reported before [13].

This PAE vs. gain tradeoff is more apparent with V_{CC} = 3.3V, as shown in Fig. 10. For the same P_{OUT} of 22.1 dBm, the ET-PA has a 2.3 dB lower gain, but now the efficiency improvement has increased to 10.4%. With V_{CC} = 4.2V, the tradeoff is even more drastic, resulting in a 2.9 dB lower gain and 15.4% higher efficiency for the ET-PA at P_{OUT} = 22.4 dBm (see Fig. 11).

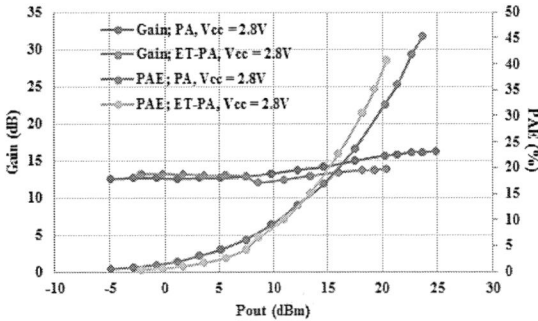

Fig. 9 Measured PAE and gain of the fixed supply PA vs. the ET-PA for V_{CC} = 2.8V with LTE 16QAM 5 MHz at 800 MHz.

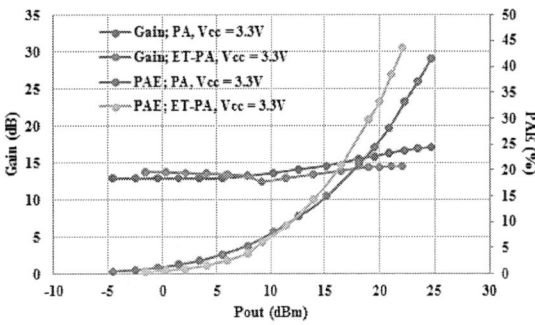

Fig. 10 Measured PAE and gain of the fixed supply PA vs. ET-PA for V_{CC} = 3.3V with LTE 16QAM 5 MHz at 800 MHz.

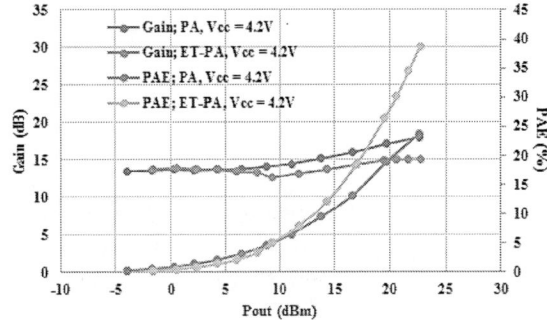

Fig. 11 Measured PAE and gain of the fixed supply PA vs. ET-PA for V_{CC} = 4.2V with LTE 16QAM 5 MHz at 800 MHz.

978-1-4799-7231-9/14 $31.00 © 2014 IEEE

Besides the higher PAE at back-off, an additional benefit of the ET-PA is the reduction of thermal issues, as observed for the case of V_{CC} = 4.2V. The data suggests that this differential SiGe ET-PA can be attractive as a MP mode PA for efficiency enhancement, which can be an important part of a dual-path multi-power-mode 3G PA. However, more work is being investigated to hopefully extend the linear P_{OUT} range of this MP ET-PA. Additional experimental data will be collected and presented at conference when available. We also suspected the device mismatches reported in the work of [12] using two different dies and TSV may have caused that pseudo-differential PA to have a higher 2nd-order distortion compared with our work here. The measured output spectra indeed supports our hypothesis as our PA's data (Fig. 7) is indeed more symmetrical than that reported in [12] for certain frequencies, and it also has a much smaller die size. Table 4 shows the ET-PA performance summary table vs. literature

Table 1 Measured Performance of the ET-PA vs. the Fixed Supply PA at a Back-Off P_{OUT} = 19.9 dBm with V_{CC} = 2.8V.

	Fixed Supply PA	ET-PA
Supply Voltage (V)	2.8	2.8
Gain (dB)	15.6	13.8
P_{OUT} (dBm)	19.9	19.9
Peak PAE (%)	31.1%	**38.4%**

Table 2 Measured Performance of the ET-PA vs. the Fixed Supply PA at a Back-Off P_{OUT} = 20.7 dBm with V_{CC} = 3.3V.

	Fixed Supply PA	ET-PA
Supply Voltage (V)	3.3	3.3
Gain (dB)	16.8	14.5
P_{OUT} (dBm)	22.1	22.1
Peak PAE (%)	33.0%	**43.4%**

Table 3 Measured Performance of the ET-PA vs. the Fixed Supply PA at a Back-Off P_{OUT} = 20.9 dBm with V_{CC} = 4.2V.

	Fixed Supply PA	ET-PA
Supply Voltage (V)	4.2	4.2
Gain (dB)	18.2	15.3
P_{OUT} (dBm)	22.4	22.4
Peak PAE (%)	23.2%	**38.6%**

Table 4 Summary and comparison of our ET-PA system with state-of-the-art ET/EER PA designs in literature

	Freq. (GHz)	V_{CC} (V)	Max. Linear P_{OUT} (dBm)	[1]Overall PAE	Signal BW (MHz)	Modulation	[2]PD	PA Technology
[4]	1.75	4.2	24.2	43%	5	LTE 16QAM	No	0.35-μm SiGe BiCMOS
[5]	1.88	3.3	23.9	34.3%	5	WiMAX 64QAM	No	2-μm InGaP/GaAs
[6]	2.4	3.3	20	28%	20	WLAN 64QAM	Yes	0.18-μm SiGe BiCMOS
[7]	2.535	3.3	25.8	32.3%	10	LTE 16QAM	No	2-μm InGaP/GaAs
[11]	2.0	3.3	19.6	22.6%	20	WLAN 64QAM	No	0.13-μm CMOS
[12]	0.7	5	28.1	42.3%	5	LTE 16QAM	No	0.35-μm SiGe with TSVs
This Work	**0.8**	**2.8**	**19.9**	**38.4%**	**5**	**LTE 16WAM**	**No**	**0.35-μm SiGe with TSVs**
	0.8	**3.3**	**22.1**	**43.4%**	**5**	**LTE 16QAM**	**No**	**0.35-μm SiGe with TSVs**
	0.8	**4.2**	**22.4**	**38.6%**	**5**	**LTE 16WAM**	**No**	**0.35-μm SiGe with TSVs**

Note: 1. The overall PAE includes the efficiency of the envelope modulator and that of the RF PA
2. PD: predistortion

IV. CONCLUSION

In this paper, a differential PA is designed in a 0.35-μm SiGe BiCMOS technology with TSV. Measurement shows that with the help of ET-PA, its PAE improves 7.3% / 10.4% / 15.4% compared with the fixed supply PA at power back-off regions, achieving PAE of 38.4% / 43.4% / 38.6% at 800 MHz for LTE 16QAM 5 MHz without predistortion. This SiGe PA shows promise to provide power savings at the back-off region for 3G/4G PA applications as a MP PA.

ACKNOWLEDGEMENT

We wish to acknowledge Ms. D. Wang and Dr. A. Joseph at IBM for IC fabrication. We are deeply grateful to the generous funding support by DoD and the TTU Keh-Shew Lu Regents Chair Endowment.

REFERENCES

[1] V. Krishnamurthy, et al., IEEE Radio Freq. Integrated Circuits (RFIC) Symp. Dig. Papers, 2010, pp. 569-572.

[2] B. Koo, et al., IEEE Trans. Microw. Theory Tech., vol. 60, no. 2, pp. 340-351, Feb. 2012.

[3] R. Wu, et al., Proc. 12th Topical Meeting on Silicon Monolithic Integrated Circuits in RF Systems (SiRF), 2012, pp. 69-72.

[4] Y. Li, R. Zhu, D. Prikhodko and Y. Tkachenko, IEEE ICSICT, China, 2010

[5] Y. Li, et al., IEEE Trans. Microw. Theory Tech., vol. 59, no. 10, pp. 2525-2536, Oct. 2011.

[6] J. Choi, et al., IEEE Trans. Microw. Theory Tech., vol. 57, no.7, pp. 1675-1686, July 2009.

[7] F. Wang, et al., IEEE J. Solid-State Circuits, vol. 42, no. 6, pp. 1271-1281, 2007.

[8] J. Choi, et al., IEEE Trans. Microw. Theory Tech., vol. 59, no. 7, pp. 1796-1802, July 2011.

[9] Yan Li, et al., IEEE Trans. Circuits Syst .I, vol. 58, no. 5, pp.893–901, May 2011

[10] F. Wang, et al., IEEE Trans. Microw. Theory Tech., vol. 54, no. 12, pp. 4086–4099, Dec. 2006.

[11] J.S. Walling, et al., IEEE J. Solid-State Circuits, vol. 44, no. 9, pp. 2239-2347, Sept. 2009.

[12] R. Wu, Y. Li, J. Lopez and D.Y.C. Lie, , Proc. IEEE BCTM, pp. 45-48, Sept. 30-Oct.3, Portland OR, USA (2012)

[13] R. Wu, Y. Li, J. Lopez and D.Y.C. Lie, , IEEE J. Solid-State Circuits, 48, 9, pp. 2030-2040, Sept. (2013)

978-1-4799-7231-9/14 $31.00 © 2014 IEEE

W-band SiGe Power Amplifiers

Peter Song, Ahmet Çağrı Ulusoy, Robert L. Schmid, Saeed N. Zeinolabedinzadeh, and John D. Cressler

School of Electrical and Computer Engineering, Georgia Institute of Technology
777 Atlantic Drive N.W. Atlanta, GA 30332-0250 USA

Abstract—This paper presents W-band cascode power amplifiers implemented in a 90 nm SiGe BiCMOS technology. The one-way cascode PA achieves a saturated output power of 18.0 dBm with 17.2% peak PAE at 93 GHz. The two-way power-combined cascode PA achieves a saturated output power of 20.8 dBm with 14.5% peak PAE at 93 GHz. The one-way PA and two-way power-combined SiGe PAs occupy an area of 0.29 mm² and 0.57 mm² without pads, respectively. To the authors' best knowledge, this work demonstrates the highest PAEs and associated output powers achieved by W-band PAs in any silicon-based technology to date.

Keywords—millimeter-wave integrated circuits, power amplifier, silicon-germanium, SiGe, W-band

I. INTRODUCTION

Advances in SiGe BiCMOS technology have continued to increase the performance of silicon-based millimeter-wave circuits and systems. Recently, 4th generation SiGe HBTs have demonstrated modest output power densities at W-band frequencies [1], [2]. This is significant, as the power amplifier (PA) is the primary limitation of wireless systems implemented at any frequency in silicon-based technologies, especially when considering medium- to long-range applications. Constrained by low breakdown voltages, silicon designers often utilize current combining with many transistors in parallel, leading to inefficient on-chip power-combining due to resistive losses. This leads to a reduction in power added efficiency (PAE), which is an issue for all power-constrained wireless systems.

Considering this fundamental design constraint, some silicon designers have focused on efficient, lower power PAs that can be spatially power combined in free space using a grid or a phased array configuration [3], [4]. Performing power combining in free space ideally has a gain relation of $10 \times \log_{10}$ (N) dB, where N is the number of elements. For this reason, the PA design should emphasize minimizing the tradeoff of PAE for output power. However, it is still important to design for high output power in order to reduce the overall number of elements. This paper reports two W-band SiGe PAs which achieve 18.0 dBm P_{SAT} with 17.2% peak PAE and 20.8 dBm P_{SAT} with 14.5% peak PAE at 93 GHz, respectively.

II. TECHNOLOGY

The PAs are fabricated in IBM's 90 nm SiGe BiCMOS technology (IBM 9HP). The process features high-speed SiGe HBTs with f_T/f_{MAX}/BV_{CEO} of 300 GHz/350 GHz/1.5 V. The back-end-of-line (BEOL) consists of four Cu digital metals, four Cu intermediate metals, a Cu thick RF metal, and a thick Al top metal, as shown in Fig. 1.

The microstrip transmission lines were formed over a 10 x

10 μm² M1-M4 gridded ground plane which uses a via-connected cross-stitch pattern. This allows designs to meet the density and stress design rules of the stricter digital metals while creating an excellent ground return path. In addition, the lower metals can be used for bias routing below the top ground plane metals to isolate DC and RF signals. Low characteristic impedances for quarter-wave ($\lambda/4$) matching networks were attained by using a lower signal metal in the back-end-of-line stack-up to increase the capacitance per unit length of the microstrip transmission lines.

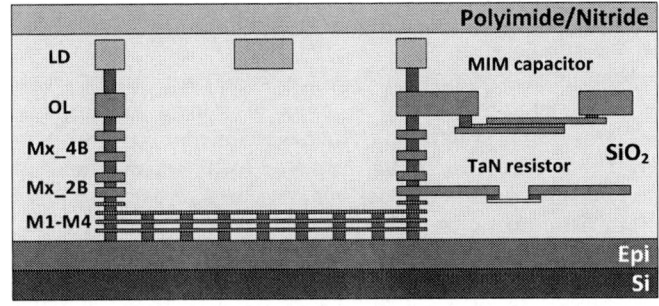

Fig. 1. Cross-sectional diagram of IBM 9HP's BEOL with the 10 metal-layer option.

III. CIRCUIT DESIGN

The circuit schematic of the one-way cascode PA without power combining is shown in Fig. 2.

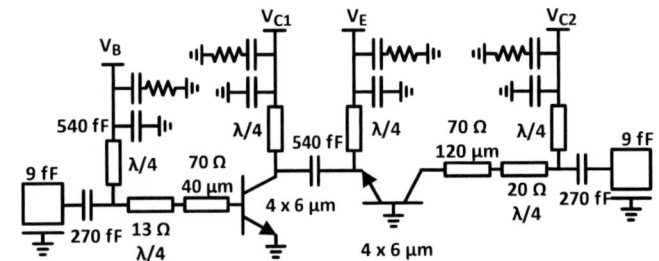

Fig. 2. Schematic diagram of the one-way cascode SiGe PA design.

The cascode topology is chosen for three primary reasons. The limited output voltage swing is increased to near BV_{CBO} of the common-base device, the load-line impedance of the output stage is increased, and the high reverse isolation of the cascode versus the common-emitter topology greatly simplifies the requisite iterative input and output matching design procedure.

However, the cascode topology presents issues with regards to its implementation at high frequencies. The base node of the common-base stage of the cascode is especially sensitive to the base termination network, and can exhibit in-band self-

978-1-4799-7231-9/14 $31.00 © 2014 IEEE

resonance if it is not designed carefully. In-band oscillations may occur when the resonance of the base termination coincides with the negative impedance looking into the base node at high current densities [5]. This issue is exacerbated at millimeter-wave frequencies since the inductance of the via-stack leading up from the device to the metal-insulator-metal (MIM) capacitor is fixed due to the BEOL of the technology. In addition, the physical size of MIM capacitors do not allow for a compact PA cell layout as each base node individually requires a low impedance RF path to ground. This issue can lead to matching networks which are disproportionately large and lossy, as well as transistor feeds which are phase-imbalanced.

For these reasons, a cascode topology in which the base nodes of the common-base stage are physically grounded is chosen, in a similar fashion as the work by Reed *et al.* [6]. As a result, the common-emitter and common-base stages of the cascode must be separated by a series capacitor, and all nodes requiring biases are biased via $\lambda/4$ transmission lines using the highest copper metal trace for low-loss and high current handling. The major disadvantages of this cascode configuration is that it requires an additional bias supply over the typical cascode topology, and the emitter bias of the common-base stage requires a negative bias which sinks high current. While this approach precludes this topology from achieving high integration at the system level, high electron mobility transistors (HEMT) in III-V compound semiconductor technologies are often depletion-mode devices and also require negative biases for transistor operation.

The SiGe HBTs used in the cascode PA have two base and two collector stripes per emitter stripe (CBEBC). The SiGe HBTs are arranged in parallel with four 6 µm emitter length devices in each stage of the cascode, with an output emitter periphery of 24 µm. Two series self-resonant 270 fF MIM capacitors DC block the common-emitter stage from the common-base stage. The layout of the cascode PA core is shown in Fig. 3.

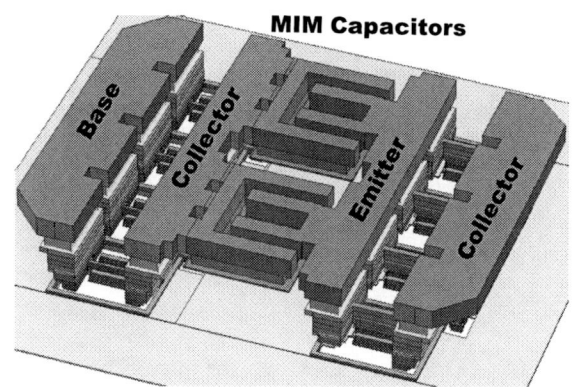

Fig. 3. Sonnet EM 3D view of the 4 x 6 µm cascode PA core layout.

The input and output matching networks consist of short, high-impedance (70 Ω) transmission lines (TL$_{ind}$) used to resonate out the capacitance of the PA cell and low-impedance (13–20 Ω) $\lambda/4$ transmission lines to match the real portion of the source and load targets to 50 Ω. The signal pads are absorbed into the matching networks as they contribute a shunt capacitance of 9 fF. The simulated loss of the output matching network from 90–100 GHz is 1.35–1.50 dB. A Smith chart

illustrating the load target of the PA cell including EM simulated device interconnect parasitics at 94 GHz and the output matching network response is shown in Fig. 4.

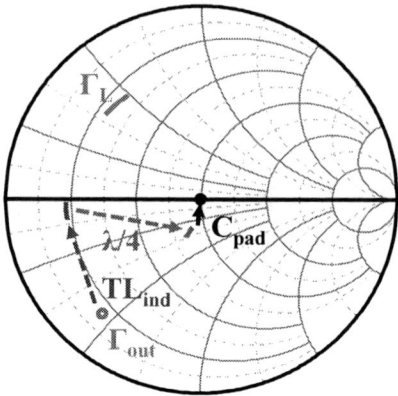

Fig. 4. Smith chart diagram of the load target and the output matching network response from 90-100 GHz.

The RF short and subsequent bypass network which terminate each of the $\lambda/4$ biasing transmission lines consists of an initial 540 fF of shunt MIM capacitors followed by a bank of resistively-terminated MIM capacitors.

Two cascode PAs are also power-combined using a low-loss, two-way Wilkinson power combiner designed using the highest metal in the stack-up. The $\lambda/4$ transmission lines with Z_0 of 70.7 Ω are implemented using 8 µm wide metal, and a 100 Ω TaN resistor is used for the odd-mode termination with evacuated ground plane underneath it to minimize the parasitic shunt capacitance. The simulated versus measured insertion loss results of the standalone Wilkinson power-combiner structures are shown in Fig. 5. The 0.2–0.3 dB difference in measured versus simulated insertion loss is due to the contact resistances of the BeCu probes to Aluminum pads (1–2 Ω), which has not been de-embedded in these measurements.

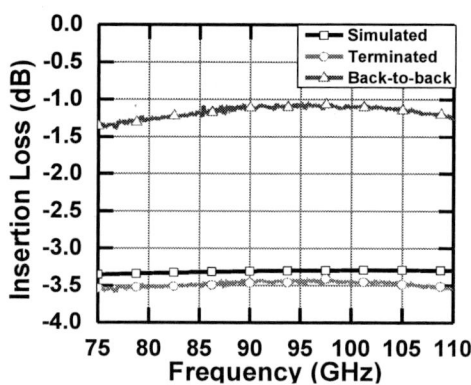

Fig. 5. Measured versus simulated insertion loss of the Wilkinson power combiner with one port terminated into an on-chip 50 Ω resistor, as well as measured insertion loss in a back-to-back configuration.

All of the passive networks, including device interconnects, are EM simulated in a piecemeal fashion using Sonnet. The simulation setups are sized such that box resonances do not occur for up to 10 harmonics. This minimizes the errors from large-signal harmonic balance simulations in Agilent ADS.

978-1-4799-7231-9/14 $31.00 © 2014 IEEE

The one-way and two-way power amplifiers are biased in deep Class-AB and Class-B, respectively, which increases the efficiency and stability of the power amplifiers due to the low current density. The low current density also allows for operating the transistors in the region of weak avalanche for increased output power and PAE as described by Sun *et al.* [7]. When biasing in region of weak avalanche, the class of biasing is especially important to ensure out-of-band stability of the amplifiers. The cascode collector supply voltage was set to 2.4–2.5 V such that the common-base output stage can swing from saturation to near BV_{CBO}. This is about a factor of 3x higher than BV_{CEO}, but does not cause breakdown due to the low DC resistance presented to the base of the transistors.

IV. MEASURED RESULTS

The chip micrographs of the fabricated SiGe W-band PAs are shown in Fig. 6. The dimensions of the one-way and two-way power-combined PAs without pads are $650 \times 450\ \mu m^2$ and $850 \times 675\ \mu m^2$, respectively.

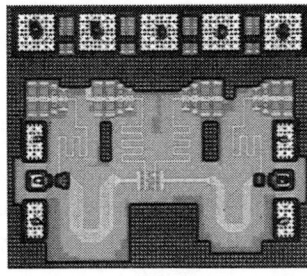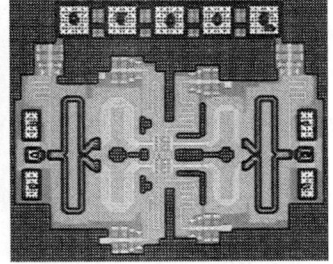

Fig. 6. Die micrographs of the one-way and two-way power-combined PAs.

An Anritsu ME7808C VNA with mm-wave extenders and 1-mm coax probes is used to measure S-parameters from 0.05-110 GHz.

As a result of the deep Class-AB and Class-B biasing, the PAs must be driven with RF power to be turned on; thus, their small-signal gain is negligible. The one-way cascade PA is biased with 0.85 V on V_B, 1.5 V on V_{C1}, -0.76 V on V_E, and 2.4 V on V_{C2}, and the two-way power-combined PA is biased with 0.75 V on V_B, 2.5 V on V_{C1}, -0.75V on V_E, and 2.5 V on V_{C2}. Additional decoupling capacitance is provided off-chip both on the GGB DC probes and on each BNC cable for low frequency bypassing. The small-signal results for the one-way and two-way power combined PAs are shown in Fig. 7.

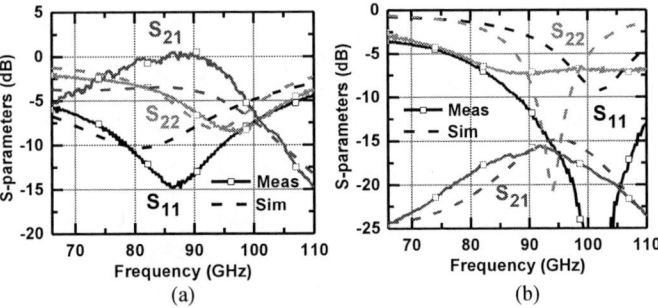

Fig. 7. Measured versus simulated S-parameters across W-band of the (a) one-way PA and the (b) two-way power-combined PA.

The large-signal measurement setup consists of an OML

S10MS 6x multiplier, WR-10 isolators, a Millitech voltage controller attenuator, 80–86 and 90–96 GHz Millitech instrumentation PAs, an Agilent WR-10 power sensor, 1-mm coax probes, and HP WR-10 to 1-mm coax converters. An instrumentation PA was not available to fully characterize the DUT above 96 GHz. The large-signal results of the one-way cascode PA are shown in Fig. 8.

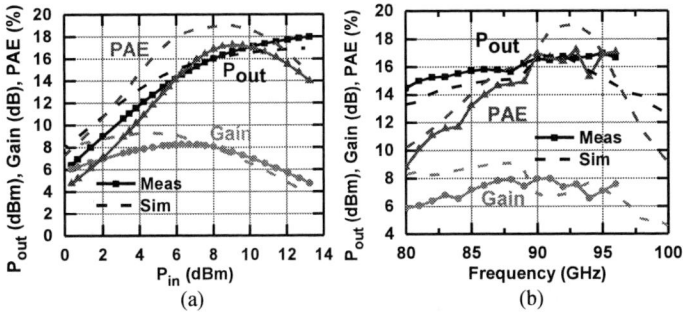

Fig. 8. Measured versus simulated large-signal performance of the one-way PA for a (a) power sweep at 93 GHz and at (b) P_{in} for peak PAE across frequency.

The peak PAE of the one-way PA at 93 GHz is measured to be 17.2%, occurring at an input power of 9 dBm with an associated output power and gain of 16.6 dBm and 7.6 dB, respectively. The total power consumption at the peak PAE at 93 GHz is 221 mW. The saturated output power is measured to be 18 dBm at 93 GHz. The peak PAE and associated output power are measured to be greater than 15% and 16 dBm from 90–96 GHz, respectively.

The large-signal results of the two-way power-combined PA are shown in Fig. 9.

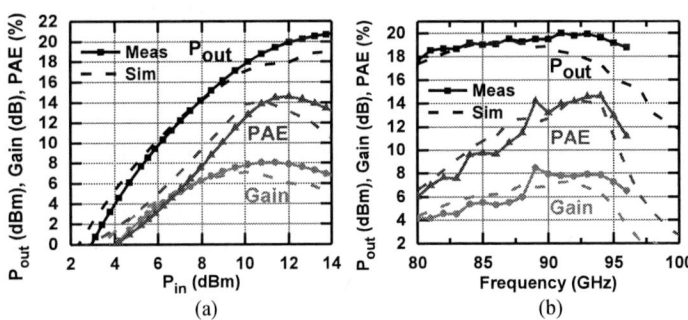

Fig. 9. Measured versus simulated large-signal performance of the two-way power combined PA for a (a) power sweep at 93 GHz and at (b) P_{in} for peak PAE across frequency.

The two-way power-combined PA obtains a peak PAE at 93 GHz of 14.6% at an input power of 12 dBm. The associated output power and gain at the peak PAE are 19.9 dBm and 7.9 dB, respectively. The total power consumption at the peak PAE at 93 GHz is 564 mW. With increased input drive, the saturated output power is measured to be 20.8 dBm at 93 GHz. The peak PAE and associated output power are greater than 11% and 19 dBm from 88–96 GHz, respectively.

The measured and simulated large-signal results are in reasonable agreement once the aforementioned probe-to-pad contact resistances are considered in the simulations.

978-1-4799-7231-9/14 $31.00 © 2014 IEEE   153

TABLE I.   COMPARISON OF STATE-OF-THE-ART W-BAND POWER AMPLIFIERS

Reference	[8] 2011	[9] 2012	[10] 2013	[11] 2013	[12] 2013	[13] 2014	[4] 2011	This Work	This Work
Technology	0.15 μm GaN HEMT	0.14 μm GaN HEMT	0.25 μm InP HBT	65 nm CMOS	45 nm SOI CMOS	45 nm SOI CMOS	0.13 μm SiGe HBT	90 nm SiGe HBT	90 nm SiGe HBT
Topology	1-mm output CS	1.2-mm output CS	2-way balun CE	16-way CS	Stacked FET	Multi-drive Stacked FET	40-μm output CE	24-μm output Cascode	2-way Cascode
Frequency (GHz)	91	93.5	86	90	89	91	94	93	93
P_{SAT} (dBm)	30.8	31.7	23.1	18.3	17	19.2	13	18.0	20.8
PAE_{peak} (%)	20	20.5	30.4	9	9	14	7.2	17.2	14.6
P_{OUT} (dBm) at PAE_{peak}	30.8	31.7	23.1	18	15	18	-	16.6	19.9
Gain (dB) at PAE_{peak}	13	10	11	8	5	7	-	7.6	7.9
Area (mm²)	2.25	-	0.67	0.85	-	0.228	-	0.29	0.57

V.   SUMMARY

Two state-of-the-art W-band PAs implemented in a 90 nm SiGe BiCMOS technology were presented. A comparison of this work to other state-of-the-art W-band PAs operating above 85 GHz is shown in Table I. While the work by Chang et al. [14] showed good results for fully-differential measurements, it assumes ideal 3-dB combining. For this reason, it has been omitted from Table I, in which the reported results include combining losses. To our best knowledge, this work demonstrates the highest PAEs and associated output powers reported for W-band PAs in any silicon-based technology to date.

ACKNOWLEDGMENT

The authors are grateful to D. Harame, J. Pekarik, and the IBM 9HP team; L. Kuo, M. Gilbert, C. Hornbuckle of Semtech; J. Kotce, R. Leoni, N. Kolias of Raytheon Microelectronics Engineering and Technology; and T. Thrivikraman, G. Sadowy of NASA Jet Propulsion Laboratory, for their support of this work. This work was supported by a NASA Office of the Chief Technologist's Space Technology Research Fellowship.

REFERENCES

[1]   I. Hasnaoui, E. Canderle, P. Chevalier, D. Gloria, and C. Gaquiere, "94-GHz load pull measurements of SiGe HBT by extracting output power density in W-Band," in *European Microwave Integrated Circuits Conference*, 2013, pp. 400–403.

[2]   A. Pottrain, T. Lacave, D. Ducatteau, D. Gloria, P. Chevalier, and C. Gaquière, "High power density performances of SiGe HBT from BiCMOS technology at W-Band," *IEEE Electron Device Letters*, vol. 33, no. 2, pp. 182–184, Feb. 2012.

[3]   Y. A. Atesal, B. Cetinoneri, R. A. Alhalabi, and G. M. Rebeiz, "Wafer-scale W-band power amplifiers using on-chip antennas," in *IEEE Radio Frequency Integrated Circuits Symposium*, 2010, pp. 469–472.

[4]   Y. A. Atesal, B. Cetinoneri, M. Chang, R. Alhalabi, and G. M. Rebeiz, "Millimeter-wave wafer-scale silicon BiCMOS power amplifiers using free-space power combining," *IEEE Transactions on Microwave Theory and Techniques*, vol. 59, no. 4, pp. 954–965, Apr. 2011.

[5]   R. L. Schmid, C. T. Coen, S. Shankar, and J. D. Cressler, "Best practices to ensure the stability of SiGe HBT cascode low noise amplifiers," in *IEEE Bipolar/BiCMOS Circuits and Technology Meeting*, 2012, pp. 1–4.

[6]   T. B. Reed, M. J. W. Rodwell, Z. Griffith, P. Rowell, M. Field, and M. Urteaga, "A 58.4mW solid-state power amplifier at 220

GHz using InP HBTs," in *IEEE MTT-S International Microwave Symposium Digest*, 2012, pp. 1–3.

[7]   Y. Sun, G. G. Fischer, and J. C. Scheytt, "A compact linear 60-GHz PA with 29.2% PAE operating at weak avalanche area in SiGe," *IEEE Transactions on Microwave Theory and Techniques*, vol. 60, no. 8, pp. 2581–2589, 2012.

[8]   A. Brown, K. Brown, J. Chen, K. C. Hwang, N. Kolias, and R. Scott, "W-band GaN power amplifier MMICs," in *IEEE MTT-S International Microwave Symposium Digest*, 2011, pp. 1–4.

[9]   M. Micovic, A. Kurdoghlian, A. Margomenos, D. F. Brown, K. Shinohara, S. Burnham, I. Milosavljevic, R. Bowen, A. J. Williams, P. Hashimoto, R. Grabar, C. Butler, A. Schmitz, P. J. Willadsen, and D. H. Chow, "92–96 GHz GaN power amplifiers," in *IEEE MTT-S International Microwave Symposium Digest*, 2012, pp. 1–3.

[10]   H. Park, S. Daneshgar, J. C. Rode, Z. Griffith, M. Urteaga, B. Kim, and M. Rodwell, "30% PAE W-Band InP power amplifiers using sub-quarter-wavelength baluns for series-connected power-combining," in *IEEE Compound Semiconductor Integrated Circuit Symposium*, 2013, pp. 1–4.

[11]   Z.-M. Tsai, Y.-H. Hsiao, H.-C. Liao, and H. Wang, "A 90-GHz power amplifier with 18-dBm output power and 26 GHz 3-dB bandwidth in standard RF 65-nm CMOS technology," in *IEEE MTT-S International Microwave Symposium Digest*, 2013, pp. 1–3.

[12]   J. Jayamon, A. Agah, B. Hanafi, H. Dabag, J. Buckwalter, and P. Asbeck, "A W-band stacked FET power amplifier with 17 dBm Psat in 45-nm SOI CMOS," in *IEEE Silicon Monolithic Integrated Circuits in RF Systems*, 2013, pp. 156–158.

[13]   A. Agah, J. A. Jayamon, P. M. Asbeck, L. E. Larson, and J. F. Buckwalter, "Multi-Drive Stacked-FET Power Amplifiers at 90 GHz in 45-nm SOI CMOS," *IEEE Journal of Solid-State Circuits*, vol. 49, no. 5, pp. 1148–1157, May 2014.

[14]   M. Chang and G. M. Rebeiz, "A wideband high-efficiency 79–97 GHz SiGe linear power amplifier with > 90 mW output," in *IEEE Bipolar/BiCMOS Circuits and Technology Meeting*, 2008, pp. 69–72.

A Low-Power and Ultra-Compact W-band Transmitter Front-End in 90 nm SiGe BiCMOS Technology

Taiyun Chi, Jong Seok Park, Robert L. Schmid, A. Cagri Ulusoy, John D. Cressler, and Hua Wang

School of Electrical and Computer Engineering, Georgia Tech, Atlanta, GA 30332-0250 USA

Abstract— This paper presents a W-band direct-conversion transmitter front-end implemented in SiGe BiCMOS technology. The transmitter consists of an up-conversion mixer, a power amplifier (PA), and all the required on-chip passive matching networks. A differential cascode structure with capacitive neutralization is used in the PA design for improved gain and reverse isolation. W-band on-chip transformers are extensively utilized to realize low-loss and ultra-compact passive matching networks. This design strategy enables the whole transmitter core to fit within a 600 μm × 250 μm die area. The transmitter achieves +7.2 dBm peak output power and 18 dB conversion gain at 87 GHz, with 73.5 mW total power consumption. Modulation tests with QPSK and 16QAM signals (both at 50 MSym/s) were also performed. Average EVM values of 5.9% / 6.7% have been achieved at 7 dB output power back-off for QPSK/16QAM signals, respectively.

Index Terms — Power amplifier, SiGe BiCMOS integrated circuits, transmitter front-end, up-conversion mixer, W-band.

I. INTRODUCTION

There has been increasing research interest recently in developing high-performance W-band transceiver systems (75 GHz to110 GHz) for satellite communications, military radars, vehicular radars, and millimeter-wave imaging and sensing applications [1]-[5]. Advanced SiGe BiCMOS processes represent an attractive candidate to implement such W-band systems due to the RF capabilities of the SiGe HBT (f_T/f_{max}) and its compatibility with low-cost CMOS technology and which is needed for baseband signal processing. In particular, two applications have shown promising potential for the W-band wireless systems. First, ultra-high speed wireless links, e.g. chip-to-chip communications, is a natural application for W-band systems. The W-band carrier frequency provides a correspondingly large available bandwidth for high-speed data transmission, and which may serve as a potential solution to the existing interconnect bottleneck. In addition, W-band phased-array radar systems are also highly attractive due to their potential high spatial-temporal resolution compared to radars operating at lower frequencies. Array operation enables electronic beam-forming and steering, and helps achieve improved signal-to-noise ratio (SNR), effective isotropic radiated power (EIRP), and spatial filtering.

To fully exploit the benefits of the W-band spectrum, it is highly desirable to leverage the concurrent operation of many W-band wireless links in parallel. This enables Multiple-Input and Multiple-Output (MIMO) architectures for ultra-high speed wireless communications and large scale phased arrays for radar applications. Therefore, it is critical to achieve both low-power and ultra-compact sized W-band front-end circuits to realize such large scale system integration.

In this paper, we present a SiGe W-band direct-conversion transmitter front-end which achieves both low-power and an ultra-compact footprint. On-chip transformer-based passive networks have been judiciously designed to perform low-loss and compact matching at W-band. The transmitter includes the up-conversion mixer, the PA, and all the matching networks, all of which are implemented monolithically in a full 90 nm SiGe BiCMOS platform (IBM 9HP). This W-band transmitter solution naturally lends itself to large-scaled W-band communications/radar systems.

This paper is organized as follows: Section II describes the active circuit designs, including the mixer and the PA. Section III presents the W-band on-chip transformer-based passive networks and their modeling. Detailed measurement results are shown in section IV.

II. ACTIVE CIRCUIT DESIGN

The simplified schematic of the transmitter front-end is shown in Fig. 1.

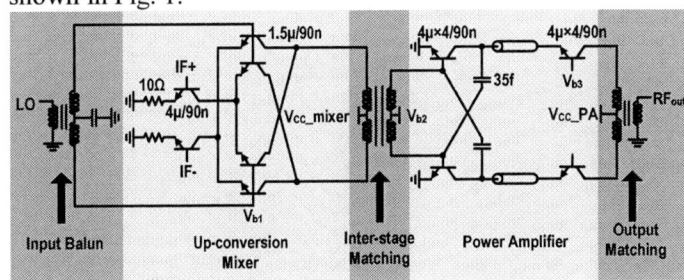

Fig. 1. Transmitter front-end circuit schematic.

The transmitter consists of an up-conversion mixer and a PA. The mixer adopts a double-balanced Gilbert Cell topology, with source degeneration resistors to improve its linearity. The PA is designed as a one-stage cascode amplifier with neutralization capacitors to boost the gain. Note that due to the low loss of the transformer-based matching networks, no additional amplification stage is needed, resulting in a significant power savings and footprint reduction.

A. W-band Power Amplifier Design

One major challenge in designing W-band PAs in silicon-based platforms is the limited gain provided by the active device. This is often exacerbated by the interconnection parasitics and losses, leading to overall low gain and low output power.

In order to harvest the maximum output power from a PA stage, a pseudo-differential cascode topology with cross-

978-1-4799-7231-9/14 $31.00 © 2014 IEEE

coupled neutralization capacitors is adopted. Compared with single-ended designs, differential PAs naturally offer 3dB more output power and are inherently more robust against the parasitics associated with the ground terminal. In addition, the cascode structure helps improve the reverse isolation of the PA, thus enhancing the amplifier stability and desensitizing the effect of the antenna load mismatch. The layout of the PA has been carefully EM simulated with extracted parasitics of the local transistors (Fig. 2).

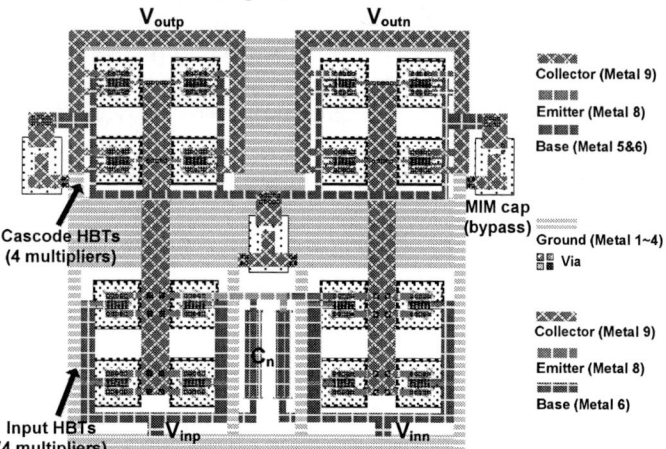

Fig. 2. Layout of the differential cascode PA.

To further enhance the gain, capacitive neutralization is adopted [6]. Based on the Miller effect [7], a cross-coupled neutralization capacitor C_n between the base and the collector of the opposite-side transistor functions as a negative capacitor in the PA's half-circuit. This negative capacitor cancels the device parasitic base-to-collector capacitor C_{bc} and thus improves the reverse isolation and gain. Moreover, it also offers capacitive cancellation at the device base input to ease the design of the driving circuit. Fig. 3 shows the simulation results of the gain and the differential stability factor (k) of the cascode PA structure, both with and without neutralization capacitors. Based on the simulations, with 35 fF neutralization capacitor, the PA can achieve around 1.8 dB more gain, and the stability factor can be improved significantly.

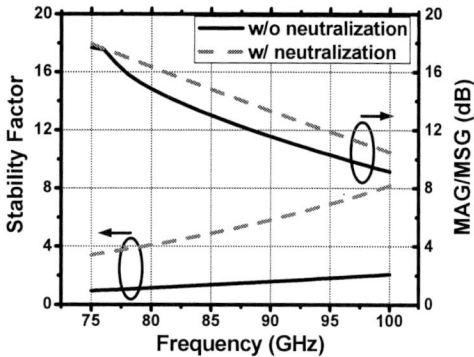

Fig. 3. Simulated stability factor and maximum power gain of the differential cascode PA device, both with and without a neutralization capacitor.

B. W-band Up-Conversion Mixer

To achieve sufficient conversion gain and suppress the LO leakage, the Gilbert Cell double-balanced mixer topology is adopted for the up-conversion mixer design. 10 Ω source degeneration resistors are used to improve the mixer linearity. Fig. 4 shows the simulation results of the input referred 1 dB compression point (P_{in-1dB}) of the mixer. The differential IF signal (up to 10 GHz) is generated off-chip and fed to the chip by a GSGSG probe. The W-band LO signal is provided by a W-band GSG probe and converted to a differential signal using an on-chip balun (Fig. 1). The mixer output and the PA input are coupled using a differential transformer as the inter-stage matching.

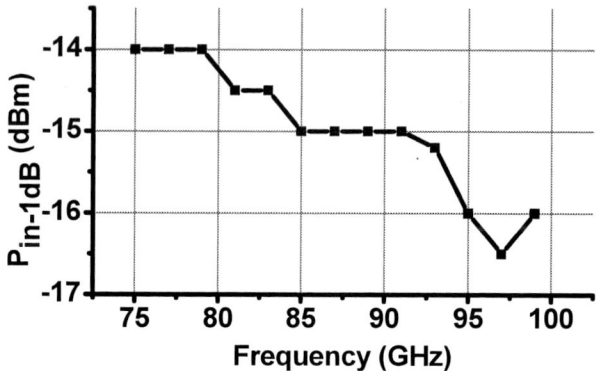

Fig. 4. Simulated P_{in-1dB} of the up-conversion mixer.

III. TRANSFORMER-BASED PASSIVE NETWORK DESIGNS

Transmission-line based matching is a common practice for mm-wave integrated circuit design due to its ease of modeling [1][2]. On the other hand, on-chip transformer-based passive networks have merged as a competitive solution to address high-frequency matching and signal routing. Transformer-based networks typically offer broad bandwidth, low-loss, and small footprints. This matching strategy has been fully exploited in this W-band transmitter design to achieve its compactness.

The PA output matching network is based on a single W-band transformer, which simultaneously offers impedance conversion, power combining and differential to single-ended conversion. Top aluminum metal layer (4μm thickness) and the second copper metal layer (3μm thickness) are used in a vertical coupling configuration to maximize the magnetic coupling. Due to the high reverse-isolation of the cascode PA, only output load-pull optimization is sufficient for performance optimization. The PA supply V_{CC} is fed through the transformer center tap. To maintain a well-defined common-mode return current path and good ground reference for all of the terminals, a ground ring is implemented to surround the primary and secondary coils. Fig. 5 shows the EM drawing of the output transformer and its simulated passive loss.

The inter-stage matching adopts a fully differential transformer structure. The design ensures that the PA and mixer device parasitics are resonated out. The coupling coefficient is designed to provide broadband matching [8].

At the W-band transmitter input, a transformer-based balun provides the input matching of the LO port for the up-

978-1-4799-7231-9/14 $31.00 © 2014 IEEE

conversion mixer and converts the single-ended W-band LO input signal to a differential driving signal. Special design considerations are needed to help ensure the balance of the differential LO signal, since it impacts the proper operation of a double-balanced mixer. The primary coil of the W-band LO balun is first rotated 90° to better fit the floor plan and to partially compensate the imbalance between the two differential output ports. To further improve the differential balancing, the primary and secondary coils are offset by 8 μm. Fig. 6 shows the EM model of the LO balun as well as the simulated amplitude and phase mismatch.

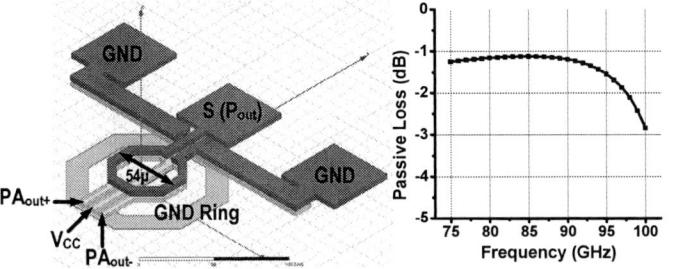

Fig. 5. The EM model of the output matching network for the PA with output GSG pads. The simulated passive loss is shown in the right inset.

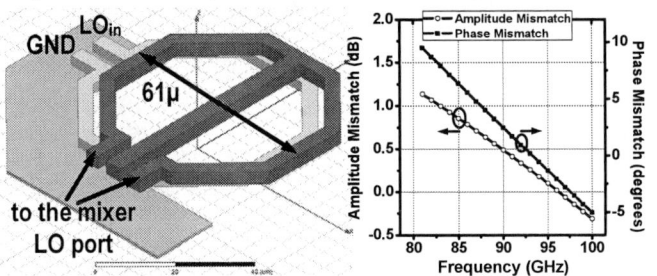

Fig. 6. The EM model of the W-band LO balun with the simulated amplitude and phase mismatch.

IV. MEASUREMENT RESULTS

The transmitter front-end was fabricated in the IBM 9HP SiGe BiCMOS process technology with a typical f_T / f_{max} of 300 / 350 GHz (Fig. 7). Due to the compact transformer-based passive networks, the core area of the transmitter is only 600 μm by 250 μm. The total chip area is only 700 μm by 700 μm including all pads.

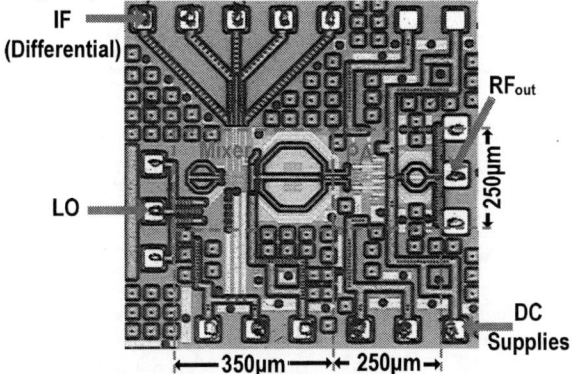

Fig. 7. Chip microphotograph.

The transmitter front-end was measured using on-wafer probing. The measurement setup is shown in Fig. 8. All of the losses of the cables and the waveguide connections are characterized by the power meter across the entire W-band. In the continuous-wave (CW) measurement, the LO signal is generated by a W-band source module (×6 multiplier) and the resulting LO power is within 2 to 5 dBm across W-band at the probe tip. The output power is measured by a power meter and a W-band power sensor. Fig. 9a shows the measured saturated output power (P_{sat}) and the conversion gain when the LO frequency is swept (IF frequency = 10 MHz). The transmitter center frequency appears at 87 GHz with the peak P_{sat} of +7.2 dBm with a PA drain efficiency of 13.3%. The peak conversion gain is 18 dB at 87 GHz with a 3 dB bandwidth from 84 GHz to 91 GHz. Note that this large conversion gain is achieved with only a one-stage PA in this design.

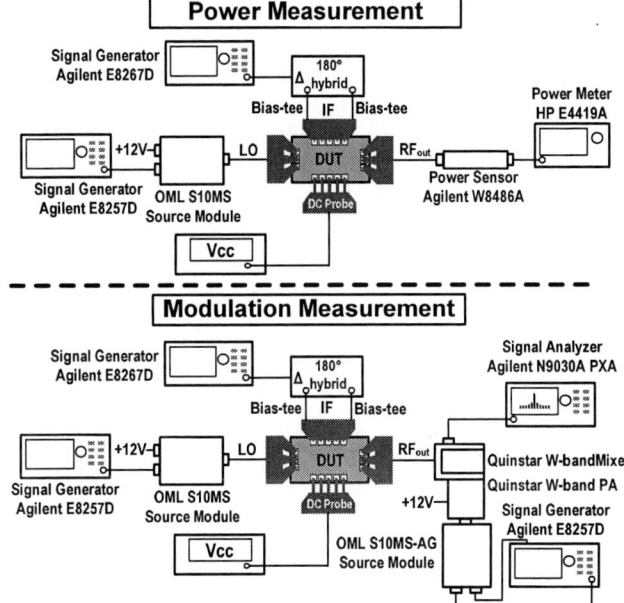

Fig. 8. Measurement setups.

To verify the wireless data transmission capability of the W-band transmitter, a digital modulation test has also been performed by using both QPSK and 16QAM signals. The IF complex modulation signal was generated using an Agilent vector signal generator. The W-band output of the chip is directly down-converted by a Quinstar W-band balanced mixer and then measured with a spectrum analyzer. To provide enough power to drive the W-band mixer, a Quinstar W-band PA was connected after the W-band source module. The two E8257D signal generators are synchronized together. Limited by the test equipment, the maximum symbol rate is 50 MSym/s for both QPSK and 16QAM signals. Fig. 9b shows the average EVM result as a function of the average output power level. Since 16QAM is more sensitive to both AM-AM/AM-PM distortion and has a higher Peak-to-Average-Power-Ratio (PAPR) compared to the QPSK signals, its measured EVM is higher than that of the QPSK signal at the same average output power level. Fig. 9b also shows that the EVM results of the QPSK and the 16QAM signals are 5.9% and 6.7% at 7 dB power back-off from the saturated transmitter output power, respectively. Example measured

978-1-4799-7231-9/14 $31.00 © 2014 IEEE   157

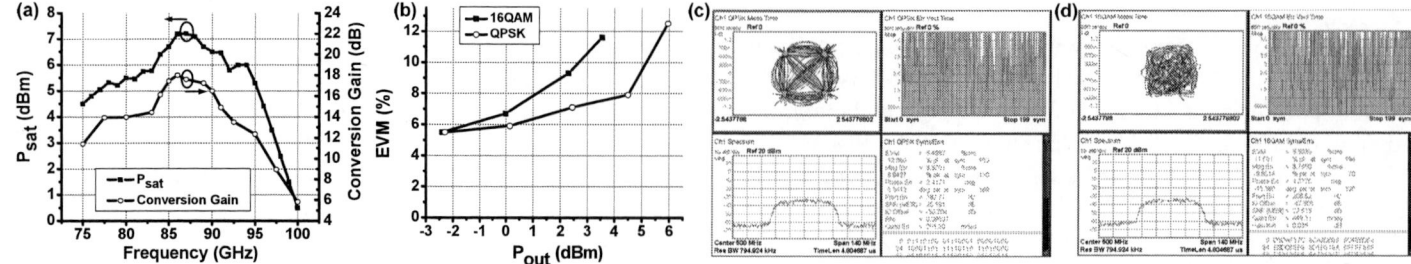

Fig. 9. (a) Measured P_{sat} and conversion gain vs. LO frequency. (b) Measured EVM of 16QAM and QPSK signals vs. average output power. (c) Constellation of the measured QPSK signal. (d) Constellation of the measured 16QAM signal.

constellations of the demodulated QPSK and 16QAM signals are shown in Fig. 9c and Fig.9d.

V. CONCLUSION

A low-power and ultra-compact W-band SiGe transmitter front-end, including an up-conversion mixer, a PA, and all the passive matching networks, is implemented in an IBM SiGe 9HP BiCMOS process technology. By utilizing transformer-based passive networks, the core area of the entire transmitter is only 600 μm by 250 μm. Measurement results show 7.2 dBm P_{sat} and 18 dB conversion gain at 87 GHz. Modulation tests were performed for both QPSK and 16QAM signals. Such a transmitter is very conducive to the implementation of large-scaled W-band MIMO wireless links or W-band phased-array radars.

ACKNOWLEDGMENT

The authors are grateful to D. Harame, J. Pekarik, and the IBM 9HP SiGe team, and L. Kuo, M. Gilbert, and C. Hornbuckle of Semtech, for their support of this work. The authors also wish to acknowledge the members of the Georgia Tech Electronics and Micro-System (GEMS) group, GT SiGe Group, and GT MIRCTECH group, and the Georgia Electronic Design Center (GEDC) for helpful technical discussions and measurement support.

REFERENCES

[1] D. Sandstrom, M. Varonen, M. Karkkainen, and K. A. I. Halonen, "A W-band 65nm CMOS transmitter front-end with 8GHz IF bandwidth and 20dB IR-ratio," *IEEE ISSCC Dig. Tech. Papers*, pp. 418-419, Feb. 2010.

[2] W. Shin, B.-H. Ku, O. Inac, Y.-C. Ou, G. M. Rebeiz, "A 108–114 GHz 4×4 wafer-scale phased array transmitter with high-efficiency on-chip antennas," *IEEE J. Solid-State Circuits*, vol. 48, no. 9, pp. 2041–2054, Sept. 2013.

[3] S. Shahramian, Y. Baeyens, N. Kaneda, and Y.-K. Chen, "A 70–100 GHz direct-conversion transmitter and receiver phased array chipset demonstrating 10 Gb/s wireless link," *IEEE J. Solid-State Circuits*, vol. 48, no. 5, pp. 1113–1124, May 2013.

[4] I. Sarkas, M. Khanpour, A. Tomkins, P. Chevalier, P. Garcia, and S. P. Voinigescu, "W-band 65-nm CMOS and SiGe BiCMOS transmitter and receiver with lumped I-Q phase shifters," in *Proc. IEEE Radio Frequency Integrated Circuits (RFIC) Symp.*, pp.441-444, Jun. 2009.

[5] Z. Xu, Q. J. Gu, and M.-C. F. Chang, "A W-band current combined power amplifier with 14.8dBm P_{sat} and 9.4% maximum PAE in 65nm CMOS," in *Proc. IEEE Radio Frequency Integrated Circuits (RFIC) Symp.*, Jun. 2011.

[6] M. Boers, "A 60GHz Transformer Coupled Amplifier in 65nm Digital CMOS" in *Proc. IEEE Radio Frequency Integrated Circuits (RFIC) Symp.*, pp.343-346, May 2010.

[7] T. H. Lee, *The design of CMOS radio-frequency integrated circuits*, 2nd edition, Cambridge University Press, 2003.

[8] H. Wang, C. Sideris, and A. Hajimiri, "A CMOS broadband power amplifier with a transformer-based high-order output matching network," *IEEE J. Solid-State Circuits*, vol. 45, no. 12, pp. 2709–2722, Dec. 2010.

TABLE I
PERFORMANCE SUMMARY AND COMPARISON

Ref.	Center Frequency	3-dB Bandwidth (GHz)	P_{sat}(dBm)	Conversion Gain (dB)	P_{DC} (mW)	Average EVM	Core Area (mm²)	Technology
This work	**87GHz**	**75-95 (P_{sat})** **84-91 (Gain)**	**7.2**	**18**	**73.5**	**5.9% / 6.7%** **(QPSK/16QAM, 50MS/s)**	**0.15**	**IBM 9HP SiGe BiCMOS (90 nm)**
[1]	85GHz	75-95 (Gain)	6.6	8.5	120	N/A	1.2	65 nm CMOS
[2]	110GHz	100-133 (Gain)	1.0/channel	13.5*	42.8	N/A	0.12	0.18 μm SiGe BiCMOS
[3]	90GHz	70-100 (P_{sat})	8.5/channel	>25	500†	3% (256QAM, 5MS/s)	3.4**	0.18 μm SiGe BiCMOS
[4]	90GHz	80-94 (Gain)	3.0	3.8*	142	N/A	0.08	65 nm CMOS

† Includes both TX and RX power consumption because no specific TX power consumption number is found in the paper.
* No on-chip mixer, the gain is the RF gain.
** Include all pads.

978-1-4799-7231-9/14 $31.00 © 2014 IEEE

Comparison Between MOS and Bipolar mm-Wave Power Amplifiers in Advanced SiGe Technologies

A. Serhan, *Student Member, IEEE*, E. Lauga-Larroze, S. Bourdel, J.-M. Fournier, N. Corrao

IMEP-LAHC, Université de Grenoble, Minatec, 38000 Grenoble, France

Abstract— This article provides a comparison between the performance of MOS and bipolar single stage power amplifiers (PA) in silicon germanium SiGe BiCMOS 55 nm technology from STMicroelectronics. The comparison is made in the same technology node and under similar design conditions (bias current, supply voltage, class of operation and silicon area). Moreover, slow wave coplanar waveguides (S-CPW) were used for matching network in order to reduce the impact of the passive components on the overall performances. Measurement results prove the superiority of bipolar PA in terms of power gain (8.2 dB against 5.5 dB for MOS), and power added efficiency (PAE) (16 % against 12 % for MOS). The output compression point (OCP_{1dB}) and saturation power (P_{sat}) (7 dBm and 10 dBm respectively) are similar for both amplifiers. These results are clarified through a brief theoretical study. To our best knowledge, the presented bipolar PA has the highest figure of merit (FOM) when compared to the state of art of single stage, common source, class-A, 60 GHz power amplifiers.

Keywords— *bipolar, comparison, CMOS, power amplifier, millimiter-wave.*

I. INTRODUCTION

In the past, bipolar devices were specially developed for high-power/high-frequency applications, while CMOS were developed mainly for low power digital circuitry. Hence, comparison between MOS and bipolar transistor was difficult to done. This situation is changed with the fast growing of CMOS technologies, in the sub-100 nm range, where MOS devices provide comparable performance in terms of transition frequency (f_t) and maximum oscillation frequency (f_{max}) which are the key characteristics for RF/mmWave design. Furthermore, the low cost of CMOS technology creates a serious competition between MOS and bipolar especially when talking about the system on chip approach (SoC).

Afterward, SiGe BiCMOS technologies offer both bipolar and MOS transistors, on same substrate, and allow an optimal choice between the two devices since the cost is no more a relevant factor. Works found in the literature, at millimeter wave frequencies, compare bipolar and MOS devices from different technology nodes and different foundries and/or at different operation conditions (power consumption, circuit topology) which may lead the reader to an incorrect conclusion or choice [1][2]. This explains our motivation to compare these two devices, benefiting from the availability of the BiCMOS 55 nm technology.

The comparison done in this paper aims to qualify the advantage of using either bipolar or MOS devices for power amplifiers at mmWave frequencies. In fact, the thin gate oxide, low breakdown voltage, and high substrate loss of CMOS technology result in the mm-Wave power amplifier begin the most challenging to implement in CMOS [3].

Several techniques were applied to CMOS power amplifier in order to enhance gain and output power by using the multi-stage cascade and/or power combining techniques. All these techniques increase the power consumption and reduce the amplifier efficiency.

Taking into account the challenge on power consumption and efficiency, the comparison made in this article uses same bias conditions and class of operation for both amplifiers. In addition, the contribution of the passive structure was reduced by the careful choice of the passive interconnect devices.

The article is organized as follow: Section II presents the technology features and the choice of the passive devices. The design methodology and conditions are explained in section III. Finally, measurement results are presented in section IV and are followed by a short theoretical discussion in section V.

II. PASSIVE AND ACTIVE STRUCTURES

The power amplifiers are fabricated in an industrial SiGe BiCMOS 55 nm process (B55) from STMicroelectronics. The technology process uses a millimeter wave dedicated back-end of line (BEOL) with eight copper metal layers and one aluminum cap-layer on the top as shown in the Fig .1.

The main specification of the B55 technology is the availability of a high speed NPN hetero-junction bipolar transistor (HBT) with f_t/f_{max} of 320 GHz/370 GHz. The transistor model is a level-two high current model (HICUML2) with self heating and non-quasi static effect options. Details regarding the model can be found in [4]. The MOS transistor is a shrunken version of the ST CMOS 65 nm bulk model with a shrink factor of 0.9. The f_t/f_{max} are 180 GHz/220 GHz for low power NMOS devices. The description of the model is provided in [3].

For passive structures, slow wave Coplanar Waveguides (S-CPW) transmission line (TL) is the best candidate for impedance matching network due to its high quality factor Q as demonstrated in [5]. However, micro-strip transmission line is still important for short interconnect and discontinuities as it will be explained in the PAs design section.

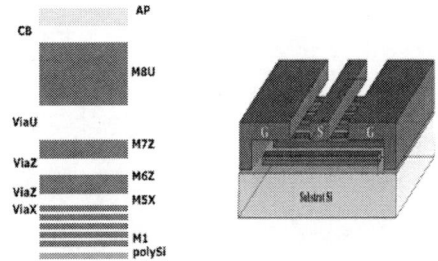

Fig. 1. BEOL of the B55 technology (left), SCPW TL structure (right).

This work has been performed in the RF2THZ SiSoC project of the EUREKA program CATRENE in which the G-INP partner is funded by the DGCIS, France.

978-1-4799-7231-9/14 $31.00 © 2014 IEEE

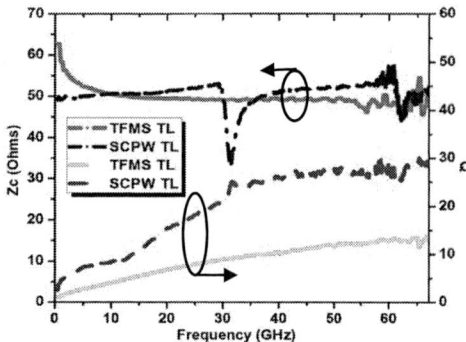

Fig. 2. Measured SCPW TLs characteristics

Fig. 2 shows the measured characteristic impedance and the quality factor of the S-CPW TL and TFMS TL designed and used in this paper. The results show a Z_c of around 50 ohms for both TLs. The quality factor are 28 and 13 for S-CPW TL and TFMS TL respectively.

Finally, the process provides 3D multifinger Metal-oxide-Metal capacitors. The capacitor structure and model are described in [7]. For the power amplifier designed in this work, inductive effects of the capacitor access were extracted using HFSS and added to the original model.

III. POWER AMPLIFIERS DESIGN CONSIDERATION

To provide a good comparison between MOS and bipolar power amplifiers common conditions were used: (1) same bias current, (2) same supply voltage of 1.2 V, (3) transmission lines with same characteristic impedance and propagation loss for input and output matching, (4) they are fabricated on the same die, (5) both power amplifiers are common source, single stage, and should operate in class-A at 60 GHz. This implies that the load resistance presented at the collector of HBT and the drain of MOS is the same.

For MOS power amplifier, we used the methodology presented in [3][5]. The power transistor has a total width of 72 μm and is biased at 0.38 $mA/\mu m$, which corresponds to the peak f_t. This results in a DC drain current of 27 mA and an optimum load, for class-A operation, R_{load} of 45 ohms. The total number of fingers is 48 and is chosen as to respect the electromigration rules at 125 °C. Minimum gate length is used in order to achieve the highest possible f_t.

Once the transistor size and bias point are selected, the output matching network is designed to transform the 50 ohms external load to $Z_{load} = R_{Load} + jX_{load}$ where X_{laod} is an inductive part to cancel out the drain capacitor of the transistor.

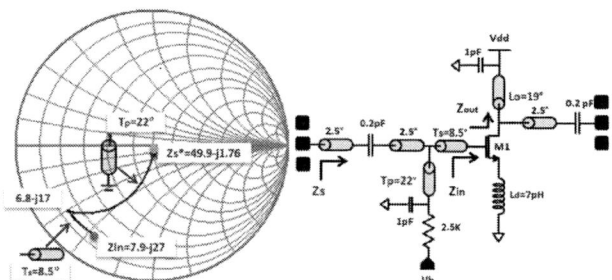

Fig. 3. simplified schematic of the MOS PA (right), input matching network for High-Q input of MOS transistor (left).

Finally, the input impedance Zin=7.9-j27 ohms is transformed to 50 ohms using a single stub matching network,

with $T_s = 8.5°$ for the series TL and $T_p = 22°$ for the parallel shorted stub. Fig. 3 shows the schematic of the MOS PA and the simplified trajectory of the input matching network.

For bipolar PA, we fixed the DC current to the value obtained during the MOS PA design. Next, the transistor area is determined by dividing the total current by the appropriate value of current density (the current density for maximum f_t). For bipolar transistor, in the B55 technology, the peak ft is around 9 $mA/\mu m^2$ which results in a total emitter area of 3.1 μm^2. Finally, CBEBC layout configuration was used to improve f_{max} and f_t due to the reduction in base and collector resistances [8]. The Emitter width was kept to its minimum value (0.2 μm) for maximum ft.

The input impedance (Z_{in}=11-j5.8 ohms) of the bipolar PA has lower quality factor ($Q_{in_{Bip}} \approx 0.5 < 1$) than the MOS input quality factor ($Q_{in_{MOS}} \approx 3.4 \gg 1$). To increase the quality factor of the bipolar PA, a series capacitor C_s is used. As shown in Fig. 4, the capacitor C_s is chosen to have a Q_{in} of 2 so that the 50 ohms impedance can be achieved using a shorted stub ($T_p = 27°$). The alternate solution, without series capacitor, was evaluated. The series capacitor is replaced by a TL ($Ts = 32°$), and the shorted parallel stub was replaced by an open-ended stub ($T_P = 59°$). These large TLs in the second input matching network topology (IMN with Ls) cause additional loss and increase the layout area. Hence the first topology was retained.

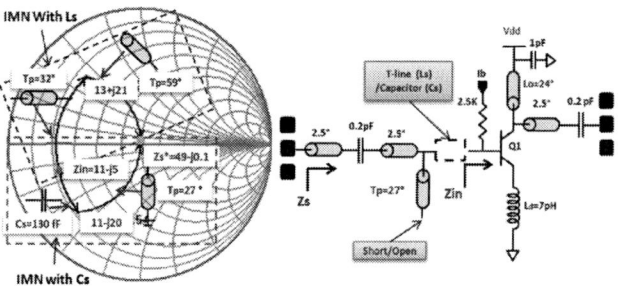

Fig. 4. simplified schematic of the bipolar PA (right), input matching network for Low-Q input of bipolar transistor (left).

For both amplifiers, a 7 pH degeneration inductor (L_d) was artificially used to ensure the in-band unconditional stability (Fig.3 and Fig. 4). The effect of this inductor on the PAs performances will be discussed in the last section. Moreover, as mentioned in [5], the S-CPW TLs have a large lateral dimension which necessitates the use of T-shaped micro-strip TLs to facilitate the layout connection between the S-CPW stubs and the transistors. Noting that the ground plan of the micro-strip TLs should be connected to the ground of S-CPW TLs to ensure a perfect return current path and avoid additional losses.

IV. POWER AMPLIFIERS MEASUREMENTS

S-parameter measurements are performed using an Anritsu ME7808C Broadband Vector Network Analyser and a semi-automatic Cascade S300 Station. The VNA was calibrated using the Line Reflect-Reflect Match (LRRM) calibration method. For large signal measurement, an external power amplifier was used to achieve the input power requirements.

Source power calibrations are done using Agilent V8486 power sensor coupled with E4418B power-meter. The RF probes used are Cascade Microtech Infinity Probe in a ground-signal-ground (GSG) configuration.

 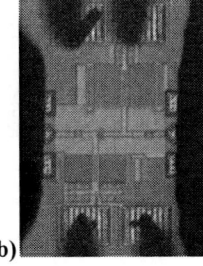

a)   b)

Fig. 5. Die photo of the PA using bipolar (left), and MOS (right)

The die photos of the amplifiers under test are shown in Fig. 5.a and Fig. 5.b for bipolar and MOS PA respectively. The active sizes are 238x265 μm^2 and 200x275 μm^2 for bipolar and MOS respectively. These similar areas for the two PAs make the comparison more representative.

S-parameters measurement and simulation results of MOS and bipolar PAs are shown in Fig. 6 and Fig. 7 respectively. For MOS PA the results show a good agreement between measurement and simulation around 60 GHz. The measured S21 is 6 dB which are 0.2 dB higher than the simulated one. The input reflection coefficient is centered at 60 GHz with -12 dB of return loss.

For bipolar transistor, the maximum measured gain is around 8.2 dB at 60 GHz; which is 0.75 dB lower than the simulated value. This can be explained by the shifted S11 characteristic from 60 GHz to 64 GHz. This frequency shift is attributed to the tentative version of post extracted bipolar transistor model.

The amplifier are both unconditionally stable over the entire measured frequency band, from 40 MHz to 110 GHz, with stability factor k greater than unity and absolute value of the S-parameters determinant |Δ| smaller than unity.

Fig. 8 shows the measured large signal power gain for both MOS and bipolar PAs measured at 60 GHz. An 8.2 dB gain for bipolar amplifier was obtained versus 5.5 dB for MOS amplifier. The output saturation power (P_{sat}) is approximately similar since the output load is the same for both PAs.

As shown in Fig. 9, the peak value of the power added efficiency (PAE) is 16 % for the bipolar PA and 12 % for the MOS PA. We can also notice from Fig. 8 and Fig. 9 that the input compression point of the MOS PA is about 2.5 dB higher than the one for bipolar transistor.

We defined the efficiency factor (E_{Factor}) as the ratio between the PAE of the two PAs:

$$E_{Factor} = \frac{PAE_{Bipolar}}{PAE_{MOS}} = \frac{\left(\frac{P_{out}-P_{in}}{P_{dc}}\right)_{Bipolar}}{\left(\frac{P_{out}-P_{in}}{P_{dc}}\right)_{MOS}} \quad (1)$$

Benefiting from class-A operation linearity, we made the assumption that the evolution of the DC current in function of input power is the same in both amplifiers, so that the E_{Factor} can be simplified to:

$$E_{Factor} = \frac{(G_p-1)_{Bipolar}}{(G_p-1)_{MOS}} \quad (2)$$

This factor compares the PAE for a given input power. As we can notice from (2), for amplifiers with same power consumption and class of operation, the difference in PAE value is referred only to the difference in the power gain.

Fig. 9 shows the PAE of the two PAs and the enhancement factor as a function of input power. The E_{Factor} is approximately linear and equal to 2 at low input power (deep linear amplification region), and tends to one when both amplifiers enter in their saturation regions.

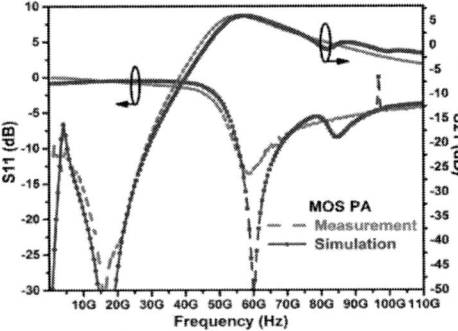

Fig. 6. Measured and simulated S-parameters of MOS PA.

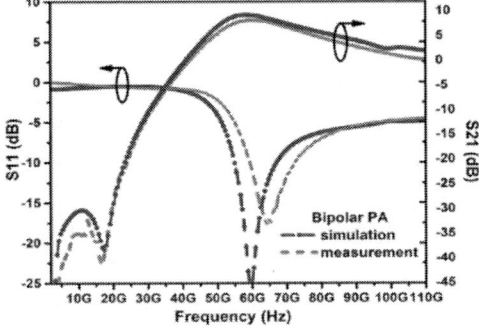

Fig. 7. Measured and simulated S-parameters of bipolar PA.

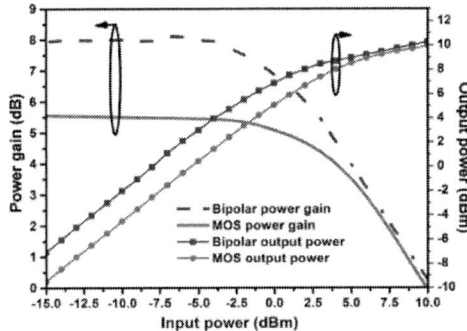

Fig. 8. Measured output power and power gain of both amplifiers.

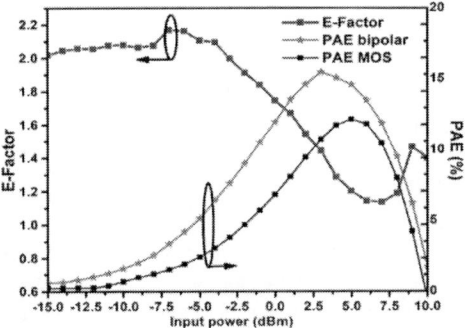

Fig. 9. Measured power added efficiency and E_{factor} of both amplifiers.

978-1-4799-7231-9/14 $31.00 © 2014 IEEE

TABLE I. COMPARISON BETWEEN THE FABRICATED PAs AND THE STATE OF ART FOR SINGLE STAGE, COMMON SOURCE, CLASS-A 60 GHZ PAs.

PA type	Technology	S_{11}(dB)	G_p(dB)	I_{DC} / V_{dd}	OCP_{1dB} (dBm)	PAE (%)	P_{sat} (dBm)	FOM^t
MOS [3]	CMOS 65 nm	-13	4.5	20 mA -1.2 V	6.4	26	9.2	22
MOS [5]	CMOS 40 nm	-15	5.6	22 mA -1.2 V	7	16	10	20
MOS [9]	CMOS 65 nm	x	4.5	27 mA - 1.2 V	6	8.5	9	7.3
This work/MOS	BiCMOS 55nm	-12	5.5	27 mA - 1.2V	7	12	10	15
This work/Bipolar	BiCMOS 55nm	-18	8.2	27 mA - 1.2V	6.8	16	9.9	37

^{t}FOM = G_p × P_{sat} × f^2 × PAE [10]

V. RESULTS DISCUSSION

The linear power gain of a class-A single stage PA can be approximated by the relation:

$$G_p = \frac{P_{out}}{P_{in}} = \alpha \frac{\left(gm^* V_{in} \frac{R_0 R_L}{R_0 + R_L}\right)^2}{R_L \, \Re(Y_{in}) V_{in}^2} = \alpha \frac{\left(gm^* \frac{R_0 R_L}{R_0 + R_L}\right)^2}{R_L \, \Re(Y_{in})} \quad (3)$$

where α represents the total loss in the matching networks. $\Re(Y_{in})$ and R_0 are the real part of the input admittance and output impedance of the device respectively. R_L is the optimum load for a given maximum output power, while gm^* is the effective linear transconductance of the device after source/emitter degeneration:

$$gm^* = gm_i / \sqrt{1 + (gm_i \, L_d w_0)^2} \quad (4)$$

where gm_i is the intrinsic transconductance, and L_d is the source/emitter inductance mentioned in section III.

By considering the same value of α and R_L for the two PAs (Table II), the ratio between the power gains is given by:

$$G_{ratio} = \frac{G_{P_{Bip}}}{G_{P_{MOS}}} = \frac{\Re(Y_{in})_{MOS}}{\Re(Y_{in})_{Bip}} \left(\frac{gm^*_{Bip} \frac{R_{0_{Bip}}}{R_{0_{Bip}} + R_L}}{gm^*_{MOS} \frac{R_{0_{MOS}}}{R_{0_{MOS}} + R_L}} \right)^2 \quad (5)$$

Equation (5) was used to estimate the performances of the amplifiers using the data provided in table II. As we can notice, the bipolar PA dominates the MOS PA in terms of R_0 and gm^*. However, the bipolar device has higher input admittance than the MOS PA, resulting in a moderate power gain ratio between the two devices.

Finally, table I provides a comparison of the PAs developed in this work with previously reported 60GHz PAs in terms of the PA FoM. All compared PAs are single stage, common source topology, and they are supposed to operate in class-A. It can be seen that the bipolar PA marks the highest PA FoM while also having the highest large signal power gain.

TABLE II. PAs PARAMTERS SUMMARY

	$\Re(Y_{in})$ $m\Omega^{-1}$	R_0 Ω	gm^* $m\Omega^{-1}$	RL Ω	α dB	Estimated G_p (dB)	Measured G_p (dB)
MOS	10.5	90	60	45	-3	**5.4**	5.5
Bipolar	70	180	205	45	-3	**9.2**	8.25

VI. CONCLUSION

Comparison between MOS and bipolar power amplifiers is described. The amplifiers were fabricated in an industrial SiGe BiCMOS 55 nm process and compared under the same conditions. In addition, the use of S-CPW TLs with high quality factor reduces the contribution of the passive structure on the overall performances. Measurement results prove the superiority of bipolar PA in terms of power gain (8.2 dB against 5.5 dB), and PAE (16 % against 12 %). The output compression point OCP_{1dB} and maximum saturated power P_{sat} (7 dBm and 10 dBm respectively), are the same for both amplifiers which is expected since they operate in the same class and under the same DC current. Concerning the power gain of the amplifiers, the theoretical study demonstrates that the advantage of the high transconductance of the bipolar transistor is limited by its high input admittance.

ACKNOWLEDGMENT

The authors would like to acknowledge the RF2THZ project for financial support, and Mr. Daniel Gloria from STMicroelectronics for technology access and technical support.

REFERENCES

[1] Avenier, G.; et al., "0.13μm SiGe BiCMOS technology for mm-wave applications," Bipolar/BiCMOS Circuits and Technology Meeting, 2008. BCTM 2008. IEEE , vol., no., pp.89,92, 13-15 Oct. 2008

[2] Voinigescu, S.P.; Dickson, T.O.; Beerkens, R.; Khalid, I.; Westergaard, P., "A comparison of Si CMOS, SiGe BiCMOS, and InP HBT technologies for high-speed and millimeter-wave ICs," Silicon Monolithic Integrated Circuits in RF Systems, 2004. Digest of Papers. 2004 Topical Meeting on , vol., no., pp.111,114, 8-10 Sept. 2004

[3] Quemerais, et al., "A CMOS class-A 65nm power amplifier for 60 GHz applications," Silicon Monolithic Integrated Circuits in RF Systems (SiRF), 2010 Topical Meeting on , vol., pp.120,123, 11-13 Jan. 2010

[4] Pawlak, A.; Schroter, M.; Krause, J.; Céli, D.; Derrier, N., "HICUM/2 v2.3 parameter extraction for advanced SiGe-heterojunction bipolar transistors," Bipolar/BiCMOS Circuits and Technology Meeting (BCTM), 2011 IEEE , vol., no., pp.195,198, 9-11 Oct. 2011

[5] X.L. Tang, E. Pistono, P. Ferrari, and J.-M. Fournier "Enhanced Performance of 60-GHz Power Amplifier by using Slow-wave Transmission Lines in 40 nm CMOS Technology ", International Journal of Microwave & Wireless Technology, Vol. 4, pp. 93-100,Feb 2012.

[6] Franc, A.; Pistono, E.; Meunier, G.; Gloria, D.; Ferrari, P., "A Lossy Circuit Model Based on Physical Interpretation for Integrated Shielded Slow-Wave CMOS Coplanar Waveguide Structures," Microwave Theory and Techniques, IEEE Transactions on , vol.61, no.2, pp.754,763, Feb. 2013.

[7] Quémerais, T; Moquillon, L; Benech, P; Fournier, J-M; Pruvost, S; "CMOS 45-nm 3D metal-oxide-metal capacitors for millimeter wave applications",Microwave and Optical Technology Letters,Volume 53, Issue 7, pages 1476–1478, July 2011

[8] Jae-Sung Rieh; et al., "SiGe heterojunction bipolar transistors and circuits toward terahertz communication applications," Microwave Theory and Techniques, IEEE Transactions on , vol.52, no.10, pp.2390,2408, Oct. 2004

[9] Valdes-Garcia, A.; Reynolds, S.; Plouchart, J.-O., "60 GHz transmitter circuits in 65nm CMOS," Radio Frequency Integrated Circuits Symposium, 2008. RFIC 2008. IEEE , vol., no., pp.641,644, June 17 2008-April 17 2008.

[10] International Technology Roadmap for Semiconductors, 2007 edition.

978-1-4799-7231-9/14 $31.00 © 2014 IEEE

On-Wafer Small-Signal and Large-Signal Measurements up to Sub-THz Frequencies

Invited Paper

Viktor Krozer, Ralf Doerner, Franz-Josef Schmückle,
Nils Weimann, Wolfgang Heinrich

Ferdinand-Braun-Institut (FBH),
Leibniz-Institut für Höchstfrequenztechnik
Berlin, Germany
viktor.krozer@fbh-berlin.de

Andrej Rumiantsev[*,$]

[*] MPI Corporation,
Chu-Pei City, Hsinchu County, Taiwan
[$] Brandenburg University of Technology (BTU),
Cottbus-Senftenberg, Cottbus, Germany
a.rumiantsev@ieee.org

Marco Lisker, Bernd Tillack

IHP GmbH
Innovations for High Performance Microelectronics
Leibniz-Institut für innovative Mikroelektronik
Frankfurt (Oder), Germany

Abstract — **Recent advances in MMIC technology have opened the possibilities for circuit operation in the THz range. There are numerous examples of BiCMOS and III-V compound device technologies with demonstrated performance beyond 600 GHz. Characterization of such MMIC are predominantly performed on-wafer in a planar environment. However, on-wafer characterization facilities do not fully keep pace with MMIC development in terms of frequency and power. The paper discusses issues involved in on-wafer calibration at mm-wave frequencies, which is the basis for accurate measurements and characterization of active and passive device. Subsequently, the paper discusses mm-wave interconnect characterization. Low-loss interconnects are important for mm-wave MMIC, especially in case of heterogeneous integration. Finally, a novel heterogeneous integration approach of bipolar technologies, using both BiCMOS and InP DHBT processes is presented. This approach heavily relies on low-loss interconnects and accurate device modelling. It will be shown that accurate large-signal models can be efficiently extracted from well-calibrated on-wafer multi-bias small-signal measurements, but verification is difficult due to calibration difficulties at mm-wave frequencies.**

Keywords — *on-wafer calibration techniques, on-wafer characterization, mm-wave and sub-mm-wave circuits, calibration methodologies, sub-mm-wave transmission lines, interconnects*

I. INTRODUCTION

Recent advances in MMIC technology have opened the possibilities for integrated circuit operation in the THz range. There are numerous examples of BiCMOS and III-V compound device technologies, with demonstrated performance beyond 600 GHz. The design of such MMIC is generally based on a detailed knowledge of EM effects of the passives and interconnects, and on good large-signal and small-signal device modelling. Characterization of such MMIC is predominantly performed on-wafer in a planar environment and requires accurate calibration techniques.

But on-wafer characterization facilities do not fully keep pace with MMIC development in terms of frequency of operation and power. Especially critical is the accurate characterization and modelling of devices operational up to THz frequencies. This owes to the fact that accurate on-wafer calibration is difficult at higher frequencies, due to excitation of higher order modes, signal coupling and a multi-mode operation as well as coupling effects corrupting the calibration data. The situation becomes even more difficult with the newly implemented processes utilizing transfer substrates and heterogeneous integration, which exhibit many transitions to a multitude of transmission line technologies. Characterization of passives like interconnects requires calibration up to THz frequencies, while device model extraction necessitates measurements at somewhat lower frequencies. An example, the attenuation of a coplanar waveguide (CPW) line is presented in Fig. 1 as a function of frequency. It shows that corrupted or inaccurate data obtained during calibration leads to inaccuracies in the device under test (DUT) measurement results. The paper will show below that the origin of these issues have to be determined and overcome in order to accurately determine the DUT performance at THz frequencies.

Of particular importance is the choice of the transmission line technologies (e.g. CPW, standard, thin-film and inverted microstrip) and the according interconnects. In particular, the paper discusses transmission line and interconnect performances, using simulations verified with experiments at millimeter-wave frequencies. It is shown that thin-film microstrip or stripline lines are especially suitable for line calibration standards and interconnects at THz frequencies. Further, the paper shows that although active device modelling generally utilize data up to 60 GHz, device model verification requires measurements up to THz frequencies due to the high cut-off and maximum oscillation frequencies of modern bipolar devices. One important aspect for useful measurements is the control of output power, which has to be sufficiently low for transistor characterization and hence limits considerably the dynamic range of today's measurement equipment. Another aspect is the positioning accuracy of the probes at high frequencies.

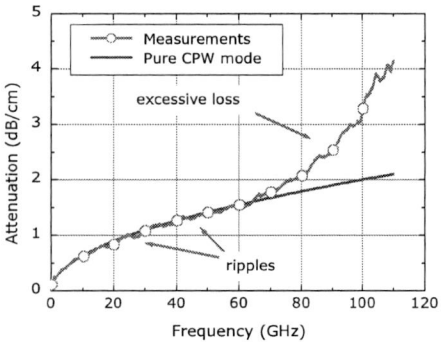

Fig. 1: CPW attenuation versus frequency for a CPW line extracted from measurements using multi-line TRL. The theoretical CPW attenuation is also indicated [3].

Thin-film transmission lines and interconnects are built on top of a substantial dielectric stack on top of the active device layers. In this paper we go a step further and demonstrate that such stacks can be efficiently combined using BiCMOS and III-V technologies at mm-wave frequencies. The paper will demonstrate interconnect and device characterization in this wafer-level heterogeneous integration processes. It is based on a newly established InP-on-BiCMOS process, available now through the SciFab IHP foundry. The transitions from the BiCMOS circuits to the InP circuits and vice versa as well as active device performance will be discussed in detail using simulated and measured results.

II. Pitfalls in On-Wafer Calibration at MM-Wave Frequencies

It becomes indispensable to perform accurate on-wafer device characterization at mm-wave frequencies, due to the increasing transit and maximum oscillation frequencies of the devices employed in such circuits. Generally, results rely to date on extrapolations from low-frequency measurements. The difficulties in on-wafer measurements have many sources, such as for example: crosstalk between probes, crosstalk with other structures on-wafer, parasitic mode excitation with probes, on-wafer calibration errors, probe misalignment, multi-mode propagation on the calibration standards, signal losses, limited dynamic range of equipment due to low signal levels, radiation into the substrate, parallel plate mode excitation in CPW structures etc. Another difficulty appearing at frequencies above 110 GHz is the necessity for re-calibration of the system at each band with the according re-occurrence of all the above difficulties.

The key components for understanding the above effects are the on-wafer probes, the transmission line properties and the calibration procedures employed. Calibration procedures will be treated in the following section, while the other two will be discussed below.

The impact of the on-wafer probe can be understood by considering that the outer structure dimensions are large compared to the on-wafer calibration standard dimensions. This enables the excitation of additional propagating modes on-wafer, which are not related to the desired CPW mode. Firstly, the coupling between probe ground and on-wafer ground creates a parasitic propagating signal. Secondly, in the vicinity of the pad structure a mode is excited similar to the parallel-plate mode between the CPW ground contacts and backside contact. Finally, there is a mode which is guided between the ground of the coaxial probe and the wafer backside metallization. Fig. 2 shows these effects obtained from 3D EM simulations of calibration standards employing microstrip lines. The interaction of the probe with the surrounding structures is clearly visible.

Fig. 2: Left: Electrical field E_y in the longitudinal cross-section (y is the vertical axis, blue color means negative, red color positive and green color zero field values, respectively); Right: Electrical field E_y in the top view cross section directly under the ground metallization of the microstrip lines (y is the normal direction). Inset: The exciting probe positioned over a short neighboring microstrip line [3].

The impact of this coupling and excitation of parasitic modes on the calibration procedure can be evaluated, when simulating different calibration standards in microstrip technology [4] and CPW environment [1], [2], [5], [6], respectively. As an example, Fig. 3, shows simulated results for S_{21} of three different transmission line lengths, which are typically used during calibration. It can be concluded from the figure that coupling to a neighboring line gives rise to transmission dips at frequencies related to the length of this line. Coupling to several lines on both ends of the DUT will give rise to several dips in the frequency characteristic. These dips are then responsible for measurement inaccuracies in the final measurements after calibration, independent of the calibration methodology used.

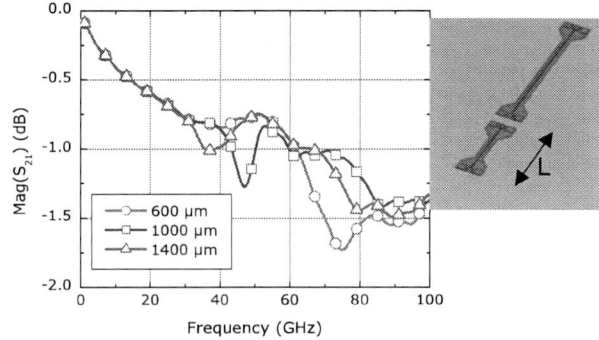

Fig. 3: Signal transmission for neighboring short microstrip lines of three different lengths (L = 600 µm, 1000 µm and 1400 µm) corresponding to pronounced resonance points (75, 48 and 37 GHz) [4].

It is common practice to employ CPW waveguides as calibration standards on-wafer in addition to the calibration substrates. These calibration standards are prone to parallel-plate mode propagation and consequently to substrate mode excitation. Fig. 4 illustrates the coupling effect for a substrate placed on top of a ceramic mimicking an infinite substrate and the same situation with a metallized back-plane. One can clearly see the strong coupling into a parallel-plate mode in case of backside metallization.

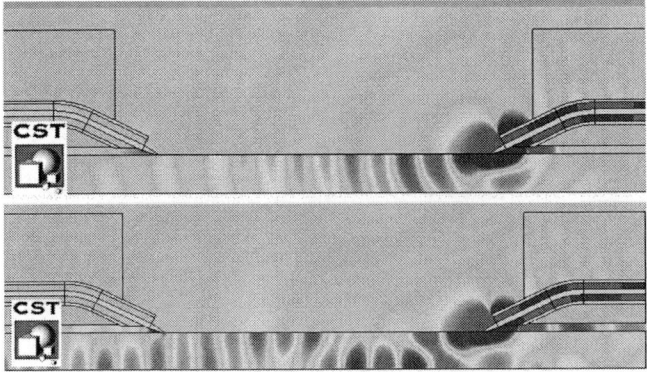

Fig. 4: Field plots from CST Microwave Studio simulations with the CPW substrate placed on: ceramic (top) and metal (bottom). Open boundary condition is applied at the bottom for ceramic CPW substrate, while an electric wall boundary is applied for metal CPW substrate [22].

III. ADVANCED CALIBRATION PROCEDURES

A. On-Wafer Calibration

Scattering (S)-parameter measurements of devices fabricated on semiconductor wafers are usually calibrated using a two-step procedure: first the measurement reference plane is set close to the RF probe tips using suitable calibration substrates and then parallel and serial parasitic impedances are characterized by measurements of the device de-embedding elements. As it was demonstrated in [13]–[15], the on-wafer S-parameter calibration with *in-situ* standards is the preferable strategy for accurate device characterization at sub-mm-wave frequencies, in particular on silicon. The measurement reference plane is shifted close to the device terminals (Fig. 5). Application of commercial planar calibration standards assures consistent level of calibration accuracy for the wafer-level S-parameter measurement system [11]. This procedure remains accurate only if the above coupling effects on the calibration substrate can be ignored. In addition, de-embedding of the device back end of line (BEOL) parasitics becomes a challenge at mm-wave and sub-mm-wave frequencies due to the increased complexity of the BEOL equivalent circuit and de-embedding algorithms, involving more than five measurement steps (e.g. [12]).

While a great variety of S-parameter calibration methods are available nowadays, the multiline thru-reflect-line (mTRL) [16] and the transfer thru-match-reflect (TMR, or LRM) [17] are relevant for implementation. The convincing advantage of mTRL is its capability to directly measure the S-parameters of *in-situ* transmission lines at a well-defined single-mode reference plane setting the calibration reference impedance Z_{REF} to the characteristic impedance of the line

standard [20], [21]. Thin-film lines provide line standards with characteristic impedance close to 50 Ω and reduced dispersion region using modern semiconductor technologies. Therefore, the transformation of the calibration reference impedance to the system of 50 Ω reference impedance (i.e. to "pseudo S-parameters", [18]) may not be required for the majority of measurement tasks. The remaining practical limitation of the mTRL, however, is the relatively big size of the test chip taken by calibration standards and the need to re-position wafer probes during calibration procedure. It is important to note, that the minimal size of the TRL chip is reverse proportional to measurement frequencies and is less relevant at sub-mm-wave frequencies, omitting low frequencies.

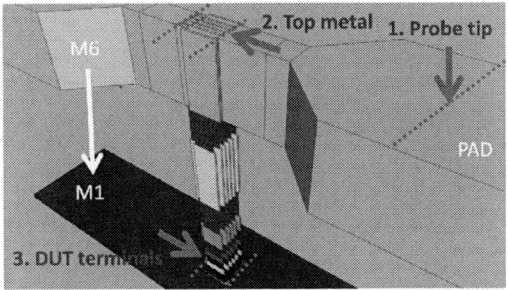

Fig. 5: Three different locations of the reference plane: probe tip, top metal and DUT terminals

The transfer TMR approach has a remarkable advantage: it requires only three standards with the same geometry taking the minimum space of a test chip. The TMR can be easily automated and runs without the need of operator interaction on a conventional wafer probe system. However, the accurate determination of the calibration reference impedance of TMR at sub-mm-wave range is difficult (same holds for any other lumped-standard based calibration method). While [19] suggested several methods applicable for W-band range, the solution for higher frequencies is still underway.

B. Calibration Reference Plane

Designing *in-situ* broadband thin-film transmission lines close to the DUT terminals necessitates the utilization of the top metal level, for instance M6 for the ST Microelectronics' BiCMOS9MMW process (Fig. 6), or the LB metal for IBM's SOI12S0 process (Fig. 7). Therefore, the optimal position of the *in-situ* calibration reference plane is at the top metal level and can be shifted from the top metal to the intrinsic device terminals by a conventional de-embedding approach. Because the parasitic impedance of the contact pads and interconnect lines are already included into the systematic calibration error model, the equivalent impedance of the de-embedding elements should be mostly of pure lumped nature [20], [21].

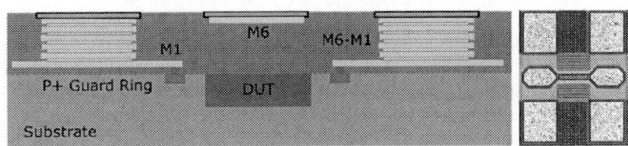

Fig. 6: Cross-section (left) and a photograph (right) of the thru standard implemented in the ST Microelectronics' Si/SiGe:C BiCMOS9MMW process(taken from [20]).

The work in [15] demonstrated an implementation of the *in-situ* mTRL calibration up to 750 GHz frequencies with smooth and consistent propagation constant suitable for transistor device characterization. The calibration lines were realized in a thin-film microstrip design on a thin bisbenzocyclobutene-based (BCB) monomers film. Later, a similar experiment was performed on calibration lines realized by IBM's SOI12SO process (Fig. 8). These results confirm that suitable sub-mm-wave transmission lines for calibration purposes can be realized using thin-film transmission lines and mTRL is applicable.

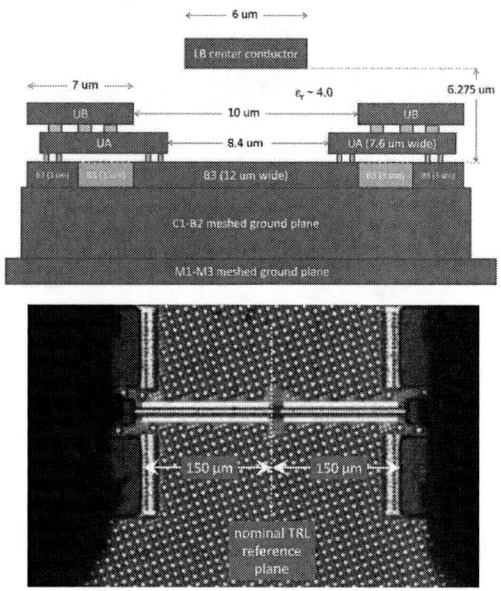

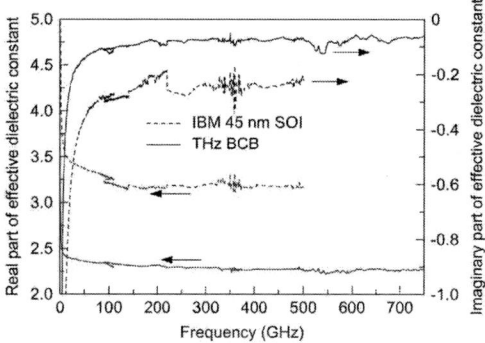

Fig. 7: Cross-section of the transmission-line (top) and test structure (bottom) implemented in IBM 45-nm complementary metal-oxide-semiconductor (CMOS) silicon-on-insulator SOI12S0 integrated-circuit process [21].

Fig. 8: Comparison of measured effective dielectric constant of the line standard implemented in Teledyne BCB and IBM's SOI process [21]).

C. Correction for Crosstalk and Coupling

With increase of measurement frequency to the sub-THz range, the impact of parasitic effects described above can significantly contribute to the calibration residual errors. Then, the conventional approach of modeling the systematic measurement errors of a wafer probe measurement system by twelve error terms becomes insufficient. A comprehensive study of two different approaches, the 16-term and the two-tier

interior crosstalk models was reported recently in [22]. It was demonstrated that correction for the parasitic coupling and crosstalk errors become relevant already at upper W-band frequencies (Fig. 9). Several useful recommendations on optimization of on-wafer standards were given there, such as minimizing of the signal-conductor width and the gap to the ground conductors for CPW standard design and keeping the access lines short.

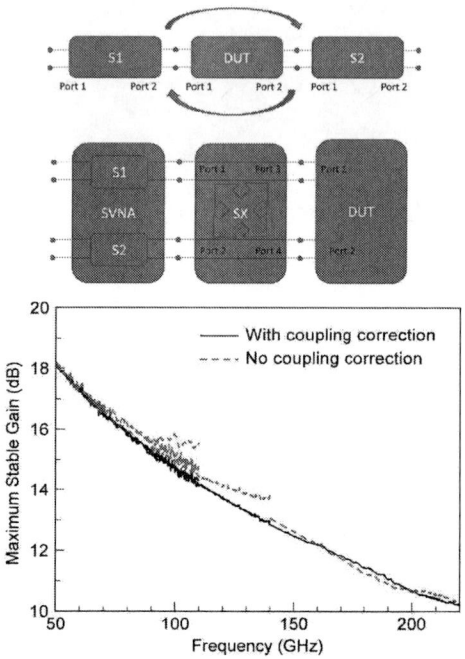

Fig. 9: Two-tier interior crosstalk model (top) and the maximum stable gain of a transistor corrected with standard TRL and with TRL augmented with coupling corrections (bottom) [22].

IV. MM-WAVE DEVICE ON-WAFER CHARACTERIZATION

A. Interconnects and Transmission Lines

Mm-wave and sub-mm-wave MMIC require low-loss interconnects and low-loss transmission lines. As indicated in the previous section, thin-film and inverted microstrip lines have demonstrated suitability for MMIC applications up to THz frequencies, due to their advantageous characteristics [21].

Thin-film microstrip lines require, however, an appropriate dielectric stack with the according interconnects. Fig. 10 shows a comparison of simulated and measured transmission coefficient up to 220 GHz for a thin-film microstrip line in BiCMOS, with interconnect transitions to a BCB thin-film microstrip line and pads in InP technology. These results confirm the broadband capabilities of low-loss interconnect transitions employed in thin-film transmission line technology. In fact, the losses of each transition from the BCB to the BiCMOS stack are lower than 0.5 dB at 220 GHz. Such low-loss transition structures can be employed in heterogeneous integration, as described below or for chip-to-chip transitions. Measurement results for similar lines are presented in Fig. 11 for different line lengths. These results demonstrate the

978-1-4799-7231-9/14 $31.00 © 2014 IEEE   166

usefulness of such lines and interconnects for calibration and circuit purposes.

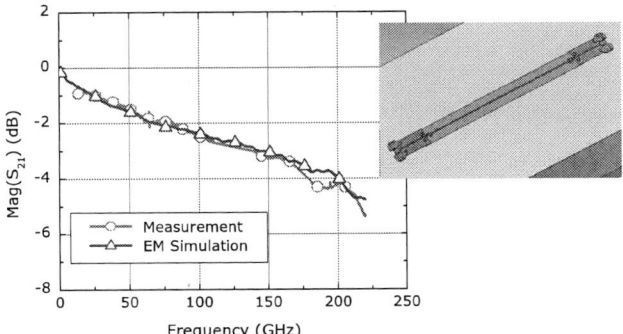

Fig. 10: Comparison of simulated and measured transmission losses of a microstrip line on BiCMOS probed through a microstrip line on InP heterogeneously integrated on top of the BiCMOS wafer. The results include line losses and losses due to two transitions between the two technologies [8].

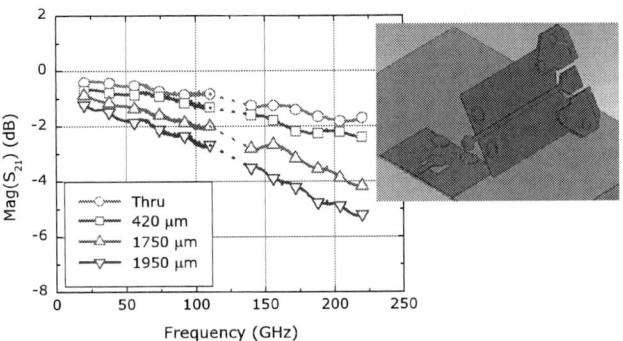

Fig. 11: Comparison of simulated and measured transmission losses of a microstrip line on BiCMOS probed through a microstrip line on InP heterogeneously integrated on top of the BiCMOS wafer. The results include line losses and losses due to two transitions between the two technologies.

Thin-film transmission lines exhibit also a relatively small phase constant as can be seen in Fig. 12 with a respective attenuation constant of around 1 dB/mm at 200 GHz.

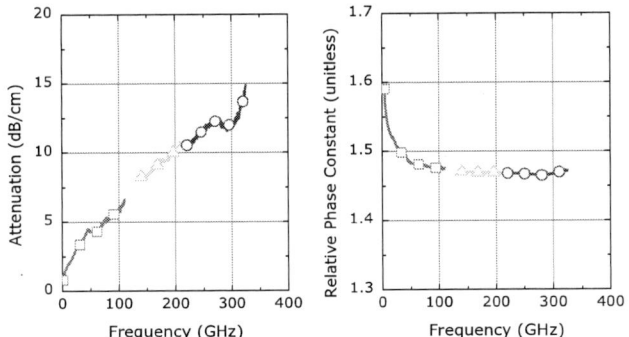

Fig. 12: Measured attenuation constant (left) and relative phase constant (right) frequency characteristics of a thin-film microstrip line in BCB.

B. Transistor Devices and MMIC

Wafer-level device characterization technology made a substantial progress in the last decade: RF probe frequency

capabilities increased exponentially, new probe technologies came to place and the list of probe manufactures expanded [23]. Recently, micromachined probes designed to facilitate the development of TMICs in the 750 GHz to 1.1 THz frequency range have been introduced. This probe demonstrated an insertion loss of less than 7 dB and a return loss of greater than 15 dB, as can be observed in Fig. 13 [24].

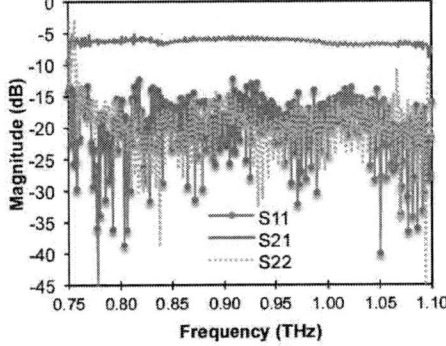

Fig. 13: Measured S-parameters of the 1.1 THz wafer probe [24].

While small-signal measurement capabilities of modern systems have already achieved the THz frequency range, the more sophisticated wafer-level large-signal and noise measurements still remain a great challenge. Load-pull and RF noise measurement systems for the W-band are seldom available. The traditional load-pull techniques that require integration of the electro-mechanical impedance tuners on the wafer probe system suffer from high losses (e.g. larger than 2 dB at W-band) and other limitations at millimeter-wave frequencies and beyond. As a result, the maximum reachable reflection coefficient of such systems is lower than 0.6.

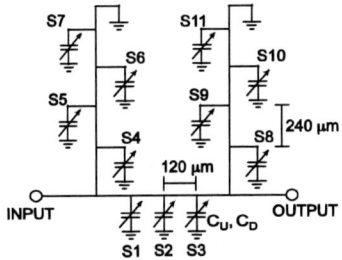

Fig. 14: The schematic diagram (top) and the photograph (bottom) of the W-band double-stub MEMS switch impedance tuner [26].

A novel concept of an active real-time load-pull W-band system with frequency up/down conversion technique was reported in [25]. The authors claimed to reach load reflection coefficients as high as 0.95 at 94 GHz. Nearly full coverage of

978-1-4799-7231-9/14 $31.00 © 2014 IEEE   167

the Smith chart enables verification of the large-signal device models and accurate characterization of microwave integrated circuits the sub-THz frequency range. An alternative approach to minimize the setup losses is to integrate a reconfigurable impedance tuner into the wafer probe [26]. The schematic circuit and a micrograph are shown in Fig. 14. Realizing double and triple switching topologies and using MEMS switches, maximum load reflection coefficient lower than 0.92 and 0.82 were achieved for 75 GHz and 100 GHz frequencies, respectively.

A further step toward reducing the setup losses is the integration of the tuner on-wafer, e.g. the *in-situ* impedance tuning. In [27] the design of an *in-situ* impedance tuner and measurement results of the device noise parameters obtained at D-band have been presented. As it was originally discussed in [28], the tuner is based on 50 Ω transmission lines and digital tunable capacitances (DTC) connected in parallel. It can provide 32 different impedance points, optimized for characterization of a particular device, as illustrated in Fig. 15. As it is claimed in [27] the developed *in-situ* impedance tuning technique is an efficient solution for extracting the device noise parameters in the entire D-band range.

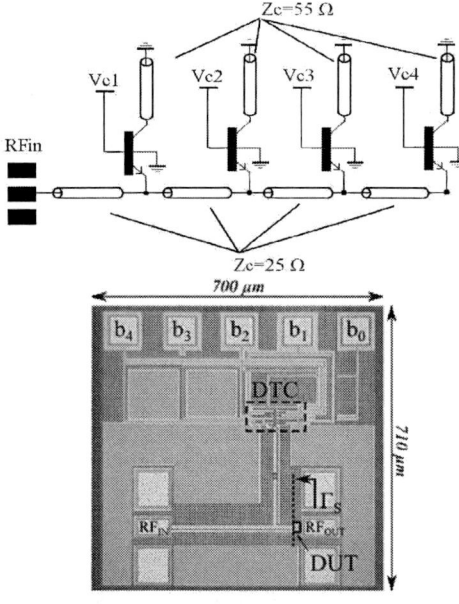

Fig. 15: The schematc of the DTC (top) and a photograph of the D-band source-pull *in-situ* impedance tunners (bottom). Pictures from [27], [28].

C. Large-Signal Device Characterization

The small-signal equivalent circuit parameters for transistor devices in two-terminal and three-terminal configurations are determined by employing a direct parameter extraction methodology. The accuracy of this extraction relies entirely on the accuracy of the on-wafer calibration and on the accuracy of the reference plane shift towards the intrinsic device terminals. Assuming that the calibration issues described above have been avoided and a suitable calibration procedure has been applied allows for direct parameter extraction using multi-bias S-parameter transistor measurements. As an example, results for InP III-V

based HBTs are presented here based on [29]. After transferring the elements found from small-signal extraction to the large-signal model, the parameters affecting the DC and thermal characteristics of the device are determined. The large-signal model can accurately predict large-signal performance including device scaling, as illustrated in Fig. 16.

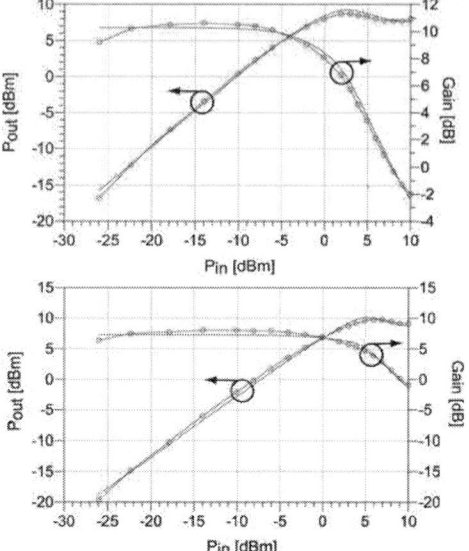

Fig. 16: Measured (solid line w. symbols) and simulated (solid line) large-signal performance at 77 GHz for single-finger device biased at $V_{ce} = 1.4$ V with a quiescent current of $I_{cq} = 22.7$ mA (top) and two-finger device biased at $V_{ce} = 1.4$ V with a quiescent current of $I_{cq} = 29.6$ mA (bottom) [29].

These results demonstrate a successful sequence of steps comprised of on-wafer system calibration, low-loss interconnect transitions and single mode transmission lines, reference plane shift, small-signal model parameter extraction and finally large-signal parameter extraction. This procedure is subsequently applied to the characterization of devices in the heterogeneous integration technology described below.

V. HETEROGENEOUS MMIC DEVICE TESTING

Wafer-level integration of heterogeneous technologies has recently attracted a lot of attention, due to technological developments in several large-scale projects, such as e.g. COSMOS [5], [6], or SciFab [7]–[10]. The major driving force is to provide high-speed III-V semiconductor devices on Si platforms, providing not only speed improvement without trade-off in signal amplitude, but also functional complexity to high-speed circuits.

Such technology developments put high demands on device characterization. Fig. 17 illustrates schematically the cross-section of the InP-on-BiCMOS process. One challenge is the device characterization of devices after heterogeneous wafer-level integration. The characterization of the impact of such a technology on device performance is facilitated by introducing pad and interconnect structures, which interconnect the buried BiCMOS devices with top-level pads on InP.

The impact of the additional layers on transistor performance in BiCMOS and InP technologies, respectively,

978-1-4799-7231-9/14 $31.00 © 2014 IEEE

are shown in Fig. 18 and Fig. 19. Fig. 18 shows the extrapolated transit and maximum oscillation frequencies, respectively, from measurements of BiCMOS transistor devices on completely processed InP-on-BiCMOS wafers. The various curves represent values for devices across a 3-inch wafer area. The horizontal lines in Fig. 18 indicate the maximum values for the transit and maximum oscillation frequencies for devices before heterogeneous integration. These results employ low-loss interconnect via technology. It can be concluded from the figure that the impact of the heterogeneous integration can be estimated to cause a decrease in both values of only around 5 GHz and is not limited by the interconnects.

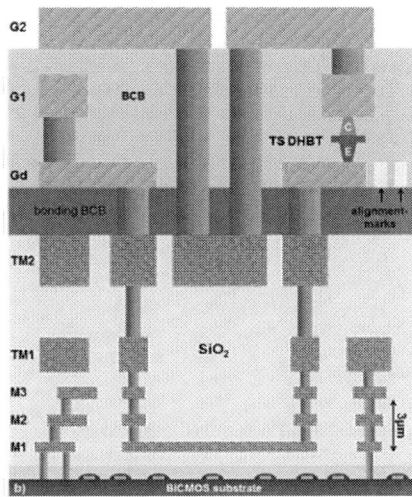

Fig. 17: Cross-section of the InP-on-BiCMOS substrate transfer process. The interconnects between TM2 with G2 and Gd with M1, respectively, form a thin-film microstrip interconnection. Top-level pads are on metal level G2 [9].

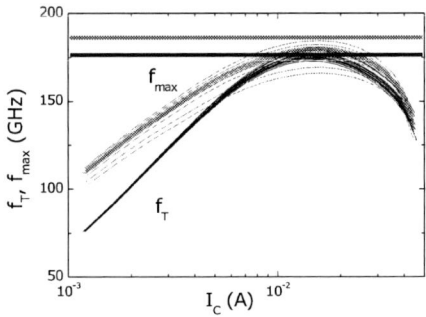

Fig. 18: Interpolated transit frequency f_T and maximum oscillator frequency f_{max} as a function of collector current I_c measured at 30 GHz and an emitter-collector voltage $V_{ce} = 1.5$ V for eight NPN HBTs in parallel. Horizontal lines indicate the respective values before heterogeneous integration [9].

It has been shown in [10] that device small-signal and large-signal performance of the BiCMOS and InP HBT devices, respectively, remains essentially unaffected by the hetero-integration InP-on-BiCMOS process. Small-signal measurements provided in Fig. 19 indicate that there is only a minor impact by fillers and thermal vias in SiGe BiCMOS on InP DHBT performance. This is most probably due to the fact

that the filler structures, as well as the thermal via structures are small and non-interconnected, which has been confirmed by 3D EM simulations.

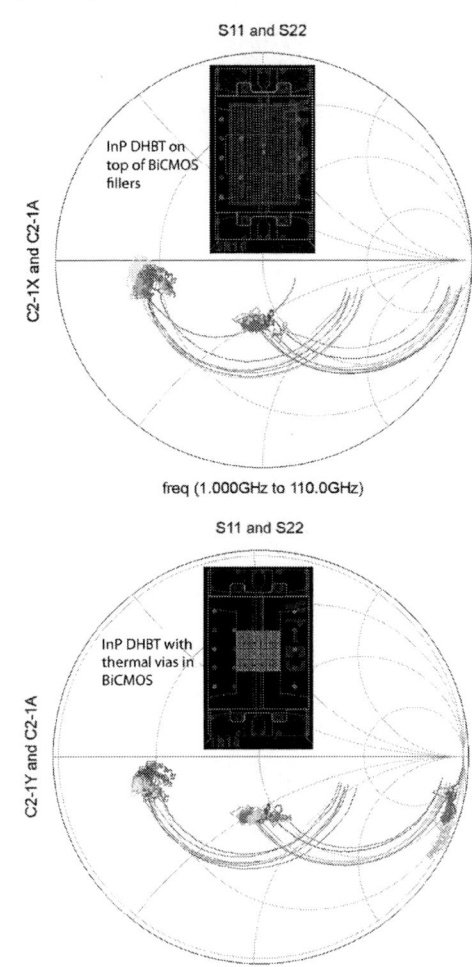

Fig. 19: Small-signal input and output S-parameters of single-finger InP DHBT in a transferred-substrate (TS) process and the same device with fillers in the InP-on-BiCMOS process (top) and implementing a heat-sink underneath the transistor in SiGe (bottom). Measurements are in the range 1 GHz–110 GHz. Cyan and blue lines are results for the standard TS process, red and magenta are results for InP HBT transistors in InP-on-BiCMOS process. Insets show the transistor layout for the two cases, respectively [10].

These results are important for circuit integration in both technologies, which requires dissipation of thermal energy from the top-level InP DHBT devices operating under large-signal conditions.

VI. SUMMARY

It is shown that mm-wave and sub-mm-wave on-wafer characterization presents a challenge from calibration, on-wafer measurement, measurement system and circuit design point of view. The methodologies to overcome these challenges include modified calibration standards avoiding probe coupling and coupling between structures in on-wafer calibration substrates.

These also include the suppression of multi-mode propagation using thin-film transmission lines, as well as appropriate transitions between the levels. The suggestions discussed in the paper lead to improved calibration standards suitable for TRL and LRM calibration methodologies.

It is further shown, that successful attempts exist for small-signal characterization of devices and MMICs up to sub-mm-wave frequencies using thin-film transmission lines. The impact of standard calibration procedures on device characterization is discussed. It is shown that accurate device characterization is important for the verification of the extracted models. Finally, characterization results for heterogeneously integrated devices and MMIC are presented.

ACKNOWLEDGMENT

Viktor Krozer thanks Oerlikon AG for partial financial support.

REFERENCES

[1] E. M. Godshalk, "Surface wave phenomenon in wafer probing environments," in *40th ARFTG Microwave Measurement Conf. Dig.*, Orlando, FL, USA, Fall 1992, pp. 10–19.

[2] A. Lewandowski and W. Wiatr, "Errors in on-wafer measurements due to multimode propagation," in *Proc. 15th Int. Conf. Microwave, Radar and Wireless Communication (MIKON-2004)*, Warsaw, Poland, 2004, pp. 759–763.

[3] F.-J. Schmückle, R. Doerner, G. N. Phung, *et al*, "Radiation, Multimode Propagation, and Substrate Modes in W-Band CPW Calibrations," in *Proc. 41st European Microwave Conf. (EuMC)*, Manchester, U-K., 2011, pp. 297–300.

[4] G. N. Phung, F.-J. Schmückle, W. Heinrich, "Parasitic Effects and Measurement Uncertainties in Multi-Layer Thin-Film Structures," in *Proc. 43rd European Microwave Conf. (EuMC)*, Nuremberg, Germany, 6-11 Oct. 2013, pp. 318–321.

[5] W. K. Liu, D. Lubyshev, J. M. Fastenau, *et al.*, "Monolithic integration of InP-based transistors on Si substrates using MBE," *J. Crystal Growth*, vol. 311, no. 7, pp. 1979–1983, March 2009.

[6] A. Gutierrez-Aitken, P. Chang-Chien, D. Scott, *et al.*, "Advanced Heterogeneous Integration of InP HBT and CMOS Si Technologies," in *Proc. IEEE Compound Semiconductor Integrated Circuit Symp. (CSICS)*, Monterey, CA, USA, 3-6 Oct. 2010, pp. 1–4.

[7] T. Jensen, T. Al-Sawaf, M. Lisker, *et al.*, "A 164 GHz Hetero-Integrated Source in InP-on-BiCMOS Technology," in *Proc. 8th European Microwave Integrated Circuits Conf. (EuMIC)*, Nuremberg, Germany, 6-8 Oct. 2013, pp. 244–247.

[8] T. Kraemer, I. Ostermay, T. Jensen, *et al.*, "InP-SiGe BiCMOS Technology with fT/fmax of 400/350 GHz for Heterogeneous Integrated Millimeter-Wave Sources," *IEEE Trans. Electron Devices*, vol. 60, no. 7, pp. 2209–2216, July 2013.

[9] M. Lisker, A. Trusch, A. Krueger, *et al.*, "InP-Si BiCMOS Hetero-integration Using a Substrate Transfer Process," *ECS J. Solid State Sci. Technol.*, vol. 3, no. 2, pp. 17–20, 2014.

[10] T. Jensen, T. Al-Sawaf, M. Lisker, *et al.*, "Millimeter-wave hetero-integrated sources in InP-on-BiCMOS technology," *Int. J. Microwave and Wireless Technologies*, vol. 6, no. 3-4, pp. 225–233, June 2014.

[11] A. Rumiantsev, P. Sakalas, N. Derrier, *et al.*, "Influence of probe tip calibration on measurement accuracy of small-signal parameters of advanced BiCMOS HBTs," in *Proc. IEEE Bipolar/BiCMOS Circuits and Technology Meeting (BCTM)*, Atlanta, GA, USA, 2011.

[12] F. Pourchon, C. Raya, N. Derrier, *et al.*, "From measurement to intrinsic device characteristics: Test structures and parasitic determination," in *Proc. IEEE Bipolar/BiCMOS Circuits and Technology Meeting (BCTM)*, Monterey, CA, USA, 2008, pp. 232–239.

[13] N. Derrier, A. Rumiantsev, and D. Celi, "State-of-the-art and future perspectives in calibration and de-embedding techniques for characterization of advanced SiGe HBTs featuring sub-THz fT/fMAX," in *Proc. IEEE Bipolar/BiCMOS Circuits and Technology Meeting (BCTM)*, Portland, OR, USA, 2012, pp. 92–99.

[14] D. F. Williams, P. Corson, J. Sharma, *et al.*, "Calibrations for millimeter-wave silicon transistor characterization," *IEEE Trans. Microwave Theory Tech.*, vol. 62, pp. 658–668, March 2014.

[15] D. F. Williams, A. C. Young, and M. Urteaga, "A prescription for sub-millimeter-wave transistor characterization," *IEEE Trans. Terahertz Science and Technology*, vol. 3, no. 4, pp. 433–439, 2013.

[16] R. B. Marks, "A multiline method of network analyzer calibration," *IEEE Trans. Microwave Theory Tech.*, vol. 39, pp. 1205–1215, 1991.

[17] H. J. Eul and B. Schiek, "A generalized theory and new calibration procedures for network analyzer self-calibration," *IEEE Trans. Microwave Theory Tech.*, vol. 39, pp. 724–731, 1991.

[18] D. Williams, "Traveling waves and power waves: Building a solid foundation for microwave circuit theory," *IEEE Microwave Mag.*, vol. 14, no. 7, pp. 38–45, Nov. 2013.

[19] A. Rumiantsev, "On-wafer calibration techniques enabling accurate characterization of high-performance silicon devices at the mm-wave range and beyond," Fakultät für Maschinenbau, Elektrotechnik und Wirtschaftsingenieurwesen, BTU Cottbus, Cottbus, 2014.

[20] A. Rumiantsev, F. Pourchon, N. Derrier, *et al.*, "Designing on-wafer calibration standards for advanced high-speed BiCMOS technology," in *Proc. 41st European Microwave Integrated Circuits Conf. (EuMIC)*, Manchester, U-K., 2011, pp. 57–60.

[21] D. F. Williams, P. Corson, J. Sharma, *et al.*, "Calibration-kit design for millimeter-wave silicon integrated circuits," *IEEE Trans. Microwave Theory Tech.*, vol. 61, no. 7, pp. 2685–2694, July 2013.

[22] D. F. Williams, F.-J. Schmückle, R. Doerner, *et al.*, "Crosstalk corrections for coplanar-waveguide scattering-parameter calibrations," *IEEE Trans. Microwave Theory Tech.*, vol. 62, no. 8, pp. 1748–1761, Aug. 2014.

[23] A. Rumiantsev and R. Doerner, "History and selected topics of on-wafer S-parameter measurements at mm-wave frequencies," *IEEE Microwave Mag.*, vol. 14, no. 7, pp. 46–58, Nov. 2013.

[24] M. F. Bauwens, N. Alijabbari, A. W. Lichtenberger, *et al.*, "A 1.1 THz micromachined on-wafer probe," in *IEEE MTT-S Int. Microwave Symp. (IMS) Dig.*, Tampa, FL, USA, 2014, pp. 1–4.

[25] V. Teppati, H. Benedickter, D. Marti, *et al.*, "A W-band on-wafer active load-pull system based on down-conversion techniques," *IEEE Trans. Microwave Theory Tech.*, vol. 62, no. 1, pp. 148–153, Jan. 2014.

[26] T. Vähä-Heikkilä, J. Varis, J. Tuovinen, *et al.*, "W-band RF MEMS double and triple-stub impedance tuners," in *IEEE MTT-S Int. Microwave Symp. (IMS) Dig.*, Long Beach, CA, USA, 2005, pp. 4.

[27] M. Deng, L. Poulain, D. Gloria, *et al.*, "Millimeter-wave in situ tuner: An efficient solution to extract the noise parameters of SiGe HBTs in the whole 130-170 GHz range," *IEEE Microwave and Wireless Components Letters*, vol. PP, 2014.

[28] T. Quemerais, D. Gloria, S. Jan, *et al.*, "Millimeter-wave characterization of Si/SiGe HBTs noise parameters featuring fT/fMAX of 310/400 GHz," in *Proc. IEEE Radio Frequency Integrated Circuits Symposium (RFIC)*, 2012, pp. 351–354.

[29] T. K. Johansen, M. Rudolph, T. Jensen, *et al.*, "Small- and Large-Signal Modeling of InP HBTs in Transferred-Substrate Technology," *Int. J. Microwave and Wireless Technologies*, vol. 6, no. 3-4, pp- 243–251, June 2014.

Temperature Impact on the *In-Situ S*-Parameter Calibration in Advanced SiGe Technologies

Andrej Rumiantsev [1), 2)]
[1)] MPI Corporation,
Chu-Pei City, Hsinchu County, Taiwan
[2)] Brandenburg University of Technology (BTU),
Cottbus-Senftenberg, Cottbus, Germany
a.rumiantsev@ieee.org

Ralf Doerner
Ferdinand-Braun-Institut (FBH),
Leibniz-Institut fuer Hoechstfrequenztechnik,
Berlin, Germany
ralf.doerner@fbh-berlin.de

Falk Korndoerfer
Innovations for High Performance Microelectronics (IHP)
Leibniz-Institut fuer innovative Mikroelektronik,
Frankfurt (Oder), Germany
korndoerfer@ihp-microelectronics.com

Abstract—This paper analyzes the *in-situ S*-parameter multiline TRL and the transfer TMR calibration methods for the sensitivity to the thermal variation of electrical characteristics of calibration standards. The standards were realized in IHP's SG13 130 nm SiGe:C BiCMOS process. The measurement experiment was performed for the frequency range up to 110 GHz. We demonstrate that the calibration error caused by thermal instability of electrical characteristic of standards is in order of magnitude of the system drift error and, thus, negligible.

Keywords—S-parameter calibration, device characterization, mm-wave measurements, BiCMOS.

I. INTRODUCTION

Increasing demand for more content, higher data transfer rates and cost reduction pushes the development of advanced SiGe technologies operating at mm-wave and sub-THz ranges. Optimization of device models over extremely wide frequency ranges and across multiple temperatures as well as verification of the results with highest level of accuracy and confidence becomes more and more challenging. Therefore, accurate and consistent *in-situ S*-parameter calibration procedures covering wide frequency and temperature ranges are the critical success factor for development of advanced SiGe technologies.

The state-of-the-art *in-situ S*-parameter calibration techniques for Silicon processes were reviewed in [1]. It recommended implementation of two comparable calibration concepts: 1) the probe-tip calibration performed on the commercially available Alumina substrate, followed by the advanced six-step de-embedding procedure, and 2) straightforward *in-situ* multiline Thru-Reflect-Line (TRL) or transfer Thru-Match-Reflect (TMR, also known as LRM+). However, [1] presented results for the room temperature only. Possible variation of electrical characteristics of calibration standards with the change of the measurement temperature,

and, as a result, the accuracy of the over-temperature calibration remained unclear.

Later, the work in [2] addressed this topic for the first calibration concept. It proposed a method for accuracy analysis of the over-temperature probe-tip calibration. It was demonstrated that the temperature variation has a marginal impact on the electrical characteristic of the evaluated commercially available Alumina calibration standards. In this paper, we present the investigation results for the second calibration concept applying custom on-wafer standards and the multiline TRL and the transfer TMR. The objective of this work is to identify standards parameters that are most sensitive to the temperature variation as well as to propose a practical method for compensation of possible calibration error.

II. ANALYSIS METHOD

Multiline TRL and TMR calibration methods differently define calibration reference impedance Z_{REF}. For the multiline TRL, the reference impedance is set to the characteristic impedance Z_{LINE} of the line standard. Assuming that the conductive loss of the microstrip-designed transmission line is negligible in the frequency range of interest, we calculated Z_{LINE} from the measured capacitance per unit length C' and propagation constant γ, as proposed in [3, 4]. Once Z_{LINE} is known, the calibration reference impedance can be transformed to the desired value of $Z_{REF} = 50\,\Omega$.

The Z_{REF} for TMR depends on the impedance of the match (load) as well as γ and Z_{LINE} of the thru. Therefore, multiline TRL and the transfer TMR may show different calibration residual error across the temperature range.

As it was already proposed in [2], the maximum error bounds for measured *S*-parameters of a passive device are a reliable figure of merit (FoM) of the calibration accuracy, calculated by the calibration comparison technique at given temperature. The benchmark calibration conditions were established by using pre-characterized temperature-dependent electrical properties of standards. The test calibration was performed at the test temperature, but using electrical models of standards extracted at the room temperature. Thus, the maximum error bounds can be calculated for each temperature point and calibration method used.

III. EXPERIMENTAL SETUP

A. Calibration Standards

The test structures were realized in IHP's SG13 process. It is a 130 nm SiGe:C BiCMOS process with 5 thin and 2 thick aluminum metallization layers [5]. The top metallization layers have 2 μm and 3 μm thickness, respectively. Two types of bipolar transistors are available in the process. The high speed HBTs feature f_T/f_{MAX} of 240 GHz/330 GHz at a breakdown voltage BV_{CEO} of 1.7 V. The high voltage HBTs BV_{CEO} of 3.7 V with cut-off frequencies f_T/f_{MAX} of 50 GHz/130 GHz.

We designed test structures consisting of open, short, load, thru, and 10 microstrip lines (Fig. 1). Their lengths vary from 0.4 mm to 15.1 mm. The length ratio between two lines is always non-integer. We decided for microstrip lines to circumvent influences of substrate losses in the silicon wafer.

The signal line is designed in the topmost metallization layer (TM2) and the ground plane in the lowest (M1). The signal line is 15 μm wide. The ground plane has a width of 90 μm. The insulator between signal line and ground plane is 9.8 μm thick with a dielectric constant of $\varepsilon_r = 4.1$. The expected line impedance is about 50 Ω with that dimensions.

All structures are placed in GSG pads with 100 μm pitch. We added 65 μm feeding lines between contact pad and lines to allow a homogenization of the fields.

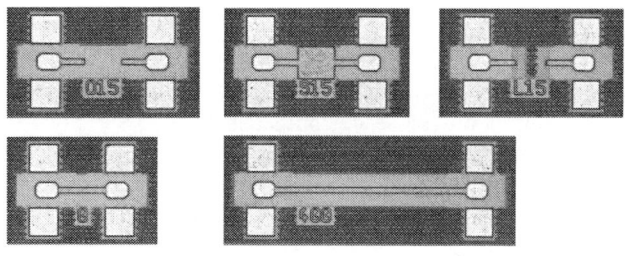

Fig. 1. Selected layouts of the designed calibration standards (from left to right: open, short, load, thru, and line).

B. Measurement Setup

The experimental setup included a semi-automatic thermal probe system PA200 and Agilent PNA 67 GHz vector network analyzer (VNA). Experiments were conducted for three temperature points: 233 K, 300 K, and 398 K covering typical device modeling temperature range. All measured data of calibration standards and the device under test were acquired uncorrected, in one series, and saved for further analysis. Thus, uncertainty caused by contact repeatability is minimized.

IV. RESULTS AND DISCUSSION

First, the propagation constant γ (Fig. 2) and the capacitance per unit length C' were determined using the multiline TRL method for every temperature point. The extracted characteristic impedance of the line Z_{LINE} (Fig. 3), as well as other parameters of the calibration standards are given in the Table I.

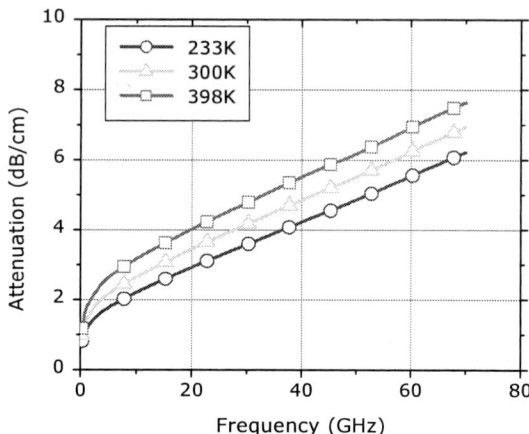

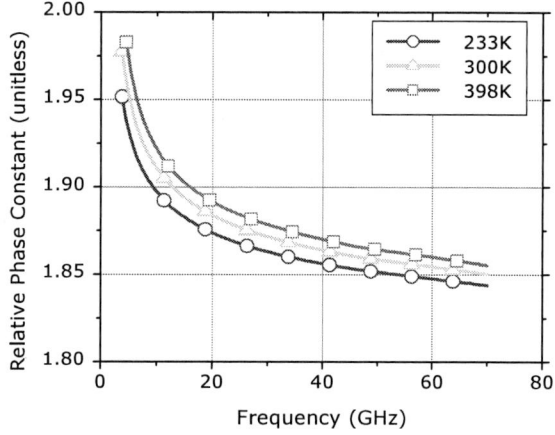

Fig. 2. Attenuation constant (top) and the relative phase constant (bottom) of the line standard measured across the temperature.

With the measurement temperature variation from 233 K to 398 K, the maximum variation of key standard parameters compared to room temperature (300 K) is: 1.8% for the load resistance, 33% for the M1 sheet resistance, 1.4% for the line capacitance per unit length C', 14% for the attenuation constant α, and is marginal (0.4%) for the relative phase constant β/β_0. As a result, the maximum variation of the characteristic impedance Z_{LINE} is 1.7% for its real and 15% for its imaginary part respectively (for 40 GHz frequency). We attributed measured variations in the electrical parameters of calibration standards to the temperature impact on the sheet resistance and the permittivity of dielectric. Once $Z_{LINE}(T)$, $R_{LOAD}(T)$ and $\gamma(T)$ are defined, the impact of each of them on the calibration accuracy of the multiline TRL and the transfer TMR can be estimated.

978-1-4799-7231-9/14 $31.00 © 2014 IEEE

TABLE I. STANDARD PARAMETERS VARIATION OVER TEMPERATURE

Parameter	Temperature, K		
	233	*300*	*398*
α @ 40 GHz, dB/cm	4.22	4.85	5.53
β/β_0	1.856	1.864	1.870
C', pF/cm	1.248	1.241	1.224
$\Re(Z_{LINE})$ @ 40 GHz, Ω	49.61	50.10	50.96
$\Im(Z_{LINE})$ @ 40 GHz, Ω	-1.55	-1.79	-2.06
$R_{LOAD,P1}$, Ω	51.64	50.77	49.99
$R_{LOAD,P2}$ Ω	51.33	50.43	49.76
M1 sheet resistance, $\Omega/\square$	0.0835	0.1086	0.1439
TM2 sheet resistance, $\Omega/\square$	0.0082	0.0111	0.0151

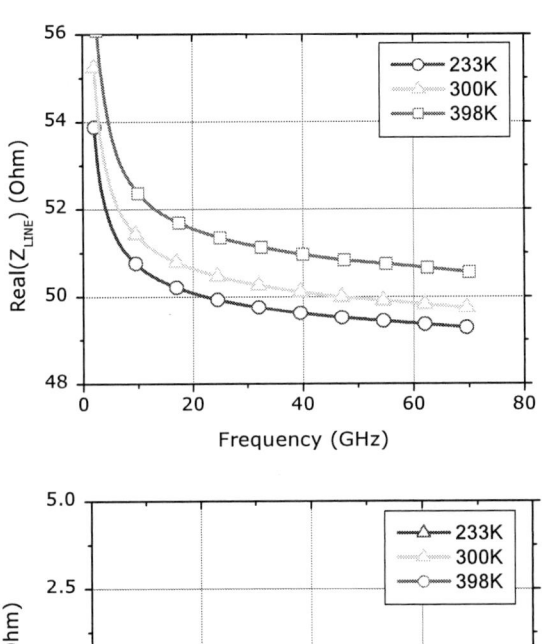

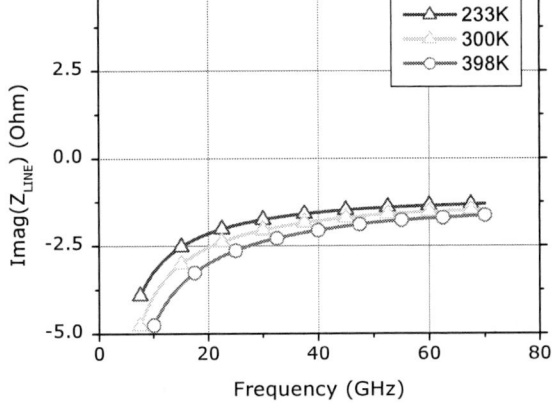

Fig. 3. Real (top) and imaginary (bottom) parts of the characteristic impedance Z_{LINE} of the line standard measured across the temperature.

We calculated the maximum error bounds for both calibration methods for simplified ($T = 300\,K$) and characterized ($T = T_{TEST}$) models of calibration standards for every temperature point T_{TEST} using the method from [2]. The maximum error bounds are 0.02 for the multiline TRL for $T_{TEST} = 398\,K$ and 0.027 for the transfer TMR for $T_{TEST} = 233\,K$ at 67 GHz, respectively. It is lower than the

typical drift of the on-wafer setup and, thus, negligible (Fig. 4). The setup drift was obtained from a separate experiment. Therefore, the extensive characterization of the electrical properties of both the lumped and the distributed calibration standards is not required. This founding significantly simplifies implementation of the over-temperature *in-situ* calibration for advanced SiGe technologies.

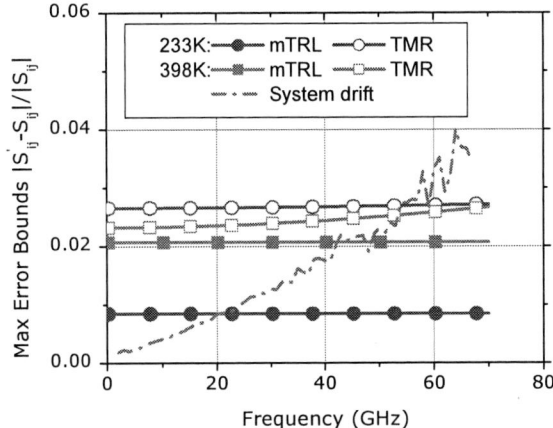

Fig. 4. Maximum error bounds calculated for multiline TRL, and transfer TMR for $T_{TEST} = 233\,K$ and $T_{TEST} = 398\,K$ with benchmark and worst case definition of standard properties. Typical experimental drift of a comparable setup is added for reference.

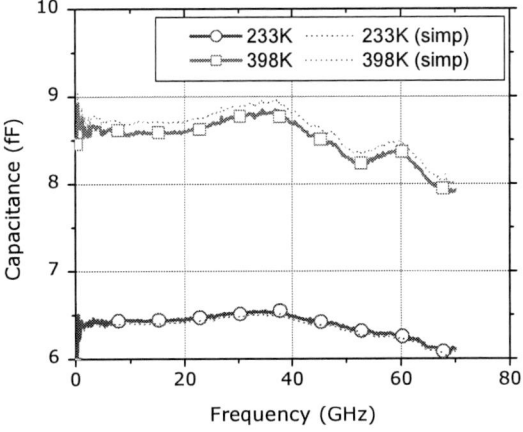

Fig. 5. C_{BE} of a test DUT extracted from the cold S-parameter conditions for the benchmark and simplified multiline TRL at $T_{TEST} = 233\,K$ and $T_{TEST} = 398\,K$. The results extracted for the transfer TMR are similar.

The calibration residual error of the simplified multiline TRL stems from the error of the characteristic impedance of the line ΔZ_{LINE} that we found to be frequency independent. Thus, we observed constant maximum error bound $|S'_{ij} - S_{ij}|/|S_{ij}|$. The transfer TRM residual errors are the sum of errors of the characteristic impedance ΔZ_{LINE} and the propagation constant $\Delta\gamma$ of the line, as well as of the impedance of the load ΔZ_{LOAD}. The individual impact of each of the errors depends on the frequency range: The error in the load resistance and the line characteristic impedance contribute the most at lower frequencies, while the impact of errors in the load reactance and line propagation constant predominate at higher frequencies.

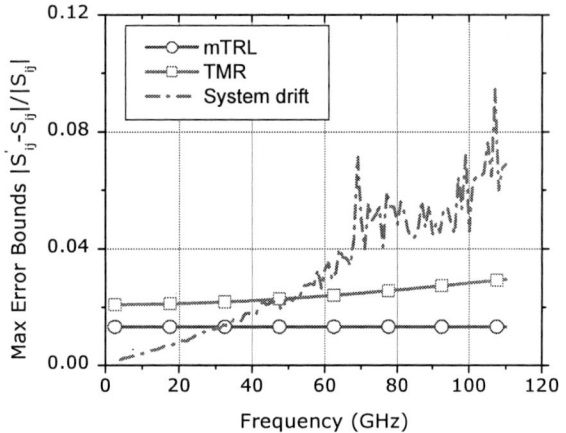

Fig. 6. Maximum error bounds calculated for multiline TRL, and transfer TMR for the 110 GHz setup and for $T_{TEST} = 358\,K$ with benchmark and worst case definition of standard properties. Typical experimental drift of a comparable setup is added for reference. Measurement data were obtained on the second experimental system.

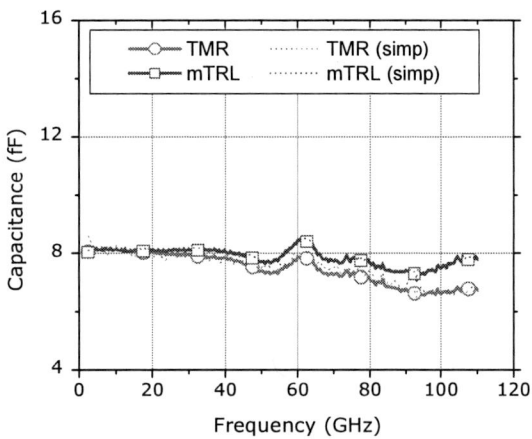

Fig. 7. C_{BE} of a test DUT extracted from the cold S-parameter conditions for the benchmark and simplified multiline transfer TMR at $T_{TEST} = 358\,K$. Measurement data were obtained on the second experimental system.

Finally, the capacitance C_{BE} of a test HBT at $T_{TEST} = 233\,K$ and $T_{TEST} = 398\,K$ was extracted from the cold S-parameters corrected with the benchmark and the simplified calibrations (Fig. 5). In both cases, the difference between the extracted value from the benchmark and the simplified calibration is negligible. Fig. 5 shows the results for the multiline TRL calibration. The results for the transfer TMR calibration are similar.

We repeated the experiment for the same device on a 110 GHz system, which allowed a temperature variation from 300 K to 358 K. The obtained results are, in general, comparable with those for the 67 GHz probe system (Fig. 6, 7). However, we observed a minor deviation of the extracted C_{BE} from the expected value above 60 GHz for the TMR calibration (up to 13% at 110 GHz). This error can be attributed to insufficiencies of the load standard model. It can be decreased by an appropriate description of the load impedance [6] what was out of the scope of this experiment.

V. CONCLUSION

For the first time, sensitivity analysis of the *in-situ* over-temperature calibration was performed for the multiline TRL and the transfer TMR implemented in the advanced SiGe BiCMOS process up to 110 GHz. The obtained results demonstrated that temperature variations of the electrical characteristics of load and line standards lead to different calibration residual errors depending on the calibration method used.

For the considered experimental setup, we found that the multiline TRL calibration method was less sensitive to temperature. However, the worst-case error bound for both methods were three times less than the typical drift of the measurement setup. Therefore, complicated and time-consuming experiments for characterization of the temperature coefficients of custom calibration standards are not required. Extracted C_{BE} of a test HBT proved this statement. Results of this investigation significantly simplify implementation of the advanced *in-situ* calibration methods into the conventional characterization workflow of advanced SiGe devices.

REFERENCES

[1] N. Derrier, A. Rumiantsev, and D. Celi, "State-of-the-art and future perspectives in calibration and de-embedding techniques for characterization of advanced SiGe HBTs featuring sub-THz fT/fMAX," in *Bipolar/BiCMOS Circuits and Technology Meeting (BCTM), 2012. IEEE*, Portland, OR, 2012, pp. 92-99.

[2] A. Rumiantsev, G. Fisher, and R. Doerner, "Sensitivity analysis of wafer-level over-temperature RF calibration," in *Microwave Measurement Symposium (ARFTG)-Fall, 80th*, San Diego, CA, USA, 2012, pp. 1-3.

[3] R. B. Marks and D. F. Williams, "Characteristic impedance determination using propagation constant measurement," *IEEE Microwave and Guided Wave Letters*, vol. 1, pp. 141-143, June 1991.

[4] D. F. Williams and R. B. Marks, "Transmission line capacitance measurement," *Microwave and Guided Wave Letters, IEEE*, vol. 1, pp. 243-245, 1991.

[5] H. Rucker, B. Heinemann, W. Winkler, *et al.*, "A 0.13 um SiGe BiCMOS technology featuring ft/ fmax of 240/330 GHz and gate delays below 3 ps," *Solid-State Circuits, IEEE Journal of*, vol. 45, pp. 1678-1686, 2010.

[6] A. Rumiantsev, "On-Wafer calibration techniques enabling accurate characterization of high-performance silicon devices at the mm-wave range and beyond," Fakultät für Maschinenbau, Elektrotechnik und Wirtschaftsingenieurwesen, BTU Cottbus, Cottbus, 2014.

A Simple and Accurate Method for Extracting the Emitter and Thermal Resistance of BJTs and HBTs

Andreas Pawlak[1], Steffen Lehmann[1], Michael Schroter[1],[2]

[1]Technische Universität Dresden, Germany
[2]University of California San Diego, USA

Abstract — **A simple yet accurate extraction method for the emitter and thermal resistance of bipolar transistors is presented. Only DC measurements taken on a thermally controlled wafer prober are required. The knowledge of the collector resistance is preferable, but not mandatory. The method yields excellent results when applied to advanced HBT technologies.**

Index terms — **Parameter extraction, emitter resistance, compact bipolar transistor models, self-heating, thermal resistance, external collector resistance.**

I. INTRODUCTION

Knowledge of the emitter resistance R_E is crucial for accurate models of bipolar transistors because both the DC and the RF behavior are strongly affected by the parasitic voltage drop and the resulting negative feedback.

A large variety of methods exists for extracting R_E for a given equivalent circuit. Extraction methods based on small-signal parameters were shown to give acceptable results, e.g. [1]-[3]. However, except for the requirement of small-signal measurements, which is not a limiting factor for a standard characterization infrastructure, selecting the suitable frequency range can be tricky ([4]). In contrast, DC methods are strongly limited by electrothermal effects (e.g. [5]) or may lead to device destruction and require a special measurement setup (e.g. the open-collector method [6]). A DC method incorporating electrothermal effects while using a standard measurement setup was published in [7].

The extraction method presented here is also completely based on DC-measurements and always yields a unique solution. It uses similar assumptions as in [7] but a different optimization strategy for making the method more stable. In contrast to other DC methods, self-heating of the device is *not* a limiting effect.

The theory and equations of the extraction method are given in section II, while its application is presented in section III. The accuracy of the method is demonstrated in section IV first based on synthetic data and then on experimental results of transistors with different geometries.

II. THEORY

In the forward operating region at medium injection, i.e. for negligible recombination in the base-emitter space charge region and the neutral base (caused by the collector heterojunction barrier), the base current I_B in absence of breakdown can be described by

$$I_B = I_{BEs}\exp\left(\frac{V_{BEi}}{m_{BE}V_T}\right) \tag{1}$$

with the saturation current I_{BEs}, the non-ideality factor

m_{BE}, the thermal voltage V_T and the internal base emitter voltage V_{BEi}. The latter reads for a DC current gain β much larger than 1 and thus $R_E \gg R_B/(\beta+1)$ [1]

$$V_{BEi} = V_{BE} - I_E R_E \tag{2}$$

with the terminal voltage V_{BE} and the emitter current I_E. In (1) and (2), I_{BEs}, V_T and R_E are functions of the temperature T, while the temperature dependence of m_{BE} can be neglected. Since the terminal values V_{BE}, I_B and I_E can be measured by standard equipment, T can be determined from solving (1) with (2) for known values of $R_E(T)$ and $I_{BEs}(T)$.

Neglecting avalanche breakdown, the temperature increase ΔT in the device with respect to the ambient temperature T_0 reads

$$\Delta T = T - T_0 = I_C V_{CEi} R_{th}, \tag{3}$$

where V_{CEi} is the internal collector-emitter voltage and R_{th} is the (temperature dependent) thermal resistance. V_{CEi} differs from the terminal voltage V_{CE} by voltage drops across the emitter and collector resistance

$$V_{CEi} = V_{CE} - (I_E R_E + I_C R_{Cx}). \tag{4}$$

Combining (3) and (4) yields

$$\Delta T = (V_{CE} - I_C R_{Cx} - I_E R_E)I_C R_{th}. \tag{5}$$

Keeping I_C constant in (5) and sweeping only V_{CE} then allows to determine R_E from known ΔT as described next. Assuming that I_C is kept sufficiently low such that the variation of the temperature dependent parameters (e.g. R_{th}, R_E) is small, results in an almost ideal linear behavior when plotting ΔT as a function of V_{CE} using (5) in the desired operating range. This is verified in Fig. 1 based on data from a complete compact model (HICUM/L2 [8]) with a realistic and partially nonlinear T dependence of its parameters (including R_{th} and R_E). The actual values of ΔT show a linear increase with V_{CE} that is followed very accurately by ΔT calculated from (1) with (2). Fig. 1 also indicates the suitable V_{CE} range, which is limited by saturation towards low V_{CE} and by avalanche breakdown towards high V_{CE}.

When extrapolating ΔT vs. V_{CE} (cf. Fig. 1), finite series resistances R_E and R_{Cx} lead to a zero crossing at V_{CE0}. According to (5) the corresponding value can be expressed by the series resistances as

$$V_{CE0} = I_E R_E + I_C R_{Cx}. \tag{6}$$

---

[1]Note that advanced SiGe HBTs typically exhibit peak DC current gains of at least several 100.

978-1-4799-7231-9/14 $31.00 © 2014 IEEE

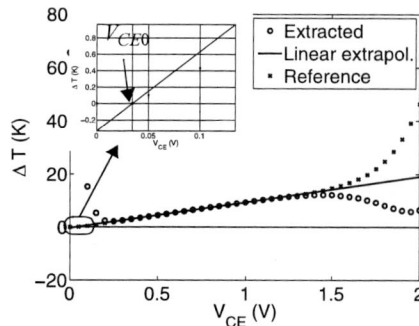

Fig. 1: Device temperature increase ΔT vs. CE terminal voltage V_{CE}. Comparison between actual (reference) values obtained from a complete compact model (crosses) and ΔT calculated from (1) with (2) (circles). In addition, the corresponding linear extrapolation from the forward active region is shown (solid line). The inset shows a zoom-in of the region around the zero-crossing, giving V_{CE0}.

R_E can now be determined iteratively based on the temperature sensor (1) with (2) and (6), as described next.

III. EXTRACTION METHOD

A. Base current parameters

In (1), the parameters for the base current, i.e. $I_{BEs}(T)$ and m_{BE}, are required. These are extracted using standard forward Gummel curves at $V_{BC} = 0$ V and sufficiently low injection for a large range of temperatures.

B. Measurement setup

As explained in sec. II, a constant value of I_C as a function of V_{CE} is desired to enable a linear extrapolation of (5) with V_{CE}. Three methods can be applied to achieve this.

- First, using a user-written measurement software, one can regulate V_{BE} for each V_{CE} to result in the desired value of I_C. This method may lead though to noisy V_{BE} data caused by the limited accuracy of the equipment.
- Second, a constant value of I_E can be forced into the device combined with a sweep of V_{BC}. This has the advantage of resulting in an almost constant I_C and an increase of the usable operating range. However, this cannot be directly applied to devices in RF-GSG-pads.
- The third method, which was also applied in sec. IV, is a simple forced I_B output sweep. This can be used with standard measurement software and equipment and also with standard device contact configurations. It is suitable for devices with a small temperature coefficient of the current gain, which is the case for most of the advanced SiGe technologies.

The extraction itself is performed for several ambient temperatures, allowing to extract the temperature coefficient of the emitter resistance. Also, the extraction should be performed with different values of constant I_C or I_B, respectively, in order to reduce the impact of numerical and measurement noise.

C. Extraction procedure

Starting from (6), an emitter resistance value is calculated from V_{CE0} by

$$R_{E,\,VCE0} = \frac{V_{CE0} - I_C R_{Cx}}{I_E}. \qquad (7)$$

V_{CE0} in turn depends on the correct calculation of ΔT from I_B, which requires a known value $R_{E,IB}$ used in eq. (2). Therefore, the nonlinear equation

$$R_{E,\,IB} - R_{E,\,VCE0} = 0 \qquad (8)$$

is solved. Note that the knowledge of R_{th} is not required for this calculation.

As shown in Tab. 1, by applying a bisection method, $R_{E,VCE0}$ is decreasing when increasing $R_{E,IB}$ and vice versa. This is expected, since a too large value of R_E is compensated by a large ΔT, which leads to a smaller (and possibly negative) V_{CE0}.

It #	$R_{E,\text{IB}}$	$R_{E,\text{VCE0}}$	$R_{E,\text{IB}}$	$R_{E,\text{VCE0}}$
1	0	404.85	0	404.85
2	1	221.58	2.59	-73.43
3	2	38.22	2.33	-21.67
4	3	-143.07	2.25	-8.12
5	2.5	-54.35	2.17	8.13
...				
8	2.18	2.62	2.19	2.19
...				
28	2.19	2.19		

Tab. 1: Numerical solution of (8) using the bisection method (left columns) and the Newton method (most right columns).

Another possible method is the Newton algorithm, where in each iteration step the value of $R_{E,IB}$ is varied in a small range in order to obtain the value of the numerical derivative. The application of this method is visualized in Fig. 2, and an example for the convergence behavior is given in Tab. 1. While $R_{E,IB}$ does not vary much, $R_{E,VCE0}$ converges quickly from a quite distant starting solution.

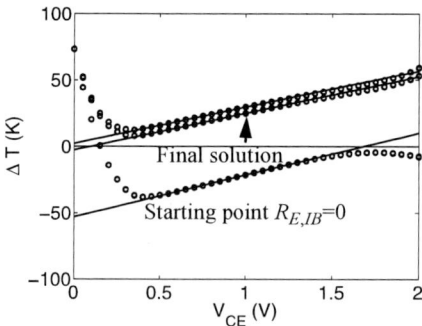

Fig. 2: Iteration towards the solution of (8), based on simulation data.

D. Thermal resistance

After convergence of the iteration, the thermal resistance of the device is also extracted according to (5) from the slope $I_C R_{th}$ of $\Delta T(V_{CE})$. This value for the thermal resistance represents a lumped self-heating model (3), which is common in compact models.

E. Collector resistance

In the previous extraction step a known value of R_{Cx} was assumed. Since this may not always be the case, the extraction method can be extended to also include the extraction of this value.

The extraction is performed for two different values of I_{C1} and I_{C2} at the same ambient temperature. Two operating points V_{CE1} and V_{CE2} with the same dissipated power exist. Then also the temperature increase in both operating points is the same and therefore also the value of temperature dependent values. It follows

$$I_{C1}(V_{CE1} - (I_{E1}R_E + I_{C1}R_{Cx}))R_{th} \\ = I_{C2}(V_{CE2} - (I_{E2}R_E + I_{C2}R_{Cx}))R_{th}. \quad (9)$$

This is rewritten with respect to R_{Cx}

$$R_{Cx} = \frac{I_{C1}V_{CE1} - I_{C2}V_{CE2} - R_E(I_{C1}I_{E1} - I_{C2}I_{E2})}{I_{C1}^2 - I_{C2}^2}. \quad (10)$$

Therefore, additionally to (7) for R_E an equation for R_{Cx} exists. A global optimization can be used to extract both values. This optimization has been implemented here sequentially. First, using an initial value of R_{Cx} (e.g. 0), R_E is extracted. Then, using this value of R_E and (10) leads to R_{Cx}, which again can be used to extract an updated value of R_E. The operating points $I_{C1}(V_{CE1})$ and $I_{C2}(V_{CE2})$ should be chosen in the center between saturation and breakdown.

IV. VERIFICATION

A. Simulation (synthetic) data

Synthetic data, i.e. reference data produced by a compact model, are suitable for (i) estimating the maximum accuracy that can be achieved by an extraction method and (ii) evaluating the valid operating range of the method. Results for two SiGe HBT technologies are shown here, comprising a low-cost medium speed [9] and a high-speed [10] technology. A much more comprehensive comparison of this method for a wide range of technologies along with other methods will be published in [4].

The application of the extraction method on the low-cost technology is shown for a selected device in Fig. 3. Deviations of about 10% are observed. These originate from a relatively high value of I_B that is already affected by the BC barrier effect, since self-heating is less prominent for these devices so that they have to be operated at higher current density levels.

In contrast, the results for the high-speed technology are very accurate for both R_E and R_{th} as shown in Fig. 4. This is also directly related to the much stronger self-heating in these devices, allowing to use operating ranges far enough below the onset of parasitic effects in the base current.

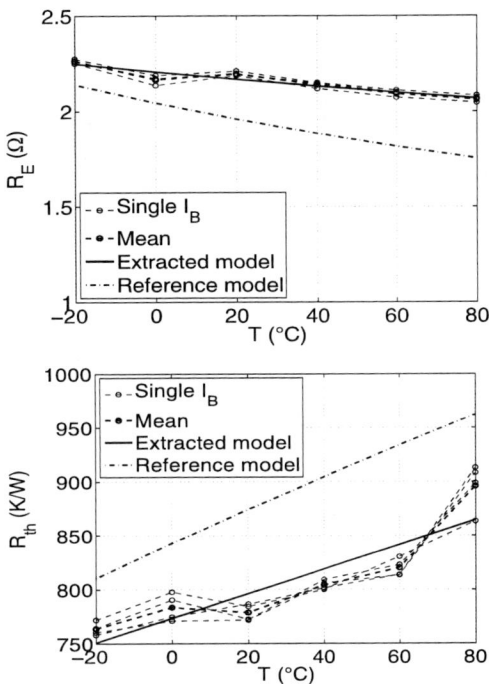

Fig. 3: Extraction results for the technology in [9] for the emitter and thermal resistance as a function of ambient temperature. The extraction was performed for different values of forced I_B. A mean value was used for extracting the parameters. For reference, the curve from the original model is included, showing its slightly nonlinear T dependence.

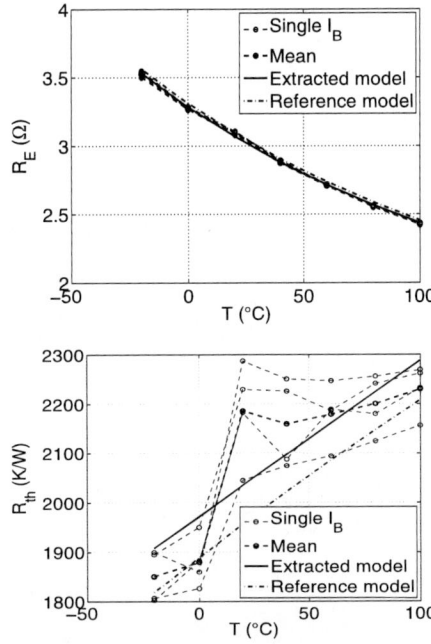

Fig. 4: Extraction results for the high-speed technology in [10] for R_E and R_{th} as a function of temperature based on synthetical data.

All results shown above use known values for the external collector resistance. Extraction using (10) was performed on several model cards but did not converge in most cases since even small errors in ΔT (compared to the reference) and therefore V_{CE1} and V_{CE2} can lead to a completely incorrect (and even negative) R_{Cx}.

A sensitivity analysis for the model cards from Fig. 3 and 4 was performed. The inserted value of R_{Cx} was swept from 0 to 150% of the original value. As shown in Fig. 5, only a weak dependence of R_E and almost no dependence of R_{th} on R_{Cx} used in (5) is observed, making the extraction method rather independent of R_{Cx} and the availability of corresponding specific test structures for extraction of the latter.

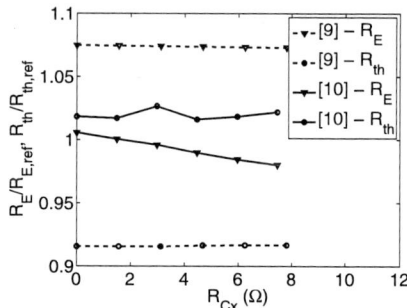

Fig. 5: Sensitivity of extracted R_E and R_{th}, normalized to their reference value from the model card, with respect to different values for R_{Cx} inserted in (5).

B. Experimental data

A verification on experimental data was performed here based on the expected physical behavior of the extracted values in terms of geometry scaling. Results are shown in Fig. 6, where different ways of emitter size scaling were applied, namely the variation of the emitter contact area and of the number of emitter fingers. During extraction of these results, R_{Cx} had already been extracted using special test structures.

The method was also successfully applied on InP DHBTs as demonstrated in [11].

V. Conclusion

An accurate method for extracting the emitter resistance of bipolar transistors requiring only DC measurements was presented. This method incorporates electrothermal effects leading to accurately extracted values for the thermal resistance as well. The comparison to synthetic data shows excellent agreement when using a suitable bias range. Extraction of the external collector resisance using dedicated test structures is preferred but not mandatory for improving the accuracy of the extracted R_E and R_{th} values. The experimental results of the method show the expected scaling of R_E with the lateral emitter size.

Acknowledgements

This work has been supported by the European Commission within DOTSEVEN (FP7-IP ICT-316755), the German Research Foundation (DFG) in the Collaborative Research Center 912 "Highly Adaptive Energy-Efficient Computing" and the German

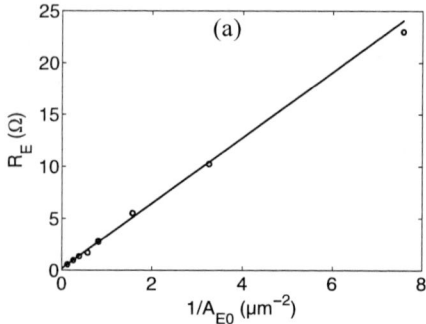

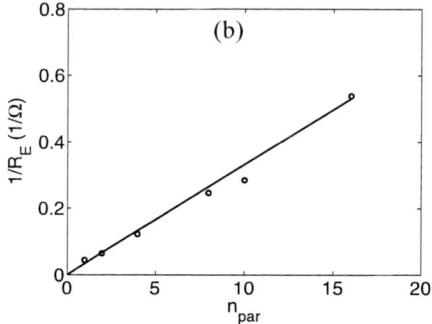

Fig. 6: Extracted values of the emitter resistance: (a) for geometry scaled devices varying both emitter width and emitter length, and (b) for structures with a different number of parallel devices.

Ministry of Research and Education (BMBF) within the RF2THz project.

References

[1] S. Maas, B.L. Nelson, D.L. Tait, "Intermodulation in Heterojunction Bipolar Transistors", IEEE Trans. Microwave Theory Tech., Vol. 40, 1992, pp. 442-448.

[2] Z. Huszka and E. Seebacher, "Extraction of RE and its temperature dependence from RF measurements", Working Group Bipolar (AKB) Meeting, 2009.

[3] C. Raya, B. Ardouin, Z. Huszka, "Improving Parasitic Emitter Resistance Determination, Methods for Advanced SiGe:C HBT transistors", IEEE BCTM 2011, pp. 191-194.

[4] J. Krause, M. Schroter, "An evaluation of methods for determining the emitter resistance of SiGe HBTs", to be published.

[5] T. H. Ning, D. D. Tang, "Method for Determining the Emitter and Base Series Resistances of Bipolar Transistors", IEEE Trans. Electron Dev., Vol. ED-31, No. 4, 1984, pp. 409-412.

[6] R. Gabl and M. Reisch, "Emitter Series Resistance from Open-Collector Measurements - Influence of the Collector Region and the Parasitic pnp Transistor," IEEE Transactions on Electron devices, Vol. 45, No. 12, 1998.

[7] H. Tran, M. Schroter, D.J. Walkey, D. Marchesan, T.J. Smy, "Simultaneous Extraction of Thermal and Emitter Series Resistances in Bipolar Transistors," in Proc. Bipolar/BiCMOS Circuits and Technology Meeting, Minneapolis, MN, 1997, pp. 170-173.

[8] M. Schroter and A. Chakravorty, "Compact Hierarchical Bipolar Transistor Modeling With HiCUM". World Scientific Publishing Co., Inc., 2010.

[9] D. Knoll et al., "A flexible, low-cost, high performance SiGe:C BiCMOS process with a one-mask HBT module," Internation Electron Devices Meeting, IEDM, 2002, pp. 783 – 786.

[10] H. Rucker, B. Heinemann, A. Fox, "Half-Terahertz SiGe BiCMOS technology," IEEE 12th Topical Meeting on Silicon Monolithic Integrated Circuits in RF Systems (SiRF), 2012, pp. 133-136.

[11] T. Nardmann, J. Krause, M. Schroter, "An evaluation of extraction methods for the emitter resistance for InP DHBTs", accepted at CSICS 2014.

A study on transient intra-device thermal coupling in multifinger SiGe HBTs (Student)

R. D'Esposito, M. Weiß, A. K. Sahoo, S. Fregonese and T. Zimmer

Laboratoire IMS, CNRS - UMR 5218, Université de Bordeaux

351 Cours de la Libération - 33405 Talence Cedex, France

rosario.desposito@ims-bordeaux.fr

Abstract—**This paper presents a study of transient mutual thermal coupling occurring between the fingers of trench isolated SiGe HBTs. Three-dimensional thermal TCAD simulations have been carried out to obtain the temperature evolution in transient operation in a multifinger HBT structure. The same behavior has been simulated using a netlist-based model, which provides an accurate representation of the substrate thermal coupling between active device areas. On-wafer measurements in pulsed conditions have been conducted on specially designed test structures that permit to determine the thermal coupling between the different fingers of a 5x(CBEBC) SiGe HBT; the results from the measurements are found to be in good agreement with a simulation in which the thermal coupling network has been added to the thermal nodes of five HiCuM transistor models.**

Keywords—*Heterojunction Bipolar Transistors, Silicon germanium, Multifinger, Transient thermal coupling, Intra-device, Mutual coupling, Thermal capacitance*

I. INTRODUCTION

The thermal analysis of solid-state devices is a topic of increasing interest in the last years, due to the current trend in nanoelectronics to shrink device dimensions and increase the integration levels, thus yielding very high power densities. A common case where electro-thermal problems arise is in high performance silicon-germanium (SiGe) heterojunction bipolar transistors (HBTs), which are becoming largely used for millimeter-wave applications, due to their high performances in terms of transconductance and cutoff frequency. These high speed performances are achieved with an aggressive shrinking of device dimensions and the introduction of deep trench isolation (DTI) and shallow trench isolation (STI) that help reducing parasitic capacitances. Shrinking the vertical and lateral dimensions of these components results in high internal electric fields and high current densities and thus to an increase in power density. For these devices, the region where there is the most significant heat dissipation can be approximated by the base-collector space charge region, and its dimensions are close to the effective emitter area (L_E x W_E). Multi-finger transistors (MFTs) are often used in PA cells since they are more suitable for high frequency combined with high power operation due to the small base access resistance. In these devices, the center fingers are considered to be the most critical ones from an electro-thermal point of view, since, due to the thermal coupling with the neighboring fingers, they reach the highest temperature; this effect is made stronger by the presence of the DTI, which prevents the heat to flow laterally. These thermal inhomogeneities can be quite significant: as it

has been previously shown by means of TCAD simulations in [1], a temperature difference of even 15°K can be observed between center and side fingers when there is a power dissipation of 30mW. For these reasons, to avoid thermal runaway and subsequent destruction of the MFT component, it is necessary to simulate the right temperature distribution in the structure during the design process; to accomplish that, a distributed approach can be used, in which one transistor model per finger is considered and the thermal node is connected to a network that is able to predict the self and mutual heating effects arising in the device. Concerning a single finger transistor, several research studies were performed to study self-heating effects in steady-state, like [2] and [3] and in transient conditions [4], [5] to obtain thermal resistance *Rth* and thermal capacitance *Cth*. In [6] mutual thermal coupling in multifinger SiGe:C HBTs is modeled under DC operation, using a distributed transistor model that considers self-heating as well as thermal coupling between devices. The aim of this paper is to study and model the thermal evolution in pulsed conditions of operation for multifinger HBTs. The characterization of the transient thermal response in these devices is of critical importance in cases of pulsed power dissipations (like in switching applications), where the transient temperature doesn't reach the steady state value.

II. THE MODEL PROPOSED FOR TRANSIENT INTRA-DEVICE THERMAL COUPLING

In compact models a common way to take into account electro-thermal effects is to provide a representation of the device electrical characteristics as function of its operating temperature, while providing a prediction of its temperature according to the power dissipated in the device itself. To calculate the temperature of a single device due to self-heating, a simple parallel RC network is typically used, but when two devices are operating at the same time in close proximity, an additional temperature increase must be considered, due to thermal coupling. In fact the spread of the heat flux through the substrate causes device 1 to give a certain contribution in the increase of the temperature of device 2; an efficient way to take into account this thermal coupling effect can be found in [7]. If we consider self-heating and mutual thermal coupling, we can express the temperature rise above the ambient temperature for two active devices as shown in (1), where Rth_i and Pd_i are respectively the thermal resistance and the power dissipated of each component; the terms c_{ij} instead are the thermal coupling coefficients between device i and device j and take into account how much of the temperature of device j is coupled to that of device i (see Fig. 1). This coupled temperature can thus be simulated by the introduction of a voltage-controlled-voltage-source (VCVS) in series with the thermal resistance.

$$T_1 = Rth_1 Pd_1 + Rth_{12} Pd_2 = Rth_1 Pd_1 + c_{12} Rth_2 Pd_2 =$$
$$= Rth_1 Pd_1 + c_{12} \Delta T_2$$
$$T_2 = Rth_{21} Pd_1 + Rth_2 Pd_2 = c_{21} Rth_1 Pd_1 + Rth_2 Pd_2 =$$
$$= c_{21} \Delta T_1 + Rth_2 Pd_2$$

(1)

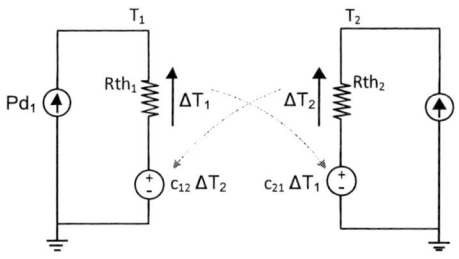

Fig. 1. Extended thermal network for self-heating and DC thermal coupling for two heat sources [7].

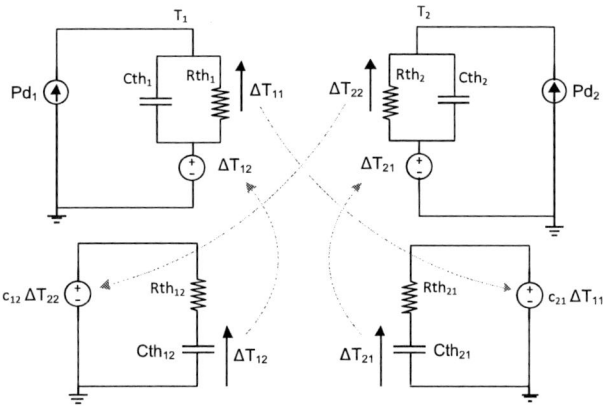

Fig. 2. Extended thermal network for self-heating and transient thermal coupling for two heat sources [8].

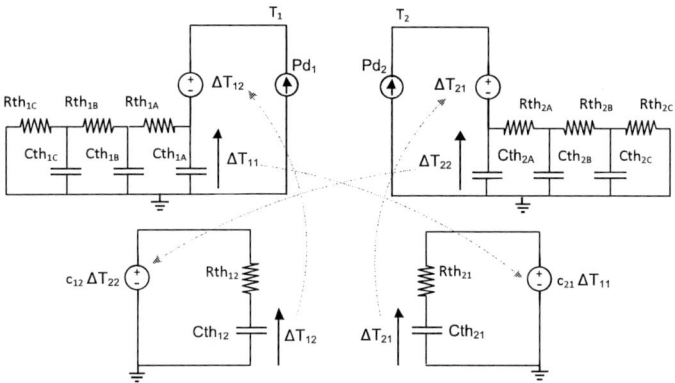

Fig. 3. The proposed 3 poles thermal network for self-heating and transient thermal coupling considering two sources.

In [8], this approach has been further extended to model the thermal coupling between two heat sources in case of transient operation. As it is possible to see in Fig. 2, an additional separate RC network has been introduced, in order to take into account the delay in the time response. This concept can be easily extended to multiple devices and heat sources, in our case the fingers of a MFT. As already shown in [5] and [9] a single pole network is not able to approximate the actual thermal response to a step in power, however for compact modeling it is suggested to limit the number of RC subcircuit nodes for computational reasons. Hence a three time constants RC circuit has been chosen as a good compromise between accuracy to reproduce the transient thermal response and computational effort; the Rth and Cth of each finger in Fig. 2 have been replaced by a recursive Cauer type network (which has already been demonstrated to give good accuracy in [10]). Fig. 3 presents the network that we propose to model the intra-device mutual thermal coupling inside multifinger SiGe HBTs in pulsed operation conditions (to simplify the image, just 2 heat sources are shown); in our case we will consider a five emitters component.

III. VALIDATION OF THE MODEL WITH TCAD SIMULATIONS

In order to test the validity of this substrate representation, a 3D TCAD electro-thermal simulation is realized on the structure in Fig. 4, which reproduces a five-emitter SiGe HBT component. The 5 emitters are assumed to be the heat sources and each one has a drawn emitter area $A_E = 5 \times 0.18$ μm². The device was embedded on the top of a 300μm (thickness of the wafer) x 500μm (lateral dimension >> than device dimensions) Si-substrate block.

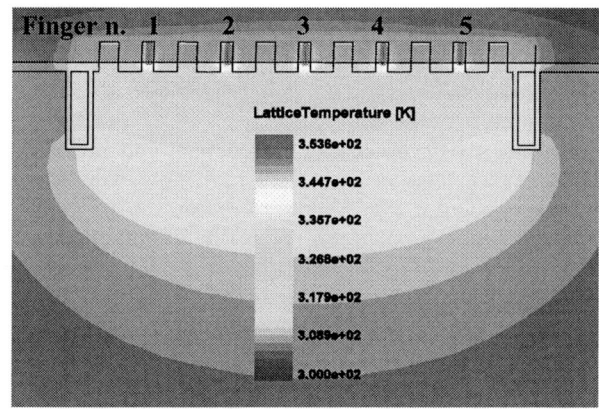

Fig. 4. The structure of the five emitters HBT.

The thermal behavior of the component is first analyzed under DC conditions: a constant power is applied to the emitter stripes one at a time, to obtain the Rth of each finger and evaluate the coupling with the other fingers; besides it is also possible to evaluate the impact of the temperature dependent thermal conductivity. Next, the transient thermal behavior of the structure is simulated by applying a pulse of 10mW per finger. The results from the proposed model are also presented in Fig. 5, using a logarithmic scale on the time axis, to show the accuracy achieved during the transient (since the structure is symmetrical the temperature of finger 4 and 5 is the same as finger 1 and 2 respectively). The maximum deviation of the model from the TCAD results is around 1°K. From the results of Fig. 5 it is possible to notice that the thermal evolution of

978-1-4799-7231-9/14 $31.00 © 2014 IEEE

finger 1 and 5 during the transient differs from that of the other fingers due to the proximity of the DTI, acting as a vertical adiabatic surface. In fact, a lower temperature rise is expected because finger 1 and 5 have only one neighboring heat source. To further study the transient behavior of the component another scenario is presented, where a pulse of 10mW is applied to just one emitter stripe.

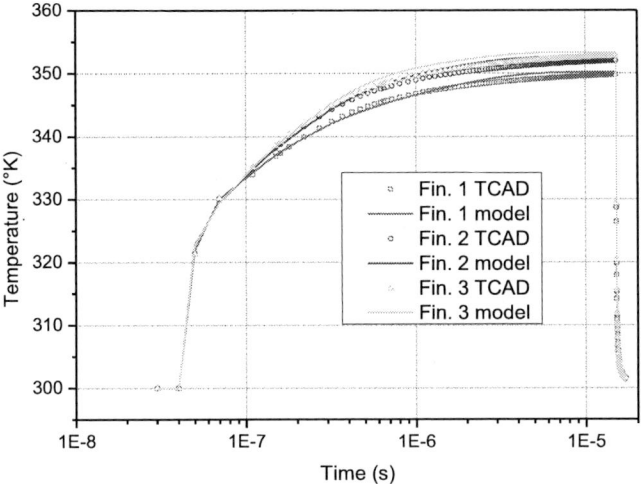

Fig. 5. Temperature results from the TCAD simulation and the compact model of a 5 finger HBT when a pulse of 10 mW power dissipation is applied to all fingers ($T_{finger1}=T_{finger5}$, $T_{finger2}=T_{finger4}$).

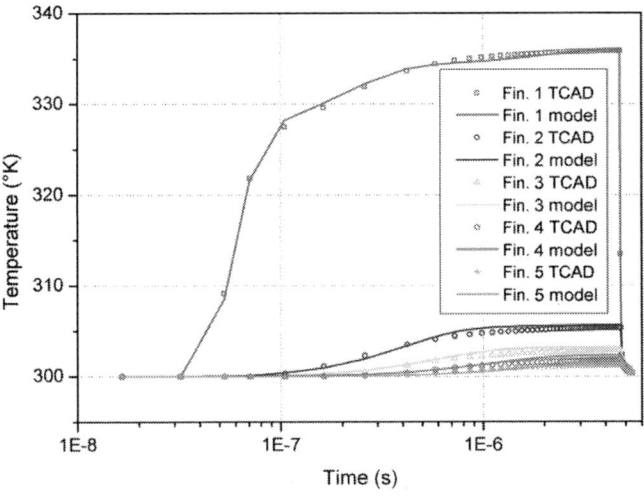

Fig. 6. Temperature results from the TCAD simulation and the compact model of a 5 finger HBT when a pulse of 10 mW power dissipation is applied at finger 1.

As expected, a lower temperature will be sensed on fingers that are placed at a larger distance from the heat source, as can be seen on Fig. 6; in addition, it is possible to notice a slower time response when the distance between the heat source and the sensing finger increases, this is due to the longer time that the heat needs to propagate in the substrate (this effect is modeled by the delay RC network in Fig. 3). In this scenario the lower accuracy to simulate the temperature on the sensing fingers can be justified by the inability of a single pole network to take into account this delay in the heat propagation.

IV. ON-WAFER MEASUREMENTS

To further evaluate the effect of the intra-device transient coupling, an on-wafer test structure (equivalent to the one already presented in [6]) has been measured in pulsed operation conditions. The test structure has been fabricated in STMicroelectronics B55 technology and consists of a 5 fingers SiGe HBT, where each finger is thermally coupled, but electrically separated; the 5 transistors have a common collector contact and each one of the 5 emitters is accessible, while the bases are all grounded (see Fig. 7). This configuration allows to use each finger as a heater, by simply turning it on, or as a temperature sensor, by measuring the shift in its I_E (V_{BE}) characteristics. In [11] a measurement system has been presented that uses a low-distortion function generator, a lock-in amplifier, and a parameter analyzer for the extraction of the thermal resistance and capacitance in bipolar devices. In this work a new setup for the measurement of the thermal response is used, that allows to measure the response of each finger of the multifinger SiGe HBT to a voltage pulse applied to a neighboring finger.

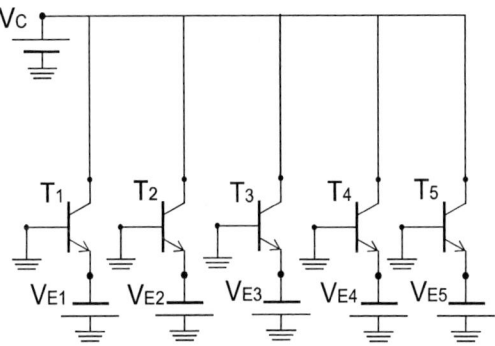

Fig. 7. The equivalent schematic of the test structure used for the thermal coupling measurements.

The measurements were carried out on-wafer at 27°C using a SUSS MicroTec probing station equipped with a thermal chuck. Probing was done with two differential GSGSG probes (SUSS MicroTec) and two GSG probes (Picoprobe). The DC analyzer Keithley 4200 containing a 4225-SCS module has been used to generate the pulses on the heating finger and measure the response on the sensing finger. While the common collector is constantly biased to 0.5V using the HP4155A DC analyzer, a negative pulse of -0.95V is applied to the emitter finger that acts as a heater, to forward bias its BE junction, while the emitter finger that senses the temperature is biased to a constant negative voltage of -0.85V; in these conditions a current increase can be noticed on the sensing finger, due to the temperature rise of the heater. In Fig. 8 are shown the currents of the sensing fingers 1, 4 and 5 when finger 2 acts as a heater; the fingers are sensed one by one, while the heater is always finger 2. The compact model simulation results are also added. It can be noticed that if the distance between heating and sensing finger increases, the variation of current that is measured on the sensing finger is smaller due to its lower temperature increase. This behavior is qualitatively similar to the TCAD simulation results (Fig. 6). The TCAD simulation shows a faster time constant (of a factor 2); in fact during

TCAD simulation the back end of line (BEOL) is not considered; the BEOL has an important impact on the thermal capacitance [12]. It is also interesting to show the current measured on the heater, since it will have a small increase due the power dissipation on the sensing finger (in our case no more negligible), which is constantly biased (see Fig. 9). To simulate the electro-thermal behavior of the measured multifinger HBT, the thermal nodes of five lumped HiCuM transistor models are connected to the distributed thermal network proposed in section 2; in Fig. 8 and 9 the results of the simulation of the distributed model are compared to the measurements, showing a good agreement.

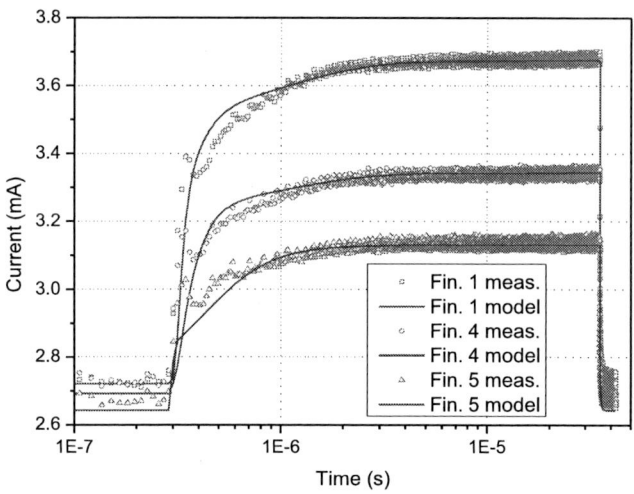

Fig. 8. Measured current (red) and simulated current (blue) on the sensing fingers; from top: current sensed on finger 1, finger 4 and finger 5.

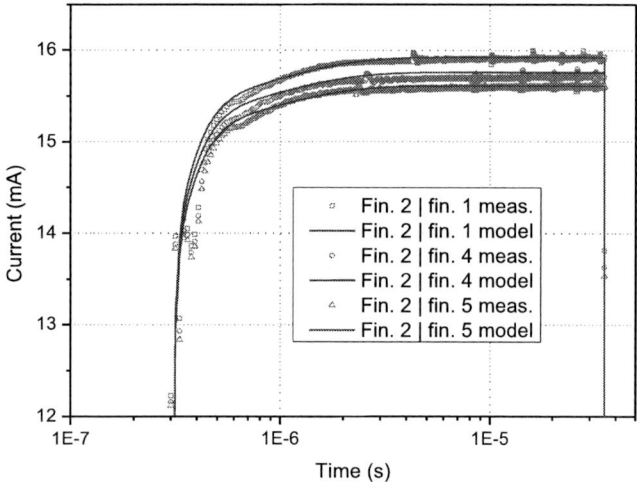

Fig. 9. Measured current (red) and simulated current (blue) on the heating finger 2 in the case when (from the top): finger 1, finger 4 and finger 5 are sensing.

V. CONCLUSIONS

The effects of intra-device transient thermal coupling in trench isolated SiGe HBTs have been characterized with the aid of thermal TCAD simulations and with on-wafer pulsed measurements. The model reported uses a simple netlist-based

method, which allows to accurately reproduce the transient thermal effects described in an electrical simulator; for these reasons it can be useful to simulate in a more realistic way the behavior of multifinger SiGe HBTs during the design process. In addition, it can be used to choose the ballast resistors in the PA cell for a balanced heat distribution.

ACKNOWLEDGMENTS

This work is part of the RF2THZ SiSoC project supported by the European EUREKA Program CATRENE and the French Ministry for Economics and Industry and of the Dotseven project supported by the European Commission through the Seventh Framework Program for Research and Technological Development. The Authors would like to thank STMicroelectronics for providing the B55 wafer.

REFERENCES

[1] M. Weiss, A. K. Sahoo, C. Raya, M. Santorelli, S. Fregonese, C. Maneux, e T. Zimmer, «Characterization of intra device mutual thermal coupling in multi finger SiGe:C HBTs», in *2013 IEEE International Conference of Electron Devices and Solid-State Circuits (EDSSC)*, 2013, pagg. 1–2.

[2] N. Rinaldi, «Thermal analysis of solid-state devices and circuits: an analytical approach», *Solid-State Electronics*, vol. 44, n. 10, pagg. 1789–1798, ott. 2000.

[3] R. Menozzi, J. Barrett, e P. Ersland, «A new method to extract HBT thermal resistance and its temperature and power dependence», *IEEE Transactions on Device and Materials Reliability*, vol. 5, n. 3, pagg. 595–601, set. 2005.

[4] N. Rinaldi, «On the modeling of the transient thermal behavior of semiconductor devices», *IEEE Transactions on Electron Devices*, vol. 48, n. 12, pagg. 2796–2802, dic. 2001.

[5] A. K. Sahoo, S. Fregonese, M. Weis, N. Malbert, e T. Zimmer, «A Scalable Electrothermal Model for Transient Self-Heating Effects in Trench-Isolated SiGe HBTs», *IEEE Transactions on Electron Devices*, vol. 59, n. 10, pagg. 2619–2625, ott. 2012.

[6] M. Weiss, A. K. Sahoo, C. Maneux, S. Fregonese, e T. Zimmer, «Mutual thermal coupling in SiGe:C HBTs», in *2013 Symposium on Microelectronics Technology and Devices (SBMicro)*, 2013, pagg. 1–4.

[7] D. J. Walkey, T. J. Smy, R. G. Dickson, J. S. Brodsky, D. T. Zweidinger, e R. M. Fox, «Equivalent circuit modeling of static substrate thermal coupling using VCVS representation», *IEEE Journal of Solid-State Circuits*, vol. 37, n. 9, pagg. 1198–1206, set. 2002.

[8] Y. Zimmermann, "Modeling of spatially distributed and sizing effects in high-performance bipolar transistors", MSEE/Diploma thesis, Chair for Electron Devices and Integrated Circuits, TU Dresden, 2004

[9] D. J. Walkey, T. J. Smy, D. Marchesan, H. Tran, C. Reimer, T. C. Kleckner, M. K. Jackson, M. Schroter, e J. R. Long, «Extraction and modelling of thermal behavior in trench isolated bipolar structures», in *Bipolar/BiCMOS Circuits and Technology Meeting, 1999. Proceedings of the 1999*, 1999, pagg. 97–100.

[10] Sahoo, A.K., S. Fregonese, M. Weiss, N. Malbert, and T. Zimmer. "Electro-Thermal Dynamic Simulation and Thermal Spreading Impedance Modeling of Si-Ge HBTs." In *2011 IEEE Bipolar/BiCMOS Circuits and Technology Meeting (BCTM)*, 45–48, 2011.

[11] Nenadovic, N., S. Mijalkovic, L.K. Nanver, L.K.J. Vandamme, V. D'Alessandro, H. Schellevis, and J.W. Slotboom. "Extraction and Modeling of Self-Heating and Mutual Thermal Coupling Impedance of Bipolar Transistors." *IEEE Journal of Solid-State Circuits* 39, no. 10 (October 2004): 1764–72.

[12] A. K. Sahoo, S. Fregonese, M. Weis, C. Maneux, N. Malbert, and T. Zimmer, "Impact of back-end-of-line on thermal impedance in SiGe HBTs," in *2013 International Conference on Simulation of Semiconductor Processes and Devices (SISPAD)*, 2013, pp. 188–191.

A Low Phase Noise Signal Generation System for Ka–Band P2P Applications based on an Injection-Locked Frequency Tripler

D. Cabrera[1], JB Begueret[1], Y. Deval[1], O. Tesson[2], P. Gamand[2], O. Mazouffre[1], T. Taris[1].

[1]IMS Laboratory, University of Bordeaux, France.

[2]NXP Semiconductors, Caen, France.

Abstract—A signal generation system composed by a subharmonic VCO followed by an injection-locked frequency tripler (ILFT) is designed in a 0.25 μm BiCMOS SiGe:C technology. The ILFT implements a cascoded current-biased common emitter configuration that exploits the second harmonic of the VCO to enhance the efficiency in the generation of the injecting signal responsible for the ILFT locking. At 30.8 GHz, the system achieves a phase noise of -112 dBc/Hz at 1 MHz offset. The total current consumption is 38 mA for a supply voltage of 2.5 V.

Index Terms—phase noise; millimeter wave frequency generation; subharmonic VCO; frequency tripler.

I. INTRODUCTION

The recent advancements in silicon-germanium (SiGe) heterojunction transistors (HBTs) have enabled the development of circuits and methods for oscillators working near the transit frequency of the transistors. These signal sources have allowed integrated solutions for radar and imaging applications in millimeter-wave frequencies [1], [2]. The phase noise performances of millimeter wave signals continue to be significantly challenging for integrated implementations of new communication standards at these frequencies [3]. The Colpitts topology, in fundamental or push-push mode, has gained wide acceptance for low-noise frequency generation. In this topology (and for any other LC VCO working at these frequencies) the trade-off between phase noise and frequency tuning range becomes more evident as the tank's quality factor is dominated by the quality factor of the varactors, which decreases exponentially with frequency. This fundamental trade off can be treated separately by the use of subharmonic oscillators followed by frequency multipliers [4]. In these systems, the phase noise specification is met by the subharmonic oscillator and the tuning range depends on the bandwidth of the frequency multiplier.

In this work, a signal generation system composed by a sub-harmonic VCO followed by an injection-locked frequency tripler (ILFT) for a low-IF point-to-point fixed radio system in the Ka-Band band is proposed. From a system analysis, a phase noise at 1 MHz better than -107 dBc/Hz was found to be sufficient for the targeted application. In particular, the proposed ILFT exploits the second harmonic of the VCO to enhance the generation efficiency of the injecting signal responsible for the locking of the ILFT. The proposed system is represented in Fig. 2a.

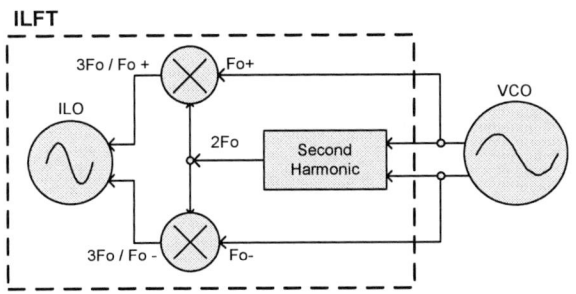

Fig. 1. Conceptual block diagram of the proposed system. To generate a low phase noise signal, a subharmonic VCO followed by an injection-locked frequency tripler is proposed. The second harmonic of the VCO is used to enhance the generation of the signal responsible to lock the ILO.

II. CIRCUIT ANALYSIS AND DESIGN

A. 12 GHz VCO

Fig. 2.a shows the schematic of the VCO. Considering the frequency band, a classical cross-coupled NPN-only topology is used. The VCO's frequency tuning capability is achieved using a mixed scheme of a diode varactor and switched capacitor bank. The oscillating signal is buffered using a pair of common collector transistors. The bias voltage BIASOSC sets the current of the VCO core.

Special attention was given to the $1/f^2$ phase noise of this VCO as it will set the phase noise of the entire system. Based on a good understanding of the noise contribution of each component on the oscillating output signal, different strategies for phase noise minimization have been proposed [5]. These methodologies suggest relating the noise generated by each component of the circuit to the phase noise generated by the parallel resistance of an equivalent RLC circuit. This parallel resistance will also determine the amplitude of the oscillating signal. The phase noise of a conceptual parallel RLC oscillator can be expressed as [5]:

$$\mathcal{L}(f_{off}) = 10\log_{10}\left[\frac{K_B T f_o}{Q_{FL} C V_o{}^2 \Delta f_{off}{}^2}\right] \quad (1)$$

where f_{off} is the offset frequency; f_o is the oscillation frequency; K_B is the Boltzmann constant; T is the temperature

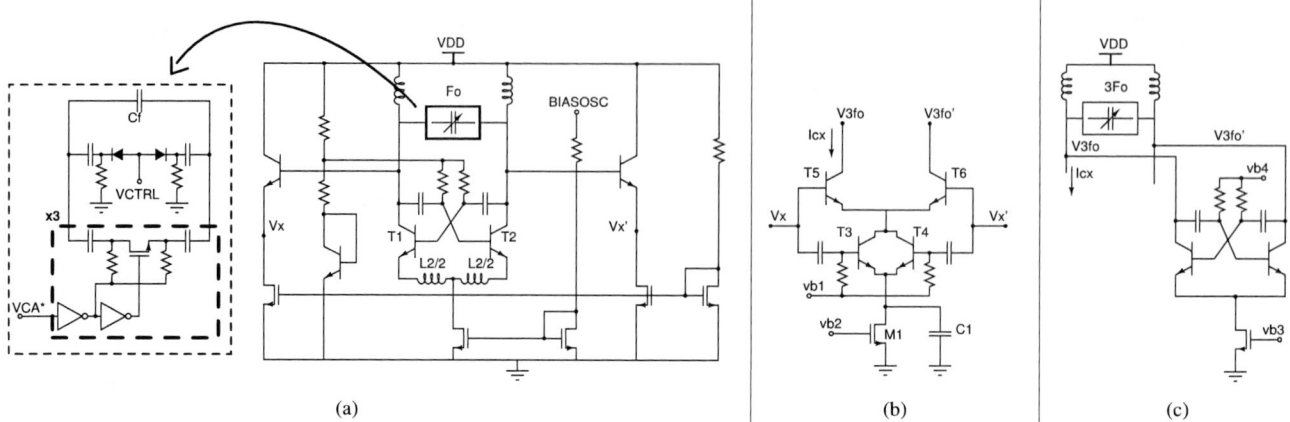

Fig. 2. Schematic diagram of the blocks composing the system: a) VCO. b) Harmonic current generator. c) ILO.

in Kelvins; Q_{FL} is the tank's quality factor; C is the equivalent parallel capacitance and V_o the signal amplitude.

The strategy used for the phase noise minimization of the VCO is mainly derived from Eq. 1: for a given oscillation frequency, a high equivalent capacitance (low inductance) and a high tank's quality factor are desirable for a low phase noise. Indeed, when lowering the inductance of top-metal spiral inductors its own quality factor tends to decrease, but if the increment on the capacitance (C) is higher than the resulting degradation on the tank's quality factor (Q_{FL}), then the phase noise will tend to decrease.

On other hand, to provide a frequency tuning of 10% while minimizing the degradation on Q_{FL}, special attention was given in the design of each component of the capacitance-varying scheme: varactors and switched-caps. Considering that the quality factor of a lossy capacitor bank is mainly determined by the quality factor of the highest capacitance contributor [6], the fixed capacitance Cf was implemented with a high Q MIM (Metal-Insulator-Metal) capacitor. The small size varactor and switched capacitors were designed to provide the needed frequency range. The size of the MOS switches was optimized to minimize its ON-resistance, and to not degrade the given tuning range when in the off-state. Regarding to the core transistors (T1, T2), its noise contribution is mainly given by its base resistances and bias current. A lower base resistance was achieved with the lower emitter width for a given bias current. Finally, the inductor L2 was added to filter high frequency noise coming from the bias current.

B. Injection Locked Frequency Tripler (ILFT)

The ILFT is based on the current-pulses generator of Fig. 2b and the injection-locked oscillator (ILO) of Fig. 2c, which is tuned at $3 \times f_{VCO}$. If $3 \times f_{VCO}$ lies in the locking range of the ILO, the third harmonic component (I_{3fi}) of the pulsed current Icx will lock the ILO. To adjust the frequency range where the ILFT will be locked to the input signal, a diode varactor was used to set its free running frequency. Additionally, the magnitude of I_{3fi} can be modified by changing the DC current

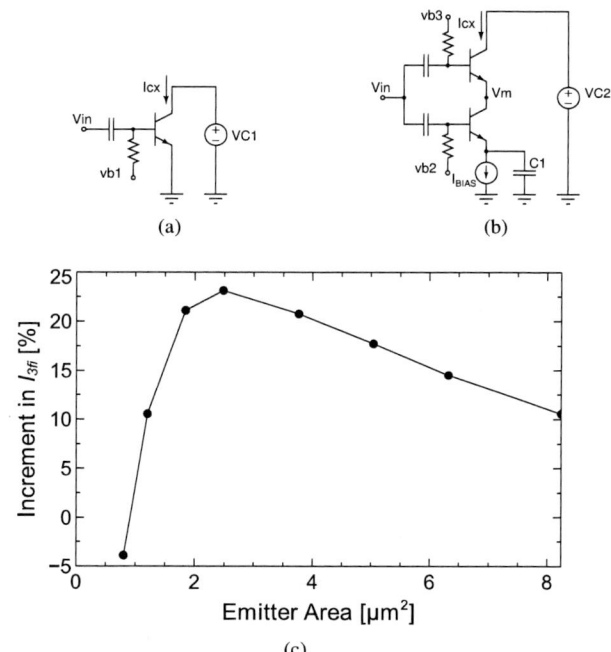

Fig. 3. Simulation-based comparison of the third harmonic of Icx by using: (a) voltage-biased common emitter (*conf1*) and (b) cascoded current-biased common emiter (*conf2*). (c) The plot shows the relative increment of I_{3fi} by using *conf2* over the *conf1* (all transistors have the same size).

trough M1. These capabilities provide flexibility to set the locking range.

The pulsed current generator is based on the current-biased emitter/collector coupled transistors (T3-T4) biased in class A/B. Current pulses at twice the frequency of the input signal $(2 \times f_{VCO})$ are addressed by transistors T5-T6 to the ILO at f_{VCO}. Compared to a voltage-biased single transistor in common emitter configuration (*conf1* showed in Fig. 3a), the current-biased topology was found to be more effective to generate a higher 3rd harmonic component of the current for a given input voltage and frequency. The cascoded version of

978-1-4799-7231-9/14 $31.00 © 2014 IEEE   184

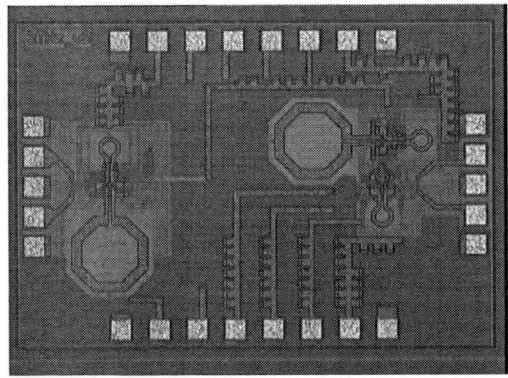

Fig. 4. Photograph of the fabricated chip. Chip size: 800 μm by 1200 μm.

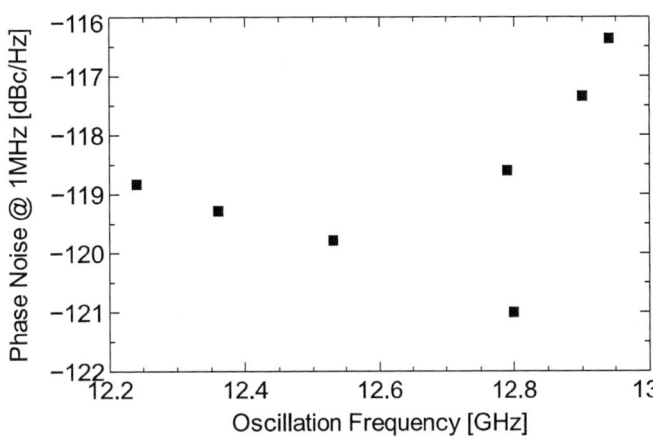

Fig. 6. Measured phase noise at 1 MHz over the frequency range of the VCO

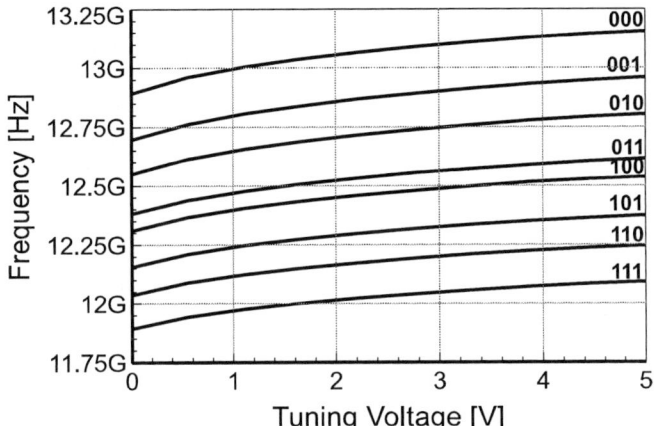

Fig. 5. Measured tuning Range of the VCO.

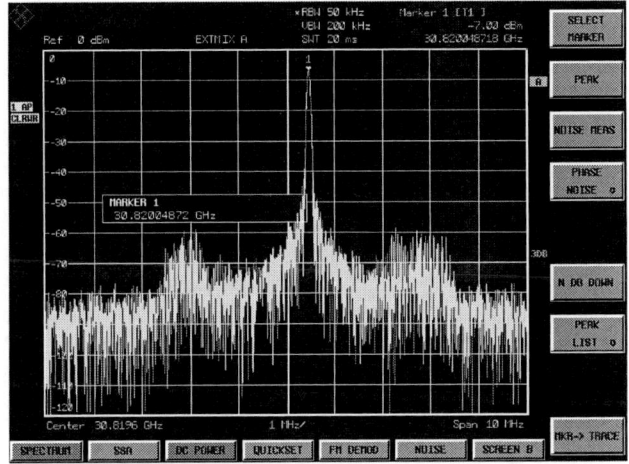

Fig. 7. Measured spectrum of the ILFT when locked to the VCO.

the current-biased transistor, *conf2*, is shown in Fig. 3b.

The relative increment on the magnitude of I_{3fi} by using the circuit *conf2* over the circuit *conf1* is plotted for several emitter areas (Fig. 3c). A maximum increment of 25% was verified. These results, which are based on harmonic-balance simulations, were obtained using the MEXTRAM 504.7 model of the transistor provided by the foundry [7]. In the simulations, the same input signal was applied and all transistors have the same size. The input signal is composed by a sinusoidal voltage at 12 GHz and its 2nd harmonic at 24 GHz with voltage magnitudes of 400 mV and 50 mV respectively. These values correspond to the simulated output voltage (single ended) of the VCO in Fig. 2a. Finally, the average of *Icx* for both topologies was kept the same by changing the voltage vb1 on *conf1*. Note that the working principle for the generation of pulsed currents is different for both configurations. For *conf2* the current pulses are generated by the charging/discharging process of the capacitor $C1$ and its better efficiency is explained by the mixing action of the cascoded transistor. A similar concept was already presented in [8].

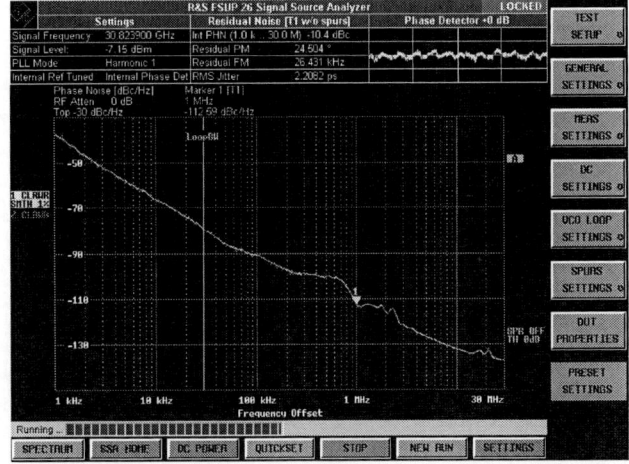

Fig. 8. Measured phase noise at 1 MHz of the ILFT when locked to the VCO.

978-1-4799-7231-9/14 $31.00 © 2014 IEEE

III. IMPLEMENTATION AND MEASUREMENTS

The system was realized in the 0.25 μm SiGe:C BiCMOS process QUBIC4XI from NXP featuring NPN transistors with transit frequency (f_t) of 216 GHz and f_{max} of 177 GHz [9]. Fig. 4 shows the chip micrograph. This prototype contains also a standalone version of the VCO. Compared to the schematic in Fig. 2a, the standalone version has an additional stage of output buffers for measurement purpose and due to the capacitive load difference between this buffer and the input of the ILFT, a frequency shift is expected. A R&S®FSU26 spectrum analyzer was used for the phase noise and frequency tuning characterization. The measures were performed on die using a GSGSG probe.

The standalone version of the VCO was characterized first. The whole VCO consumed 28 mA for a supply voltage of 2.5 V. The core consumes 18 mA (bias circuit 3 mA) and the output buffer 10 mA. Fig. 5 shows the tuning range for all values of the switching capacitors. The tuning voltage was varied from 0 V to 5 V. Fig. 6 shows the measured phase noise at 1 MHz offset over the tuning range. A FOM[1] of 184.7 at 12.8 GHz is obtained. This result compares well with state-of-the-art VCOs (Table I).

In a second step, the system composed of the VCO and the ILFT was characterized. An initial measurement test was developed in a probe station, using the same network analyzer and a down conversion harmonic mixer. The ILFT consumed 10 mA from a 2.5 V supply. The frequency range of the locked system goes from 30.72 to 33.6 GHz. The power spectrum and the phase noise of the locked ILFT at 30.88 GHz is shown in Fig. 7 and Fig. 8; its measured phase noise at 1 MHz is -112 dBc/Hz. When compared to the unlocked mode in the same frequency range, a phase noise improvement around 25 dBc/Hz is observed.

IV. CONCLUSIONS

In order to address the trade-off between phase noise and tuning range present in millimeter wave oscillators, this work presents a low noise signal generation system from 30.72 to 33.6 GHz in a SiGe:C BiCMOS technology for P2P applications. The system is based on a subharmonic VCO followed by an injection-locked frequency tripler. The system achieved a measured phase noise at 1 MHz from a 30.8 GHz carrier frequency of -112 dBc/Hz. An efficient frequency tripler has been realized and presented which allows a low phase noise implementation of the signal generation system.

V. ACKNOWLEDGMENT

This work is part of the "RF2THz" project supported by the European CATRENE program. The authors would like to thank G. Varin from NXP Semiconductors for the support in the layout-finishing steps.

[1]Defined as: $FOM = -PN + 20 \times \log_{10}(f_o/f_{off}) - 10 \times \log_{10}(P[mW])$

Ref.	Process	Freq [GHz]	TR	Pdiss [mW]	PN@1 MHz [dBc/Hz]	FOM
[4]	0.18μm BiCMOS	32	9.5%	35	-103	184.6
[10]	0.25μm BiCMOS	13.1	5.1%	93	-123	185.6
[11]	0.13μm BiCMOS	28	30%	10.5	-104	182.7
[12]	0.25μm BiCMOS	17.4	4.6%	30	-113	183
This Work VCO only	0.25μm BiCMOS	12.8	7.5%	70	-121	184.7
This Work VCO+ILFT	0.25μm BiCMOS	30.8	9%	95	-112	182.2

TABLE I
COMPARED TO OTHER WORKS

REFERENCES

[1] A. Tomkins, E. Dacquay, P. Chevalier, J. Hasch, A. Chantre, B. Sautreuil, and S. P. Voinigescu, "A study of SiGe signal sources in the 220–330 GHz range," in *Proc. IEEE Bipolar/BiCMOS Circuits Tech. Meeting*, Sept 2012, pp. 80–83.

[2] O. Momeni and E. Afshari, "High Power Terahertz and Millimeter-Wave Oscillator Design: A Systematic Approach," *IEEE J. Solid-State Circuits*, vol. 46, no. 3, pp. 583–597, March 2011.

[3] ETSI, "Fixed Radio Systems; Point-to-point equipment; Derivation of receiver interference parameters useful for planning fixed service point-to-point systems operating different equipment classes and/or capacities," European Telecommunications Standards Institute, Tech. Rep., 2003, ETSI TR 101 854 V1.2.1. [Online]. Available: http://www.etsi.org

[4] C.-C. Wang, Z. Chen, and P. Heydari, "W-Band Silicon-Based Frequency Synthesizers Using Injection-Locked and Harmonic Triplers," *IEEE Trans. Microwave Theory Tech.*, vol. 60, no. 5, pp. 1307–1320, May 2012.

[5] A. Fard and P. Andreani, "An analysis of 1/f2 phase noise in bipolar colpitts oscillators (With a digression on bipolar differential-pair LC oscillators)," *IEEE J. Solid-State Circuits*, vol. 42, no. 2, pp. 374–384, 2007.

[6] S. Toso, A. Bevilacqua, A. Gerosa, and A. Neviani, "A thorough analysis of the tank quality factor in LC oscillators with switched capacitor banks," in *Proc. IEEE Int. Symp. Circuits Syst.*, May 2010, pp. 1903–1906.

[7] R. Van der Toorn and W. Paasschens, J.C.J.; Kloosterman. (2008) The mextram bipolar transistor model: level 504.7. [Online]. Available: http://mextram.ewi.tudelft.nl

[8] M. Bao, R. Kozhuharov, and H. Zirath, "A High Power-Efficiency D-Band Frequency Tripler MMIC With Gain Up to 7 dB," *IEEE Microwave Wireless Compon. Lett.*, vol. 24, no. 2, pp. 123–125, Feb 2014.

[9] W. Van Noort, A. Rodriguez, H. Sun, F. Zaato, N. Zhang, T. Nesheiwat, F. Neuilly, J. MelaI, and E. Hijzen, "BiCMOS technology improvements for microwave application," in *Proc. IEEE Bipolar/BiCMOS Circuits Tech. Meeting*, Oct 2008, pp. 93–96.

[10] Y. Wang, D. Leenaerts, E. Van der Heijden, and R. Mahmoudi, "A low phase noise Colpitts VCO for Ku-band applications," in *Proc. IEEE Bipolar/BiCMOS Circuits Tech. Meeting*, Sept 2012, pp. 1–4.

[11] Q. Wu, T. Quach, A. Mattamana, S. Elabd, S. Dooley, J. McCue, P. Orlando, G. Creech, and W. Khalil, "A 10mW 37.8GHz current-redistribution BiCMOS VCO with an average FOMT of -193.5dBc/Hz," in *IEEE ISSCC Dig. Tech. Papers*, Feb 2013, pp. 150–151.

[12] E. van der Heijden, A. Farrugia, R. Breunisse, C. Vaucher, and R. Pijper, "Colpitts VCOs for low-phase noise and low-power applications with transformer-coupled tank," in *Proc. IEEE Radio Frequency Integrated Circuits Symp.*, June 2008, pp. 653–656.

A Digitally-Controlled Seven-State X-Band SiGe Variable Gain Low Noise Amplifier

Robert L. Schmid and John D. Cressler

School of Electrical and Computer Engineering, 777 Atlantic Drive, N.W.
Georgia Institute of Technology, Atlanta, GA 30332-0250 USA

Abstract—**This work describes the design of a digitally-controlled seven-state variable gain low noise amplifier. The amplifier utilizes separately biased transistor cores to activate additional transistor area and change the amplifier gain. Variable gain low noise amplifiers enable greater dynamic range in RF receivers and provide amplitude control for phased-array systems. The amplifier was fabricated in a 180 nm SiGe BiCMOS technology platform featuring SiGe HBTs with an f_T/f_{max} of 240/260 GHz. When all transistor cores are activated, the amplifier achieves 16.5 dB of gain and a noise figure of 1.6 dB at 10 GHz. A gain variation of 7 dB from the maximum to the minimum state is demonstrated. In addition, the noise figure, linearity, and gain all improve with increasing transistor area and power dissipation, indicating an optimum state can be selected to meet receiver requirements while minimizing power dissipation.**

Keywords—*SiGe HBT, silicon-germanium, low noise amplifier, LNA, variable gain, VGLNA, X-band*

I. INTRODUCTION

Variable-gain low noise amplifiers (VGLNAs) play an important role in receivers. VGLNAs have the ability to reduce the gain of the system when a high power signal is received at the input and extend the dynamic range of the system. Variable gain amplifiers (VGAs) can be placed after down-conversion mixers and operate at low frequencies without stringent noise figure requirements [1], but this often creates difficult linearity requirements for the mixer. Utilizing a VGLNA early in the receiver chain can relax the linearity requirements for all subsequent components. In addition, VGLNAs can be used in phased-array systems to provide amplitude control without additional attenuators or VGAs, which would necessarily increase size and power consumption.

One method commonly used to develop a VGLNA is to add a current stealing device into a cascode core [2], [3]. In this topology the first transconductance stage converts the RF signal into a current. The output current is split between two common-base or common-gate devices. One device amplifies the current into the output load and the other "current stealing" device dumps the RF signal into the voltage rail. By controlling the bias of the current stealing device, the gain can be attenuated. However, in this topology, the current and power dissipation of the current stealing device is wasted and does not contribute to the gain of the amplifier. Another simple method to achieving gain control is to vary the gate or base bias of an amplifying transistor [4]. This approach is convenient because it requires no modification to the core amplifier, but it requires analog input voltages. In order to interface the bias nodes with digital control circuitry, additional digital-to-analog converters

(DACs) must be included, which increases the amplifier footprint and power consumption [5].

In the present work, a digitally-controlled VGLNA with seven states is designed by activating additional transistor area. The amplifier is fabricated on the TowerJazz 180 nm SBC18H3 SiGe BiCMOS technology, which features a 130 nm SiGe HBT with a peak f_T/f_{max} of 240/260 GHz. When all of the transistor area is activated, the amplifier achieves a gain of 16.5 dB with a noise figure of 1.6 dB at 10 GHz. A gain control of 7 dB is achieved by reducing the active transistor area, while maintaining a noise figure of less than 3.25 dB. In Section II, the digitally-controlled SiGe core is described. Section III discusses how the digitally-controlled core is incorporated into a seven-state VGLNA. In Section IV, measurement results are presented and compared to other published VGLNAs.

II. DIGITALLY-CONTROLLED SiGe CORE

One technique to reconfigure the RF performance of an amplifier is to activate additional transistor area. While this can be achieved by using RF switches [6]-[8], any switches in the RF path reduces the gain of the amplifier and increases the noise figure. An alternative method to activate transistor area is shown in Fig. 1, where multiple current mirrors are used to separately activate different transistor cores.

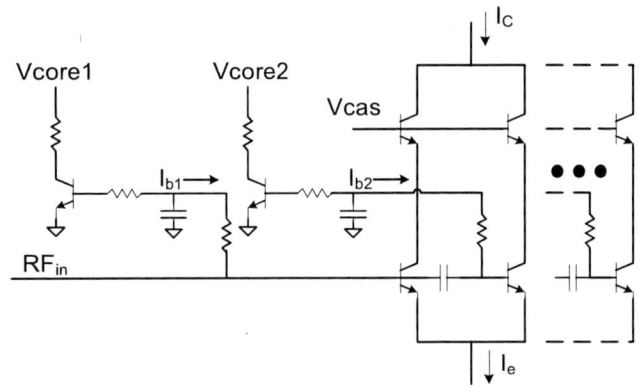

Fig. 1. Schematic diagram of a cascode core with separate resistive current mirrors for digital activation.

In Fig. 1, the DC voltage at the base node of each transistor core is isolated by using blocking capacitors. The blocking capacitors are made large in size to reduce the phase difference between amplifier cores. The resistive current mirrors enable each transistor core to be activated with a 1.0 V bias. Each core can be deactivated with 0 V bias, which prevents current from flowing through the cascode pair. Large resistors are used in

978-1-4799-7231-9/14 $31.00 © 2014 IEEE

the current mirrors to present a high impedance to the RF path and reduce the impact of the current mirrors on the RF matching.

A cascode structure is utilized to provide high gain and high isolation. The high isolation between input and output is particularly beneficial for maintaining a good output match while changing the active transistor area at the input. The open-circuit voltage gain of a cascode amplifier can be derived as given in (1) [9], where g_{m1} is the transconductance of the input transistor, $r_{\pi2}$ is the input resistance of the cascode transistor, g_{m2} is the transconductance of the cascode device, and r_{o2} is the output resistance of the cascode transistor.

$$A_{vo} = -g_{m1}r_{\pi2}g_{m2}r_{o2} \qquad (1)$$

When the active transistor area is changed using the current mirrors in Fig. 1, the effective transconductance of the common-emitter device (g_{m1}) is changed, allowing the gain of the amplifier to be controlled.

III. SEVEN-STATE VGLNA DESIGN

The digitally-controlled transistor core was incorporated into an X-band VGLNA. The VGLNA has three amplifier cores that are binary weighted in size with one, two, and four 6 µm emitter length SiGe HBTs, providing seven different states of operation. The transistor core blocking capacitors, C_{block}, are chosen to be 550 fF to balance the tradeoff between the phase delay and physical size of the capacitor. The current mirrors use 5 kΩ resistors, R_b, to inject the base current into the amplifying SiGe HBTs.

The schematic of the seven-state X-band VGLNA is shown in Fig. 2. The size of the transistor core is chosen to move the optimum noise impedance toward 50 Ω. The degeneration inductor, Le, is chosen to balance the tradeoff between achieving a good input match and providing a high gain. The output network is matched at 10 GHz using a 500 Ω resistor, 625 pH inductor and 165 fF capacitor.

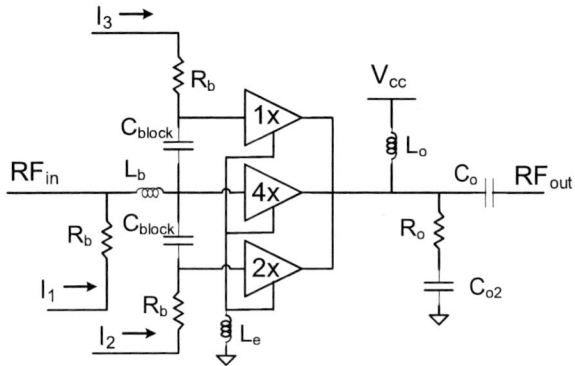

Fig. 2. Schematic diagram of the seven-state LNA. Each amplifier core represents a cascode pair as shown in Fig. 1.

The input impedance of a degenerated LNA is given in (2), where C_{be} is the base-emitter capacitance and L_b is the base inductance.

$$Z_{in} = g_{m1}\frac{L_e}{C_{be}} + j*\left[\omega\left(L_b + L_e\right) - \frac{1}{\omega C_{be}}\right] \qquad (2)$$

When extra transistor area is activated, the effective transconductance g_{m1} of the amplifier increases. However, the equivalent base-emitter capacitance remains roughly the same whether the transistor cores are activated or deactivated. As a result, the real part of the input impedance increases as additional transistor area is activated, while the imaginary part remains roughly the same. This characteristic may be used to adjust the amplifier to account for process variation or changes in the antenna impedance. Fig. 3 shows the simulated S_{11} and optimum noise impedance, Γ_{opt}, of the VGLNA at 10 GHz as the active transistor area is increased. In this case, the S_{11} has been partially compromised to achieve good noise matching.

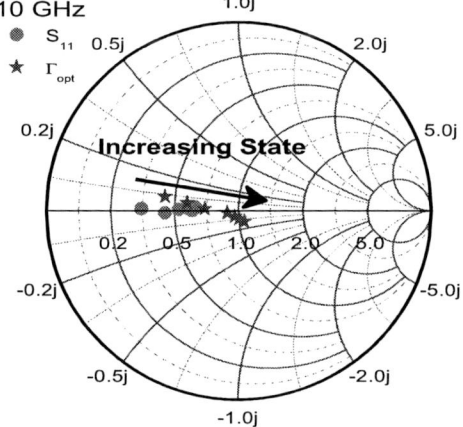

Fig. 3. Simulated S_{11} and Γ_{opt} normalized impedances at 10 GHz as a function of the state of the transistor core.

The layout of the seven-state LNA is shown in Fig. 4. The final circuit is 800 µm by 950 µm including pads and 500 µm by 700 µm without pads. The digitally-controlled SiGe core is only 80 µm by 80 µm. It is clear from the die photo that the added functionality of the digitally-controlled transistor core does not increase the overall size of the circuit.

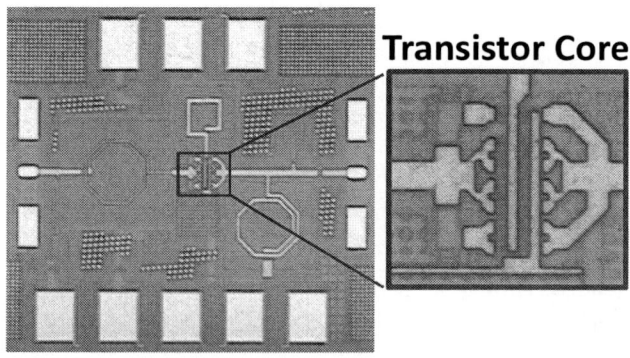

Fig. 4. Die photograph of the seven-state VGLNA with a zoomed in picture of the transistor core.

978-1-4799-7231-9/14 $31.00 © 2014 IEEE

IV. MEASUREMENT

The S-parameters of the seven-state VGLNA were measured on an Agilent E8351A PNA from 1-40 GHz. The gain of the VGLNA can be digitally controlled from 9.5 to 16.5 dB, as shown in Fig. 5. The output return loss remains the same between different states, but the input return loss degrades for small transistor areas due to the small g_{m1}, as expected from (2). In future designs, the input match could be improved at the cost of a slightly higher noise figure by modifying the device size and degeneration inductance. However, the input return loss is still better than 10 dB at 10 GHz for four of the seven states.

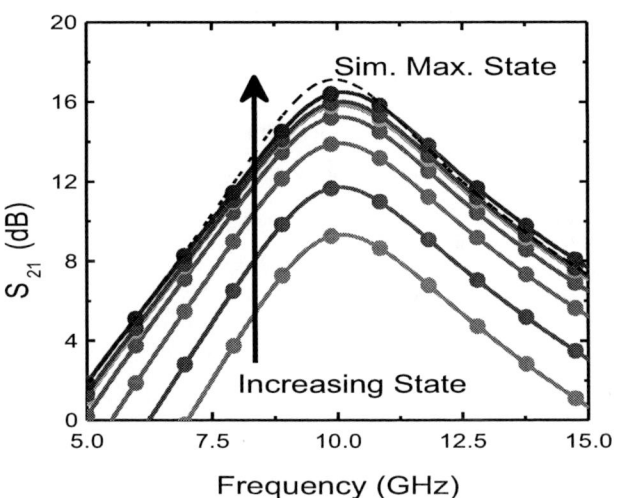

Fig. 5. Measured gain of the seven-state VGLNA as a function of the state of the transistor core.

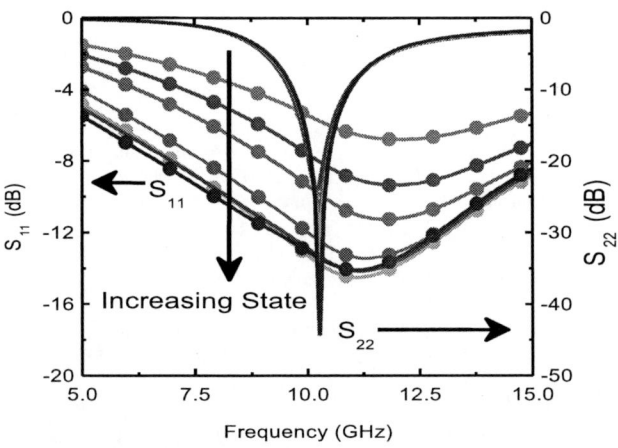

Fig. 6. Measured S_{11} and S_{22} of the seven-state VGLNA as a function of the state of the transistor core.

The noise figure and linearity of the seven-state VGLNA was measured using a custom-built integrated S-parameter, noise figure, and load-pull probing station which provides RF switching between the network analyzer, the signal sources, and the spectrum analyzer.

Fig. 7 compares the two-tone linearity of the amplifier with the smallest and largest transistor area activated, demonstrating the additional transistor area can also provide benefits in terms of linearity. The OIP3 improves from 5.5 dBm in the smallest area state to 18.5 dBm in the maximum area state.

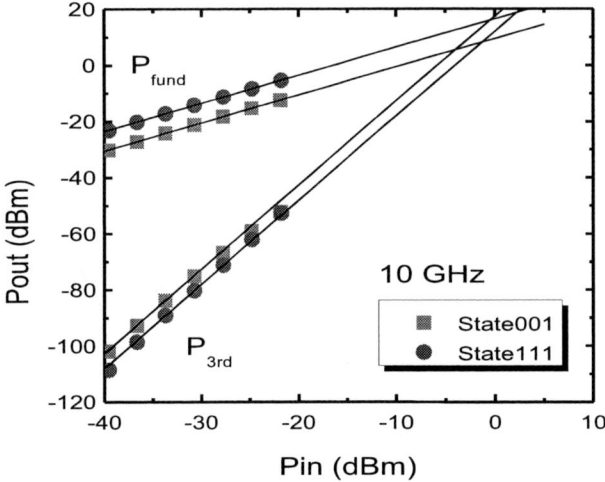

Fig. 7. A comparison of the measured two-tone linearity of the VGLNA in the maximum and minimum transistor area states.

The digitally-controlled SiGe transistor cores enable the transconductance of the amplifier to be changed by activating additional transistor area. However, the transconductance can also be increased by maintaining the same transistor area, and increasing the bias point of the amplifier. To compare these two techniques, the performance of the VGLNA across the seven states is compared to the LNA with all transistors activated and the bias voltage modified to achieve the same current density as the digitally-controlled VGLNA.

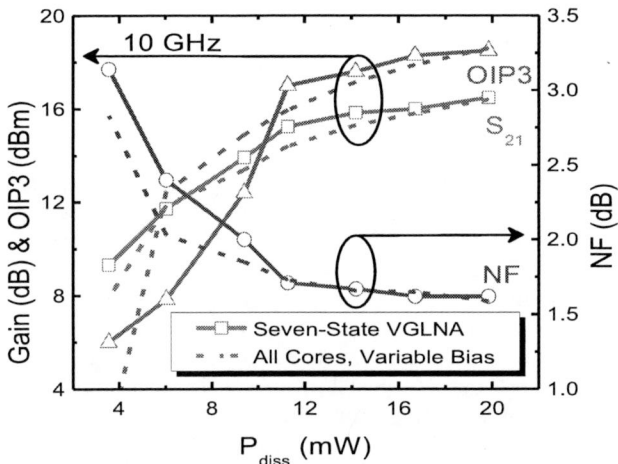

Fig. 8. Measured gain, two-tone linearity, and noise figure for the seven-state VGLNA. The performance of the VGLNA using the digitally-controlled transistor core (solid lines) is compared to having all transistors turned on and adjusting the bias for the same current consumption (dashed lines).

The digitally-controlled VGLNA and variable bias LNA show similar performance. However, there are several benefits

TABLE I. PERFORMANCE COMPARISON OF MICROWAVE VGLNAS

Reference	Technology	Topology	Freq (GHz)	Gain (dB)	NF (dB)	OIP3 (dBm)	Pdiss (mW)	Area mm²
This work	130 nm SiGe BiCMOS	Digital core	10	9.5 - 16.5	3.3 - 1.6	5.5 - 18.5	3.5 - 19.9	0.76
[2]	120 nm SiGe BiCMOS	Digital current stealing	32.0 – 34.0	9.0 - 20.0	4.3 - 3.4	-10.0 -1.0	26.3	0.13[a]
[3]	180 nm SiGe BiCMOS	Analog current stealing	8.0 – 16.0	7.5 – 15.0	8.1 - 4.5[b]	2.0 - 16.0[b]	36.0	1.17
[10]	250 nm SiGe BiCMOS	Analog bias control	1.0 – 12.0	-17.0 – 18.0	12.0 – 2.2	-	2.5 - 55	0.99
[8]	130 nm CMOS	Switchable Core	2.4 – 5.4	12.0 - 24.0	5.0 - 2.2	0.0 – 8.0	4.6	0.49[a]

[a] Without pads   [b] Not measured over gain states   - Not reported

to the digitally-controlled VGLNA topology. First, in highly-integrated systems that require reconfigurable RF circuitry or self-healing capabilities, the variable bias LNA will be controlled by some digital logic and require a digital-to-analog converter (DAC). In addition, the digitally-controlled SiGe transistor core offers improved reliability. Since each transistor core is biased separately, the amplifier can still function even if one of the transistor cores is damaged from radiation, breakdown, or fabrication failures.

The seven-state VGLNA also achieves excellent noise figure, with a noise figure of 1.6 dB when all transistor cores are activated and less than 2 dB noise figure for five of the seven states. This demonstrates that using resistive current mirrors in a digitally-controlled transistor core can provide lower noise figure than topologies which use switches to activate transistor area [6]-[8]. A clear tradeoff between the LNA performance and power dissipation is also shown in Fig. 8. Based on this tradeoff, a state can be selected to meet the receiver requirements while minimizing power consumption.

Table 1 summarizes the performance of published state-of-the-art VGLNAs at microwave frequencies. The digital core VGLNA presented in this work shows excellent noise figure in comparison to other publications. In addition, the power consumption is reduced in comparison to the current stealing designs.

V. SUMMARY

A method of activating additional transistor area using separately bias transistor cores has been investigated as means to change the gain of an RF amplifier. This technique has been incorporated into a SiGe X-band VGLNA. By utilizing three different size transistor cores, seven distinct gain states are achieved. The VGLNA demonstrates 7 dB of gain control while maintaining excellent noise figure. The gain, noise figure, and linearity of the VGLNA can all be improved by increasing the active transistor area at the cost of higher power consumption. Thus, the state of operation can be selected to meet design requirements while minimizing power dissipation. This demonstrates the possibility of using addition transistor area to reconfigure RF performance.

ACKNOWLEDGMENT

The authors wish to acknowledge the members of the SiGe Devices and Circuits team at Georgia Tech, T. Quach and the circuit design team at the AFRL, and S. Jordan and Tower Jazz for fabrication support.

REFERENCES

[1] P.K. Saha, D.C. Howard, S. Shankar, R. Diestelhorst, T. England, and J.D. Cressler, "A 6–20 GHz adaptive SiGe image reject mixer for a self-healing receiver," *IEEE Journal of Solid-State Circuits*, vol. 47, no. 9, pp. 1998-2006, Sept. 2012.

[2] B.-W. Min and G.M. Rebeiz, "Ka-Band SiGe HBT Low Phase Imbalance Differential 3-Bit Variable Gain LNA," *IEEE Microwave and Wireless Components Letters*, vol. 18, no. 4, pp. 272-274, April 2008.

[3] D.C. Howard, P.K. Saha, S. Shankar, R.M. Diestelhorst, T.D. England, N.E. Lourenco, E. Kenyon, and J.D. Cressler, "An 8–16 GHz SiGe low noise amplifier with performance tuning capability for mitigation of radiation-induced performance loss," *IEEE Transactions on Nuclear Science*, vol. 59, no. 6, pp. 2837-2846, Dec. 2012.

[4] F. Ellinger, H. Jackel., "Low-cost BiCMOS variable gain LNA at Ku-band with ultra-low power consumption," *IEEE Transactions on Microwave Theory and Techniques*, vol. 52, no. 2, pp. 702-708, Feb. 2004.

[5] S.M. Bowers, K. Sengupta, K. Dasgupta, and A. Hajimiri, "A fully-integrated self-healing power amplifier," *IEEE Radio Frequency Integrated Circuits Symposium*, pp. 221-224, June 2012.

[6] R.L. Schmid, T. Quach, A. Mattamana, and J.D. Cressler, "The analysis of SiGe HBT switchable transistor cores of low noise amplifiers," *GOMAC-Tech Conference*, pp. 115-118, Mar. 2012.

[7] Y.-H. Wang, K.-T. Lin, T. Wang, H.-W. Chiu, H.-C. Chen, and S.-S. Lu, "A 2.1 to 6 GHz tunable-band LNA with adaptive frequency responses by transistor size scaling," *IEEE Microwave and Wireless Components Letters*, vol. 20, no. 6, pp. 346-348, June 2010.

[8] C.-T. Fu, C.-L. Ko, C.-N. Kuo, and Y.-Z. Juang, "A 2.4–5.4-GHz Wide Tuning-Range CMOS Reconfigurable Low-Noise Amplifier," *IEEE Transactions on Microwave Theory and Techniques,* vol. 56, no. 12, pp. 2754-2763, Dec. 2008.

[9] J.C.H. Poh, P. Cheng, T.K. Thrivikraman, and J.D. Cressler, "High gain, high linearity, L-band SiGe low noise amplifier with fully-integrated matching network," *IEEE Silicon Monolithic Integrated Circuits in RF Systems*, pp. 69-72, Jan. 2010.

[10] W. Chang, S. Lee, J. Mun, E. Nam, "Differential variable-gain LNA for UWB system," *Microwave Integrated Circuits Conference*, 2012, pp. 377-380, Oct. 2012.

978-1-4799-7231-9/14 $31.00 © 2014 IEEE

Ultra-Low Noise and Low Power 18.7 GHz Radiometer LNAs in a 0.5 THz SiGe Technology Utilizing Back-Side Etched Inductors

Christopher T. Coen, Robert L. Schmid, John D. Cressler, Mehmet Kaynak[1], and Bernd Tillack[1]

School of Electrical and Computer Engineering, Georgia Institute of Technology, Atlanta, GA 30332 USA

[1]IHP Microelectronics, Im Technologiepark 25, 15236 Frankfurt (Oder), Germany

Abstract — This paper presents two 18.7 GHz low-noise amplifiers (LNAs) for radiometer applications designed in a SiGe technology featuring HBTs with peak f_T / f_{MAX} of 300/500 GHz. Back-side substrate etching is utilized to reduce inductor losses and improves the noise figure (NF) of the LNAs by an average of 0.12 dB across the measured band. At 18.7 GHz, the first LNA achieves 1.10 dB NF, 17.1 dBm OIP_3, and 8.6 dB gain while consuming only 5 mW of power. The second LNA achieves 1.48 dB NF, 11.5 dBm OIP_3, and 13.9 dB gain while consuming 10 mW of power. To the authors' best knowledge, these amplifiers have the lowest measured NF of all silicon-based LNAs at this frequency and are competitive with the best III-V LNAs.

Index Terms — Inductors, low-noise amplifiers, noise figure, radiometers, silicon germanium.

I. INTRODUCTION

Rapidly advancing silicon-germanium (SiGe) BiCMOS technologies have made significant inroads into numerous high-frequency commercial applications in recent years. The use of SiGe in demanding scientific applications such as remote sensing and radio astronomy, however, has been limited due to the traditionally superior performance of III-V technologies. Due to the advent of >300 GHz platforms, SiGe technologies are now gaining interest for use in these highly performance-constrained applications, especially those where large numbers of components are required and cost, reliability, and integration are important considerations.

SiGe technologies are specifically well-suited for use in microwave radiometers [1]. Radiometers are highly sensitive receivers which precisely measure incoherent radiated electromagnetic power within a defined bandwidth. The power measured by a radiometer is typically interpreted as the equivalent noise temperature of a matched resistor. The measurement precision of a total power radiometer (often referred to as radiometric resolution, or sensitivity) is given by

$$\Delta T = \frac{(T_{ANT} + T_{REC})}{\sqrt{B\tau}} \quad (1)$$

where T_{ANT} is the antenna noise temperature, T_{REC} is the receiver noise temperature, B is the bandwidth, and τ is the integration time. Remote sensing radiometers are often limited to narrow bandwidths and short integration times, so ultra-low noise LNAs are required to obtain highly precise data.

The present work demonstrates SiGe LNAs for 18.7 GHz snow and ice measurement radiometers. Proposed satellite-based snow and ice measurement concepts require hundreds of highly-integrated ultra-low power radiometer receivers [2]. Integrated SiGe receivers would enable cost reductions and

improve the feasibility of such mission concepts. The LNAs presented here leverage IHP's SG13G2 HBTs [3] (f_T / f_{MAX} of 300/500 GHz) to achieve unprecedented noise performance at 18.7 GHz with very low power consumption. To further improve NF, IHP's localized backside etching (LBE) module was utilized to remove the conductive silicon substrate under the inductors and suppress substrate-induced losses. Versions of the LNAs both with and without LBE were compared to demonstrate the performance improvements enabled by LBE.

In this paper, we report two 18.7 GHz SiGe LNAs with noise figures (NF) of 1.10 dB and 1.48 dB which consume 5 mW and 10 mW of power, respectively. To the authors' best knowledge, these LNAs have the lowest NF of all published silicon-based LNAs at this frequency to date.

II. AMPLIFIER DESIGN

Two single-stage LNAs were designed to demonstrate the potential of SG13G2 for use in this application. The first LNA was designed to demonstrate the absolute minimum NF achievable using this SiGe technology. This LNA was implemented using a common-emitter (CE) topology without a lossy on-chip input bias tee, as shown in Fig. 1(a). A supply voltage of 1.0 V was chosen to minimize power dissipation while keeping the SiGe HBTs biased in forward-active mode.

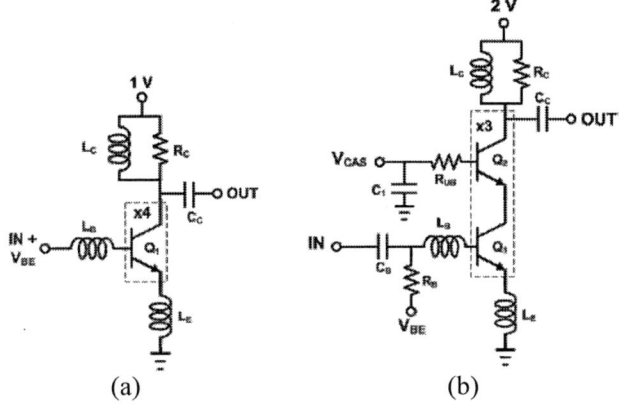

Fig. 1.   Simplified schematics of the (a) common-emitter and the (b) cascode LNAs.

The second LNA was designed to demonstrate the NF which can be achieved by a more practical LNA that does not utilize the NF reduction techniques used in the first design. A cascode core was utilized to obtain higher gain and reverse isolation than the common-emitter design, and a lossy input bias tee was included on chip to enable practical use. A schematic is shown in Fig. 1(b). Both the supply voltage and the upper base bias were set to 2.0 V to balance gain and NF and to keep both SiGe HBTs in forward-active mode. The input biasing is injected through a 5 kΩ resistor to minimize loss. To ensure stable operation, a 15 Ω resistor was added in series with the upper base to reduce the reflection coefficient at that node [4].

Both LNAs utilized the simultaneous gain and noise matching procedure outlined in [5]. The current density for each LNA core was selected to balance NF, gain, and power dissipation, and the emitter area was scaled to set the optimum noise resistance to 50 Ω. Candidate transistor cores were laid out and parasitic extracted, and this process was iterated until sufficient performance was attained. After optimization, the CE LNA core consisted of four 0.12 μm x 0.96 μm x 5-emitter SiGe HBTs operating at a current density of 2.2 mA/μm^2. The cascode LNA core used three cascode pairs consisting of 6-emitter SiGe HBTs biased at 2.4 mA/μm^2.

Emitter degeneration inductance was utilized in each design to match the input resistance to 50 Ω and improve linearity. This inductance was implemented as a thin metal trace over the substrate. A series inductor at each common-emitter base node resonates out the input capacitance and completes the input match. Each core output was matched to 50 Ω using a shunt-L series-C network, with a shunt resistor to ensure stability. Due to the finite reverse isolation of each core, this process was iterated until sufficient matching was attained.

Particular emphasis was placed on the design of the inductors. Custom square spiral inductors were designed, and all inductors used wide signal traces where possible to minimize resistive losses. The foundry's LBE module was utilized to remove the silicon substrate beneath the inductors, virtually eliminating capacitive substrate coupling and eddy current-induced losses. All inductors were EM simulated in Sonnet following the procedure in [6]. The EM-simulated quality factor (Q) for the 380 pH common-emitter LNA input inductor with and without LBE is shown in Fig. 2. LBE-induced Q improvements become apparent above 10 GHz, where capacitive losses become significant. At 18.7 GHz, LBE improves the Q from 18.0 to 20.5—a 14% improvement.

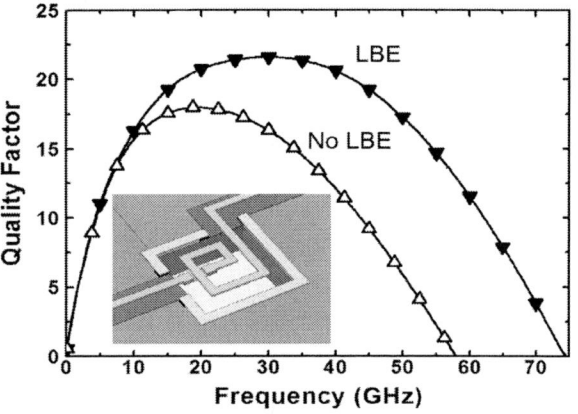

Fig. 2.   EM-simulated quality factor of the 380 pH input inductor used in the common-emitter LNA, with and without LBE. The inset shows a 3-D model of the inductor.

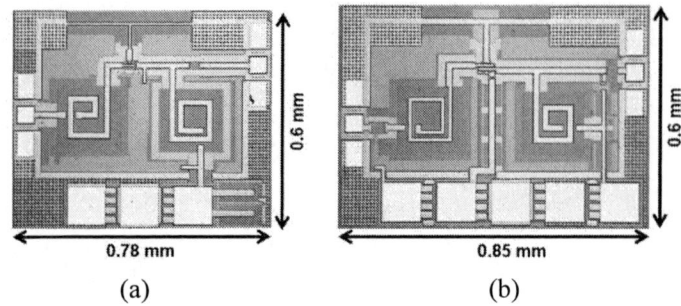

Fig. 3.   Die photographs with annotated dimensions of the (a) common-emitter and (b) cascode LNAs.

Die photographs of the fabricated LNAs are shown in Figs. 3(a) and 3(b). The bias current of each LNA is 5 mA, so the common-emitter LNA consumes 5 mW from a 1.0 V supply and the cascode LNA consumes 10 mW from a 2.0 V supply.

III. MEASUREMENTS

Each of the LNAs was characterized on-wafer, both with and without LBE, in an RF-shielded room. The measured S-parameters of the common-emitter LNA with and without LBE are shown in Fig. 4, along with the simulated S-parameters of the LNA with LBE. The measured S_{12} was greater than predicted by simulation, which was primarily due to incomplete modeling of the HBT substrate network. The reverse isolation of this common-emitter LNA is inherently low due to the Miller effect, so this S_{12} change induced a frequency shift. At 18.7 GHz, the LNA with LBE has an S_{21} of 8.6 dB as compared to 8.1 dB for the LNA without LBE. Due to the relatively low gain and reverse isolation, this LNA would be most practically used as the first stage of a multi-stage LNA in a radiometer receiver.

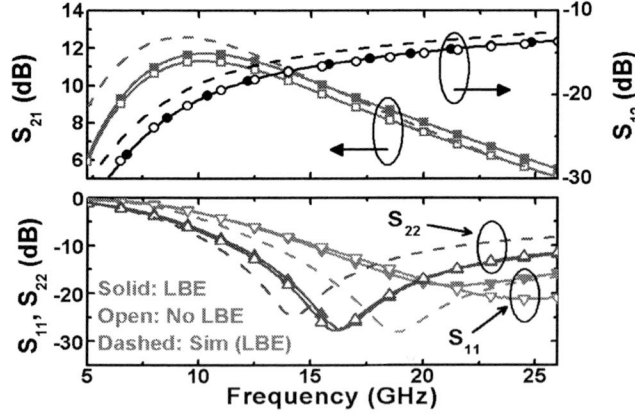

Fig. 4.   Measured S-parameters of the common-emitter LNA with and without LBE and simulated performance with LBE.

The measured performance of the cascode LNA with and without LBE versus the simulated performance with LBE is shown in Fig. 5. The measured S_{12} for this LNA was also greater than predicted by simulation. Cascode pairs have much higher reverse isolation than common-emitter HBTs, so despite the S_{12} change, a much smaller frequency shift from

978-1-4799-7231-9/14 $31.00 © 2014 IEEE

simulation is observed in this LNA. The return loss is greater than 12 dB at both the input and output, and the use of LBE only slightly shifts the matching. At 18.7 GHz, the S_{21} of this LNA is 14.0 dB with LBE and 13.6 dB without LBE—a gain improvement similar to that observed in the common-emitter LNA, as expected. The measured S_{21} is more narrowband than simulated, although the LNA is well-centered near 18.7 GHz.

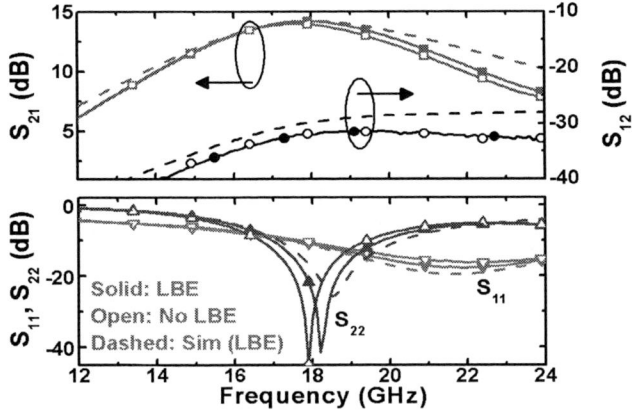

Fig. 5. Measured S-parameters of the cascode LNA with and without LBE and simulated performance with LBE.

The NF of the LNAs was measured on-wafer using an accurate cold-source noise parameter measurement method controlled by Focus Microwaves' Load-Pull Explorer software [7]. A block diagram of the measurement setup is shown in Fig. 6. A matched termination at ambient temperature is used as a cold noise source, and the output noise power is measured across a narrow bandwidth using a precisely calibrated noise receiver. Great care was taken to minimize the NF of the noise receiver, which directly impacts measurement precision. The switched-in network analyzer enables accurate vector calibration of all setup losses and impedance mismatches. The LNA noise parameters were extracted from NF measurements at 22 source impedances for each frequency, from which the reported 50 Ω NF values were obtained.

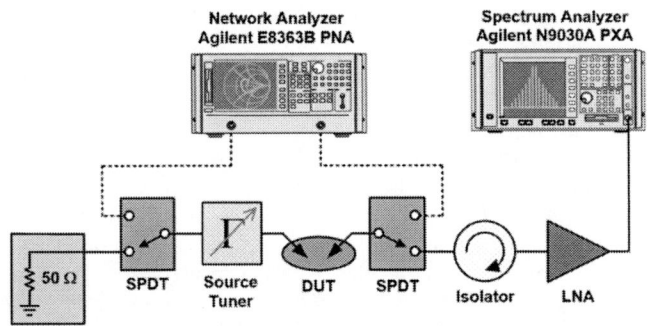

Fig. 6. Simplified block diagram of the cold-source noise measurement setup.

The measured noise performance of both LNAs with and without LBE is shown in Fig. 7. The data exhibits some ripple due to imperfect calibration and reflections in the measurement path; however, all measurements were highly repeatable and the data trends are accurate. We therefore

conservatively estimate an NF uncertainty of ±0.1 dB. The NF of the common-emitter LNA at 18.7 GHz is 1.10 dB (84 K noise temperature) with LBE, in comparison to 1.24 dB (96 K) without LBE. The cascode LNA has a measured 18.7 GHz NF of 1.48 dB (118 K) with LBE and 1.60 dB (129 K) without LBE. LBE improves the NF and noise temperature of these LNAs by averages of 0.12 dB and 11.0 K, respectively, across the band. While this LBE-induced NF reduction may seem small, this represents a roughly 10% reduction in linear noise temperature and is significant for radiometers (see eqn. (1)).

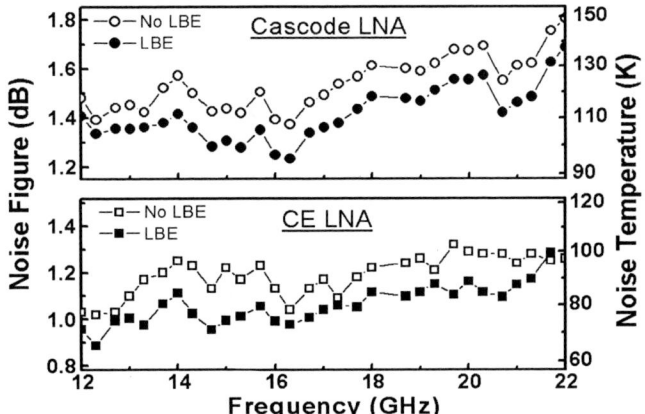

Fig. 7. Measured noise performance of both LNAs with and without LBE.

Although radiometers measure small input power levels, they operate in a crowded spectral environment and are often subjected to spurious RF interference. High linearity is desirable to prevent interference-induced compression and intermodulation distortion. The two-tone linearity of each LNA was measured at 18.7 GHz with an 8 MHz offset, and the results for each LNA with LBE are shown in Fig. 8. The extrapolated input- and output-referred third-order intercept points (IIP_3 and OIP_3) are +8.5 and +17.1 dBm for the common-emitter LNA and -2.4 and +11.5 dBm for the cascode LNA. Without LBE, the measured OIP_3 was +16.4 dBm for the common-emitter LNA and +11.2 dBm for the cascode LNA. These values are sufficiently high for typical radiometer applications.

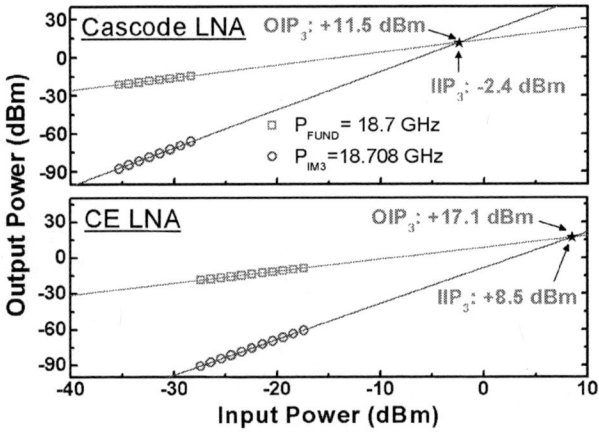

Fig. 8. Measured two-tone linearity of both LNAs with LBE.

TABLE I

COMPARISON WITH SIMILAR LOW-NOISE AND LOW-POWER LNAS

Reference	Frequency (GHz)	Noise Figure (dB)	Gain (dB)	P_{DC} (mW)	OIP_3 (dBm)	Technology
This work	**18.7**	**1.10**	**8.6**	**5.0**	**17.1**	**300 / 500 GHz SiGe HBT**
This work[1]	**18.7**	**1.48**	**13.9**	**10.0**	**11.5**	**300 / 500 GHz SiGe HBT**
[5]	10	1.36	19.5	15	20.3	200 / 285 GHz SiGe HBT
[8]	10	1.98	10	2	10	200 / 285 GHz SiGe HBT
[9][2]	33	2.9	23.5	11	4	200 / 285 GHz SiGe HBT
[10]	24	3.2	13	4.1	8	40 nm CMOS
[11][1]	10	1.60	8.7	0.60	-	0.1 µm InP pHEMT
[12][2]	16	1.06	38	16	-	70 nm GaAs mHEMT

[1]Full input bias tee included on-chip [2]Multi-stage design

The performance of these LNAs is compared with other state-of-the-art ultra-low noise and low-power LNAs in Table 1. It should be noted that many of the listed results did not utilize a lossy on-chip input bias tee unlike the cascode LNA presented here. To the authors' best knowledge, these LNAs have the lowest NF of all published silicon-based LNAs at this frequency and are competitive with the best III-V results.

IV. SUMMARY

The design and performance of 18.7 GHz radiometer LNAs in a 0.5 THz SiGe technology has been presented. These LNAs achieve record-setting noise performance for a silicon-based technology at this frequency, with very low power consumption and sufficient linearity for radiometers. We have also shown the ability of LBE to reduce inductor losses and improve circuit performance. At 18.7 GHz, LBE improved the gain of the two LNAs by 0.4/0.5 dB, the NF by 0.12/0.13 dB, and the OIP_3 by 0.3/0.7 dBm. These LNAs show that best-in-class SiGe technologies now offer competitive performance to the best III-V technologies and are suitable for use in exceptionally performance-demanding applications.

ACKNOWLEDGEMENT

The authors are grateful to IHP Microelectronics for fabrication support. We would like to thank Jeff Piepmeier from NASA Goddard Space Flight Center and John Papapolymerou at Georgia Tech for their contributions. This work was supported by a NASA Office of the Chief Technologist's Space Technology Research Fellowship.

REFERENCES

[1] C. T. Coen, J. R. Piepmeier, and J. D. Cressler, "Integrated Silicon-Germanium Electronics for CubeSat-Based Radiometers," in *IEEE International Geoscience and Remote Sensing Symposium (IGARSS)*, 2013, pp. 1286-1289.

[2] D. Cline *et al.*, "Parametric Evaluation of Cold-Land Processes Measurement Technologies, Final Report To NASA Earth Science Technology Office," NASA Earth Science Technology Office, Goddard Space Flight Center, 2003.

[3] H. Rucker, B. Heinemann, and A. Fox, "Half-Terahertz SiGe BiCMOS Technology," in *IEEE 12th Topical Meeting on Silicon Monolithic Integrated Circuits in RF Systems (SiRF)*, 2012, pp. 133-136.

[4] R. L. Schmid, C. T. Coen, S. Shankar, and J. D. Cressler, "Best Practices to Ensure the Stability of SiGe HBT Cascode Low Noise Amplifiers," in *IEEE Bipolar/BiCMOS Circuits and Technology Meeting (BCTM)*, 2012, pp. 194-197.

[5] W.-M. L. Kuo, Q. Liang, J. D. Cressler, and M. A. Mitchell, "An X-Band SiGe LNA With 1.36 dB Mean Noise Figure for Monolithic Phased Array Transmit/Receive Radar Modules," in *IEEE Radio Frequency Integrated Circuits (RFIC) Symposium*, 2006, pp. 395-398.

[6] F. Korndorfer, M. Kaynak, and V. Muhlhaus, "Simulation and Measurement of Back Side Etched Inductors," in *European Microwave Conference*, 2010, pp. 1631-1634.

[7] C. Tsironis, "AN-19: On-Wafer Noise Parameter Measurements Using Cold-Noise Source and Automatic Receiver Calibration," Focus Microwaves, Inc., 1999.

[8] T. K. Thrivikraman *et al.*, "A 2 mW, Sub-2 dB Noise Figure, SiGe Low-Noise Amplifier for X-Band High-Altitude or Space-Based Radar Applications," in *IEEE Radio Frequency Integrated Circuits (RFIC) Symposium*, 2007, pp. 629-632.

[9] B.-W. Min and G. M. Rebeiz, "Ka-Band SiGe HBT Low Noise Amplifier Design for Simultaneous Noise and Input Power Matching," *IEEE Microwave and Wireless Components Letters*, vol. 17, no. 12, pp. 891-893, 2007.

[10] M.-H. Tsai *et al.*, "An Ultra-Low Power K-Band Low-Noise Amplifier Co-Designed With ESD Protection in 40-nm CMOS," in *IEEE International Conference on IC Design & Technology (ICICDT)*, 2011, pp. 1-4.

[11] L. Liu, A. R. Alt, H. Benedickter, and C. R. Bolognesi, "InP-HEMT X-Band Low-Noise Amplifier With Ultralow 0.6-mW Power Consumption," *IEEE Electron Device Letters*, vol. 33, no. 2, pp. 209-211, 2012.

[12] A. H. Akgiray *et al.*, "Noise Measurements of Discrete HEMT Transistors and Application to Wideband Very Low-Noise Amplifiers," *IEEE Transactions on Microwave Theory and Techniques*, vol. 61, no. 9, pp. 3285-3297, 2013.

K-Band Digitally Controlled Oscillator with Integrated Divide-by-16 Divider Chain for ADPLL Applications

Christopher Maxey and Sanjay Raman
Bradley Department of ECE, Virginia Tech, Arlington, VA, USA
cmaxey@vt.edu

Abstract— **This paper presents the design and measurements of a differential K-band digitally-controlled oscillator (DCO) utilizing a tunable-dielectric resonator structure with an integrated divide-by-16 divider chain for use in an all-digital phase locked loop. The prototype differential oscillator exhibits 256 digitally-controlled states with the LSB corresponding to ~700 kHz of frequency tuning. At a 1 MHz offset, the simulated divider output phase noise is -104 dBc/Hz. The divider chain includes four divide-by-two stages in series for a final division ratio of 16. To our knowledge, this is the first published tunable-dielectric DCO and divider chain integrated onto a single chip and implemented in a SiGe BiCMOS process.**

Keywords—all-digital phase-locked loop, digitally controlled oscillator; Miller divider; SiGe BiCMOS; static divider

I. INTRODUCTION

In recent years, all-digital phase locked loops (AD-PLLs) have been explored as local oscillator (LO) sources for size- and power-constrained applications such as cellular phones and mobile wireless devices [1]. As the name suggests, the majority of the components in an AD-PLL are digital in nature, including the loop filter, and they can typically be synthesized from CMOS standard cells using hardware description languages. However, AD-PLLs still rely on several analog or hybrid analog/digital circuit blocks, particularly in high frequency regimes. A K-band AD-PLL block-level schematic is shown in Figure 1.

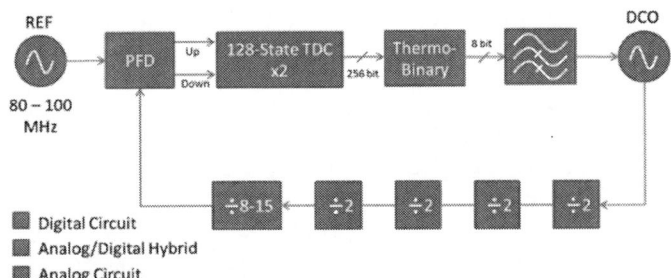

Fig. 1. Schematic of K-band AD-PLL identifying the types of circuit comprising each block

The AD-PLL has several advantages compared to more conventional analog PLLs, such as elimination of the power-consumptive charge pump and radical size reduction of the loop filter, that have become more important as advances in CMOS technology continue to drive device sizes into the deep sub-micron regime. Despite recent process advancements, AD-PLLs have to-date been primarily limited to cellular frequency ranges due to high spurious levels generated by non-linearities and finite voltage level quantization of digital circuit blocks. The goal of this research is to investigate the feasibility of extending ADPLLs into microwave/mm-wave frequency ranges by using a BiCMOS process to realize high-frequency oscillator cores and divider chains in combination with high-speed digital circuits.

In this work, the K-band signal is generated by a digitally-controlled oscillator (DCO) – a variant of the traditional voltage-controlled oscillator (VCO) that uses a digital tuning word to control the output frequency rather than an analog voltage across a varactor or series of varactors. Many previous DCO designs relied on banks of switched capacitors to tune the output center frequency in discrete steps limited by the smallest quantifiable unit of capacitance available in the chosen process. This implementation faces capacitor matching issues due to process variability and potentially high least-significant bit frequency tuning quantization depending on process ground rules for the available capacitors.

An alternative approach for implementing the oscillator tank circuit that scales well to the mm-wave regime uses a transmission line resonator fabricated in the top metal layer with a mechanism to discretely adjust the line's effective dielectric constant integrated into the underlying metal stack (DiCAD) [2]. A similar structure has been recently utilized in oscillator designs [3] in SiGe technology; this work focuses on the integration of more digital tuning bits and the divider chain.

The chain of divider circuits follows the DCO and is responsible for shifting the DCO frequency down so that it can be compared to a low frequency (and low phase noise) reference oscillator in the phase-frequency detector. In cellular band versions of AD-PLLs, these dividers can be implemented in digital logic cells (typically D flip-flops in a negative feedback configuration). At K-band, however, the f_t of the digital process (130 nm node in the case of this work) is insufficient for this approach. Analog dividers such as regenerative-mode and emitter-coupled logic latches are thus required for the first stages of the divider.

II. DIGITALLY CONTROLLED OSCILLATOR DESIGN

A. Tunable Dielectric Structure

Figure 2 shows a Sonnet 3D view of the tunable transmission line resonator used in this work. The resonator is implemented in the thick top metal (AM) of the IBM 8HP process' seven total metal layers. Strips in the next two underlying metal layers (LY and MQ) are used to control the effective dielectric constant of the transmission line. (Consequently, these strips are referred to as an 'artificial dielectric'.) Neighboring strips alternate between LY and MQ in order to maximize the strip density underneath the top-metal transmission line. By turning on and off a three transistor pi-switch situated between the two sides of the resonator along the virtual ground (see inset Figure 2), strips can be shorted together or isolated respectively.

In the all-shorted state, the effective dielectric constant is highest (the metal line crates a large effective dipole induced by the differential voltage on the line), and the resonant frequency is at its lowest value. Conversely, in the all-open state, the resonant frequency is at its highest value due to the low effective dielectric constant.

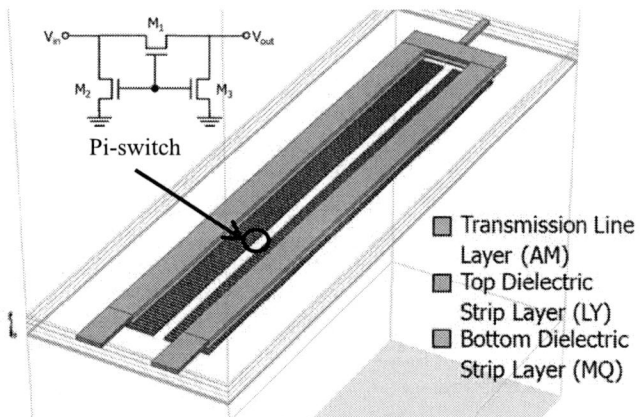

Fig. 2. Sonnet 3D view of the 256-state tunable dielectric used as the tunable element in the DCO.

As in the case of the capacitor bank DCO, a digital word applied to a series of these switches enables digital oscillator tuning between these two extremes. However, in this case, the quantization is limited to the smallest possible dielectric constant change which is a function of the separation between the line and the strips and on the width of the strips. Ultimately, the limit is based on the lithographic limits of the process in these metal layers. Spurious tones in AD-PLLs can be reduced by incorporating as many quantization states in the DCO control as possible [4].

B. DCO Design

A schematic of the prototype DCO used in this work is shown in Figure 3. The DCO comprises a cross coupled SiGe HBT pair (Q_1 and Q_2) biased by a MOSFET current source. The DCO employs the 256-state (8 bit) tunable dielectric resonator tank with control lines supplying 256 separate switches (S_1–S_{256}). For simplicity, the strips are connected together in a binary-weighted control line tree, i.e. 128 of the switches are tied to the most-significant bit of the control word, 64 to the next-most significant bit and so on. Since the strips in the LY layer are closer to the transmission line and thus have more effect on the oscillation frequency, the specific switches tied to each control line are chosen carefully so as to linearize the dielectric constant change across the tuning range. This was done to simplify actuation with DC probes; in an actual AD-PLL implementation, each switch would be independently modulated by a thermometer-encoded digital word generated by the on-chip filter, as indicated in Fig. 1. The DCO is buffered with simple common collector amplifiers (Q_3 and Q_4) before routing the output to both the divider chain and to a pair of output RF probe pads.

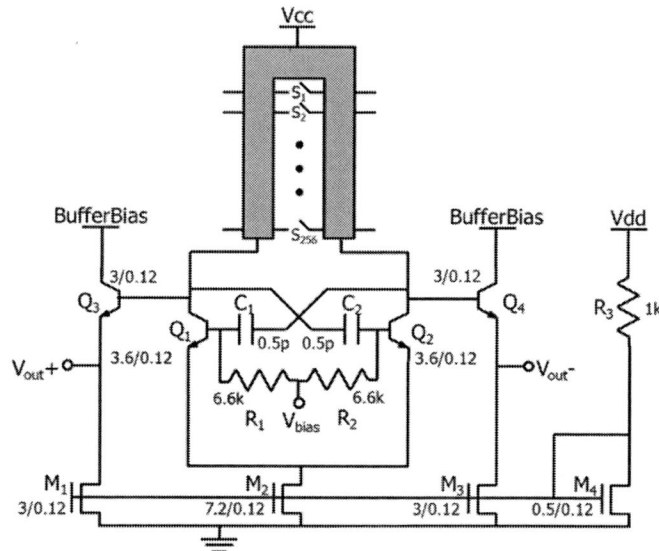

Figure 3: Schematic of 256-state digitally-controlled oscillator

III. DIVIDER CHAIN DESIGN

The divider chain is divided into two distinct sections. The first two analog stages are implemented using Miller regenerative dividers given the high frequency of the output of the DCO. The third and fourth stages are implemented using static emitter-coupled logic dividers to help lower the overall circuit power consumption.

A. Miller Regenerative Divider

The Miller regenerative divider has been a popular choice for high speed dividers since its inception in the late 1930s [5]. Advances in silicon BiCMOS technology have pushed the performance of this type of divider to 160 GHz [6]. The basic concept of the Miller divider is shown in Figure 4.

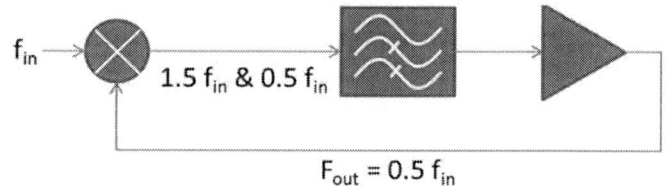

Fig. 4. High-level schematic of Miller divider

To generate the half-frequency tone, a mixer is used in a feedback configuration with the IF output routed back as the LO input. This will result in two possible main frequency components at the output of the mixer: $1.5*f_{in}$ and $0.5*f_{in}$. Since only the $0.5*f_{in}$ term is desired, a low-pass filter is integrated into the feedback path to remove higher order harmonics. An amplifier is necessary to maintain the division at a suitable amplitude unless an active mixer is used as the input stage (as was done for this work). A schematic of the divider used in the first two stages of the divider chain is shown in Figure 5.

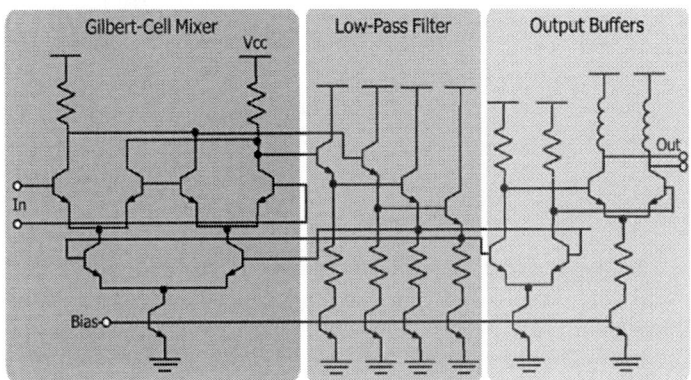

Fig. 5. Schematic of first two divider blocks.

An active Gilbert-cell mixer using the SiGe BiCMOS HBTs as the RF and LO cores is used as the divider input mixer. The output of the mixer (the 'IF') is passed through a low-pass filter network of cascaded emitter follower stages before it is routed back as the LO input to the mixer. A two-stage differential buffer amplifier is employed the output of the low-pass filter to ensure impedance compatibility with the following stages. The inductor load in the second stage of the buffer amplifier extends the bandwidth of the buffers to accommodate the high frequency of the divided output.

B. Emitter-Coupled Static Divider

Given that the first two stages have reduced the output frequency of the DCO by 4x, it is unnecessary (and prohibitively power consumptive) to also implement the third and fourth stages as Miller dividers. Instead, a lower power approach using emitter-coupled logic (ECL) latches was employed for the final two stages of the divider chain. A schematic of the ECL divider is shown in Figure 6.

As with a voltage-mode D-flip-flop divider, the emitter-coupled logic divider employs two D-latches in a master-slave configuration with an input clock buffer representing the input signal. The second latch output is fed back to the input of the first latch in a negative feedback loop. Effectively, the circuit acts as a binary two-bit counter, only changing state when two cycles of the input signal have passed, thus dividing the frequency by two. This topology does not require the two-stage buffer amplifier or low-pass filter as in the case of the Miller divider and thus has lower power consumption.

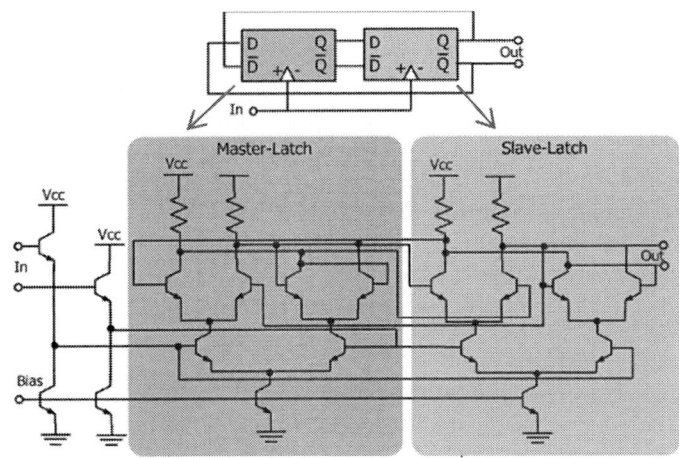

Fig. 6. Emitter-coupled latch-based static divider schematic

The final division-by-16 ratio is achieved by cascading two Miller divider stages divider with two ECL divider stages.

IV. MEASUREMENT RESULTS

The integrated BiCMOS DCO and divide-by-16 divider chain were wafer-probed and measured with an HP Agilent 8562A 9 kHz - 50 GHz spectrum analyzer to confirm proper operation of both blocks. The buffered DCO output is routed to a set of pads A photograph of the die under test is shown in Figure 7. The large inductors are located in the buffer amplifiers for the first two Miller divider stages. The die size is 1.5 mm x 1.1 mm.

Fig. 7: Photograph of die under test. The die size is 1.5 mm x 1.1 mm.

The DCO control lines and bias voltages were supplied using two 9-pin DC probes; the RF outputs from the DCO and divider chain were measured using two GSGSG RF probes. The control lines for tuning the DCO were manually switched between 0 and 1.2 V (the digital positive rail voltage) using an external breadboard. The DCO tunes between 26.63 GHz and 26.813 GHz depending on the tuning word supplied to the control lines. The LSB of the tuning word corresponds to ~700 kHz of frequency tuning. The tuning range of the divided output is 16x lower (1.664 GHz – 1.675 GHz). For this work, the goal was not to demonstrate wide tuning range, but rather to demonstrate the divider chain integration towards a subsequent ADPLL design. Wider band frequency control could alternatively be implemented with switched capacitor networks in addition to the DiCAD fine tuning structure.

978-1-4799-7231-9/14 $31.00 © 2014 IEEE

Figure 8 shows the measured phase noise of the output of the DCO compared to simulated values from Spectre for the DCO and divider stages. The output of the DCO was measured at the nodes before the input to the divider, thus the DCO output phase noise is somewhat degraded by the loading of the divider input relative to [3]. Also, the probe station measurement setup DC biasing paths could have contributed to higher than expected noise coupling (particularly for the tail current biasing lines). The cause of the flat noise response at low offsets is currently under investigation

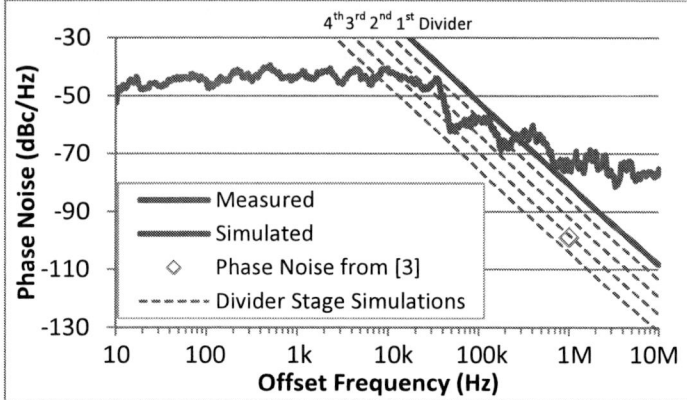

Fig. 8. Phase noise plot for DCO output.

Generally speaking, the phase noise at the output of a frequency divider can be given as:

$$\varphi_{out}^2 \approx \left(\frac{\varphi_{in}}{N}\right)^2 \qquad (1)$$

where N is the total division ratio. In the case of this work, the total reduction is ~24 dB from the 4 division stages. This is an optimistic estimate because it does not include white or flicker noise components from the divider itself; however, the simulated values for the divider phase noise match closely with this estimate. At a 1 MHz offset, the simulated divider output phase noise is -104 dBc/Hz. Measurement issues are currently preventing characterization of the phase noise at the divider output; this is currently under investigation

Figure 9 shows the measured spectrum for the DCO output at the highest operating frequency. Figure 10 shows the output spectrum of the divider output identifying the prominent peak at one sixteenth the DCO frequency; the inset of Figure 10 zooms-in on this peak.

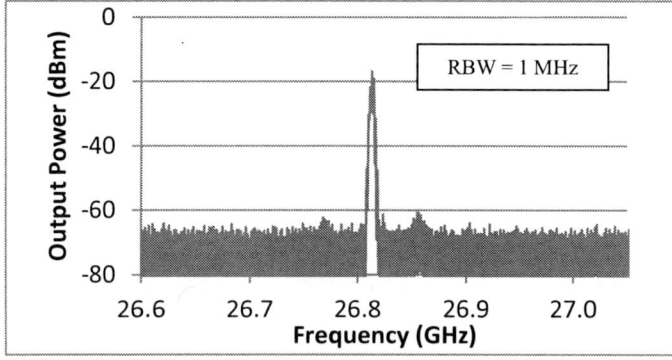

Fig. 9. DCO Output spectrum at 26.813 GHz

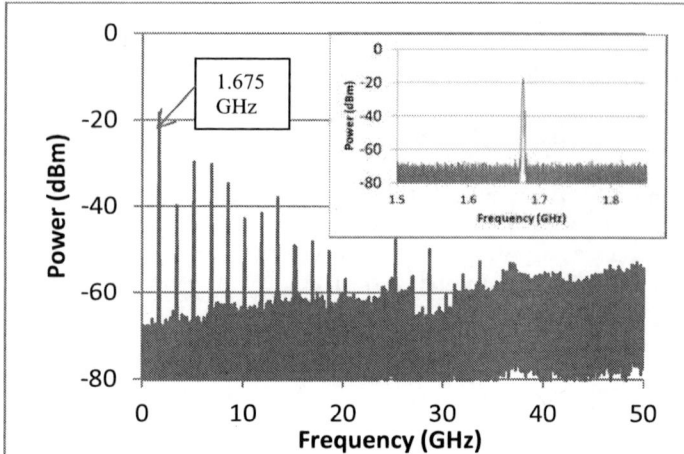

Fig. 10: Divider output spectrum (Inset: zoom-in on divide-by-16 output at 1.675 GHz)

CONCLUSIONS

This paper presented measurements of a digitally-controlled oscillator employing an artificial dielectric resonator integrated with at divide-by-16 chain with two types of frequency divider architectures. The total current consumption of the circuit is 36 mA from a 3.0 V supply that is dominated by the first two divider stages. The presented work represents a significant building block for an all-digital K-band PLL.

ACKNOWLEDGEMENTS

The authors wish to thank the generous support of Tony Quach, Kari Groves, Aji Mattamana, Len Orlando and Vipul Patel from the Air Force Research Laboratory who funded this work via a Berrie-Hill Research Corp. Task Order. The authors also wish to thank Ravi Raghunathan at Virginia Tech in Blacksburg, VA for measurement assistance.

REFERENCES

[1] R.B. Staszewski, "State-of-the-Art and Future Directions of High-Performance All-Digital Frequency Synthesis in Nanometer CMOS," *IEEE Transactions on Circuits and Systems I: Regular Papers*, vol. 58, no. 7, pp. 1497-1510, July 2011.

[2] T. LaRocca, T. Sai-Wang, D. Huang, Q. Gu, E. Socher, W. Hant, and F. Chang, "Millimeter-wave CMOS digital controlled artificial dielectric differential mode transmission lines for reconfigurable ICs," *2008 IEEE MTT-S International Microwave Symposium Digest*, pp. 181-184, 15-20 June 2008.

[3] C. Maxey and S. Raman, " K-Band Differential and Quadrature Digitally-Controlled Oscillator Designs in SiGe BiCMOS Technology," *Proceedings 2014 Topical Meeting on Silicon Monolithic Integrated Circuits in RF Systems*, 20-22 Feb. 2014.

[4] R.B. Staszewski and P.T. Balsara, *All-Digital Frequency Synthesizer in Deep-Submicron CMOS*, Hoboken, NJ: Wiley, 2006.

[5] S. Trotta, H. Knapp, T.E. Meister, K. Aufinger, J. Bock, B. Dehlink, W. Simburger, S.L. Scholtz, "A New Regenerative Divider by Four up to 160 GHz in SiGe Bipolar Technology," *2006. IEEE MTT-S International Microwave Symposium Digest*, pp.1709,1712, 11-16 June 2006.

A Wide Tuning Triple-Band Frequency Generator MMIC in 0.18μm SiGe BiCMOS Technology

Hechen Wang[1], Feng Zhao[1], Fa Foster Dai[1], Guofu Niu[1], Bogdan Wilamowski[1],
Jun Fu[2], Wei Zhou[2], and Yudong Wang[2]

1. Dept. of Electrical and Computer Eng., Auburn University, Auburn, AL 36849
2. Inst. of Microelectronics, Tsinghua University, Beijing, China

Abstract— **This paper presents a wide-tuning frequency generation scheme based on bottom-series coupled quadrature VCO (QVCO). The low band and middle band frequency signal is generated from the QVCO while the high band signal is obtained from a gilbert mixer. The tuning-range enhancement technique allows a three band frequency generation in the range from 2.4GHz to 8.4GHz without penalizing its phase noise. The VCO monolithic microwave integrated circuit (MMIC) is implemented in a 0.18 μm SiGe BiCMOS technology with 1.8 mm² area. The measured frequency range is 15.4% for the three bands centered at 2.5GHz, 5GHz, and 7.5GHz. The measured phase noise are -124.4dBc/Hz, -119.1dBc/Hz and -108.8dBc/Hz at 1 MHz offset for the low, middle, high band signals, separately. The measurement results demonstrate that proposed frequency generation technique can achieve wide tuning range capability as well as the low phase noise with very compact circuit.**

Keywords—quadrature VCO; RF; voltage-controlled oscillator; frequency doubler; tripler; Gilbert mixer; phase noise; tuning range

I. INTRODUCTION

Multi-standard wireless transceivers are in high demand for emerging applications such as smart phones and multi-function PDAs. However, wireless standards allocate channel frequencies across wide-spread spectrum from 900MHz to 5.8GHz. Ultra-wide band (UWB) requires even wider frequency band from 3GHz to 13 GHz. Radar transceivers also requires wide-tuning frequency synthesis to cover the specific radar band such as C-band (4-8GHz) and X-band (8-12GHz) [1]. Among these multi-band communication transceivers, the frequency generation block usually costs large area and power consumption since each band needs its own oscillator core. The requirement of noise optimization for each VCO core adds the design complexity. As a result, a frequency generation scheme which can provide wide frequency range from a compact circuit without several inductors becomes a valid technique to address those demands.

Frequency range and phase noise are two crucial parameters for VCO designs. Following traditional design approach, the target of low phase noise and wide frequency range usually cannot be achieved simultaneously with one VCO core since it requires large varactor size but this will deteriorate the quality factor Q of the tank and thus leading to poor phase noise performance. Moreover, large varactor will

increase the sensitivity of noise to amplitude fluctuations due to AM-FM conversion mechanism.

As a result, enlarging varactors dimensions are unacceptable when the bandwidth requirement extends to a certain level. Therefore, a compact frequency generation scheme that can provide both wide range and low phase noise at the same time will relax such design challenges.

This paper proposes a frequency generation scheme that achieves wide tuning range without penalizing phase noise performance from a low-noise QVCO circuitry. Section II will describe the proposed technique and main functional blocks. The implementation and measurement results will be given in Section III. Section IV draws the conclusion.

II. PROPOSED FREQUENCY GENERATION MODULE

The system diagram of the proposed frequency scheme is given in Fig. 1. The low-band quadrature signals are generated at the output of QVCO core. The differential middle band signals are obtained at the common-mode node of the cross-coupled differential pair. Mixing the differential fundamental frequency signals (f_0+ and f_0-) with its second harmonic ($2f_0$) produces an upper sideband and a lower sideband ($2f_0 \pm f_0$) at the output nodes of the differential mixer. By filtering the lower frequency terms with the bandpass filter (BPF), signal with a frequency $3f_0$ can be produced at the output of BPF block. The BPF-load is formed by an inductor and two capacitors that peaks at frequency $3f_{vco}$ which provides filtering effect to rest frequency bands. As a result, triple band signal generation is achieved with one QVCO core and a mixer. The QVCO core not only functions as low-band frequency generation but also works as frequency doubler without extra circuit.

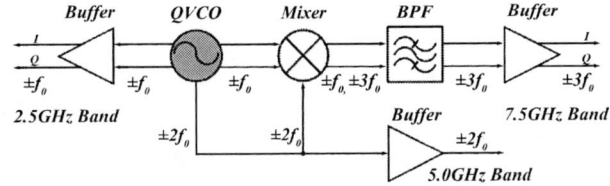

Fig. 1. Proposed triple band frequency generation scheme.

A. Bottom-Series QVCO Core

Fig. 2(b) illustrates the QVCO core circuit. SiGe bipolar

978-1-4799-7231-9/14 $31.00 © 2014 IEEE

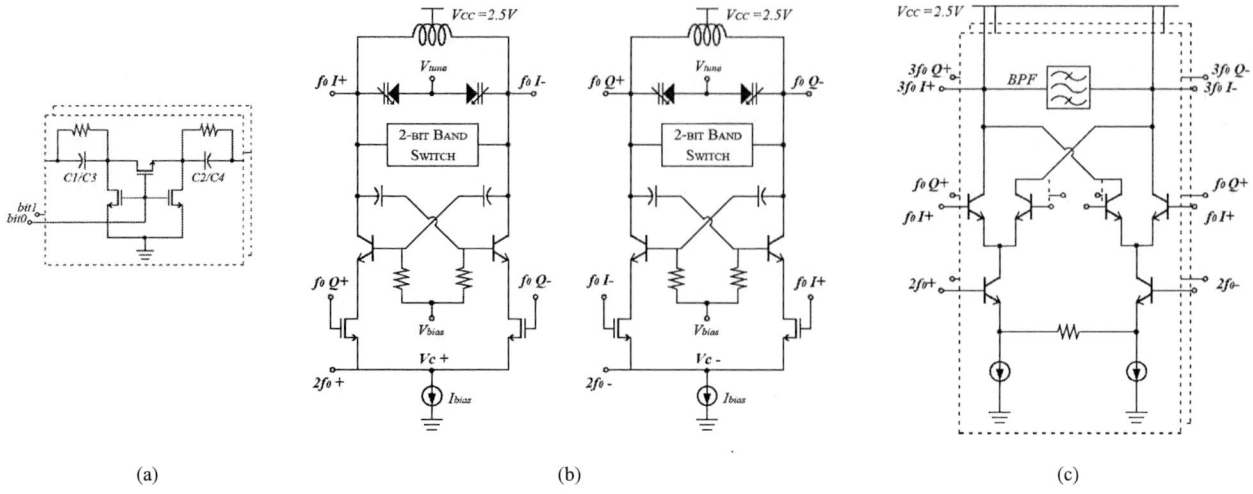

Fig 2. Proposed circuit. (a) band switches. (b) circuit schematic of the proposed QVCO. (c) Gilbert cell based frequency tripler.

transistors are used for oscillation transistors in order to obtain a better current efficiency. Signal swing of the oscillator output was maximized to improve the phase noise performance. A tradeoff between signal swing and avoiding BJT's saturation are carefully balanced through the transistors' size, the bias voltage and current optimization. If the main oscillation transistors are operated in the saturation region, noise coupling to the substrate will degrade the overall VCO phase noise greatly.

The resonant tank is built by using a three turns differential inductor with optimized Q factor at 2.5GHz and two varactors with reasonable size to improve the phase noise performance. This work uses bottom-series MOSFET coupling (BS-QVCO) to achieve a relatively low phase noise and lower power consumption [2].

B. 2-Bits Capcitor Array Band Switch

As mentioned before, the tradeoff between tuning range and phase noise eventually determines the size of the varactors. As a result we cannot arbitrarily increase VCO's tuning range by simply increasing the size of varactors. In order to widen the tuning range without affecting overall phase noise, this work introduces a 2-bit band switch controlled capacitor array to the LC-tank resonator to switch the resonant frequency band instead of using large varactors.

The capacitor array is built using four capacitors. Binary weighted capacitance is adopted to enlarge tuning range. The implemented capacitance ratio is *(C1=C2): (C3=C4) = 1: 2*. Each bit contains two identical capacitors to compensate the process, voltage and temperature (PVT) variations. In this arrangement, two binary bits can provide the control of four different frequency sub-bands.

Conventional switches used for coarse tuning between different VCO bands consist of three NMOS transistors. However, parasitic capacitances resulted from parasitic drain (n+)-to-substrate (p+) diodes vary with the dc voltage applied to its drain, i.e. the smaller the dc voltage, the larger the parasitic capacitance. When the NMOS transistors are off, the drain of the NMOS switch would be floating and close to zero and thus large parasitic capacitance loads the LC tank. Moreover, to decrease the series resistor and increase the

quality factor of the capacitor array, the NMOS transistors for switches are usually implemented with large dimensions. Then the large tuning range is suffered from these parasitic drain-substrate diodes. In order to widen the VCO frequency tuning range, when the NMOS switches are off, a dc voltage equal to supply voltage is applied to the drain through two parallel resistors as shown in Fig. 2 (a). With the proposed frequency extending technique, the frequency tuning range can be increased to cover 15.4% tuning range without degrading the phase noise.

C. Frequency Doubling & Tripling

The double frequency band is obtained at the virtual ground (nodes V_C+, V_C- in Fig.2 (b)) of the differential pair, where odd order harmonics are cancelled and even order harmonics survived. Comparing to MOSFET transistors, which have greater parasitic capacitors than BJTs, BJT differential pair has larger 2nd order harmonic component at its common mode. And derived from differential pair nonlinear signal model, the 2nd order harmonic gains for BJT and MOSFET transistors at virtual ground are given by

$$A_{V2^{nd},MOS} = \frac{I_{tail}}{\frac{(V_{OV})^2}{2}} R_s, \qquad V_{OV} = V_{GS} - V_{TH} \qquad (1)$$

$$A_{V2^{nd},BJT} = \frac{I_{tail}}{V_T^2} R_s \qquad (2)$$

where I_{tail} is tail current of the differential pair, V_{OV} and V_T are transistor's overdrive voltage and thermal voltage, and R_S is current mirror's output resistance.

Thermal voltage is 26mV and MOSFET's overdrive voltage is around 100mV for low power design, normally. Under identical working condition, same tail current, $1/V_T^2$ is much larger than $1/(V_{OV}^2/2)$, which leads to a greater 2nd harmonic gain for BJT transistors. The amplitude of the 2nd order harmonics at the common mode node of a BJT differential pair can reach as high as 254mV$_{pp}$ or −5.95dBm in simulation.

As far as the triple frequency band concerned, many designs use single balanced mixer to construct the frequency

978-1-4799-7231-9/14 $31.00 © 2014 IEEE   200

multiplication [3], which is easy to be implemented. However, when the frequency increases, clock feed through and coupling to the substrate will degrade the mixer performance which is not desirable.

In order to achieve low noise and low spurious tones, a frequency tripler built with double-balanced Gilbert cell is utilized. The Gilbert mixer has a high common-mode rejection ratio and can suppress the clock feed through. The two input signals from the frequency doubler (virtual ground of the QVCO) are differential signals that can be directly connected to Gilbert cells' differential inputs. The triple frequency band generation mechanism is inferred as follows:

$$Signal\ I = cos(\omega t)\ cos(2\omega t) = cos(3\omega t), \quad (3)$$

$$Signal\ Q = cos\left(\omega t + \frac{\pi}{2}\right) cos(2\omega t) = cos\left(3\omega t + \frac{\pi}{2}\right). \quad (4)$$

Mixing fundamental quadrature frequency signals with double-frequency signal generates quadrature phase triple-frequency signals.

D. VCO Output Buffer

An output buffer is needed for the oscillator testing. Fig 3. shows the circuit of the output buffer. A 3-stage-buffer is implemented in this design to isolate the LC-tank from loading and to increase the output buffer's drive capability. Directly loading the oscillator LC-tank will alter oscillation center frequency and lower the loaded Q of the tank as well. The bias currents of these stages are gradually increased stage by stage. MOSFETs are used in the first two stages because they can handle larger voltage swings than BJTs. The last stage uses BJTs which have better current drive capability with higher g_m efficiency.

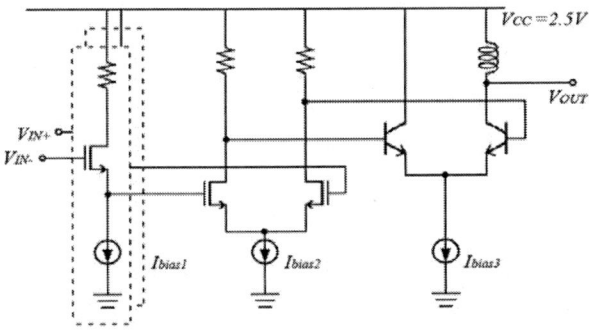

Fig. 3. Output buffer circuit.

III. IMPLEMENTATION AND MEASURED RESULTS

The prototype of this wideband frequency generator MMIC is implemented in a 0.18μm SiGe BiCMOS process. The VCO core consumes 15mW and the frequency multiplier consumes 40mW under a 2.5V supply voltage. The die photograph is shown in Fig. 4 with total area including bond pads of 3.6m². The VCO core area is 1.8x1.0 mm², and the rest on-chip area is occupied by a built-in band pass filter for 7.5GHz band output. Off-chip filter can substitute for it when chip's size is limited.

A. Measured Phase Noise Results

Phase noise measurements were performed on an Agilent E4446A spectrum analyzer with the phase noise measurement option. The phase noise is measured as -124.4dBc/Hz, -119.1dBc/Hz and -108.8dBc/Hz at 1 MHz offset in each band, respectively.

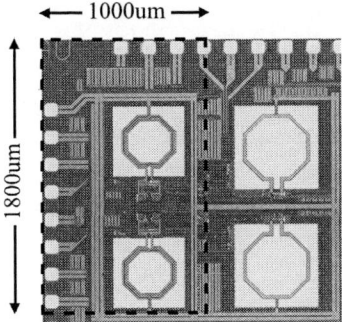

Fig. 4. Die photo of the wideband frequency generator MMIC

Fig. 5. shows the measured and simulated phase noise in the 2.5GHz band, 5GHz band, and 7.5GHz band at center point of each band's tuning curve (a 1.2 V voltage difference across the varactors). In the first two bands, measurements show good agreement with the simulation results. For 7.5GHz band, the measured result has a relatively large deviation from simulation result, mainly due to the process drifting of the built-in band pass filters.

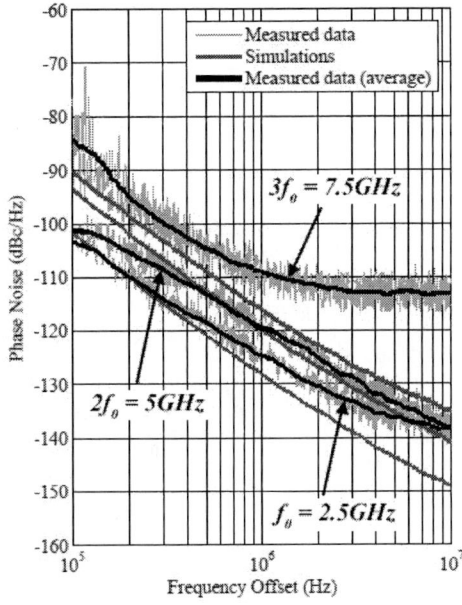

Fig. 5. Measured/simulated phase noise at 2.5, 5.0, and 7.5GHz, respectively.

B. Measured Tuning Range

The measured tuning range of the QVCO fundamental frequency band is presented in Fig. 6. According to the measured results, a 15.4% tuning range (from 2.395GHz to 2.804GHz) is achieved in the fundamental frequency band (f_0=2.5G).

Tuning ranges of other two frequency bands are derived from the fundamental band accordingly, since they are produced by mixing the fundamental frequencies. Phase noise varies within the tuning range. Fig. 7. Shows the measured phase noise of the QVCO at 1MHz offset across the entire tuning range of the fundamental frequency band. The phase noise degraded at 2.4GHz point is because varactors have been biased at the edge of forward biasing range.

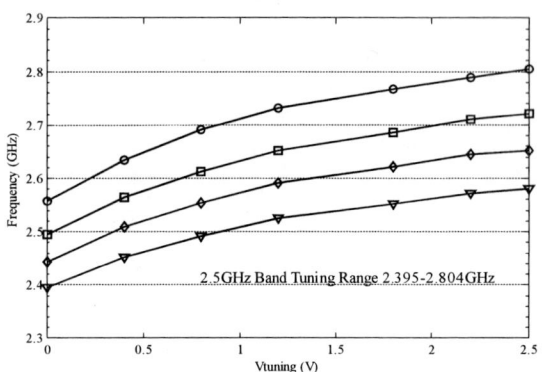

Fig. 6. Measured tuning rang of the QVCO fundamental frequency.

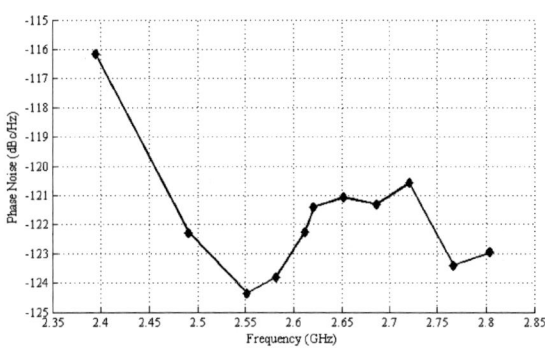

Fig. 7. Measured phase noise versus the tuning range of the proposed circuit @1MHz offset in fundamental frequency band.

C. VCO Performance Comparison

Table I summarizes the performance of the proposed QVCO and comparison with previously published QVCO work. When compared with prior art, the proposed QVCO and frequency multiplier achieves a Figure of Merit (FoM) of 182.2dB, 182.9dB and 170.4dB for each band separately, where the FoM is defined as [4]:

$$FoM = 10log\left[\left(\frac{f_0}{\Delta f}\right)^2 \frac{1mW}{P}\right] - L(\Delta f) \qquad (5)$$

In the above definition, $L(\Delta f)$ is the phase noise at the Δf offset from the oscillator frequency f_0, and P is the QVCO's core power consumption in mW. The FoM is calculated for each band. First two bands consume same amount of power. The third band's power include main core and mixer cell.

IV. CONCLUSIONS

Using a 0.18um SiGe BiCMOS technology, this paper have presented a compact integration design consisting a mixer based low phase noise wide tuning range QVCO producing three frequency bands of 2.5GHz, 5GHz, 7.5GHz respectively.

According to the tested results, the proposed QVCO with frequency multiplier demonstrated a low-phase-noise and wide tuning range performance. Each frequency band achieves 15.4% tuning range while the phase noise is -124.4dBc/Hz, -119.1dBc/Hz and -108.8dBc/Hz respectively. The QVCO uses NPN transistors for oscillation and NMOS devices for coupling. The second harmonic of QVCO is easy to obtain and utilize with SiGe BiCMOS process. The QVCO occupies a core area of 1.8mm².

TABLE-I PERFORMANCE COMPARISON OF QVCOS.

Ref./Tech	Freq. [GHz]	Tuning Range	PN@1MHz [dBc/Hz]	Power [mW]	FoM [dB]
[5] /CMOS 0.13μm	5.5	/	-117	5.28	184.58
[6] /CMOS 0.13 μm	9.6	6.6%	-121 @3MHz	9	182.6
[7] /CMOS 0.18 μm	10	15%	-95	14.4	163
[8] /CMOS 0.18 μm	4.8	/	-125	22	185
This work /SiGe 0.18 μm	2.5	15.4% all bands	-124.4	15	182.2
	5		-119.1	15	182.9
	7.5		-108.8	55	170.4

ACKNOLEGEMENT

The authors would like to acknowledge Huahong Grace Semiconductor Manufacturing Corporation for support of the IC fabrication.

REFERENCES

[1] Feng Zhao, Jianjun Yu, Joseph Cali, Fa Foster Dai, J. David Irwin, and Andre Aklian, "A 4.8-6.8GHz Phase-Locked Loop with Power Optimized Design Methodology for Dividers," *IEEE Bipolar / BiCMOS Circuits and Technology Meeting (BCTM)*, Bordeaux, France, Oct. 2013

[2] P. Andreani and X. Wang, "On the phase-noise and phase-error performance of multiphase LC CMOS VCOs," *IEEE J. Solid-State Circuits*, vol. 39, No. 11, pp. 1883-1893, Nov. 2004.

[3] Pei-Kang Tsai, Tzuen-Hsi Huang, "Integration of Current-Reused VCO and Frequency Tripler for 24-GHz Low-Power Phase-Locked Loop Applications", *IEEE transactions on circuits and systems—II*: express briefs, vol. 59, No. 4, April 2012

[4] P. Kinget, B. Soltanian, S. Xu, S. Yu, and F. Zhang, "Advanced design techniques for integrated voltage controlled LC oscillators," *Proceedings of Custom Integrated Circuits Conference*, 2007, pp. 805.

[5] C.-Y. Jeong and C.-S. Yoo, "5-GHz low-phase noise CMOS quadrature VCO," *IEEE Microw. Wireless Compon. Lett.*, vol. 16, no. 11, pp.609–611, Nov. 2006.

[6] I. R. Chamas and S. Raman, "Analysis and design of a CMOS phasetunable injection-coupled LC quadrature VCO (PTIC-QVCO)," *IEEE J. Solid-State Circuits*, vol. 44, pp. 784–796, Mar. 2003.

[7] S. Li, I. Kipnis, and M. Ismail, "A 10-GHz CMOS quadrature LC-VCO for multicore optical applications," *IEEE J. Solid-State Circuits*, vol. 38, no. 10, pp. 1626–1634, Oct. 2003.

[8] S. L. J. Gierkink,S. Levantino, R. C. Frye, C. Samori, and V. Boccuzzi, "A low-phase-noise 5-GHz CMOS quadrature VCO using superharmonic coupling," *IEEE J. Solid-State Circuits*, vol. 38, no. 7, pp. 1148–1154, Jul. 2003.

[9] Feng Zhao and Fa Foster Dai, "A 0.6-V quadrature VCO with enhanced swing and optimized capacitive coupling for phase noise reduction," *IEEE Trans. Circuits Syst. I, Reg. Papers, vol. 59, no. 8, pp. 1694–1705, Aug. 2012*

Small-Signal Modeling of the Lateral NQS Effect in SiGe HBTs

Shon Yadav, Anjan Chakravorty
Department of Electrical Engineering
Indian Institute of Technology Madras
Chennai 600036, India
Email: ee12s022@ee.iitm.ac.in, anjan@ee.iitm.ac.in

Michael Schröter
CEDIC, Department of Electrical Engineering
Dresden University of Technology
Dresden 01062, Germany
Email: schroter@iee.et.tu-dresden.de

Abstract—**Detailed formulations for DC and AC emitter current crowding are presented in view of developing an extended π-equivalent circuit (EC) model to accurately predict the lateral non-quasi-static effects in silicon germanium heterojunction bipolar transistors. Under negligible DC current crowding, the EC reduces to a simple π-model. The implementation-suitable versions of the models are also developed. Compared to state-of-the-art model formulations, the extended π-model shows better accuracy in predicting device simulated data. If desired, the high level of accuracy obtained by the extended π-model can be traded with the required extra simulation time due to one extra node.**

Keywords—*SiGe HBT, NQS Effects, Current crowding, π-model.*

I. INTRODUCTION

Silicon germanium heterojunction bipolar transistors (SiGe HBTs) have been favored for high speed and high power applications. At frequencies near the cutoff frequency (f_T), transistors show non-quasi-static (NQS) effects [1][2]. For the power HBTs, the lateral NQS effect is more dominant due to the larger emitter width yielding DC as well as AC current crowding [2][3]. Efforts were made to model the AC current crowding via small-signal analysis considering the distributed nature of the base and finally a simple lumped impedance model is obtained [2]. The low frequency approximation of this model, which is suitable for simulator implementation, yields inaccurate results above $f_T/5$. In [1], this low frequency approximated model is modified to a single-pole rational function form which improved the model accuracy but is invalid for large-signal transient simulation. A distributed π-equivalent circuit (EC) model in [4] also shows better accuracy compared to a lumped impedance model and may be better suitable for large-signal transient simulation; however the physical basis of such a π-model has not been established so far. In [5], the DC and AC current crowding effects are considered together but there exists no closed form solution. However corresponding numerical solutions are given in [6][7]. The use of an effective emitter width, obtained from the DC current crowding effect in the AC solution resulted in a more accurate and suitable model as shown in [7].

In this paper, we develop a physics-based extended π-model and demonstrate that it is more accurate than a conventional lumped impedance model and a standard model of [1]. The new model reduces to a simple π-model at negligible DC current crowding. In section II we show how DC and AC

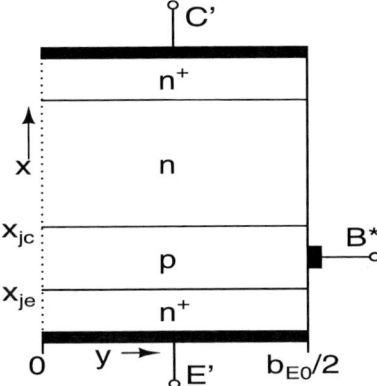

Fig. 1: *Half structure of the investigated internal structure of SiGe HBT.*

current crowding effects can be incorporated in the extended π model. We also propose the implementation suitable forms of the extended π- as well as the π-model. In section III, we present the results with discussions followed by conclusions in section IV.

II. MODEL FORMULATION

Fig. 1 shows the internal part of the investigated bipolar transistor. An extended π-equivalent circuit model for the base-emitter portion of this structure is shown in Fig. 2(a). Parameter α is used to split the total base current (I_{BEi}) between two branches. α varies between 0 and 1. A zero value for α converts Fig. 2(a) into a conventional lumped impedance model with $R_1 + R_2 = R_{Bi}$ [4] being R_{Bi} the total internal base resistance. From Fig. 2(a) we obtain

$$R_1 = \frac{V_{B^*E'} - V_{B_1E'}}{I_{BEi}} = \frac{(1-\zeta)V_{B^*E'}}{I_{BEi}}, \quad (1)$$

where the model parameter ζ relates $V_{B^*E'}$ with $V_{B_1E'}$ by $V_{B_1E'} = \zeta V_{B^*E'}$. The R_2 can be obtained as

$$R_2 = \frac{V_{B_1E'} - V_{B'E'}}{(1-\alpha)I_{BEi}} = \frac{\zeta V_{B^*E'} - V_{B'E'}}{(1-\alpha)I_{BEi}}. \quad (2)$$

978-1-4799-7231-9/14 $31.00 © 2014 IEEE

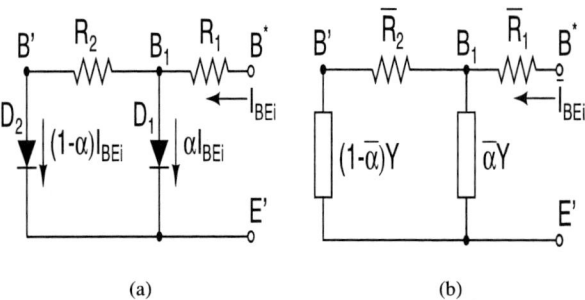

Fig. 2: *(a) Static equivalent circuit of extended π-model. (b) Small signal AC equivalent of extended π-model.*

To get rid of $V_{B'E'}$ from (2) we use the expression for the lateral base potential [4] as

$$V(y) = V_{B^*E'} - 2V_T \ln \left[\frac{\cos\left(Z \frac{2y}{b_{E0}} \right)}{\cos(Z)} \right] \quad (3)$$

with b_{E0} as the emitter width and Z is a measure of DC current crowding. Following [4] Z is expressed as

$$Z \tan(Z) = \frac{r_{SBi}}{V_T} \frac{b_{E0}}{8 l_{E0}} I_{BEi} \quad (4)$$

where r_{SBi} is the internal base sheet resistance and l_{E0} is the emitter length. From (3), we obtain $V_{B'E'}$ as the potential at $y = 0$ as

$$V_{B'E'} = V_{B^*E'} - 2V_T \ln\left(\sec(Z)\right). \quad (5)$$

Substituting $V_{B'E'}$ from (5) into (2) yields,

$$R_2 = \frac{1}{(1-\alpha)I_{BEi}}[(\zeta - 1)V_{B^*E'} + 2V_T \ln(\sec(Z))]. \quad (6)$$

Now we need to find α. From Fig. 2(a) we obtain

$$\alpha I_{BEi} = I_{SBi1} exp \left(\frac{V_{B^*E'} - I_{BEi}R_1}{V_T} \right), \quad (7)$$

$$(1 - \alpha)I_{BEi} = I_{SBi2} exp \left(\frac{V_{B'E'}}{V_T} \right) \quad (8)$$

where V_T is the thermal voltage, I_{SBi1} and I_{SBi2} are the saturation currents for the diodes D_1 and D_2, respectively. Additionally we have the expression for I_{BEi} [4] as

$$I_{BEi} = I_{SBi} exp \left(\frac{V_{B^*E'}}{V_T} \right) \frac{\cos(Z)\sin(Z)}{Z} \quad (9)$$

where $I_{SBi} = I_{SBi1} + I_{SBi2}$. Dividing (7) by (9) and using (1) we obtain

$$\alpha = \gamma \alpha' exp \left[\frac{(\zeta - 1)V_{B^*E'}}{V_T} \right], \quad (10)$$

where $\alpha' = Z/(\cos(Z)\sin(Z))$ and $\gamma \ (= I_{SBi1}/I_{SBi})$ represents the degree to which EC is distributed. To find out γ, first we divide (8) by (9) and substitute $V_{B'E'}$ from (5) to obtain

$$(1 - \alpha) = (1 - \gamma)\alpha' \cos^2(Z). \quad (11)$$

Now substituting α from (10) into (11) we obtain

$$\gamma = \frac{Z \cot(Z) - 1}{Z \cot(Z) - \alpha' exp \left[\frac{(\zeta-1)V_{B^*E'}}{V_T} \right]}. \quad (12)$$

For small-signal AC case, the diodes are replaced by their small signal admittances ($\overline{\alpha}Y$ and $(1 - \overline{\alpha})Y$) in Fig. 2(b). Also R_1, R_2 and α are replaced by a modified $\overline{R}_1$, $\overline{R}_2$ and $\overline{\alpha}$, respectively. It is important to consider the influence of DC current crowding onto the AC current crowding. However, since there is no closed form solution for the general case of DC and AC current crowding together, we introduce a modified emitter utilization factor following [8] as

$$f_{MEU} = \frac{2 l_{E0} \int_{(1-\gamma)\frac{b_{E0}}{2}}^{\frac{b_{E0}}{2}} \underline{J}_{nx}(y)dy}{\gamma b_{E0} l_{E0} \underline{J}_{nx}|_{\frac{b_{E0}}{2}}}$$

$$= \frac{\tanh(\eta_{ac})(1 - \cosh(\gamma \eta_{ac}))}{\gamma \eta_{ac}} + \frac{\sinh(\gamma \eta_{ac})}{\gamma \eta_{ac}}, \quad (13)$$

to obtain an AC solution. A unity value for γ ensures $(1 - \gamma)\frac{b_{E0}}{2} = 0$ and $f_{MEU} = f_{EU} = \tanh(\eta_{ac})/\eta_{ac}$ as detailed in [8]. Here $\eta_{ac} \ (= \sqrt{3YR_{Bi,nc}})$ is the AC current crowding factor, $\underline{J}_{nx}(y)$ is spatially varying AC emitter current density and $R_{Bi,nc} \ (= \frac{r_{SBi}b_{E0}}{12 l_{E0}})$ is the internal base resistance at negligible DC current crowding. Therefore, a modified AC $\overline{\gamma}$ can be obtained from the DC γ as

$$\overline{\gamma} = f_{MEU}\gamma. \quad (14)$$

Using this $\overline{\gamma}$ in (10) we obtain the AC $\overline{\alpha}$ and subsequently using the AC $\overline{\alpha}$ in (6) we obtain the AC $\overline{R}_2$ (in Fig. 2(b)) as

$$\overline{\alpha} = \overline{\gamma}\alpha' exp \left[\frac{(\zeta - 1)V_{B^*E'}}{V_T} \right],$$

$$\overline{R}_2 = \frac{1}{(1-\overline{\alpha})I_{BEi}}[(\zeta - 1)V_{B^*E'} + 2V_T \ln(\sec(Z))]. \quad (15)$$

In Fig. 2(b) we have also used $\overline{R}_1$ just to consider that ζ can differ from DC to AC operations.

It is clear that $\overline{\alpha}$ and $\overline{R}_2$ are frequency-dependent via f_{MEU}. Due to the complexities involved in (15), $\overline{R}_2$ and $\overline{\alpha}$ cannot be directly implemented in a compact model. Here we present a method to obtain implementation-suitable expressions for $\overline{R}_1$, $\overline{R}_2$ and $\overline{\alpha}$. The input impedance of the small signal model of Fig. 2(b) can be expressed as

$$Z_{B^*E'} = \overline{R}_1 + \frac{(1 - \overline{\alpha})Y\overline{R}_2 + 1}{\overline{R}_2 Y^2 \overline{\alpha}(1 - \overline{\alpha}) + Y} \quad (16)$$

series expansion of which leads to

$$Z_{B^*E'} = \overline{R}_1 + \frac{1}{Y}[1 + \overline{R}_2 Y(1 - \overline{\alpha})^2 - \overline{R}_2^2 Y^2 \overline{\alpha}(1 - \overline{\alpha})^3$$

$$+ \overline{R}_2^3 Y^3 \overline{\alpha}^2(1 - \overline{\alpha})^4 - \ldots]. \quad (17)$$

On the other hand, an exact solution for the input impedance given in [2] reads

$$Z_{B^*E'} = \frac{1}{Y}\eta_{ac} \coth(\eta_{ac}). \quad (18)$$

978-1-4799-7231-9/14 $31.00 © 2014 IEEE

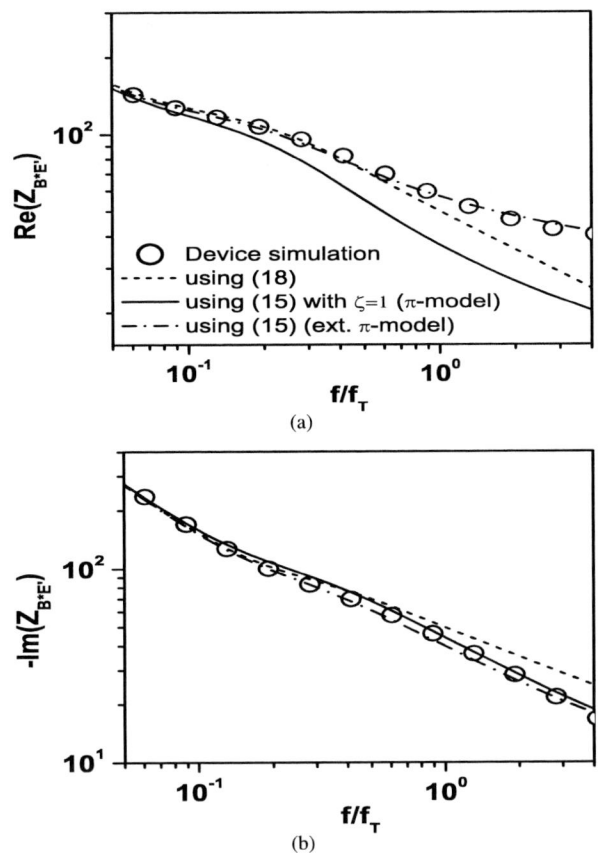

Fig. 3: *Frequency dependent (a) real and (b) imaginary parts of the input impedance corresponding to an internal SiGe HBT structure: comparison of the device simulated data with exact solution, π-model and the extended π-model. Legends of (a) are also used in (b).*

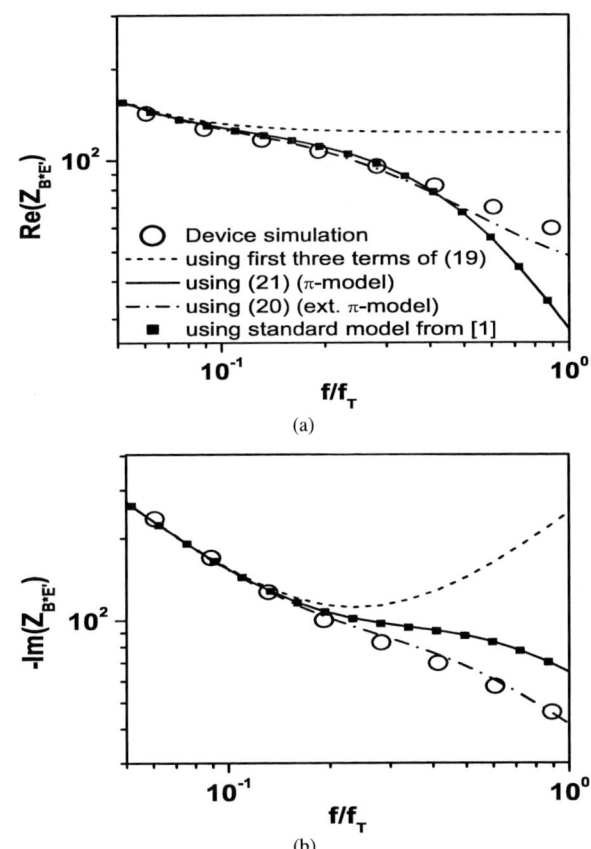

Fig. 4: *Frequency dependent (a) real and (b) imaginary parts of the input impedance corresponding to an internal SiGe HBT structure: comparison of the device simulation data with the standard model from [1] and implementation suitable versions of the exact solution, π-model and the extended π-model. Legends of (a) are also used in (b).*

which may be expanded to

$$Z_{B^*E'} = \frac{1}{Y}\left[1 + Y R_{Bi,nc} - \frac{1}{5}Y^2 R_{Bi,nc}^2 \right. $$
$$\left. + \frac{2}{35}Y^3 R_{Bi,nc}^3 - \dots \right]. \quad (19)$$

Comparisons of the same Y^n terms in (17) and (19) yield

$$\overline{R}_1 = \frac{r_{SBi}b_{E0}}{40 l_{E0}}, \overline{R}_2 = \frac{1587}{13720}\frac{r_{SBi}b_{E0}}{l_{E0}}, \overline{\alpha} = \frac{20}{69}. \quad (20)$$

Note that $\zeta = 1$ in (1) results into $R_1 = \overline{R}_1 = 0$ which essentially leads Fig. 2(b) to a π-EC model referred in [4]. Therefore, using $\zeta = 1$ in (12) and subsequently using (14) we can obtain $\overline{\gamma}_\pi$. Using this $\overline{\gamma}_\pi$ and $\zeta = 1$ in (15), we obtain $\overline{\alpha}_\pi$ and $\overline{R}_{2,\pi}$. Implementation-suitable expressions for these $\overline{\alpha}_\pi$ and $\overline{R}_{2,\pi}$ can be obtained by comparisons of the same Y^n terms in (17) with $\overline{R}_1 = 0$ and (19) leading to

$$\overline{\alpha}_\pi = \frac{1}{6}, \overline{R}_{2,\pi} = \frac{3}{25}\frac{r_{SBi}b_{E0}}{l_{E0}}. \quad (21)$$

III. RESULTS AND DISCUSSION

An internal SiGe HBT structure with half emitter width ($b_{E0}/2$) of 0.5 μm (Fig. 1) is numerically simulated using the 2-D device simulator DEVICE [9] to obtain the reference data

for model validation. The operating point is chosen at a current of I_C (corresponding to peak $f_T = 111GHz$) = $2.18mA$ at $V_{BC} = 0V$. For the model calculation, $l_{E0} = 1\mu m$ is assumed. Using $I_{BEi} = 13.4\mu A$ and $r_{SBi} = 1.5K\Omega$ obtained from the simulated device, the variable Z in (4) is iteratively calculated yielding $Z = 0.306$. Figs. 3(a) and (b) compare the models for the real and imaginary parts of the input impedance with the device simulation data. The input impedance for the extended π-model is obtained using (16) along with (1) and (15) and fitting value of $\zeta = 0.99969$. The solution for π-model is obtained by using $\zeta = 1$. It can be seen that all the three models show good match with the device simulation data up to the frequency, $f = f_T/10$. For the real part, the conventional model of (18) deviates from the device simulation data at f near to f_T and that of the π-model at a frequency significantly lower than f_T. On the other hand the extended π-model shows excellent agreement up to $f = 4f_T$. In case of imaginary part, however, both π and extended π-models show excellent accuracy up to $f = 4f_T$ while the conventional model shows slight deviation after f_T.

Figs. 4(a) and (b) compare the input impedance of the implementation suitable models with the device simulation data. Corresponding to the extended π- and the π-model, respectively, we used the model variables given in (20) and

(21). The parameter values obtained for the extended π-model (π-model) are $\overline{\alpha} = 20/69$, $\overline{R}_1 = 37.42\Omega$ and $\overline{R}_2 = 173.13\Omega$ ($\overline{\alpha}_\pi = 1/6$ and $\overline{R}_{2,\pi} = 179.6\Omega$). For the conventional model, we considered a low frequency approximation of (18), i.e., the first three terms of (19) as used in [4]. The standard model from [1] is implemented using C_{rBi} (=$C_B/5$) in parallel with the internal base resistance R_{Bi} (=$V_T ln(\alpha')/I_{BEi}$). Here C_B includes base-emitter and base-collector depletion and diffusion capacitances. It is observed that for the real part of input impedance, both the π- and standard models start deviating from the device simulation data at $f = f_T/2$ whereas the conventional model deviates at $f = f_T/10$. For the imaginary part, the π- and the standard models show deviation at $f = f_T/3$ and the conventional model at $f = f_T/5$. In comparison, the extended π-model shows high level of accuracy for both the real and imaginary parts of the input impedance up to $f = f_T$. The deviation observed in the extended π-model (see Fig. 4(a)) near f_T can be reduced by further optimizing the parameters.

IV. CONCLUSION

A physics based extended π-model is proposed to capture lateral non-quasi-static effects in SiGe HBTs. The proposed AC current crowding model is formulated in combination with the DC current crowding phenomenon. The model reduces to a simple π-EC model under negligible DC current crowding ($\zeta = 1$). Compared to the classical model of (18) and the simple π-model, the extended π-model accurately predicts the device simulation data till $f = 3f_T$. The closed-form solutions for the extended π- and the π-model are properly simplified to make them suitable for implementation. It is shown that the implementation-suitable extended π-model can predict the device simulation data up to $f = f_T$ with high level of accuracy unlike the simplified π-model, the low-frequency approximation of (18) and the standard model of [1]. The additional computational time due to the extra node of the extended π-model may be justified for significant DC current crowding. However, in advanced SiGe HBT technologies with low internal base sheet resistance and narrow emitter window widths the simple π-model and the standard model appear to be reasonable and accurate solution without increasing the simulation time.

REFERENCES

[1] M. Versleijen, "Distributed High Frequency Effects in Bipolar Transistors", in *Proc. IEEE Bipolar Circuits and Technology Meeting*, Minneapolis, 1991, pp. 85-88.

[2] R. L. Pritchard, "Two-Dimensional Current Flow in Junction Transistors at High Frequencies", *Proc. IRE*, Vol. 46, pp.1152-1160, 1958.

[3] J. R. Hauser, "The Effects of Distributed Base Potential on Emitter Current Injection Density and Effective Base Resistance for Stripe Transistor Geometry", *IEEE Trans. Electron Dev.*, Vol. ED-11, pp. 238-242, 1964.

[4] M. Schröter and A. Chakravorty, *Compact Hierarchical Bipolar Transistor Modeling with HICUM*. Sinagapore:World Scientific, 2010.

[5] G. Blasquez, J. Caminade, G. le Gac, "Analysis of the effects of current crowding on noise of transistors with a circular geometry. Application to transistor with any given geometry", *Physica 92B*, pp. 313-329, 1977.

[6] H. N. Ghosh, "A Distributed Model of the Junction Transistor and its Application in the Prediction of Emitter-Base Diode Characteristic, Base Impedance, and pulse response of the Device", *IEEE Trans. Electron Device*, Vol. Ed-12, No. 10, October, 1965.

[7] T.-Y. Lee, "Model enhancement and parameter extraction for the MM-SPICE/QBBJT model, Ph.D. thesis, U. of Florida, 1997.

[8] W. Liu, *Handbook of III-V heterojunction bipolar transistors*. John Wiley & Sons, New York, 1998.

[9] M. Schröter, "Transient and small-signal high-frequency simulation of numerical device models embedded in an external circuit," *COMPEL*, Vol. 10, pp. 377-378, 1991.

An Electrothermal PIN Diode Model with Substrate Injection

Adam W. DiVergilio, John J. Pekarik, Vibhor Jain

IBM Corporation
Essex Junction, VT 05452
awdiverg@us.ibm.com

Abstract—**This paper presents a compact model for an integrated 3-terminal PIN diode suitable for SPICE simulation. The model described within represents a significant extension to the standard 2-terminal diode model of SPICE, capturing phenomena critical to the accurate prediction of diode behavior in modern integrated process technologies. In particular, the inclusion of self-consistent electrothermal modeling and substrate injection effects is key to improving simulation accuracy for diodes operating in a forward conductive state, such as would be common for PIN diodes. The model is implemented in Verilog-A and hardware-verified results are presented for a 90nm SiGe BiCMOS technology featuring a high-performance integrated PIN diode.**

Keywords—Verilog-A; SPICE; PIN Diode; Compact Model; Semiconductor Device Modeling

I. INTRODUCTION

For many years, the spotlight in compact modeling has focused squarely on transistors, whose banner specifications often drive IC technology progression. However there are additional devices in these technologies, especially for analog / mixed-signal applications, whose performance can be a critical element of a successful IC design. One such example is the PIN diode, whose importance in RF and microwave switching applications is very high. Unfortunately, accurate modeling of such devices often takes a backseat to the more-glamorous transistor, leading the compact modeler to resort to the simple SPICE diode model [1] with perhaps a few extrinsic elements in the form of a subcircuit. In the case of a PIN diode, which is commonly used in both reverse and *forward* operation, these subcircuit implementations are often inadequate. For example, electrothermal effects can significantly impact forward-bias junction current and the apparent series resistance. Additionally, in the case of integrated silicon processes, these diodes are typically 3-terminal structures, forming a parasitic PNP with the substrate (see Fig. 1). In forward operation, the gain of this PNP is non-zero and current injected into the substrate can impact design performance. The work presented here aims to pay respect to the lowly diode by addressing these modeling deficiencies and demonstrating levels of accuracy consistent with today's application requirements.

II. TRADIONAL MODELING APPROACHES

Historically, there have been two common approaches to implementing technology-specific SPICE models: 1) reliance

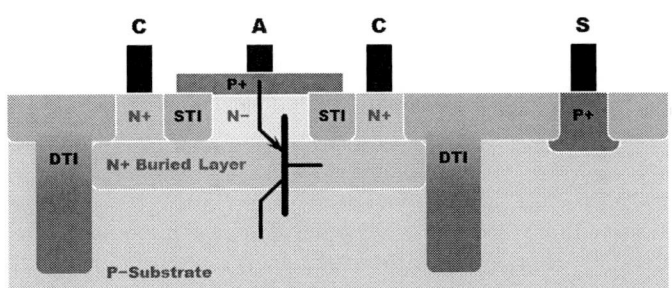

Fig. 1. Cross section of typical integrated PIN diode device on IBM's 90nm SiGe 9HP BiCMOS technology, highlighting physical PNP formation.

on parameter extraction for standard model formulations pre-defined within the simulator (e.g. VBIC, BSIM, resistor), referred to here as "core" models, or 2) construction of lumped-element subcircuit macromodels using a network comprised of the core models mentioned in 1). Both approaches are widely used in industry and can successfully cover the majority of semiconductor device modeling requirements faced today. However, there are cases where adequate standard core models do not yet exist and where phenomena cannot be properly described through the use of subcircuits. In such cases, a new core model must be written and compiled for use within the target simulator.

III. MODEL DEVELOPMENT USING VERILOG-A

The emergence of Verilog-A [2] as a standard language for disseminating new compact model equations has enabled a common path for overcoming the limitations of standard model formulations and/or subcircuits. It has been used to communicate and distribute updates to industry-standard transistor models such as HICUM [3,4], VBIC [5,6], MEXTRAM [7], and PSP [8]. In addition, it has been used to create improved models for passive devices such as MOSVARs [9] and resistors [10].

In the work presented here, Verilog-A was selected to develop a new core model for a 3-terminal PIN diode. A subcircuit-based approach using standard passive elements was not feasible, as electrothermal effects could not be accounted for properly in such a network. Similarly, an approach using existing electrothermal BJT models was deemed inappropriate due to their unnecessary complexity for the task at hand, as well as inability to provide a foundation for future device-specific enhancements (e.g. topology changes, alternate substrate coupling models, junction breakdown modeling).

978-1-4799-7231-9/14 $31.00 © 2014 IEEE

Equivalent Circuit Model

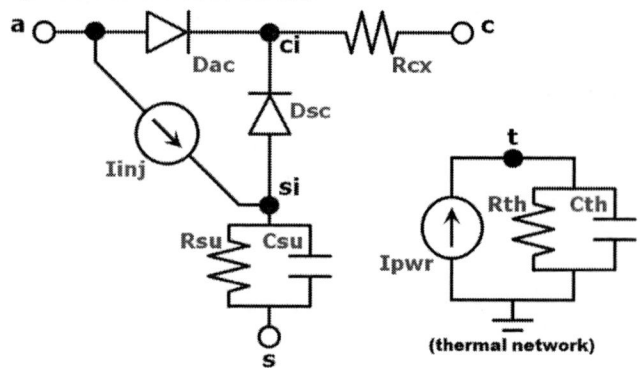

Fig. 2. Equivalent circuit model for 3-terminal electrothermal PIN diode.

Verilog-A Topological Map

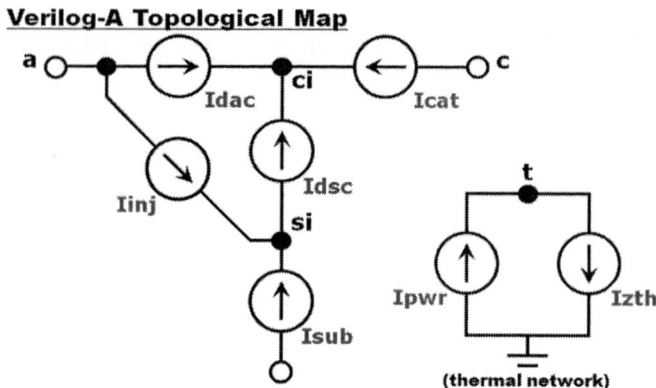

Fig. 3. Verilog-A topological map for 3-terminal electrothermal PIN diode.

IV. MODEL DESCRIPTION

Fig. 2 depicts the equivalent circuit model chosen for the PIN structure under investigation. The topological map defined for the Verilog-A code is shown in Fig. 3, provided here to aid the reader in relating the elements of the equivalent circuit model to their Verilog-A branch counterparts discussed in this work. Note that in practice the final model code has also been augmented with temperature scaling, shot and thermal noise sources, and technology-specific equations to capture geometric scaling and process statistics.

The two primary junctions, Dac and Dsc, are modeled using standard SPICE diode equations for current and charge, including depletion ($qdepa$ & $qdeps$), diffusion, and oxide terms.

$Idac$ formulation:

$$idac = ISA(T) \times (exp(Vdac/NA/vt) - 1) \quad (1)$$

$$qdac = CJA(T) \times qdepa(T) + TTA \times idac + COXA \times Vdac \quad (2)$$

$$Idac = idac + ddt(qdac) \quad (3)$$

$Idsc$ formulation:

$$idsc = ISS(T) \times (exp(Vdsc/NS/vt) - 1) \quad (4)$$

$$qdsc = CJS(T) \times qdeps(T) + TTS \times idsc + COXS \times Vdsc \quad (5)$$

$$Idsc = idsc + ddt(qdsc) \quad (6)$$

The substrate injection current, $Iinj$, is the forward transfer current of the PNP transistor formed by diodes Dac and Dsc. Similar to modern bipolar transistor models, this transfer current is modeled with its own saturation current parameters, rather than through a simple current gain parameter:

$$Iinj = ISI(T) \times (exp(Vdac/vt) - 1) \quad (7)$$

In order to preserve computational efficiency, the reverse-injection case (Dsc forward-biased) is not included in this model. The ratio $Iinj/Idac$ effectively represents the forward beta of the parasitic PNP. Ideally, one would desire this term to approach zero. In practice, it may typically be on the order of 0.1-10%, depending on the device design and technology.

Electrothermal effects, or self-heating, are added to the model using a standard first-order thermal network. This basic formulation has been demonstrated previously in several compact transistor models, including VBIC, HICUM, and MEXTRAM [3-7]. The instantaneous dissipated power is calculated and used to set current source $Ipwr$:

$$Ipwr = idac \times Vdac + idsc \times Vdsc + Iinj \times (Vdac - Vdsc)$$
$$+ Icat \times Vcat + irsu \times Vsub \quad (8)$$

This current then feeds a thermal network formed by Rth and Cth. Note that the thermal resistance, Rth, includes a temperature coefficient, providing a means for modeling the temperature dependence of the thermal conductivity of the wafer [11,12]:

$$Rth(T) = RTH \times (T/Tnom)^{XRTH} \quad (9)$$

$$Izth = V(t)/Rth(T) + ddt(Cth \times V(t)) \quad (10)$$

The voltage developed on node "t" represents the local temperature rise above ambient ($Tamb$) due to heating. Temperature-dependent model equations then reference the value of "t" dynamically, ensuring that the proper device temperature is considered at all times:

$$T = Tamb + V(t) \quad (11)$$

The cathode resistance, Rcx, is modeled as a single resistor at the cathode terminal:

$$Icat = Vcat/Rcx(T) \quad (12)$$

Finally, in order to improve the modeling of high-frequency substrate coupling, a first-order RC network formed by Rsu and Csu has been included at the substrate terminal.

$$Isub = Vsub/Rsu(T) + ddt(Csu \times Vsub) \quad (13)$$

This substrate topology is consistent with that used by the HICUM L2 model. In the case of silicon substrates, the Csu term becomes increasingly important in applications above approximately 10GHz [13,14].

978-1-4799-7231-9/14 $31.00 © 2014 IEEE

V. RESULTS

This new model has been applied to IBM's 90nm SiGe 9HP BiCMOS technology [15], which features a high-performance integrated PIN device. Excellent correlation with DC and AC characterization data has been demonstrated for this technology. Fig. 4 shows both the old and new models plotted against the same forward-bias data set, where it becomes readily apparent that the inclusion of substrate injection and electrothermal effects greatly improves the correlation with measured data. For completeness, some additional AC correlation results are provided in Fig. 5. In addition, geometric scaling and process statistics specific to the 9HP technology have been applied successfully to this model, demonstrating its applicability to a production-ready design kit.

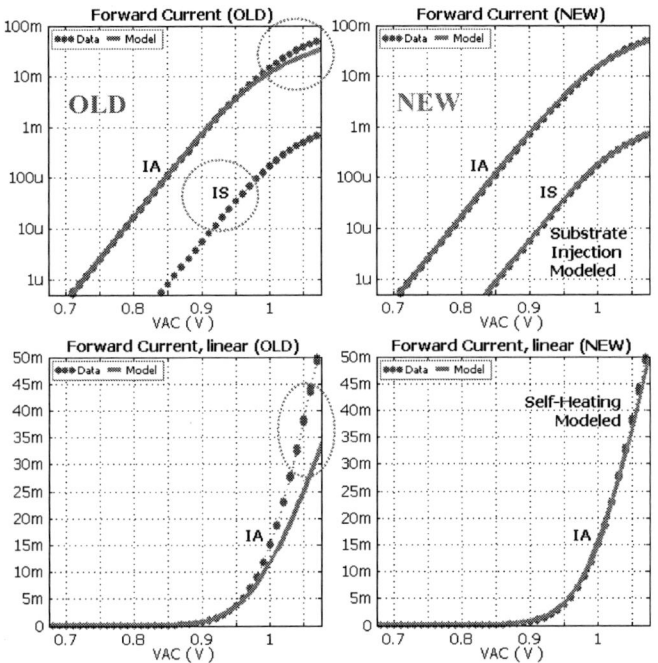

Fig. 4. Old and New models plotted against the same data set.

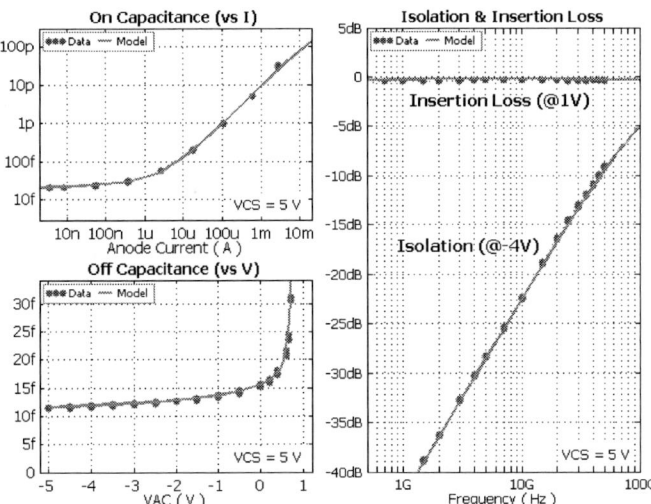

Fig. 5. AC model-hardware correlation using new PIN model.

VI. PERFORMANCE

The speed at which a model runs in a simulator is not only a function of its computational complexity, but also its code efficiency. SPICE simulation requires calculation of both the branch equations (as defined in the model description) and the derivatives of those branch equations with respect to the system variables (typically node voltages). For a complex model, the calculation time of its derivatives can far-outweigh the calculation time of its branch equations. As a result, the quality of the compiler, which generates those derivatives, can have a profound impact on the performance of the model under test.

Fortunately model compilers have improved in recent years, making the use of Verilog-A based models a more reasonable approach than previously considered. Though still not quite on par with optimized hand-coded models in commercial EDA tools, the performance gap is closing and capabilities are improving. In many circumstances, the accuracy benefits awarded through the use of a custom Verilog-A model must be weighed against the performance hit experienced in the target simulator. The ability to tradeoff speed and accuracy is an important tool throughout the design process [16].

Fig. 6 shows the results of a performance analysis conducted on the 3-terminal PIN model presented in this work. Model features and simulation times have been compared for three distinct model options of varying accuracy running in the Spectre simulator [17]. Though the Verilog-A code is slower than the other models, the improvements in accuracy and extensibility make it an important option for critical applications. Additionally, this performance gap should continue to close as compiler technology improves.

Model Feature	New PIN (Verilog-A)	Transistor (VBIC-based)	Subcircuit (simple)
Self Heating	Y	Y	N
Substrate Injection	Y	Y	N
RTH Temp Coeff	Y	N	N
Full Temperature Characterization*	Y	N	N
Extensible / Customizable	Y	limited	limited
Simulation Time (normalized)**	1.0	0.4	0.3

Pre-defined standard core models often have a limited parameter set for temperature coefficients (e.g. sharing XTI & EA for multiple currents).
**Simulation times based on large-circuit DC bias sweeps.*

Fig. 6. Simulation performance comparison. The new PIN model provides the most accuracy, but at the expense of simulation time due to its Verilog-A based implementation.

VII. CONCLUSION

We have presented a 3-terminal integrated diode model in Verilog-A which accounts for important forward-conduction effects not included in the typical 2-terminal SPICE diode model. In particular, the inclusion of electrothermal behavior (self-heating) and substrate injection are important considerations when using integrated PIN diodes, which are commonly operated in a forward-biased state. Self-heating is implemented in a manner consistent with advanced transistor models such as HICUM, MEXTRAM, and VBIC using an embedded thermal RC network. As in the case with these models, the thermal node could be exposed as a 4th terminal, allowing for mutual coupling between the diode and other electrothermal devices around it. Forward current loss through substrate injection is accurately captured through the inclusion of a parasitic PNP model. High frequency substrate coupling is accounted for using a first-order RC network, in a manner similar to that of HICUM L2. Results were presented for IBM's 90nm SiGe9HP process technology, showing good correlation between this new model and measured data.

While some of these enhancements can be realized through subcircuit macromodeling (e.g. substrate injection and HF coupling), others can only be achieved through reformulation of the underlying device equations (e.g. electrothermal effects, proper temperature modeling). The Verilog-A model presented in this work demonstrates a self-consistent solution to these issues using a framework which can be further extended to account for new topologies or the addition of other phenomena (e.g. breakdown). Such enhancements would otherwise be difficult or even impossible using traditional modeling approaches.

ACKNOWLEDGMENT

The authors wish to acknowledge the outstanding technical support of Chris Lamothe and Adnan Beganovic in the measurement of all devices used in this study.

REFERENCES

[1] L. W. Nagel, *SPICE2: A Computer Program to Simulate Semiconductor Circuits*, Memorandum no. ERL-M520, Electronics Research Laboratory, University of California, Berkeley, May 1975.

[2] http://www.eda.org/verilog-ams/

[3] M. Schroter and A. Chakravorty, *Compact heirarchical bipolar transistor modeling with HiCUM*, World Scientific Publishing Inc., 2010.

[4] http://www.iee.et.tu-dresden.de/iee/eb/

[5] C. C. McAndrew, J. Seitchik, D. Bowers, M. Dunn, M. Foisy, I. Getreu, M. McSwain, S. Moinian, J. Parker, D. J. Roulston, M. Schroter, P. van Wijnen and L. Wagner, "VBIC95: The Vertical Bipolar Inter-Company Model," *IEEE J. Solid-State Circuits*, vol. 31, no. 10, pp. 1476-1483, Oct. 1996.

[6] http://www.designers-guide.org/VBIC/

[7] http://mextram.ewi.tudelft.nl

[8] http://pspmodel.asu.edu

[9] J. Victory, Z. Yan, G. Gildenblat, C. C. McAndrew and J. Zheng, "A physically based, scalable MOS varactor model and extraction methodology for RF applicatons," *IEEE Transactions on Electron Devices*, vol. 52, no. 7, pp. 1343-1353, July 2005.

[10] L. Lemaitre and C. C. McAndrew, "Voltage-controlled-current-source-only Verilog-A resistor model for R≥0," *Behavioral Modeling and Simulation Workshop, BMAS*, 2008.

[11] J. C. J. Paasschens, S. Harmsma and R. van der Toorn, "Dependence of thermal resistance on ambient and actual temperature," *Bipolar/BiCMOS Circuits and Technology Meeting*, 2004.

[12] C. J. Glassbrenner and G. A. Slack, "Thermal conductivity of silicon and germanium from 3K to the melting point," *Physical Review*, vol. 134, no. 4A, pp. A1058-A1069, May 1964.

[13] M. Pfost, H.-M. Rein and T. Holzwarth, "Modeling substrate effects in the design of high-speed Si-bipolar IC's," *IEEE J. Solid-State Circuits*, vol. 31, no. 10, pp. 1493-1501, Oct. 1996.

[14] S. Strahle and M. Pfost, "Substrate modeling for RF and high-speed Bipolar/BiCMOS circuits," *Bipolar/BiCMOS Circuits and Technology Meeting*, 2003.

[15] J. Pekarik, et al. "A 90nm SiGe BiCMOS technology for mm-wave and high-performance analog applications," *unpublished*.

[16] A. W. DiVergilio, "Practical Modeling – When Less is More," *Bipolar/BiCMOS Circuits and Technology Meeting*, 2012.

[17] http://www.cadence.com/products/cic/spectre_circuit/

978-1-4799-7231-9/14 $31.00 © 2014 IEEE

An Investigation of f_T and f_{max} Degradation Due to Device Interconnects in 0.5 THz SiGe HBT Technology

A. Çağrı Ulusoy[1], Robert L. Schmid[1], Saeed Zeinolabedinzadeh[1], Wasif T. Khan[1], Mehmet Kaynak[2], Bernd Tillack[2,3] and John D. Cressler[1]

[1]School of Electrical and Computer Engineering, Georgia Institute of Technology,
777 Atlantic Drive, Atlanta, GA 30332-0250 USA (aculusoy@gatech.edu)
[2] IHP Microelectronics GmbH, 15236 Frankfurt (Oder), Germany
[3] Technische Universität Berlin, HFT4, 10587, Berlin, Germany

Abstract—**In this paper, the authors investigate the impact of device interconnect parasitics on the two most commonly-accepted RF small-signal figures-of-merit, the transit frequency (f_T) and the maximum frequency of oscillation (f_{max}) in state-of-the-art SiGe HBT technology. Simulations and measurement results are provided as a guideline to design an optimum device interconnect scheme to achieve a high f_{max}. Test structures were characterized with de-embedding structures providing reference planes at the device level and at the top-metal level. Measurements show an f_{max} of 450 GHz at the device level and at the top-metal level a degradation of only 4% to 430 GHz. These results demonstrate a significant advantage of the SiGe HBT technology compared to ultra-scaled CMOS technology at device speeds approaching a terahertz, and to the best of the authors' knowledge, demonstrate the highest f_{max} reported at the top-metal level in any state-of- the-art silicon technology.**

I. Introduction

Recent developments in high-speed transistors have yielded technology offerings with unprecedented device performance. Deep sub-micron CMOS and state-of-the-art SiGe HBT technologies now achieve RF figures-of-merit (FoM) for f_T and f_{max} in the range of half a terahertz [1], [2]. While this is clearly very promising for the development of ICs that can revolutionize many applications in the millimeter-wave to sub-millimeter-wave frequencies (e.g., 30 GHz to 300 GHz), the intrinsic device speed by itself is by no means a sufficient metric to determine the RF performance of a technology at the circuit level. Not only do the performance of passive components, such as transmission lines, inductors and capacitors, play a significant role in the overall performance of an RF circuit, but also the actual contact to the high-speed device and the associated parasitic elements has recently been identified as a major concern, impacting the maximum achievable performance from the device. For instance, a significant loss of up to 58% has been reported when the device interconnections are included while determining the f_{max} of a 45 nm SOI-CMOS transistor [3]. The authors in [3] report a reduction of f_{max} from 480 GHz to 200 GHz when all the interconnect parasitics are incorporated from the device level to the top-metal level, which is typically used for the implementation of RF circuits and passive elements. Similarly, a reduction of f_{max} in the

range of 20% has been reported for high performance SiGe HBTs in 90 nm (from 310 GHz to 250 GHz) and 130 nm nodes (from 480 GHz to 380 GHz) when the top-level interconnects are included [4], [5]. Clearly, this is a major limitation to the maximum RF circuit performance that can be achieved with these high-speed devices.

In this paper, we investigate the effects of device interconnects on the common RF FoMs of a state-of-the-art 0.5 THz, 130 nm SiGe HBT technology from IHP microelectronics [1]. First, we present simulation results to determine a "best-practice" design approach for the device interconnects relevant to millimeter-wave circuits. We also provide experimental results demonstrating only a small degradation (4%) in f_{max}, but a somewhat larger degradation for f_T, when the FoMs are extracted at the device level and the top-metal level de-embedding planes. We identify these results as a significant advantage of the presented SiGe HBT technology over ultra-scaled CMOS technologies for RF applications at millimeter-wave frequencies.

II. Device Interconnect Design

It is important to identify the limiting factor affecting the performance of RF circuits. In this respect, f_{max} is the more relevant FoM for actual RF circuits, as it is extracted from the maximum available gain (MAG), a quantity relevant to real applications. The f_T, on the other hand, is extracted from the forward current gain (H_{21}) when the output is a short circuit, a case which rarely occurs in RF circuits.

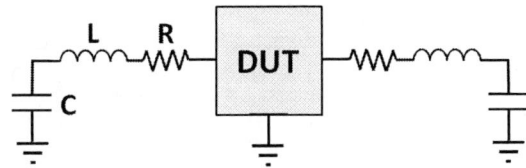

Fig. 1. Simplified schematic of the device interconnects.

In Fig. 1, a simplified but an insightful equivalent circuit is presented, representing the interconnect parasitics of a device

978-1-4799-7231-9/14 $31.00 © 2014 IEEE

with a lumped inductor, capacitor, and resistor. Here, the series inductance, L, represents top to bottom via interconnections while the series resistance, R, accounts for the via resistances. In addition to the series components, a shunt capacitance occurs due to the coupling to the ground metal as well as to the substrate. Clearly, the actual structure has a more distributed nature, but Fig. 1 is sufficient to gain a conceptual understanding.

It can be deduced from its definition that L and C in Fig. 1 will not influence the MAG of the device. The MAG is, in simple terms, the gain of the device when the input and the output are perfectly (conjugate) matched to external load impedances. Therefore, any lossless reactive component external to the device can be compensated by other lossless reactive components. From a circuit designer's point of view, these reactive components can be absorbed into the external matching network. This leads to the conclusion that the only element that has an effect on the MAG, and consequently on the transistor f_{max}, is the resistance, R, introduced by the interconnect. As a result, a simple yet effective design strategy to maximize f_{max} is to use as many via connections as possible to the device.

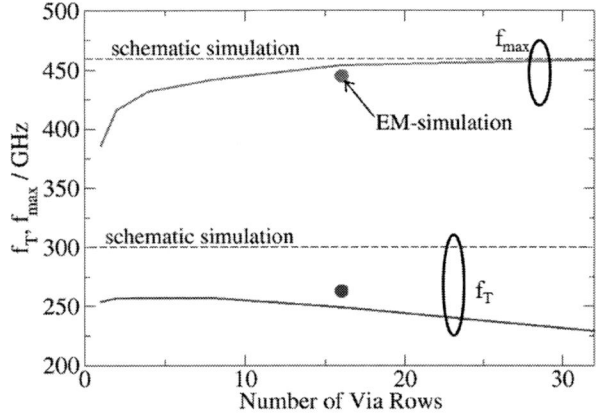

Fig. 2. Simulation of the peak f_T and peak f_{max} for a 4-finger SiGe HBT for varying numbers of M1-M5 via rows connecting to the device. The influence of the vias is determined by post-layout parasitic extraction, while each row consists of 15 vias.

In Fig. 2, the simulation of peak f_T and f_{max} for a 4-finger SiGe HBT in IHP's SG13G2 technology is presented for varying number of via rows connecting to the device. IHP SiGe technology provides seven metalization layers. In these simulations, the number of metal-1 (M1) to metal-5 (M5) layer vias are swept by increasing the number of via rows, with each row having 15 vias. The top-metal vias are kept constant in each case, as they have relatively low resistance in IHP technology and do not impact MAG. In Fig. 2, the effects of the vias are determined by post-layout parasitic extraction. As can be seen, for the worst case of a single row of vias, the f_{max} of the device degrades from 460 GHz to 380 GHz after including post-layout parasitics, a reduction of 18%. As the number of via rows is increased, the post-layout extracted f_{max} approaches the schematic simulated value, as the total series loss is reduced by parallel connection of vias. On the other hand, f_T starts to decline for increasing number of vias due to the increased parasitic capacitance. Finally, in order to

account for the distributed nature of the interconnect parasitics, a full-wave electromagnetic (EM) simulation was performed to model the device interconnects for the 16-row case. The simulated EM-model is shown in Fig. 3 and the results are presented in Fig 2. Compared to the layout extraction, the EM-simulated results show a further, but still minor, reduction in f_{max}.

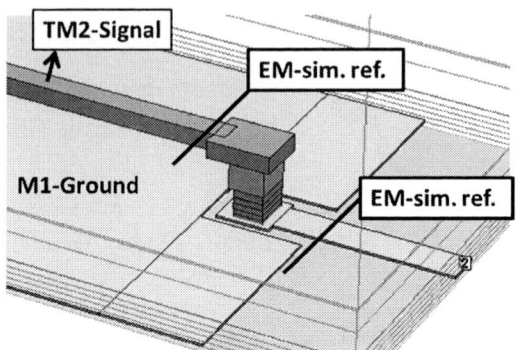

Fig. 3. Simulated EM-model of the designed via interconnects.

We conclude that for RF applications, a high number of vias is preferable for the device interconnects, thereby minimizing the series loss. The increased capacitance will impact f_T; however, when accurately modeled, this additional capacitance can be taken into account during circuit design without affecting the achievable performance of the RF circuit.

III. Experimental Results

For the experimental characterization of the impact of device interconnects on the RF FoMs, test structures were designed and fabricated with 4-finger SiGe HBTs in the common-emitter configuration with OPEN/SHORT de-embedding planes at the device and top-metal levels. This is illustrated in Fig. 4, and a micrograph of the fabricated samples is shown in Fig. 5. The top-metal to device level interconnects are designed using the 16-row case, as simulated in Fig. 2.

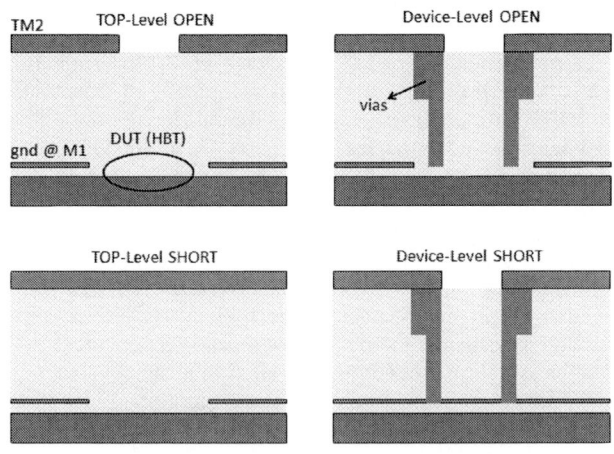

Fig. 4. Illustration of the OPEN/SHORT de-embedding structures for top-metal and device level reference planes.

The devices were characterized by S-parameter measurements up to 67 GHz using an Agilent E8351A PNA. An

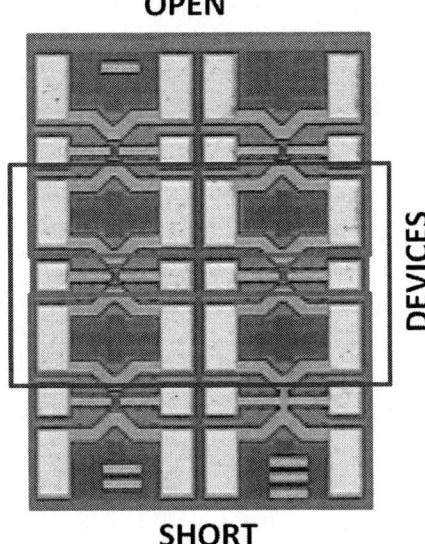

Fig. 5. Micrograph of the fabricated test structures.

SOLT calibration was performed to move the reference plane to the probe tips, and standard OPEN/SHORT de-embedding was performed to move the reference plane to the actual device at the top-metal level and the device level by using the corresponding standards.

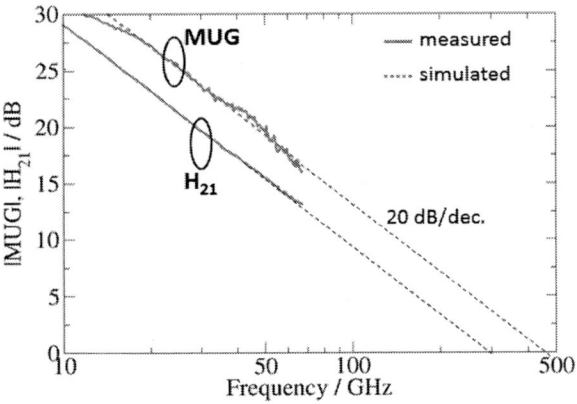

Fig. 6. Measured MUG and H_{21} of the fabricated 4-finger SiGe HBT at device level.

In Fig. 6, the extracted Mason's unilateral gain (MUG) and the H_{21} are presented when the de-embedding is performed to the device level. In general terms, f_{max} is determined by the MAG; however, within the range of the measurement frequencies the devices have a stability factor (Rollet's stability factor, k) below unity. Therefore, as commonly accepted, the MUG is preferred for f_{max} extraction instead of MAG in Fig. 6 and the following figures. The transistor was biased for peak f_{max}, at a V_{BE} of 0.92 V and a V_{CE} of 1.5 V. The simulated schematic-level MUG is provided as well, and the extracted MUG from the measurement agrees well with the transistor model. When the MUG and H_{21} are extended by a 20 dB/decade line, an f_T/f_{max} of 300 GHz/450 GHz is achieved. This result is reasonably close to the expected values

from schematic simulations using foundry-provided device models.

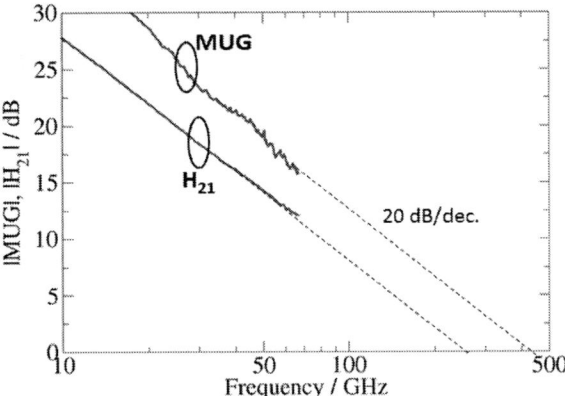

Fig. 7. Measured MUG and H_{21} of the fabricated 4-finger SiGe HBT at top metal level.

The extracted MUG and the H_{21} are presented in Fig. 7 when the de-embedding was performed to the top-metal level. In this case, the f_T/f_{max} are determined to be 260 GHz/430 GHz. The f_{max} has only slightly decreased, while f_T degradation is stronger, as predicted by simulations. This effect is presented in Fig. 8, where the f_T/f_{max} for both cases are presented as function of the 20 dB/decade extrapolation frequency from 50 to 67 GHz.

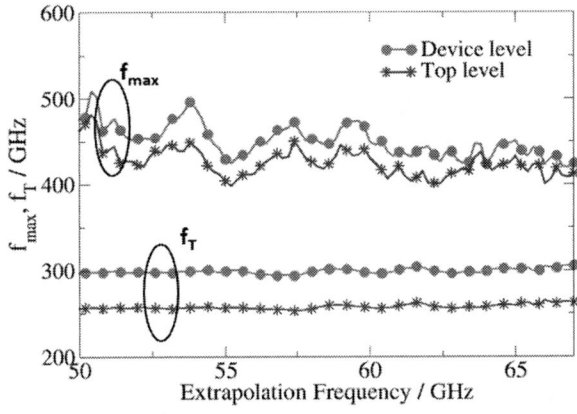

Fig. 8. Extracted f_T and f_{max} for top-metal and device level de-embedding planes for 20 dB/dec. extrapolation frequencies from 50 GHz to 67 GHz.

As it is well known, the actual value of f_{max} depends strongly on the frequency that is chosen for the 20 dB/decade extrapolation, while f_T is less dependent on this value, as can be seen in Fig. 8. On the other hand, it is clear that the extracted f_{max} for the top-metal level reference plane is consistently lower than the device-level reference plane, by almost a constant of proportionality, across all the extrapolation frequencies from 50 GHz to 67 GHz. Finally, for completeness, we report the bias dependance of f_T/f_{max} in Fig. 9 for an extrapolation frequency of 65 GHz.

As can be seen in Fig. 9, the extracted peak f_{max} at the device level equals 451 GHz, while at the top-metal level it equals 435 GHz. Similarly, the peak f_T at the device level

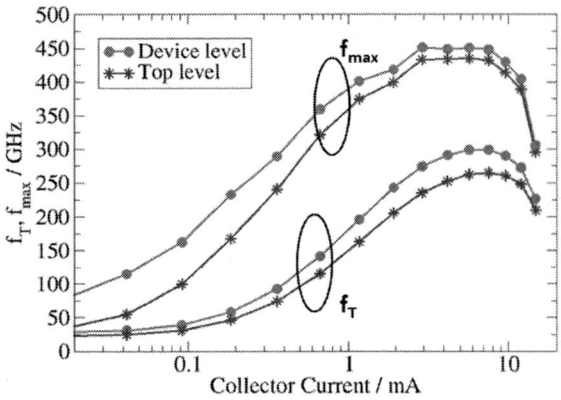

Fig. 9. Extracted f_T and f_{max} for top-metal and device level de-embedding planes for swept collector current at a V_{CE} of 1.5 V.

IV. SUMMARY

In this paper, we investigated the influence of device interconnects on the RF FoMs of a state-of-the-art 0.5 THz SiGe HBT technology. We provided a simple, yet effective approach to design the optimal RF interconnects. Experimental results show that the f_{max} degradation from device level to top-metal level can be minimized, and we quantified an f_{max} degradation of less than 4%, and an f_T degradation of 11%. We identify this feature as a significant advantage of the SiGe HBT technology under study compared to deep-scaled CMOS technologies, where an f_{max} degradation of more than 50% has been reported due to device interconnects. Furthermore, to the authors' knowledge, the results in this paper demonstrate the highest f_{max} reported at the top-metal level of any state-of-the-art silicon technology.

ACKNOWLEDGMENT

The authors are grateful to the IHP Microelectronics technology team for the fabrication of the ICs, and useful discussions.

REFERENCES

[1] B. Heinemann, R. Barth, D. Bolze, J. Drews, G. Fischer, A. Fox, O. Fursenko, T. Grabolla, U. Haak, D. Knoll, R. Kurps, M. Lisker, S. Marschmeyer, H. Rucker, D. Schmidt, J. Schmidt, M. Schubert, B. Tillack, C. Wipf, D. Wolansky, and Y. Yamamoto, "Sige hbt technology with ft/fmax of 300ghz/500ghz and 2.0 ps cml gate delay," in *2010 IEEE International Electron Devices Meeting (IEDM)*, Dec 2010, pp. 30.5.1–30.5.4.

[2] H. Li, B. Jagannathan, J. Wang, T.-C. Su, S. Sweeney, J. Pekarik, Y. Shi, D. Greenberg, Z. Jin, R. Groves, L. Wagner, and S. Csutak, "Technology Scaling and Device Design for 350 GHz RF Performance in a 45nm Bulk CMOS Process," in *2007 IEEE Symposium on VLSI Technology*, June 2007, pp. 56–57.

[3] B. Cetinoneri, Y. Atesal, A. Fung, and G. Rebeiz, "W-Band Amplifiers With 6-dB Noise Figure and Milliwatt-Level 170-200-GHz Doublers in 45-nm CMOS," *IEEE Transactions on Microwave Theory and Techniques*, vol. 60, no. 3, pp. 692–701, March 2012.

[4] Y. Yang, S. Cacina, and G. Rebeiz, "A SiGe BiCMOS W-Band LNA with 5.1 dB NF at 90 GHz," in *2013 IEEE Compound Semiconductor Integrated Circuit Symposium (CSICS)*, Oct 2013, pp. 1–4.

[5] M. Elkhouly, S. Glisic, C. Meliani, F. Ellinger, and J. Scheytt, "220 - 250-ghz phased-array circuits in 0.13-/spl mu/m sige bicmos technology," *IEEE Transactions on Microwave Theory and Techniques*, vol. 61, no. 8, pp. 3115–3127, Aug 2013.

equals 299 GHz, and 265 GHz at top-metal level. By making use of the RF device interconnect techniques highlighted in this paper, we quantify an f_{max} degradation of less than 4%, and an f_T degradation of 11% when the device interconnects from lowest metal to the top-metal layer are incorporated. These results are highly promising for RF circuit applications, as only a minor f_{max} degradation is achieved when the high-speed devices are interconnected to the RF-application relevant top-metal layer. Compared with deep sub-micron CMOS with similar intrinsic speed performance, this is a significant advantage of SiGe HBT technology, as an f_{max} degradation of from 480 GHz to 200 GHz has been reported in the literature for 45 nm SOI CMOS process [3], meaning a relative degradation of 58% compared to 4%. We would like to highlight that this result stems from a general advantage of SiGe HBTs over CMOS devices, and thus can be considered fundamental. Due to their vertical device geometry and consequently relaxed lithographic node, the device interconnects for the SiGe HBTs can be realized with lower loss, compared to a CMOS device where the via stack can result in a significant series resistance.

TABLE I. PERFORMANCE SUMMARY AND PREDICTED PERFORMANCE AT TOP-METAL LEVEL

	Schematic	Post-Layout	EM-sim.	Measured
f_T	303 GHz	252 GHz	263 GHz	265 GHz
f_{max}	458 GHz	454 GHz	445 GHz	435 GHz

Table I summarizes the results and reports the simulated f_T/f_{max} performance at the top-metal plane for the post-layout extracted and EM-simulated cases for the 16-row via interconnect. Even with the large number of via arrays, the total area of the device interconnect is still relatively small. For the 16x15 via array case, the total area of the interconnect roughly equals $7 \times 7 \times 10\ \mu m^3$. Nevertheless, it can be seen from Table I that the EM-simulated case achieves a better prediction of the interconnect parasitics, and should be preferred for accurate design of RF circuits, especially at millimeter-wave frequencies.

Analysis of the Local Extraction Method of Base and Thermal Resistance of Bipolar Transistors

Robert Setekera[1], Luuk Tiemeijer[2], Willy Kloosterman[3], Ramses van der Toorn[1],

[1]Delft University of Technology, EEMCS, Mekelweg 4, 2628 CD Delft, The Netherlands
[2]NXP Semiconductors, TSMC Research Centre, High Tech Campus 37-1.047, 5656 AE Eindhoven, The Netherlands
[3]NXP Semiconductors, Device Engineering and Characterization, Gerstweg 2, 6534 AE Nijmegen, The Netherlands

Abstract — **This paper presents an extensive method to determine the *extraction region* were the method (published earlier) that consistently accounts for self-heating and Early effect to accurately extract both base and thermal resistance of bipolar junction transistors is applicable. The method is able to determine the lower and upper limits of the extraction region (i.e., a region with very small variations of the extracted base resistance) were the method yields correct results for the extracted base and thermal resistance. A generalization of the extraction method is developed that includes devices with very small Early voltage (V_A). The method is directly applicable to transistors, thus no dedicated test structures are need. The method is demonstrated on advanced industrial SiGe HBTs.**

Index Terms – Avalanche multiplication, Base resistance, Bipolar transistor, Early voltage, HBT, Thermal resistance, Self-heating, Parameter extraction.

I. INTRODUCTION

In compact modeling and characterization of bipolar junction transistors, accurate extraction of the base resistance R_B and thermal resistance R_{TH} is of decisive importance to nowadays typical bipolar transistor applications. Successful parameter extraction by advanced methods, such as the method we presented in [1], may stand or fall with the selection of appropriate input data. These facts justify a dedicated study to solve the problem of input data selection for the parameter extraction in case.

In this paper we develop and verify a procedure to clearly identify the appropriate input data regime for the R_B- and R_{TH}- extraction method we presented earlier in [1]. This procedure (Section III) utilizes contour plots to, in practice, determine the extraction region where the underlying assumptions of the extraction method [1] are valid. We also present a generalization of the method to include cases of very small Early voltage (Section IV) and based on this generalization we study the sensitivity of the method as presented in [1] to its underlying assumption of large Early voltage.

II. EXTRACTION METHOD FOR R_B AND R_{TH}

The extraction method for the base resistance (R_B) and the thermal resistance (R_{TH}) used in this work was published earlier in [1]. Therefore, here we will only give a summary highlighting the key equations that are important for the work in this paper.
From [1], the external base-emitter voltage (V_{BE}) is given by

$$V_{BE} = V_{BEi} + I_B R_B + I_E R_E . \quad (1)$$

By forcing a constant emitter current (I_E), it ensures that the voltage drop across the emitter resistance (R_E) is constant (provided the emitter resistance can be assumed to be constant). This helps to decouple R_B and R_E. Also by considering small variations in bias and temperature space, the variations of R_E, R_B, and R_{TH} due to temperature effects are assumed to be of higher order. Employing these conditions together with $dI_B = -dI_C$ for a fixed emitter current, yields the relation

$$dV_{BE} = dV_{BEi} - R_B dI_C . \quad (2)$$

The variations in V_{BEi} (dV_{BEi}), are assumed to be due to self-heating and collector-base Early effect. These two effects are taken into account by considering the change in junction temperature and by adopting a simple model for the main forward current. After a few substitutions and simplifications (see [1] for details), the following expression is obtained

$$
\begin{aligned}
-\frac{dV_{BE}}{dI_C} &= R_B + \alpha_T R_{TH}(V_{CB} + V_A^{eff}) \\
&+ \frac{V_T}{I_C} \frac{V_A^{eff}}{V_A + V_{CB}} \left(1 - \frac{V_A + V_{CB}}{V_A^{eff}} \right), \quad (3)
\end{aligned}
$$

where $\alpha_T = -dV_{BE}/dI_T > 0$ and $V_A^{eff} = I_C(dV_{CB}/dI_C)$. By considering the regime where $I_C = I_E - I_B \approx I_E$ and $V_{CB} \ll V_A$, (3) reduces to [1]

$$-\frac{dV_{BE}}{dI_C} \approx R_B + \left[\alpha_T R_{TH} + \frac{V_T}{V_A} \frac{1}{I_E} \right] (V_{CB} + V_A^{eff}) . \quad (4)$$

Using (4) the extraction procedure for R_B and R_{TH} from DC-measurements is determined. The first step is to identify the extraction region where the approximations behind (4) actually hold and within this region, the intercept on the vertical axis of a plot of $-dV_{BE}/dI_C$ as a function of $(V_{CB} + V_A^{eff})$ gives R_B. The corresponding slope S_{TOT} is equivalent to $(\alpha_T R_{TH} + V_T/(V_A I_E))$ and it depends on the emitter current (I_E). The thermal resistance (R_{TH}) is found by plotting the slope S_{TOT} as a function of $1/I_E$. Then the intercept (γ) with the vertical axis is equivalent to $\alpha_T R_{TH}$; α_T can be measured separately. The accuracy of the extracted R_B and R_{TH} (following this extraction procedure) is strongly dependent on the choice of the extraction region where the input data is taken from; this will be discussed in detail in the next section.

III. DETERMINATION OF THE EXTRACTION REGION

To develop a method that can be used to accurately determine the extraction region for the extraction method for R_B and R_{TH} presented earlier in [1], we employ experimental data. This DC-measurement data was taken on a QUBiC4mmW SiGe HBT [2] with emitter area $A_E = 0.30$ $\mu m \times 1.0$ μm. The detailed measurement procedure can be found in [1]. In summary, at a fixed temperature ($T = 25\,^\circ\mathrm{C}$) and a sequence of constant emitter currents, we measure V_{BE}, I_B, and I_C as a function of V_{CB}. Here the constant emitter currents belong to the ideal normal forward bias regime of the Gummel measurements. The collector-base bias is chosen large enough in order to generate large avalanche current. As an example, the measurement results for the base current (I_B) are presented in Fig. 1. Detailed measurement results can be found in [1]. From Fig. 1 we can see that over the selected

extraction region is determined using a contour plot showing the extracted R_B values against the V_{CB} lower limit (lower limit of the extraction region) and V_{CB} upper limit (upper limit of the extraction region) for all possible combinations of the lower limit and upper limit containing contour lines of equal extracted R_B. For example a contour plot corresponding to a fixed emitter current $|I_E| = 6.166$ mA is presented in Fig. 2. From the figure, the suitable extraction region can clearly be observed as the large flatland area in the top-left of the contour plot with very small variations in R_B, i.e., the boundary of the lower limit of the extraction region is $V_{CB} \approx 0.4$ V and the boundary for the upper limit of the extraction region is $V_{CB} \approx 1.3$ V. Over this extraction region, variations in R_B

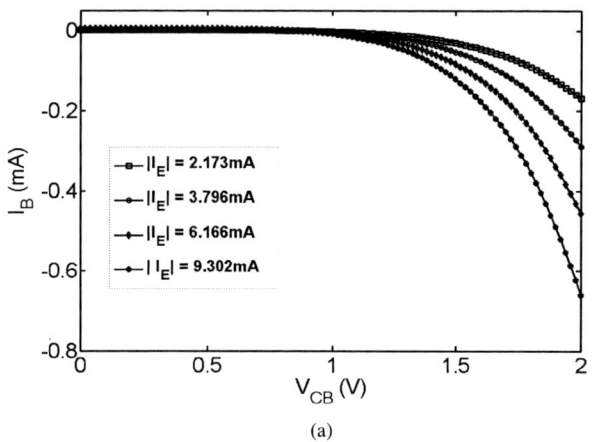

(a)

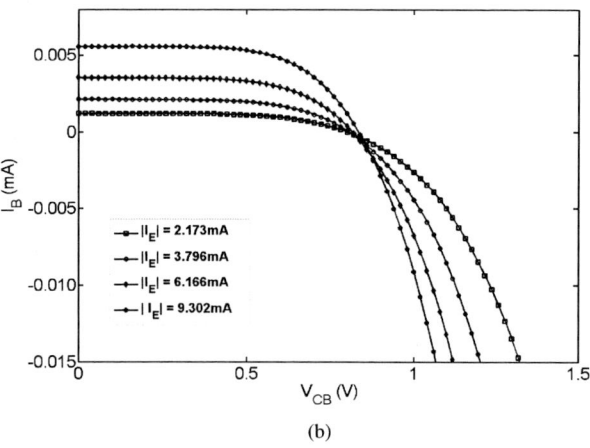

(b)

Fig. 1. Measured base current I_B (a) as a function of collector-base voltage V_{CB} for 4 different (constant) emitter currents. In (b) is a zoom in of (a) for lower values of V_{CB}. These measurements were taken on the same QUBiC4mmW SiGe HBT (with emitter area $A_E = 0.30$ $\mu m \times 1.0$ μm), in common-base configuration and at ambient temperature T $= 25\,^\circ\mathrm{C}$. I_B decreases with increasing V_{CB} and even changes sign for large V_{CB}.

V_{CB} range, we have three regions, i.e., the non-avalanche region, the weak avalanche region, and the strong avalanche region. In the derivation of (4), weak avalanche bias conditions were assumed. Therefore, we need to choose the boundaries of this weak avalanche region to have validity of the extraction method [1].

By employing the measurement data together with (4), the

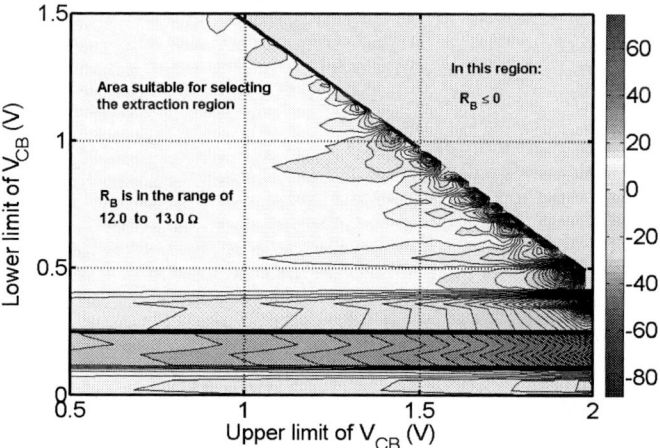

Fig. 2. Determination of the the lower and upper limits of the extraction region using a contour plot showing contour lines of equal R_B. These results correspond to a constant emitter current $|I_E| = 6.166$ mA.

values are between $12.0\ \Omega$ to $13.0\ \Omega$ (approximately less than 0.5% variation), which is really small in comparison to other complementary regions. The region below the boundary of the lower limit of the extraction region (i.e., $V_{CB} < 0.4$ V on the vertical axis Fig. 2) is the non-avalanche region (see Fig. 1) and over this region, R_B varies approximately between $2.0\ \Omega$ and $62.0\ \Omega$ (approximately 60.0% variation). The region above the boundary for the upper limit of the extraction region (i.e., $V_{CB} > 1.3$ V on the horizontal axis of Fig. 2) is the the strong avalanche region (see Fig. 1) and over this region R_B varies between negative values and $62.0\ \Omega$ (which is more than 60.0% variation).

The exact boundaries of the lower and upper limits of the extraction region can be determined using 2D plots showing the extracted R_B values against the V_{CB} lower limit (lower limit of the extraction region) and against the V_{CB} upper limit (upper limit of the extraction region). Here one of the limits is fixed, and then we determine R_B with the other limit being varied and vice verse. In Fig. 3, we present the results corresponding to the lower limit of the extraction region with fixed values of the upper limit. From the results in Fig. 3, we can clearly observe the boundary of the lower limit of the extraction region at $V_{CB} = 0.42$ V where the variations in R_B are approximately 0.08%. Below this limit, we observe large variations in R_B values. This region corresponds to the non-

978-1-4799-7231-9/14 $31.00 © 2014 IEEE

216

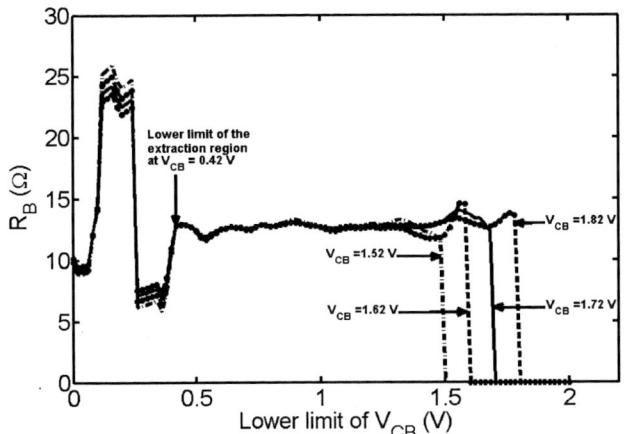

Fig. 3. Determination of the the exact boundary of the lower limit of the extraction region using a 2D plot. The upper limit is fixed to some selected values. These results correspond to a fixed $|I_E| = 6.166$ mA.

avalanche region and should be excluded from the extraction region as it yields wrong results for the extracted R_B and R_{TH}. Similarly, by taking some fixed values for the lower limit of the extraction region, we can determine the exact boundary of the upper limit of the extraction region using a 2D-plot as presented in Fig. 4. From Fig. 4, we can clearly see the

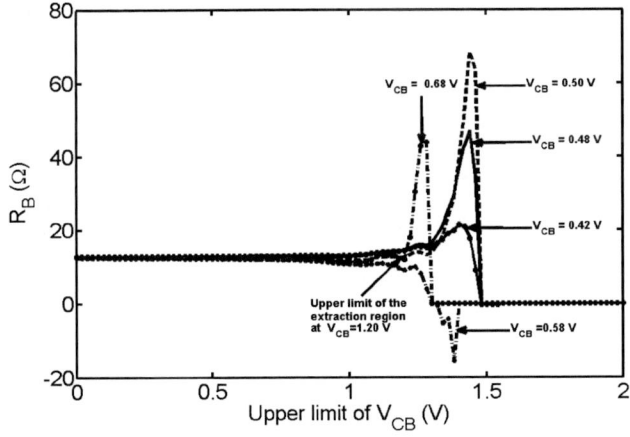

Fig. 4. Determination of the exact boundary of the upper limit of the extraction region using a 2D plot. The lower limit is fixed to some selected values. These results correspond to a fixed $|I_E| = 6.166$ mA.

exact boundary of the upper limit of V_{CB} (upper limit of the extraction region) as a point where the large variations (of more than 1.0%) in R_B start from. Beyond this point (i.e., $V_{CB} > 1.20$ V), we have the strong avalanche region and over this region it is impossible to extract correct values of R_B and R_{TH}.

In the method described above for determining the extraction region, we used DC-measurements corresponding to one constant emitter current (i.e., $|I_E| = 6.166$ mA). But as it was described in [1], to simultaneously extract both R_B and R_{TH} we need DC-measurements taken at different constant emitter

currents (see Fig. 1). Therefore, for each selected constant emitter current, the corresponding extraction region has to be separately determined. After determining the extraction regions corresponding to the sequence of the selected constant emitter currents, the extraction steps as described in [1] are followed to accurately extract both R_B and R_{TH} for a given bipolar transistor.

IV. GENERALIZATION OF THE EXTRACTION METHOD

A. Method Derivation

In the derivation of the extraction method for R_B and R_{TH} published in [1], the *Early voltage (V_A) was assumed to be much larger than the applied collector-base voltage (V_{CB})*. But in some modern bipolar transistor technologies mostly those with very high cutoff frequency, this assumption may not always hold; meaning that the extraction method [1] cannot be applied to such technologies. By relaxing this assumption, we can generalize this extraction method so that it can also be applied to technologies with low Early voltage (i.e., $V_A \leq V_{CB}$). This is done by using (3), which can be reformulated as:

$$Y = R_B + \alpha_T R_{TH}(V_{CB} + V_A^{eff}), \quad (5)$$

where

$$Y = -\left[\frac{dV_{BE}}{dI_C} + \frac{V_T}{I_C}\left(\frac{V_A + V_{CB}}{V_A^{eff}} - 1\right)\right]. \quad (6)$$

We can see that Y contains the Early voltage (V_A) and the thermal voltage (V_T). Using (5), we define an improved extraction procedure for R_B and R_{TH} in a way similar to that described in [1]. First, we determine the extraction region from the DC-measurements taken at different constant emitter currents (I_E) as described in Section III. In this region, the intercept on the vertical axis of the plot of Y as a function of ($V_{CB} + V_A^{eff}$) gives the extracted R_B and the corresponding slope ($S_{TOT_{new}}$) is equivalent to $\alpha_T R_{TH}$; where $\alpha_T = -dV_{BE}/dT$ is determined separately from V_{BE} measurements as a function of temperature at fixed I_E and constant V_{CB}.

In order to determine Y (assuming small local temperature variations), the Early voltage V_A need to be determined first. This means that another extraction method for V_A is needed. There are various methods published in literature that can be used to extract the bias dependent V_A such as those in [3], [4]. The accuracy of the extracted V_A for a given bipolar device will affect that of the extracted R_B and R_{TH} using this generalized extraction method.

B. Application of the Method

By taking arbitrary values of V_A, we extracted R_B and R_{TH} (following the above extraction steps) using the measurement data taken on the same bipolar device as the one used in Section III. The corresponding results for the extracted R_B are presented in Fig. 5. As mentioned earlier, the Early voltage affects the extracted value of the base resistance as can be seen from Fig. 5. From the figure, we observe that the extracted R_B is very large for small values of V_A and it

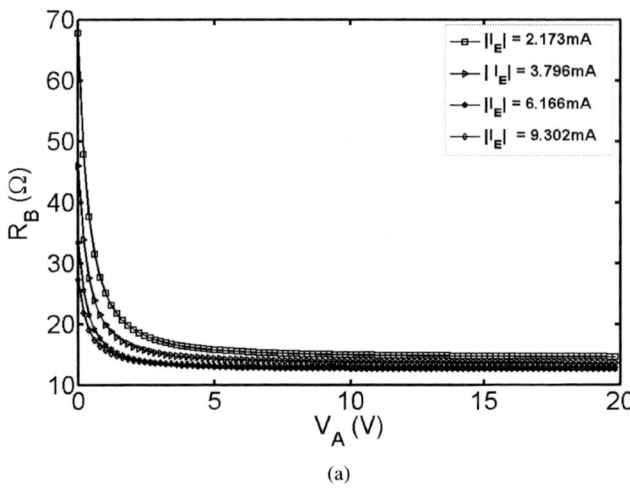

(a)

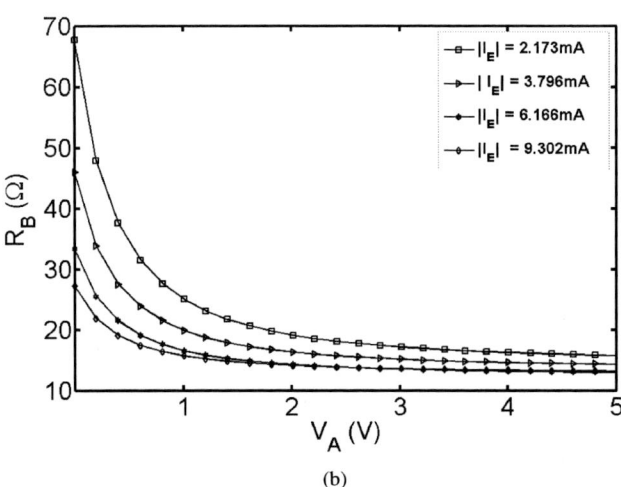

(b)

Fig. 5. Extracted base resistance (R_B) as a function of the arbitrary chosen values of the Early voltage (V_A) and constant emitter currents (I_E). (b) Zoom-in of (a) at lower values of V_A. These results correspond to the same bipolar device as that used in Section III. The applied collector-base voltage (V_{CB}) is in the range of 0.0 to 2.0V.

decreases to a constant value as V_A increases. Large values of the extracted R_B are mostly observed for $V_A \leq V_{CB}$ (for this case $0 \leq V_{CB} \leq 2$ V) as can be seen from Fig. 5(b). For $V_A \gg V_{CB}$, the extracted R_B is independent of V_A, it is only dependent on I_E (as can be seen in Fig. 5(a)). This justifies the underlying assumption of $V_A \gg V_{CB}$ employed in [1] when deriving the extraction method. From the results in Fig. 5, we also observe that the total extracted base resistance ($R_B = R_{BC} + R_{BV}$) decreases with increasing emitter current (mostly in the region where $V_A \leq V_{CB}$). This is a result of decrease in the variable part of base resistance (R_{BV}) with increase in I_E caused by the increase of the diffusion charge in the base. For very large I_E (i.e., $|I_E| = 9.302$ mA), we observe an increase in the extracted R_B. This is due to self-heating, which leads to an increase in the constant part of the extracted base resistance (R_{BC}). Due to the space limit, results corresponding to the extracted R_{TH} as a function of V_A and I_E were left out.

The above generalized extraction method need to be verified

further on measurements taken on real devices with very low V_A and either with other published methods or with compact model simulations.

V. CONCLUSION

We presented an extensive method to accurately determine the extraction region (region with very small variations in the base resistance) for the extraction method we published earlier in [1] that can be used to extract both the base resistance and the thermal resistance of bipolar junction transistors. This method utilizes contour plots showing the extracted R_B values against V_{CB} lower limit (lower limit of the extraction region) and V_{CB} upper limit (upper limit of the extraction region) for all possible combinations of the lower limit and the upper limit containing contour lines of equal extracted R_B. The desired extraction region is the region with very small variations (of less than 1.0%) of the extracted R_B. 2D-plots showing the extracted R_B against the V_{CB} lower limit and V_{CB} upper limit were used to determine the true boundaries of the extraction region. The extraction region was found to correspond to the weak avalanche regime. Accurate determination of the extraction region (where the input data is taken from) is vital in extracting correct values of R_B and R_{TH} using the method in [1].

We also presented a generalization of the extraction method [1] to include devices with low Early voltage (V_A). In the derivation of the extraction method [1], V_A was assumed to be much larger than V_{CB}, which sets a technology limit of the method. By relaxing this assumption we extended this method to bipolar technologies with very small V_A. This generalized method requires one to first apply another method to accurately extract V_A, before it is employed to extract R_B and R_{TH} for a given bipolar device. The method was demonstrated on DC-measurements taken on the advanced QUBiC4mmW devices and the results for the extracted R_B were presented. This method needs to be verified further with measurement data taken on real devices with very small V_A and with other published methods.

ACKNOWLEDGMENT

The authors would like to thank NL Agency, Dutch Ministry of Economic Affairs for the financial support. Great thanks to NXP Semiconductors for providing us with the bipolar devices used in this work and for the measurement facilities. The authors also extend their thanks to Andries J. Scholten NXP Semiconductors for the fruitful discussions during the development of this work.

REFERENCES

[1] R. Setekera, R. van der Toorn, and W. Kloosterman, "Local extraction of base and thermal resistance of bipolar transistors," in *Proc. Bipolar Circuits and Technology Meeting*, 2013, pp. 21–24.

[2] D. Leenaerts, "mmwave activities within NXP," Compus Technology Seminar, March 2012.

[3] G. Verzellesi, R. Turetta, P. Pavan, A. Collini, A. Chantre, A. Marty, and C. Canali, "Extraction of dc base parasitic resistance of bipolar transistors based on impact-ionization-induced base current reversal," *IEEE Electron Device Letters*, vol. 14, pp. 431–434, 1993.

[4] C. C. McAndrew and L. W. Nagel, "Early effect modeling in SPICE," *IEEE Journal of Solid-State Circuits*, vol. 31, no. 1, pp. 136–138, Jan 1996.

Author Index

A

Adkisson, James
Investigation of HBT layout impact on fT doubler performance for 90nm SiGe HBTs
A 90nm SiGe BiCMOS Technology for mm-wave and high-performance analog applications

Al-Eryani, Jidan
A True-RMS Integrated Power Sensor for On-Chip Calibration

Al-Sa'di, Mahmoud
An SiGe heterojunction bipolar transistor with very high open-base breakdown voltage

Aufinger, Klaus
A True-RMS Integrated Power Sensor for On-Chip Calibration

B

Bardin, Joseph
Microwave Noise Properties of Heterojunction Bipolar Transistors

Begueret, Jean-Baptiste
A Low Phase Noise Signal Generation System for Ka-Band P2P Applications based on an Injection-Locked Frequency Tripler

Bhat, Balamurali
Technologies for Very High Bandwidth Real-time Oscilloscopes

Bourdel, Sylvain
Comparison Between MOS and Bipolar mm-Wave Power Amplifiers in Advanced SiGe Technologies

Bowers, Derek
A Fast Precision Operational Amplifier Featuring Two Separate Control Loops

C

Cabrera, Dwight
A Low Phase Noise Signal Generation System for Ka-Band P2P Applications based on an Injection-Locked Frequency Tripler
A Low Phase Noise Signal Generation System for Ka-Band P2P Applications based on an Injection-Locked Frequency Tripler

Cahoon, Edward
A 90nm SiGe BiCMOS Technology for mm-wave and high-performance analog applications

Camillo-Castillo, Renata

Investigation of HBT layout impact on fT doubler performance for 90nm SiGe HBTs

High-Resistivity SiGe BiCMOS Technology Development

A 90nm SiGe BiCMOS Technology for mm-wave and high-performance analog applications

Device and circuit performance of SiGe HBTs in 130nm BiCMOS process with fT/fMAX of 250/330GHz

Candra, Panglijen

Device and circuit performance of SiGe HBTs in 130nm BiCMOS process with fT/fMAX of 250/330GHz

Cauwenberghs, Gert

A BiCMOS 50 MHz Input Bandwidth, 1-to-16 Channelizer Optimized for Low Power Analog Signal Classification

Chakravorty, Anjan

Small-Signal Modeling of the Lateral NQS Effect in SiGe HBTs

Chen, Gang

An Integrated Transmitter for LED-Based Visible Light Communication and Positioning System in A 180nm BCD Technology

Chen, Ying

A broad-band BiCMOS transmitter front-end for 27-36GHz phased array systems

Cheng, Peng

A 90nm SiGe BiCMOS Technology for mm-wave and high-performance analog applications

Device and circuit performance of SiGe HBTs in 130nm BiCMOS process with fT/fMAX of 250/330GHz

Chi, Taiyun

A Low-Power and Ultra-Compact W-band Transmitter Front-End in 90 nm SiGe BiCMOS Technology

Coen, Christopher T.

Ultra-Low Noise and Low Power 18.7 GHz Radiometer LNAs in a 0.5 THz SiGe Technology Utilizing Back-Side Etched Inductors

Corrao, Nicolas

Comparison Between MOS and Bipolar mm-Wave Power Amplifiers in Advanced SiGe Technologies

Cressler, John D.

Device-to-Circuit Interactions in SiGe Technology: Challenges and Opportunities (DP - Invited Paper)

Ultra-Low Noise and Low Power 18.7 GHz Radiometer LNAs in a 0.5 THz SiGe Technology Utilizing Back-Side Etched Inductors

Cressler, John

A 2×2, 316 GHz SiGe Scalable Transmitter Array with Novel Phase Locking Method and On-die Antennas

W-band SiGe Power Amplifiers

A Low-Power and Ultra-Compact W-band Transmitter Front-End in 90 nm SiGe BiCMOS Technology

A Digitally-Controlled Seven-State X-Band SiGe Variable Gain Low Noise Amplifier

An Investigation of fT/fmax Degredation due to Device Interconnects in 0.5 THz SiGe HBT

Technology

D

D'Esposito, Rosario
A study on transient intra-device thermal coupling in multifinger SiGe HBTs

Dai, Fa
A wide tuning triple-band frequency generator MMIC in 0.18um SiGe BiCMOS technology

Delbue, Roger
Technologies for Very High Bandwidth Real-time Oscilloscopes

Deval, Yann
A Low Phase Noise Signal Generation System for Ka-Band P2P Applications based on an Injection-Locked Frequency Tripler

Ding, Hanyi
High-Resistivity SiGe BiCMOS Technology Development

Dinh, Thanh Viet
An SiGe heterojunction bipolar transistor with very high open-base breakdown voltage

Divergilio, Adam
Investigation of HBT layout impact on fT doubler performance for 90nm SiGe HBTs

DiVergilio, Adam
A 90nm SiGe BiCMOS Technology for mm-wave and high-performance analog applications
An Electrothermal PIN Diode Model with Substrate Injection

Doerner, Ralf
Temperature Impact on the In-Situ S-Parameter Calibration in Advanced SiGe Technologies

Dong, Zongyu
An Integrated Transmitter for LED-Based Visible Light Communication and Positioning System in A 180nm BCD Technology
An Integrated Transmitter for LED-Based Visible Light Communication and Positioning System in A 180nm BCD Technology

Doshi, Kaviyesh
Technologies for Very High Bandwidth Real-time Oscilloscopes

Dunn, James
High-Resistivity SiGe BiCMOS Technology Development

E

Ellis-Monaghan, John
A 90nm SiGe BiCMOS Technology for mm-wave and high-performance analog applications

Essing, Jaap
A 27GHz, 31dBm Power Amplifier in a 0.25µm SiGe:C BiCMOS technology

F

Feng, Milton
Transistor Laser for Optical Interconnect and Photonic Integrated Circuits

Fischer, Gerhard
Degradation and Recovery of High-Speed SiGe HBTs under Very High Reverse EB Stress Conditions

Floyd, Brian
A Power-Efficient 4-element Beamformer in 120-nm SiGe BiCMOS for 28-GHz cellular communications

Fournier, Jean-Michel
Comparison Between MOS and Bipolar mm-Wave Power Amplifiers in Advanced SiGe Technologies

G

Gamand, Patrice
A Low Phase Noise Signal Generation System for Ka-Band P2P Applications based on an Injection-Locked Frequency Tripler

Gray, Peter
Investigation of HBT layout impact on fT doubler performance for 90nm SiGe HBTs
A 90nm SiGe BiCMOS Technology for mm-wave and high-performance analog applications
Device and circuit performance of SiGe HBTs in 130nm BiCMOS process with fT/fMAX of 250/330GHz

Greene, Kevin
A Power-Efficient 4-element Beamformer in 120-nm SiGe BiCMOS for 28-GHz cellular communications

Gross, B. Jeffrey
A Low-Power SiGe Feedback Amplifier with Over 110GHz Bandwidth
An Active Frequency Doubler with DC-100GHz Range
A 90nm SiGe BiCMOS Technology for mm-wave and high-performance analog applications

Gross, Blaine
Investigation of HBT layout impact on fT doubler performance for 90nm SiGe HBTs
Device and circuit performance of SiGe HBTs in 130nm BiCMOS process with fT/fMAX of 250/330GHz

H

Hannon, James
Integration Challenges for High-Performance Carbon Nanotube Logic

Harame, David
Investigation of HBT layout impact on fT doubler performance for 90nm SiGe HBTs
A 90nm SiGe BiCMOS Technology for mm-wave and high-performance analog applications

Device and circuit performance of SiGe HBTs in 130nm BiCMOS process with fT/fMAX of 250/330GHz

He, Zhong-Xiang
A 90nm SiGe BiCMOS Technology for mm-wave and high-performance analog applications

Heinemann, Bernd
Degradation and Recovery of High-Speed SiGe HBTs under Very High Reverse EB Stress Conditions

Heringa, Anco
An SiGe heterojunction bipolar transistor with very high open-base breakdown voltage

Hettrich, Horst
Linear low-power 13 GHz SiGe-Bipolar Modulator Driver with 7 Vpp differential Output Voltage Swing and on-Chip Bias Tee

I

Imai, Hisaya
Examination of Horizontal Current Bipolar Transistor (HCBT) Reliability Characteristics

Ivo, Ponky
An SiGe heterojunction bipolar transistor with very high open-base breakdown voltage

J

Jaffe, Mark
High-Resistivity SiGe BiCMOS Technology Development

Jain, Vibhor
Investigation of HBT layout impact on fT doubler performance for 90nm SiGe HBTs
High-Resistivity SiGe BiCMOS Technology Development
A 90nm SiGe BiCMOS Technology for mm-wave and high-performance analog applications
Device and circuit performance of SiGe HBTs in 130nm BiCMOS process with fT/fMAX of 250/330GHz
An Electrothermal PIN Diode Model with Substrate Injection

John, Jay
An Enhanced 180nm Millimeter-Wave SiGe BiCMOS Technology with fT/fMAX of 260/350GHz for Reduced Power Consumption Automotive Radar IC's

Joseph, Alvin
High-Resistivity SiGe BiCMOS Technology Development

K

Kamarei, Mahmoud
A 2×2, 316 GHz SiGe Scalable Transmitter Array with Novel Phase Locking Method and On-die Antennas

Kaushal, Vikas
A 90nm SiGe BiCMOS Technology for mm-wave and high-performance analog applications

Kaynak, Mehmet
A 2×2, 316 GHz SiGe Scalable Transmitter Array with Novel Phase Locking Method and On-die Antennas

Ultra-Low Noise and Low Power 18.7 GHz Radiometer LNAs in a 0.5 THz SiGe Technology Utilizing Back-Side Etched Inductors

An Investigation of fT/fmax Degredation due to Device Interconnects in 0.5 THz SiGe HBT Technology

Kerbaugh, Michael
A 90nm SiGe BiCMOS Technology for mm-wave and high-performance analog applications

Kessler, Thomas
Device and circuit performance of SiGe HBTs in 130nm BiCMOS process with fT/fMAX of 250/330GHz

Khan, Wasif
An Investigation of fT/fmax Degredation due to Device Interconnects in 0.5 THz SiGe HBT Technology

Khanna, Amarpal
Technologies for Very High Bandwidth Real-time Oscilloscopes

Khater, Marwan
A 90nm SiGe BiCMOS Technology for mm-wave and high-performance analog applications

Kirchgessner, Jim
An Enhanced 180nm Millimeter-Wave SiGe BiCMOS Technology with fT/fMAX of 260/350GHz for Reduced Power Consumption Automotive Radar IC's

Klaassen, Dick
An SiGe heterojunction bipolar transistor with very high open-base breakdown voltage

Klauk, Hagen
Low-Voltage Organic Field-Effect Transistors for Flexible Electronics

Kloosterman, Willy
Analysis of the Local Extraction Method of Base and Thermal Resistance of Bipolar Transistors

Knapp, Herbert
A True-RMS Integrated Power Sensor for On-Chip Calibration

Knierim, Dan
Ultra-Wide-Bandwidth Oscilloscope Architectures and Circuits

Knoll, Dieter
High-speed, waveguide Ge PIN photodiodes for a photonic BiCMOS process

Koričić, Marko
Examination of Horizontal Current Bipolar Transistor (HCBT) Reliability Characteristics

Korndoerfer, Falk
Temperature Impact on the In-Situ S-Parameter Calibration in Advanced SiGe Technologies

Korndörfer, Falk
High-speed, waveguide Ge PIN photodiodes for a photonic BiCMOS process

Kroh, Marcel
High-speed, waveguide Ge PIN photodiodes for a photonic BiCMOS process

Krozer, Viktor
On-Wafer Small- and Large-Signal Measurement Systems at sub-THz Frequencies

Kurps, Rainer
High-speed, waveguide Ge PIN photodiodes for a photonic BiCMOS process

L

Lachner, Rudolf
A True-RMS Integrated Power Sensor for On-Chip Calibration

Larson, Lawrence
A BiCMOS 50 MHz Input Bandwidth, 1-to-16 Channelizer Optimized for Low Power Analog Signal Classification

Lauga-Larroze, Estelle
Comparison Between MOS and Bipolar mm-Wave Power Amplifiers in Advanced SiGe Technologies

Leenaerts, Domine
A broad-band BiCMOS transmitter front-end for 27-36GHz phased array systems
A 27GHz, 31dBm Power Amplifier in a 0.25µm SiGe:C BiCMOS technology

Lehmann, Steffen
A Simple and Accurate Method for Extracting the Emitter and Thermal Resistance of BJTs and HBTs

Li, Hao
A BiCMOS 50 MHz Input Bandwidth, 1-to-16 Channelizer Optimized for Low Power Analog Signal Classification

Lie, Yu-Chun Donald
A Differential SiGe Power Amplifier Using Through-Silicon-Via and Envelope-Tracking for Broadband Wireless Applications

Lischke, Stefan
High-speed, waveguide Ge PIN photodiodes for a photonic BiCMOS process

Liu, Qizhi
Investigation of HBT layout impact on fT doubler performance for 90nm SiGe HBTs
A 90nm SiGe BiCMOS Technology for mm-wave and high-performance analog applications

Long, John R.
A Low-Power SiGe Feedback Amplifier with Over 110GHz Bandwidth
An Active Frequency Doubler with DC-100GHz Range

Lopez, Aida Vera
A 2×2, 316 GHz SiGe Scalable Transmitter Array with Novel Phase Locking Method and On-die Antennas

Lourenco, Nelson
A 2×2, 316 GHz SiGe Scalable Transmitter Array with Novel Phase Locking Method and On-die Antennas

Lu, Fei
An Integrated Transmitter for LED-Based Visible Light Communication and Positioning System in A 180nm BCD Technology

Lukaitis, Joseph
A 90nm SiGe BiCMOS Technology for mm-wave and high-performance analog applications

M

Ma, Rui
An Integrated Transmitter for LED-Based Visible Light Communication and Positioning System in A 180nm BCD Technology

Ma, Zhenqiang
Radio-Frequency Flexible Electronics: Transistors and Passives

Magnee, Peter
An SiGe heterojunction bipolar transistor with very high open-base breakdown voltage

Mahmoudi, Reza
A 27GHz, 31dBm Power Amplifier in a 0.25μm SiGe:C BiCMOS technology

Mai, Christian
High-speed, waveguide Ge PIN photodiodes for a photonic BiCMOS process

Maurer, Linus
A True-RMS Integrated Power Sensor for On-Chip Calibration

Maxey, Christopher
K-Band Digitally Controlled Oscillator with Integrated Divide-by-16 Divider Chain for ADPLL Applications

Mazouffre, Olivier
A Low Phase Noise Signal Generation System for Ka-Band P2P Applications based on an Injection-Locked Frequency Tripler

McCallum-Cook, Ian
High-Resistivity SiGe BiCMOS Technology Development

Mochizuki, Hidenori
Examination of Horizontal Current Bipolar Transistor (HCBT) Reliability Characteristics

Möller, Michael
Linear low-power 13 GHz SiGe-Bipolar Modulator Driver with 7 Vpp differential Output Voltage Swing and on-Chip Bias Tee

Morgan, Dave
An Enhanced 180nm Millimeter-Wave SiGe BiCMOS Technology with fT/fMAX of 260/350GHz for Reduced Power Consumption Automotive Radar IC's

Morita, So-ichi
Examination of Horizontal Current Bipolar Transistor (HCBT) Reliability Characteristics

N

Namarvar, Saeed Zeinolabedinzadeh
A 2×2, 316 GHz SiGe Scalable Transmitter Array with Novel Phase Locking Method and On-die Antennas

Natarajan, Arun
Device and circuit performance of SiGe HBTs in 130nm BiCMOS process with fT/fMAX of 250/330GHz

Newton, Kim
High-Resistivity SiGe BiCMOS Technology Development
Device and circuit performance of SiGe HBTs in 130nm BiCMOS process with fT/fMAX of 250/330GHz

Newton, Kimberly
A 90nm SiGe BiCMOS Technology for mm-wave and high-performance analog applications

Niu, Guofu
A wide tuning triple-band frequency generator MMIC in 0.18um SiGe BiCMOS technology

O

Ostrovskyy, Pylyp
High-speed, waveguide Ge PIN photodiodes for a photonic BiCMOS process

P

Papapolymerou, John
A 2×2, 316 GHz SiGe Scalable Transmitter Array with Novel Phase Locking Method and On-die Antennas

Park, Jong Seok
A Low-Power and Ultra-Compact W-band Transmitter Front-End in 90 nm SiGe BiCMOS Technology

Parthasarathy, Shyam
High-Resistivity SiGe BiCMOS Technology Development

Pawlak, Andreas
A Simple and Accurate Method for Extracting the Emitter and Thermal Resistance of BJTs and HBTs

Peczek, Anna
High-speed, waveguide Ge PIN photodiodes for a photonic BiCMOS process

Pei, Yu
A broad-band BiCMOS transmitter front-end for 27-36GHz phased array systems

Pekarik, John J.
A 90nm SiGe BiCMOS Technology for mm-wave and high-performance analog applications

Pekarik, John
Investigation of HBT layout impact on fT doubler performance for 90nm SiGe HBTs

Device and circuit performance of SiGe HBTs in 130nm BiCMOS process with fT/fMAX of 250/330GHz

An Electrothermal PIN Diode Model with Substrate Injection

Pupalaikis, Peter
Technologies for Very High Bandwidth Real-time Oscilloscopes

R

Raman, Sanjay
K-Band Digitally Controlled Oscillator with Integrated Divide-by-16 Divider Chain for ADPLL Applications

Rassel, Robert
High-Resistivity SiGe BiCMOS Technology Development

Rebeiz, Gabriel
Millimeter-Wave SiGe RFICs for Large-Scale Phased-Arrays

Reynolds, Scott
Device and circuit performance of SiGe HBTs in 130nm BiCMOS process with fT/fMAX of 250/330GHz

Rinaldi, Niccolò
Degradation and Recovery of High-Speed SiGe HBTs under Very High Reverse EB Stress Conditions

Rumiantsev, Andrej
Temperature Impact on the In-Situ S-Parameter Calibration in Advanced SiGe Technologies

S

Sadhu, Bodhisatwa
Device and circuit performance of SiGe HBTs in 130nm BiCMOS process with fT/fMAX of 250/330GHz

Sarkar, Anirban
A Power-Efficient 4-element Beamformer in 120-nm SiGe BiCMOS for 28-GHz cellular communications

Sasso, Grazia
Degradation and Recovery of High-Speed SiGe HBTs under Very High Reverse EB Stress Conditions

Scheit, Alexander
High-speed, waveguide Ge PIN photodiodes for a photonic BiCMOS process

Schmid, Robert L.

Ultra-Low Noise and Low Power 18.7 GHz Radiometer LNAs in a 0.5 THz SiGe Technology Utilizing Back-Side Etched Inductors

Schmid, Robert
W-band SiGe Power Amplifiers
A Low-Power and Ultra-Compact W-band Transmitter Front-End in 90 nm SiGe BiCMOS Technology
A Digitally-Controlled Seven-State X-Band SiGe Variable Gain Low Noise Amplifier
An Investigation of fT/fmax Degredation due to Device Interconnects in 0.5 THz SiGe HBT Technology

Schmidt, Nicholas
High-Resistivity SiGe BiCMOS Technology Development

Schroter, Michael
A Simple and Accurate Method for Extracting the Emitter and Thermal Resistance of BJTs and HBTs
Small-Signal Modeling of the Lateral NQS Effect in SiGe HBTs

Seo, Jung-Hun
Radio-Frequency Flexible Electronics: Transistors and Passives

Serhan, Ayssar
Comparison Between MOS and Bipolar mm-Wave Power Amplifiers in Advanced SiGe Technologies

Setekera, Robert
Analysis of the Local Extraction Method of Base and Thermal Resistance of Bipolar Transistors

Shinomura, Katsumi
Examination of Horizontal Current Bipolar Transistor (HCBT) Reliability Characteristics

Song, Peter
W-band SiGe Power Amplifiers

Srihari, Srikanth
High-Resistivity SiGe BiCMOS Technology Development

Stamper, Anthony
High-Resistivity SiGe BiCMOS Technology Development

Suligoj, Tomislav
Examination of Horizontal Current Bipolar Transistor (HCBT) Reliability Characteristics

Sureka, Anirudh
Technologies for Very High Bandwidth Real-time Oscilloscopes

T

Taris, Thierry
A Low Phase Noise Signal Generation System for Ka-Band P2P Applications based on an Injection-Locked Frequency Tripler

Tesson, Olivier
A Low Phase Noise Signal Generation System for Ka-Band P2P Applications based on an Injection-Locked Frequency Tripler

Thomas, Chris
A BiCMOS 50 MHz Input Bandwidth, 1-to-16 Channelizer Optimized for Low Power Analog Signal Classification

Tian, Xiaowei
A 90nm SiGe BiCMOS Technology for mm-wave and high-performance analog applications

Tiemeijer, Luuk
Analysis of the Local Extraction Method of Base and Thermal Resistance of Bipolar Transistors

Tillack, Bernd
High-speed, waveguide Ge PIN photodiodes for a photonic BiCMOS process
A 2×2, 316 GHz SiGe Scalable Transmitter Array with Novel Phase Locking Method and On-die Antennas
Ultra-Low Noise and Low Power 18.7 GHz Radiometer LNAs in a 0.5 THz SiGe Technology Utilizing Back-Side Etched Inductors
An Investigation of fT/fmax Degredation due to Device Interconnects in 0.5 THz SiGe HBT Technology

To, Ivan
An Enhanced 180nm Millimeter-Wave SiGe BiCMOS Technology with fT/fMAX of 260/350GHz for Reduced Power Consumption Automotive Radar IC's

Trivedi, Vishal
An Enhanced 180nm Millimeter-Wave SiGe BiCMOS Technology with fT/fMAX of 260/350GHz for Reduced Power Consumption Automotive Radar IC's

Trusch, Andreas
High-speed, waveguide Ge PIN photodiodes for a photonic BiCMOS process

Tsay, Jerry
A Differential SiGe Power Amplifier Using Through-Silicon-Via and Envelope-Tracking for Broadband Wireless Applications

U

Ulusoy, Ahmet Çağrı
W-band SiGe Power Amplifiers

Ulusoy, Ahmet Cagri
A 2×2, 316 GHz SiGe Scalable Transmitter Array with Novel Phase Locking Method and On-die Antennas

Ulusoy, Cagri
A Low-Power and Ultra-Compact W-band Transmitter Front-End in 90 nm SiGe BiCMOS Technology
An Investigation of fT/fmax Degredation due to Device Interconnects in 0.5 THz SiGe HBT Technology

V

Valdes-Garcia, Alberto
Device and circuit performance of SiGe HBTs in 130nm BiCMOS process with fT/fMAX of 250/330GHz

Vallett, Aaron
Investigation of HBT layout impact on fT doubler performance for 90nm SiGe HBTs
A 90nm SiGe BiCMOS Technology for mm-wave and high-performance analog applications

van der Toorn, Ramses
Analysis of the Local Extraction Method of Base and Thermal Resistance of Bipolar Transistors

V

Vanhoucke, Tony
An SiGe heterojunction bipolar transistor with very high open-base breakdown voltage

Vera, Leonardo
A Low-Power SiGe Feedback Amplifier with Over 110GHz Bandwidth
An Active Frequency Doubler with DC-100GHz Range

Voigt, Karsten
High-speed, waveguide Ge PIN photodiodes for a photonic BiCMOS process

W

Wang, Albert
An Integrated Transmitter for LED-Based Visible Light Communication and Positioning System in A 180nm BCD Technology

Wang, Hechen
A wide tuning triple-band frequency generator MMIC in 0.18um SiGe BiCMOS technology

Wang, Hua
A Low-Power and Ultra-Compact W-band Transmitter Front-End in 90 nm SiGe BiCMOS Technology

Wang, Li
An Integrated Transmitter for LED-Based Visible Light Communication and Positioning System in A 180nm BCD Technology

Welch, Pam
An Enhanced 180nm Millimeter-Wave SiGe BiCMOS Technology with fT/fMAX of 260/350GHz for Reduced Power Consumption Automotive Radar IC's

Wilamowski, Bogdan
A wide tuning triple-band frequency generator MMIC in 0.18um SiGe BiCMOS technology

Winzer, Georg
High-speed, waveguide Ge PIN photodiodes for a photonic BiCMOS process

Wolf, Randy
High-Resistivity SiGe BiCMOS Technology Development

Wursthorn, Jonas
A True-RMS Integrated Power Sensor for On-Chip Calibration

Y

Yadav, Shon
Small-Signal Modeling of the Lateral NQS Effect in SiGe HBTs

Yamamoto, Yuji
High-speed, waveguide Ge PIN photodiodes for a photonic BiCMOS process

Yamrone, Brian
Technologies for Very High Bandwidth Real-time Oscilloscopes

Z

Zainolabedinzadeh, Saeed
An Investigation of fT/fmax Degredation due to Device Interconnects in 0.5 THz SiGe HBT Technology

Zeinolabedinzadeh, Saeed
W-band SiGe Power Amplifiers

Zetterlund, Bjorn
Investigation of HBT layout impact on fT doubler performance for 90nm SiGe HBTs
A 90nm SiGe BiCMOS Technology for mm-wave and high-performance analog applications

Zhang, Chen
An Integrated Transmitter for LED-Based Visible Light Communication and Positioning System in A 180nm BCD Technology

Zhao, Bin
An Integrated Transmitter for LED-Based Visible Light Communication and Positioning System in A 180nm BCD Technology

Zhao, Feng
A wide tuning triple-band frequency generator MMIC in 0.18um SiGe BiCMOS technology

Zhou, Weidong
Radio-Frequency Flexible Electronics: Transistors and Passives

Žilak, Josip
Examination of Horizontal Current Bipolar Transistor (HCBT) Reliability Characteristics

Z

Zierak, Michael
High-Resistivity SiGe BiCMOS Technology Development

Zimmermann, Lars
High-speed, waveguide Ge PIN photodiodes for a photonic BiCMOS process